The Correspondence of John Tyndall

VOLUME 4

THE CORRESPONDENCE OF JOHN TYNDALL

VOLUME 4

The Correspondence, January 1853–December 1854

Edited by
Ian Hesketh and Efram Sera-Shriar

General Editors
James Elwick, Roland Jackson, Bernard Lightman,
and Michael S. Reidy

UNIVERSITY OF PITTSBURGH PRESS

Published by the University of Pittsburgh Press, Pittsburgh, Pa., 15260

Manufactured in the United States of America
Printed on acid-free paper
10 9 8 7 6 5 4 3 2 1

Cataloging-in-Publication data is available from the Library of Congress

ISBN 13: 978-0-8229-4525-3

CONTENTS

ACKNOWLEDGMENTS

This volume has been a labor of love, and it would not have been possible without the support we have received from our editorial assistants Michelle Fu, David Cross, Erin Grosjean, and Nanna Kaalund. Our thanks also extend to John Lynch who due to unforeseen circumstances had to step down as a coeditor of this volume. His efforts at the early stages of the volume, however, will not be forgotten. Thank you to Sylvia Nickerson for formatting and collating our images. We would also like to extend our gratitude to Jamie Elwick, Bernie Lightman, Michael Reidy, Roland Jackson, Peter Harrison, and Melinda Baldwin for their help and support throughout the editorial process. Michael and Roland also read through the entire manuscript and offered extremely valuable editorial guidance in preparing the volume for submission. Geoffrey Cantor, Ruth Barton, Bill Brock, Leigh Penman, Daniel Midena, Nicholas Heron, Gabriel Finkelstein, and Janet Browne have all given us invaluable information when we were unable to make sense of some of the more obscure components of the letters. Jane Harrison, the document manager at the RI, gave us wonderful assistance, particularly in the final stages when we had to order some last minute images of some key documents and letters. She was always incredibly prompt in responding to our various queries and orders. Over the years, our friends and family have been tremendously supportive of all of our academic endeavors. Special thanks, however, go to Geoff Belknap, Michaela Livingstone-Banks, Jon Livingstone-Banks, Christie Busich, and Cleo Griffin! Finally, we would like to thank our hilarious kids Magnus and Sadie respectively, who bring us boundless joy and much needed humor to our otherwise mundane academic lives.

LIST OF ABBREVIATIONS

Annal. Chim. et Phys.	*Annales de Chimie et de Physique*
BAAS	British Association for the Advancement of Science
Brit. Assoc. Rep.	*Report of the British Association for the Advancement of Science*
Brock, 'Queenwood'	William H. Brock, 'Queenwood College Revisited', Article XVII in *Science for All: Studies in the History of Victorian Science and Education* (Aldershot: Ashgate Variorum, 1996)
Brock and Meadows, *Lamp of Learning*	W. H. Brock and A. J. Meadows, *The Lamp of Learning: Taylor & Francis and the Development of Scientific Publishing* (London: Taylor & Francis, 1984)
Faraday Correspondence	Frank A. J. L. James (ed.), *The Correspondence of Michael Faraday, Volume 4, 1849–1855* (London: Institution of Electrical Engineers, 1999)
Finkelstein, *Emil du Bois-Reymond*	Gabriel W. Finkelstein, *Emil du Bois-Reymond: Neuroscience, Self, and Society in Nineteenth-Century Germany* (Cambridge: MIT Press, 2013)
FRS	Fellow of the Royal Society
Hirst, 'Journals'	Thomas Archer Hirst Journals, Bound Typescript, Volume 2, 1850–55 (RI MS JT/2/32b)
Jackson, 'Tyndall and Diamagnetism'	Roland Jackson, 'John Tyndall and the Early History of Diamagnetism', *Annals of Science* 72 (2015), pp. 435–89
Jackson, 'John Tyndall and the Royal Medal'	Roland Jackson, 'John Tyndall and the Royal Medal That Was Never Struck', *Notes and Records of the Royal Society* 68:2 (2014), pp. 151–64
Journal	John Tyndall's Journal, 1848–55, typescript (RI MS JT/2/13b)
LT	Louisa Tyndall, John Tyndall's widow
LT, 'Biography'	Louisa Tyndall's typescript biography of her husband (RI MS JT/5/9–11)
NA OS	National Archives, Kew, Ordnance Survey Records
NA WO	National Archives, Kew, War Office Records
ODNB	*Oxford Dictionary of National Biography*
OED	*Oxford English Dictionary*

Phil. Mag.	*Philosophical Magazine*
Phil. Trans.	*Philosophical Transactions of the Royal Society of London*
Poggend. Annal.	*Poggendorff's Annalen der Physik und Chemie*
RI	Royal Institution
RI MS JT	Royal Institution, London, John Tyndall Papers
Roy. Inst. Proc., vol. 1	*Proceedings of the Royal Institution of Great Britain, Vol I. 1851–1854* (London: W. Nichol, 1854)
Roy. Inst. Proc., vol. 2	*Proceedings of the Royal Institution of Great Britain, Vol. II 1854–1858* (London: William Clowes and Sons, 1858)
Roy. Soc. Proc.	*Proceedings of the Royal Society of London*
RS	Royal Society of London
Russell, *Edward Frankland*	Colin A. Russell, *Edward Frankland: Chemistry, Controversy, and Conspiracy in Victorian England* (Cambridge: Cambridge University Press, 1996)
LI	London Institution
Tyndall, 'On Diamagnetism and Magnecrystallic Action'	J. Tyndall, 'On Diamagnetism and Magnecrystallic Action', *Phil. Mag.* 2:10 (September 1851), pp. 165–88
Tyndall and Francis (eds.), *Scientif. Mem.*	J. Tyndall and W. Francis (eds.), *Scientific Memoirs: Natural Philosophy* (London: Taylor and Francis, 1853)
Trans. Roy. Soc. Edin.	*Transactions of the Royal Society of Edinburgh*
UB Mbg	University Library of Marburg

Frontispiece: Watercolor of The Royal Institution of Great Britain, Albemarle Street, *c.* 1838 (w/c on paper), by Thomas Hosmere Shepherd. There were no major revisions to the RI's exterior between 1838 and 1853, so this would have been a fairly accurate representation of what the RI looked like when Tyndall became professor of natural philosophy there in 1853. Reproduced with permission of Bridgeman Images.

INTRODUCTION TO VOLUME 4

The 329 letters in this volume cover the years 1853 and 1854 and represent a period of immense transition in Tyndall's life. The numeric data are quite compelling. Taken together the total number of surviving letters from the first four volumes of *The Correspondence of John Tyndall* is 1,025. That roughly covers a period of fourteen years from May 1840 to December 1854. Interestingly the two years covered in this volume comprise 32 percent of the total number, which is nearly a third of the letters in Tyndall's correspondence from that fourteen-year period. By contrast, the total number of letters in the nine-year period between May 1840 and December 1849, which are the years covered in the first two volumes of *The Correspondence of John Tyndall*, is 391. That amounts to approximately 38 percent of the total. There is a rise in the number of surviving letters in volume 3, which covers January 1850 to December 1852, and includes 305 letters or 29.7 percent of the total correspondence for the fourteen-year period. The spike in his extant correspondence during the early 1850s is linked to Tyndall's expanding international network, growing reputation as a leading scientific figure in Britain and abroad, and his employment at the RI. All three of these aspects of Tyndall's life are recurring themes in the letters of this volume. In early 1853, Tyndall was still without a stable scientific position, but by this time he was clearly a man on the rise. He had published several important papers on diamagnetism in respectable science journals such as *Phil. Mag.*, his presentations at recent meetings of the British Association for the Advancement of Science had caused a stir, and he was elected fellow of the Royal Society at the end of 1852.[1] Even though Tyndall was unsuccessful in his applications for several scientific positions during that year, we find him at the beginning of 1853 preparing for a much-coveted Friday evening lecture at the Royal Institution, an institution that would soon after become Tyndall's primary place of employment as professor of natural philosophy. Tyndall would also be courted by the London Institution and the Government School of Mines, both of which offered him positions as well. Indeed, by the end of the period of this volume, December 1854, Tyndall had firmly established himself as a significant man of science, complete with an influential position at the center of the British scientific establishment. Thus, the letters give us great insight into how one man of science in particular navigated a complicated course that led him into the upper echelons of the Victorian scientific world. And yet while Tyndall was no longer as anxious about his scientific future as he was in previous volumes of his correspondence, these letters show a man struggling to come to terms with his newfound status, a struggle that was often reflected in his obsession with maintaining an "inflexible integrity" that guided his actions and deeds throughout the period covered by this volume.[2]

Career Success

As our volume opens, we find Tyndall back at Queenwood College, despite training as a physicist in Germany. This time he was lecturing on mathematics and natural philosophy. While this may look like a step back, as he held a position at Queenwood before leaving for Germany, it was a strategic one because it meant that Tyndall was now closer to London and the center of scientific life in Britain. It is true that Tyndall spent much of the preceding year applying for highly sought-after scientific appointments in Sydney, Toronto, and Galway that were ultimately unsuccessful. However, as is detailed in volume 3 of the Tyndall Correspondence, the applications for these positions involved amassing testimonials from leading men of science who came to know and greatly respect Tyndall's exacting experimental work, particularly his researches into magnetism and diamagnetism. Therefore the great anxiety that Tyndall felt about his future prospects in the opening years of the 1850s, which was a central theme of some of his letters contained in volume 3, had clearly begun to evaporate by 1853 as Tyndall's reputation in Britain was more firmly established.

This is readily apparent as early as the second letter of our volume, in which Tyndall describes to his good friend Thomas Hirst that he just happened to stop in to the Royal Society to discover that the paper being read was one that was written by none other than John Tyndall. He was also at the time preparing to give an upcoming Friday evening lecture at the RI, a highly sought-after lecture that was often reserved for the most eminent men of science. Tyndall was, of course, honored to be included in the RI's program but was also filled with terror to have his name appear in between such important men of science as George Airy and George Stokes. "There I twinkle like the end of a burnt stick," he wrote to Hirst.[3] Tyndall was right to be concerned that such an important public lecture could make or break his career. But by all accounts Tyndall's lecture "On the Influence of Material Aggregation upon the Manifestation of Force" was a great success. Not only had Tyndall received rounds of "thunderclaps" from the audience as well as a very public endorsement from the RI's chief figurehead and superintendent, the physicist and leading British man of science Michael Faraday, Tyndall was made aware of various schemes "among men of influence in London" to find him a lasting scientific position there as well.[4]

Things Were about to Change

Sure enough, within just a few days of the lecture, Tyndall received word that he would likely be offered a professorship at the London Institution, and also received a letter from a manager of the RI, the chemist and physician Henry Bence Jones, that he was likely to be offered a professorship there as well.[5] Both positions were attractive ones, as the LI and RI were committed to promoting both scientific education and research and had well-established track records in doing so. The main difference, however, was the presence of Faraday at the RI. He combined an incredible ability to lecture on general science topics, while engaging in specialized research that would provide a model for Tyndall to follow. Tyndall, therefore, preferred the RI, but he was initially told that the position there was most likely to be a chair of chemistry and not of physics, which meant that Tyndall would be diverted from his "professed career," as he told Thomas Henry Huxley.[6] Huxley was at this time advising Tyndall to exploit the current interest in his work, explaining to him that there was likely a "fine opening" for him in the world of British science, if only he would avoid spoiling it "by any of your horrid modesty."[7] Another possible problem was that Tyndall would have to give a course of four lectures at the RI to be qualified as a candidate for a professorship, though luckily he was offered to give such a course in May.[8] Thus on 23 May 1853 Tyndall was officially proposed to be elected as professor of natural philosophy at the RI, a position that would indeed allow Tyndall to concentrate on physics. It was also determined that Tyndall could lecture at other institutions but only as "Professor of the Royal Institution."[9] This left him open to accept the RI offer and also accept a lecturer position at the LI, which is exactly what he did.

Many of the letters in this volume, therefore, concern speculations and negotiations about prospective scientific positions, speculations and negotiations that did not end once Tyndall had accepted his position at the RI. Most notably, after he was already professor of natural philosophy at the RI, Tyndall engaged in several negotiations with the Government School of Mines about taking up a position there as well.[10] He found himself, however, unable to get a straight answer from the geologist and founder of the Government School of Mines, Henry De la Beche, about the specific duties that would be required, and the frustrating negotiations stretched out much longer than was necessary.[11] Near the end of 1854, our volume includes letters speculating about an upcoming vacancy for the chair of natural philosophy at the University of Edinburgh.[12] While the position itself would ultimately not materialize, Tyndall found himself excited by the possibility of taking up such a highly sought-after position. His long tenure at the RI, in other words, was not necessarily a given. This volume gives us a behind-the-scenes look at this moment of immense transition in Tyndall's life and the several career opportunities that suddenly opened to him after so many years of hard work.

Controversy and Character

By the summer of 1853, Tyndall's position was greatly clarified. In early July he was officially elected professor of natural philosophy at the RI and had formally resigned his position at Queenwood College. Reflecting on this change of career circumstances in his journal, he recognized that he was now passing through the most important days of his life. Tyndall admitted to a feeling of deep satisfaction that he reached his new "position by fair means."[13] This was of utmost importance to him. Tyndall, like several other younger men of science of his generation, rejected the aristocratic and privileged nature of previous generations of gentlemanly natural philosophers.[14] He was greatly influenced by Thomas Carlyle, whose decidedly masculine philosophy stressed the need to achieve success honestly and with humility through hard work, self-discipline, and sincerity.[15] Tyndall very much relied on Carlyle as a guide in this regard. In his journal entry from 26 June 1853, he stressed how important it was to him that he achieved this newfound status by being true to this philosophy of life.

> I have stood at no man's door craving admittance, I have been asked in, every external advancement has been given, not sought. I never sought the Royal Society, still it came, I never sought the Royal Institution, but it has come. I never sought the society of the great and eminent, still I have got into such society; and this is the greatest comfort of my position, that I have been led into it by the natural guidance of circumstances, that I have scaled no walls, but entered like an honest man into the sheepfold, and in it as an honest man I shall remain, if at all.[16]

This sentiment was also expressed in Tyndall's letter to Bence Jones in negotiating the terms of the RI position.[17] He stressed that he did not seek out the position and cared not for status but merely the ability to carry out his scientific and specialized research.

This Carlylean philosophy would guide Tyndall, particularly during one of the more controversial moments of his life that is covered in this volume. In November 1853, just a few months after taking up his position at the RI, Tyndall received letters from William Francis, J. P. Gassiot, and Edward Sabine that he had been awarded the prestigious Royal Medal for his contributions to physics.[18] Tyndall was being singled out for his work on diamagnetism that he had begun in Marburg with the assistance of Hermann Knoblauch.[19] It turns out, however, that these letters of congratulations were premature. On 15 November Gassiot, acting on behalf of the Council of the Royal Society, visited Tyndall at the RI and informed him that a group of fellows of the Royal Society disapproved of the selection of Tyndall, calling attention to the fact that Tyndall had actually collaborated with Knoblauch, and the University of Berlin chemist and physicist, Gustav Magnus, in making his discoveries.[20] There were questions, therefore, about the true significance of Tyndall's singular contribution. Upon being informed of these views, Tyndall wasted little time, writing to Samuel Christie, the secretary of the Royal Society, that he declined the medal, given that it was apparently awarded while being "covered by a doubt." Tyndall stressed that he engaged in his scientific research without any consideration about the possibility of winning the Royal Medal. "It is an object at which I never aimed," Tyndall continued, "a consummation on which I never based a hope. And whatever may have been the pleasure which the award imparted to me, it was, I believe, inferior in kind to that with which I now resign all claim to the Medal, and thus restore to the Council the liberty of making a better choice."[21]

It was important for Tyndall to decline an award that was clouded by uncertainty concerning whether or not he actually deserved it.[22] And he appeared to be taking the high ground, quite in contrast to the minority of members who opposed him. As Gassiot wrote to Tyndall in response, "I would sooner hold the position you have so promptly taken than be the recipient of 20 medals."[23] Hirst also praised Tyndall for showing "the limits of your esteem for medals, and your independence of all such things. . . . To see you praised and successful is of course pleasant to me, but it is when you meet such cases as these that I am most proud of you."[24] Tyndall was particularly happy to have Hirst's approval, copying out this last sentence in his journal.[25]

If Tyndall held himself to high account, he expected much the same from others as well. As readers of volume 2 will know, Tyndall had a long and complicated relationship with George Edmondson, the headmaster of Queenwood College. Edmondson represented precisely the kind of person Tyndall wanted more than anything to avoid becoming. Tyndall believed that Edmondson was a "shortsighted little man" who would only do the right thing when the decision was forced upon him.[26] Even though Tyndall had left Queenwood College for good in July 1853, Hirst soon after began a position at Queenwood and kept Tyndall abreast of what was happening there. About the same time that Tyndall was declining the Royal Medal, Hirst told him about John Mummery, a Queenwood teacher who was seeking to purchase a German doctorate, with the seeming approval of Edmondson.[27] This was a problem for both Hirst and Tyndall because they believed that openly advertising such false credentials not only debased the character of Mummery but also the reputation of Queenwood itself and therefore anyone connected to the institution.[28] When Hirst informed Tyndall that Mummery began receiving letters at Queenwood addressed to him as "Dr. Mummery,"[29] Tyndall was particularly furious because it was made apparent to him that Edmondson "has been to some extent the abetter of the fraud."[30] This apparently called for drastic measures, and Tyndall recorded in his journal that he wrote to Edmondson with the aim of resolving the matter

> in a manner that will shake his nerves, [I] told him flatly that he was the abetter of a fraud, that his duty is to see that no immoral trick of the kind should be transacted at Queenwood for that his strength consisted in the high-toned morality of the men around him, that instead of requiring to be pushed into an honourable course of action he ought to be the pioneer of such a course, and that the men around him were not such as would respect him if they saw him waver when truth and honour were at stake, warned him that his vacillation and complicity in this dishonourable affair was loosening the very foundations of his establishment.

Tyndall believed that this was precisely how one had to deal with Edmondson, and he noted, "You can get nothing out of the man except by exciting his terror; when forced to act right he will do so, not without."[31]

This led to an interesting exchange between Tyndall and Edmondson's daughter, Jane, which also appears in the volume. Jane was understandably offended by Tyndall's treatment of her father. While her letter is not extant, Tyndall's response is, and gives us a profound example as to how Tyndall's Carlylean philosophy informed his understanding of gender. After defending his treatment of Jane's father, Tyndall suggested that while he would "leave the question to the judges who overlook earthly things" he would also leave it to those same judges "to punish a young lady who instead of writing me a Valentine or something of the kind in which she excels has misapplied her time and talents in sprinkling me with gall and bitterness."[32] It was Tyndall's view that Jane had overstepped her natural role, that she should disconnect herself from Queenwood matters and ultimately concern herself with acting in such a way that would attract a good man. As he suggested in a subsequent letter that was (perhaps wisely) never sent: "You cannot imagine what a repugnance some men feel to the interference of ladies; and moreover, these are the very men who will offer to ladies the most loyal service as long as they remain in their own sphere."[33] It is important to recognize that Tyndall's self-fashioning as a self-sufficient, disciplined, patient, and honest individual was an explicitly masculine identity that was articulated within an epistemological framework that imagined quite separate and distinct roles for men and women based on their "natural" traits.[34] While both Mummery and Edmondson were criticized by Tyndall for not fulfilling their masculine duties, Jane was criticized for transcending her "own sphere." This is a telling example of Tyndall's "inflexible integrity" at work and in concert with his Carlylean philosophy.[35]

The Correspondents

While Tyndall corresponded with fifty-seven different people in this volume, one correspondent features prominently throughout: Thomas Archer Hirst. In total there are 126 letters between the two men included here, 87 from Tyndall and 40 from Hirst. This accounts for 38 percent of the surviving correspondence for the years 1853 and 1854. By the early 1850s Tyndall and Hirst had been close for many years, a friendship that is thoroughly tracked in previous volumes of *The Correspondence of John Tyndall.* They initially met as surveyors in Halifax, and the older, more experienced Tyndall quickly became Hirst's mentor. As their friendship grew over the years included in volume 3, Hirst eventually became Tyndall's "other self."[36] Not only did Hirst travel with Tyndall to Marburg in 1850 in order to begin his own doctorate, when he completed it, he began teaching at Queenwood College just after Tyndall took up his position at the RI. If the letters in volume 3 show how the two men came to appreciate the same Carlylean philosophy, while sharing similar scientific ambitions, the letters contained in this volume give us insight into how they dealt with their careers in transition. They were both preoccupied with maintaining their integrity, particularly as career successes seemed to offer new opportunities for backsliding. They also cared deeply for one another, such that their correspondence offers a useful case study in the making of a lifelong Victorian friendship. Perhaps the other most important figure in Tyndall's life during the period of this volume is Michael Faraday. There are 15 surviving letters, accounting for 4.5 percent of Tyndall's correspondence between 1853 and 1854. Although these numbers show that Tyndall and Faraday corresponded far less than did Tyndall and Hirst, it does not mean that Faraday was less significant to Tyndall's life and career during this time. This brings to the fore an important point: when applying quantitative approaches to historical examinations, it is important to remember that numbers can be misleading. It is essential to support numerical analyses with contextual knowledge.[37] In this case, for instance, the modest exchange of letters between Faraday and Tyndall is likely because the two men worked together at the RI and therefore regularly discussed matters in person when they saw one another at work. By contrast, Hirst resided at Queenwood College near Stockbridge, Hampshire, while Tyndall was living in London. Thus the two friends had far less opportunities to speak face-to-face and relied more heavily on letter writing for keeping in touch.

In the previous volume of Tyndall's Correspondence, Faraday had become something of a mentor for Tyndall and would by 1853 become an important patron by advising the RI managers to hire Tyndall for a professorship. As we have seen, however, Tyndall had other offers available to him, but the RI was always at the top of Tyndall's list, and this primarily had to do with the presence of Faraday. Tyndall believed that he and Faraday had a similar mode of thought, such that Tyndall felt free to express "a speculative opinion" in Faraday's presence in full confidence that Faraday would grasp its meaning.[38] Indeed, Tyndall believed that it was a "great blessing to be near" Faraday, writing in his journal that "one feels one's own manhood increasing by contact with such a man."[39] Tyndall felt that he had much to learn from Faraday, particularly his "deep love of nature" that emanated from his heart and would excite the imagination of all who were around him. "I love strength," Tyndall wrote in his journal, "but let me not forget the possibility of uniting it with softness demonstrated by the character of Faraday."[40] Tyndall also began to think of Faraday as a friend, as someone he could engage in "manly conversation" while feeling entirely at home in his presence.[41] Faraday made Tyndall feel like an insider when for so long Tyndall felt himself on the margins. So when the Government School of Mines made Tyndall an attractive offer to take up a teaching position that would have allowed him to keep his position at the RI, Tyndall decided to turn it down. This was despite the fact that at the Government School of Mines Tyndall would have had the opportunity to work closely with the geologist and botanist Edward Forbes, a man Tyndall greatly admired. In the end, Tyndall recognized that even while he could keep his RI position, he did not want to divide himself between the two institutions and, perhaps more importantly, between the two men. He wrote to Forbes that as "I have plighted my faith to the Royal Institution" he could not "conscientiously accept this otherwise enviable position."[42] He was slightly more revealing in his journal, writing that "no man can serve two masters, and I am particularly incapable in this respect. It is not that I love Forbes less, for I believe he is a noble fellow, but that I love Faraday better."[43]

Tyndall also cultivated relationships with other important British men of science, most notably with Henry Bence Jones and Thomas Henry Huxley. Bence Jones brokered an invitation for Tyndall to give the Friday evening lecture at the RI in 1853 that acted as such a great catalyst for Tyndall's career. And Bence Jones was also central in securing Tyndall's position at the RI. Many of the eighteen letters between them that are included in this volume shed light on the negotiations that went on as Tyndall debated his career options. While this sum does not represent a huge percentage of the total correspondence for volume 4, the letters themselves contain crucial information on Tyndall's life during these years. It is another example of how numbers alone do not tell the whole story. Tyndall was also keen for the two men to become good friends, but he was never sure what role Bence Jones played in the fiasco over the Royal Medal. He was also irritated that Bence Jones was a great admirer of the German physiologist Emil du Bois-Reymond, whose work on animal electricity Tyndall found to be highly dubious. By the end of 1854, Tyndall worried that Bence Jones regretted that he brought him to London and wished that he had du Bois-Reymond instead. Bence Jones "now wishes to have DuBois," Tyndall wrote in his journal, "and may probably in the end find him as intractable as myself. I should however wrong myself to call myself intractable, for my wish has ever been to stand upon a cordial footing with Bence Jones."[44]

There are also twelve letters between Huxley and Tyndall in this volume that give us an insight into their developing friendship that would be so central in the shaping of British science in the years to come. That is 3.6 percent of the total correspondence for the years of 1853 and 1854. Yet within these letters we see glimpses of what Ruth Barton refers to as the "X Network"—that is, the early collaborative strategizing of future X Club members as they strove for power and leadership within the British scientific community.[45] Tyndall met Huxley in 1851.[46] As they shared similar lower-middle-class backgrounds and scientific ambitions, they became close quite quickly. By the end of 1854, it is possible to envision the two men later conspiring to shape British science while forming the nucleus of the scientific naturalists, but that is certainly not the case in early 1853. With Tyndall at Queenwood, Huxley was keen for him to take one of the pending offers in London. Huxley, though, despite being FRS, was himself without a position. The tenuousness of their situations is clear in a letter Tyndall wrote to Huxley imagining that the two of them managed to get the positions in Toronto that they both failed to get the year before. "I trust that we may weave a prediction from these omens, and that providence still intends to build us a Toronto home. That we shall be in closer proximity to each other I sincerely hope and firmly believe."[47] They were clearly looking out for one another during these years, offering strategic advice, while defending each other in public forums. For example, Huxley stood up for Tyndall at the Royal Society to ask why the latter was not present to accept the Royal Medal and demanded that his letters refusing the award be read aloud to all those present.[48] Huxley also offered to share his column on "science" with Tyndall in the *Westminster Review*, an offer that was eventually accepted.[49] In July 1854, Huxley would be appointed lecturer at the Government School of Mines, meaning that by this time both he and Tyndall had stable and secure scientific positions that could act as platforms with which to engage in larger strategic efforts to reshape British science.

Tyndall's former German colleagues corresponded with him as well, most notably Gustav Magnus (twelve letters) and Heinrich Debus (seven letters). Tyndall had worked in Magnus's Berlin lab for much of 1851. After Tyndall made the move back to England, they wrote to each other in great detail about experiments of mutual interest. Tyndall relied on Magnus to send him and his English colleagues the much more precise German apparatuses that were central to Tyndall's experimental science. Tyndall came to know Debus at Marburg where the latter was the chemical assistant in Bunsen's laboratory. During the period of this volume, Debus, like both Tyndall and Hirst, ended up at Queenwood College after his stint at Marburg, teaching chemistry there. The letters between Debus and Tyndall are rich, traversing a wide range of topics from experimental details to the politics of German and British science, as well as various incidents occurring at Queenwood College.

Tyndall also kept in contact with old friends from his surveying days with letters from the Coxes (ten), the Steuarts (two), and the Wynnes (six). Of particular note are the thirteen letters from Margaret Allen, whom Tyndall met when working with her future husband, Robert, on the Ordnance Survey of Ireland in 1840–41.[50] As the letters in this volume show, Tyndall became something of a patron to Margaret after her husband, who had drinking problems, was sent to America to reform his debased character. We glimpse a softer side of Tyndall in these letters as he sought to comfort Margaret after the loss of three of her young children to consumption. He also helped her financially, though on one occasion he did not like her impertinence in requesting some money, writing in his journal that she appeared to diverge from her "suffering" and "striving" character.[51] Tyndall therefore "asked her to sit down and write to me calmly and thus assured me that I am dealing with the same longsuffering Mrs Allen and she should have the money by return of post."[52] While Tyndall's treatment of Allen in this case is not unlike his treatment of Jane Edmonson, as both women were chastized for deviating from their more "natural" characters, Allen's sincerely apologetic response made Tyndall realize that he "had no right to let a single note disturb the estimate of a character founded upon years of acquaintanceship."[53]

Tyndall also kept in touch with a few families that he become close with while at Queenwood College. As is clear in many of his letters, as well as in his journal, Tyndall was a committed teacher and he became very interested in the future of a few of his former students, such as Alexander Cumming and Tommy Grieve. Tyndall continued to offer their families education advice, while also writing the boys letters of encouragement. Even though Tyndall wanted more than anything a position that would allow him the freedom to develop and practice his scientific research, it was bittersweet for him to take up the RI position because it meant leaving Queenwood College and the boys he had grown so fond of teaching and mentoring. In his journal entry for 6

July 1853, Tyndall made a point of noting that this would be the last entry he would make from Queenwood: "I carry many sweet thoughts away from it, I carry the thought that when I came the place was dying and it is now so vigorous that doubts are entertained of their being able to accommodate all that are coming. But I carry away with me little blossoms of affection, watered by the dew of the heart and which to me are sweetest of all, affection, not for women but for little boys."[54] This is a pertinent reminder that if Tyndall was essentially beginning an exciting new chapter in his life as this volume opens, he was also leaving an important part of his earlier life behind.

Conclusion

If much of Tyndall's time in the years covered by the previous volume of his correspondence was often taken up by translating monographs for the *Phil. Mag.* and amassing testimonials for job applications, in the years covered by this volume Tyndall's new positions at both the RI and LI meant that he spent a lot of time preparing for lectures. He recognized that a successful public lecture relied on just the right mixture of specialized competence and showmanship, and he anguished over getting that mixture just right by spending a great deal of time planning and practicing the lectures themselves. This hard work clearly paid off, as his lectures were very well attended and it became clear early on that he would make a strong, suitable replacement for Faraday when the time came.

Tyndall, however, was his own worst critic, and he rarely commented on a lecture in his journal without criticizing it in some way. He demanded a great deal from himself, both in the lecture theater and in the laboratory, and this constant work and pressure led to health problems such as exhaustion and headaches. A recurring theme throughout the letters in this volume is Tyndall's complaints about the state of his brain, which was so overtaxed that he often found himself unable to work at all. His usual remedy for these bouts of migraine and general ill health was to read Carlyle and Emerson, but he also began to discover that long "rambles" in the forest cleared his head a great deal and did wonders for his productivity upon his return. We therefore see the beginnings of Tyndall the adventurer and countryside wanderer in this volume.

If Tyndall was finding that being out in nature provided some health remedies, he was also finding that it provided some spiritual guidance as well. It is well known that Tyndall's religious views were heterodox and generally in flux in the 1850s.[55] But his visit to Germany in the summer of 1853 seems to have led to an important realization about those views, particularly after visiting with du Bois-Reymond, who Tyndall found to be too much of a materialist. "The conservation of force," Tyndall wrote in his journal, "is his sole creed."

For Tyndall, there was something more important underlying the nature of all things, which could not simply be reduced to mechanical forces.[56] As Tyndall pondered the final hours of 1854, as the clock was ticking toward a new year, he wrote, "Another year is just sliding into the past eternity—humanity in 90 minutes more will sweep by another landmark in the waste of time—hurrying on witherward? We know the destiny of individuals—we know the clod hut towards which they wander, but to what destiny does the race journey on? It is strange if nobody knows more of this than man himself, strange if he be the wisest being in the universe, and if a wiser exist what reflections does it bring! But while I ponder such a being eludes the clasp of my understanding—his presence is to be felt but not explained."[57] While Tyndall's own experimental research at this time may have been highly specialized, his thoughts here about the nature of the universe, and what may exist beyond the capacities of human explanation, make it clear that he was interested in exploring the wider implications of his science.

Notes

1. For Tyndall's election as FRS see Letter 0630 (volume 3).
2. Journal, 6 December 1853, JT/2/13b/645.
3. Letter 0698.
4. Letter 0713.
5. Journal, 16 February 1853, JT/2/13b/601; and letters 0715 and 0717.
6. Letter 0719. Also see Journal, 26 April 1853, JT/2/13b/603.
7. Letter 0721.
8. Letter 0717.
9. Letters 0762 and 0764.
10. Journal, 10 November 1853, JT/2/13b/634–35.
11. Journal, 23 December 1853, JT/2/13b/648; and letters 0858, 0859, and 0861.
12. See letters 1005, 1006, and 1008; see also Journal, 20 November 1854, JT/2/13b/699.
13. Journal, 26 June 1853, JT/2/13b/607.
14. On the influence of Carlyle on Tyndall and other scientific naturalists see Ursula DeYoung, *A Vision of Modern Science: John Tyndall and the Role of the Scientist in Victorian Culture* (London: Macmillan Palgrave, 2011), pp. 60–69; Heather Ellis, *Masculinity and Science in Britain, 1831–1918* (London: Palgrave Macmillan, 2017), pp. 117–48; and Frank M. Turner, "Victorian Scientific Naturalism and Thomas Carlyle," in *Contesting Cultural Authority: Essays in Victorian Intellectual Life* (Cambridge: Cambridge University Press, 1993), pp. 131–50.
15. This is perhaps best expressed in Thomas Carlyle, *On Heroes, Hero-Worship, and the Heroic in History* (London: James Fraser, 1841).
16. Journal, 26 June 1853, JT/2/13b/607.
17. Letter 0758.
18. See letters 0815 (Francis), 0821 (Gassiot), and 0822 (Sabine).
19. J. Tyndall, 'On Diamagnetism and Magnecrystallic Action', *Phil. Mag.* 2:10 (September 1851), pp. 165–88.
20. Journal, 15 November 1853, JT/2/13b/636.
21. Letter 0828.

22. Roland Jackson, 'John Tyndall and the Royal Medal That Was Never Struck', *Notes and Records of the Royal Society* 68:2 (2014), pp. 151–64.
23. Letter 0835.
24. Letter 0840.
25. Journal, 29 November 1853, JT/2/13b/641. This was quite in contrast to Huxley, who was apparently furious that Tyndall relinquished the award. See Journal, 26 November 1853, JT/2/13b/640.
26. Journal, 7 February 1853, JT/2/13b/657–58.
27. Letter 0833.
28. Letter 0876.
29. Letter 0879.
30. Journal, 7 February 1853, JT/2/13b/657.
31. Ibid., JT/2/13b/657–58
32. Letter 0890.
33. Letter 0893, n. 4; and Journal, 27 February 1854, JT/2/13b/663–64.
34. The classic statement of separate sphere's ideology appears in Leonore Davidoff and Catherine Hall, *Family Fortunes: Men and Women of the English Middle Class, 1750–1850* (Chicago: University of Chicago Press, 1987).
35. DeYoung, *A Vision of Modern Science*, pp. 60–69.
36. See the introduction to Volume 3.
37. Margo Anderson, "Quantitative History," in William Outhwaite and Stephen Turner (eds.), *The SAGE Handbook of Social Science Methodology* (London: Sage Publications, 2007), pp. 246–64. See also Pat Hudson, *History by Numbers: An Introduction to Quantitative Methods* (London: Bloomsbury Academic, 2000).
38. Journal, 12 October 1853, JT/2/13b/627.
39. Journal, 28 October 1853, JT/2/13b/630.
40. Journal, 5 November 1853, JT/2/13b/632
41. Journal, 22 November 1853, JT/2/13b/639.
42. Letter 0863.
43. Journal, 31 December 1853, JT/2/13b/651–52.
44. Journal, 25 December 1854, JT/2/13b/707.
45. Ruth Barton, "Huxley, Lubbock, and Half a Dozen Others: Professionals and Gentlemen in the Formation of the X Club," *Isis* 89 (1998), pp. 410–44. See also Roy, Macleod, "The X Club a Social Network of Science in Late-Victorian England," *Notes and Records of the Royal Society of London* 24 (1970), pp. 305–22; and Adrian Desmond, "Redefining the X Axis: 'Professionals,' 'Amateurs' and the Making of Mid-Victorian Biology—A Progress Report," *Journal of the History of Biology* 34 (2001), pp. 3–50.
46. Letter 0582 (volume 3).
47. Letter 0826.
48. See letters 0845 and 0848.
49. See letters 0980 and 0985. Their first shared column appeared in April 1854.
50. See volume 2 of the Tyndall Correspondence.
51. Journal, 4 March 1854, JT/2/13b/665; and letter 0896, n. 1.
52. Ibid.
53. Journal, 6 March 1854, JT/2/13b/665. Allen's response is letter 0896.
54. Journal, 6 July 1853, JT/2/13b/609.
55. See, for instance, Geoffrey Cantor, "John Tyndall's Religion: A Fragment," *Notes and Records of the Royal Society* 69 (2015), pp. 419–36.
56. Journal, 24 July 1853, JT/2/13b/612.
57. Journal, 31 December 1854, JT/2/13b/709.

EDITORIAL PRINCIPLES

Our aim is to include all extant letters to and from John Tyndall. Some of the letters are from published sources, such as letters written by Tyndall to newspapers or journals, but the majority are personal letters that are transcribed either from the original or from typescript copies produced by Louisa Tyndall from the originals, which are sometimes now missing. One central editorial principle has been to transcribe, wherever possible, from original letters rather than Louisa's typescripts. This is because of Louisa's conscious and unconscious editorial interventions, a point discussed in more detail below. We have also aimed to reproduce, as accurately as possible, the text of the letters as they were written. Idiosyncrasies of spelling and punctuation in the handwritten letters have therefore been preserved; but, because Louisa made many inadvertent errors, mistakes appearing in typescript sources have been silently corrected. But, in general, nonstandard spellings (e.g., the use of dialect) has been retained even when in a typescript-only letter (though there is the possibility that the spelling is Louisa's, not her husband's). Where the transcription is based on a typescript transcription by Louisa (or one of her assistants), this will be indicated by "LT Transcript Only." Some of the letters in this volume are transcripts made by Tyndall, indicated by "JT Transcript Only," while others are LT transcripts of letters copied from the originals by Tyndall, indicated by "LT Transcript of JT Transcript Only."

We have included illustrations that are part of the letter. In cases where it was possible to reprint the original handwritten image, we have done so. In other cases, volume editors have reconstructed drawings because it was not possible to reprint an image from the handwritten letter (due, for example, to deterioration of the original document).

The letters are presented in chronological order of writing. Where more than one letter has the same date, letters from Tyndall (in alphabetical order of addressee) come first, followed by letters to Tyndall (in alphabetical order of writer), with the exception that if the order of writing can be determined, that order takes precedence. Where letters cannot be dated precisely, they are placed at the latest likely date of writing. (In some cases, if it is possible, but unlikely, that they were written later, notes indicate such possibilities.) To indicate the beginning of a new year, the date of that year has been inserted in a large font before the first letter from that year. Readers should note that although volumes are arranged in chronological order, letters that are discovered from years already covered in published volumes will appear in one of the final volumes of the correspondence.

For non-English-language letters, both the transcription of the original and the English translation are presented. The transcription of the non-English letter appears first, followed by the English translation in a smaller font. Editorial notes are inserted in the English translation. Some foreign letters are in an original hand, while others exist only as transcriptions by Louisa or others who often did not have a full command of the language. In the latter case these letters contain frequent spelling, grammatical, and interpretation errors that do not seem to have been committed by the original author. In these typescript-only cases, and only in these cases, spelling and grammatical errors have been corrected, as have poor interpretations where a better word choice was obvious from the context.

There is a standard format for each letter. The first line lists the author, if it is to Tyndall, or the recipient, if it is from Tyndall; the date; and the letter number. The second line lists any information supplied by the writer in their salutation, such as the place, date, or time of day when the letter was written. (In some cases, though, where the writer put the date at the foot of the letter, we have left it there in order to reproduce the text accurately.) The opening salutation, the text of the letter, and closing then follow. Then come the source and editorial notes, the latter indicated in the text by Arabic numerals.

Editorial notes are intended to provide the contextualization necessary to understand the contents of each letter. There are basically six types of editorial notes. Many of them are informative notes on persons mentioned in the letters. If a person referred to in a letter is not described in a note, they have been mentioned more than twice in the volume and they will have an entry in the Biographical Register at the back of the volume. For the rest—those mentioned once or twice—biographical information will appear in the editorial note appended to the letter in which he or she is first mentioned. When there is a second reference to that person, a note will refer back to the letter in which he or she was first mentioned. Biographical notes on well-known figures such as Faraday or Huxley stress what that person was doing at the time when the letter was written rather than providing a full overview of the person's life.

A second type of editorial note identifies allusions and quotations. In the case of quotations, where possible, the first edition of the work is cited unless a different edition is mentioned in the letter. If a quote is not in English, a translation is given. Glosses on obscure places, abstruse words, and words used in an unfamiliar sense comprise the third type of editorial note.

The fourth type provides information about the context of the letter, drawing on such sources as contemporary publications, scholarly sources, Tyndall's personal journals and field notes, and other letters in the Tyndall Correspondence. The fifth type of editorial note is cross-references to other letters in the Correspondence, referring to the letter number. The last type is information on significant textual changes that Tyndall made to the original that is thought to be relevant to the meaning of the letter.

There are a series of conventions for indicating when editors have given additional information about the text of a letter:

When Louisa Tyndall has made an annotation or marginal insertion that the editors have deemed important enough to retain, we have included her annotation in an editorial footnote.

When editors other than LT have made insertions, they are indicated by unitalicized square brackets, for example, [word].

When certain words are ambiguous and we have had to conjecture their meaning they are indicated by italicized square brackets, for example, *[*words*]*; when we have considered such words illegible the number of illegible words are indicated in full italicization and square brackets, for example, *[3 words illeg]*.

If a word or series of words are not present or have been destroyed—by inkblots, holes, or mold, for example—they are indicated as missing in italicized angular brackets, for example, <*3 words missing*>.

In extremely rare cases where it is clear that Louisa herself has intentionally destroyed a few words, they are indicated as excised in italicized angular brackets, for example, <*3 words excised*>.

Finally, a brief word about Louisa. She was responsible for collecting together many of Tyndall's letters in order to write a biography of her late husband. This biography was not completed. It is important to recognize that Louisa was not always reliable as a transcriber. She often threw out the manuscript letter when she had made her transcription, but some original manuscripts remain, which indicate her editing processes. Like any transcriber, she was sometimes inaccurate, but corrections show that she proofread her transcriptions. However, she also edited Tyndall's writing to make his letters seem more consistent or grammatically correct. For example, she made punctuation more formal, corrected grammar, and, occasionally, corrected quotations. She consistently corrected Tyndall's usual (deliberate) lower case for "god" to "God." Her typewriter was unable to insert the superscripts that Tyndall used in abbreviations, or the "&c" symbols which stood for an address at the foot of the letter. (Modern typefaces are unable to deal with all the variant scrawls that the "etc" symbol took, and they are all formalized here to "&c.") Louisa also left out details that she considered too private or too salacious for public consumption. She equated postmark and date of writing and overlooked the time taken to write some letters. When she inferred dates from internal evidence, her reasoning was sometimes faulty. Similarly, her identifications of people are sometimes in error. Independent evidence has always been sought for her dates and identifications. Lastly, the patterns of punctuation differ between LT sources and manuscript sources. Tyndall himself used dashes more often than commas, semicolons, and stops, and he used dashes of many different lengths that are all standardized here to em-dashes. It is often difficult to distinguish between very short dashes and commas or very short dashes and stops. Similarly, capitalization is often ambiguous; distinguishing S/s for "science," for example, is a particular problem. Therefore, scholars for whom punctuation and capitalization are significant will need to consult the manuscripts. Users should also be aware that throughout this project the transcribers and editors have generally worked from scans and photographs rather than the original manuscripts.

NOTE ON MONEY

In the nineteenth century the currency used in England was the pound sterling, usually abbreviated to £ (or, occasionally, to L). A pound consisted of 20 shillings (abbreviation, s) and each shilling contained 12 pennies (d). A guinea was 21 shillings, that is £1 1s. A crown was 5 shillings (5s), and half a crown equalled two shillings and sixpence (2s 6d). The smallest coin was a farthing, one quarter of a penny (¼d).

In the notes to this volume the editors have used the following format: £3 4s. 5d. Several forms of representation were in contemporary use. Sometimes the numbers for pounds, shillings, and pence were separated by dots, forward slashes, or spaces; thus £3 4s. 5d. could be written as £3.4.5, £3/4/5, or £3 4 5. A zero could be represented by a dash; thus 7 shillings was often written as 7/- or just 7/. A half crown was often written as 2/6.

German currency is occasionally mentioned in this volume. The currency of the northern German states was the thaler (Thaler in German, usually abbreviated Th) or Reichsthaler (usually abbreviated Rthlin), the groschen or silbergroschen (Silbergroschen, abbreviated SG or Sgr), and pfennig (Pfennig, abbreviated Pf). A thaler consisted of 30 groschen, a groschen of 12 pfennig. This currency was introduced in Prussia in 1821 and spread to other northern German states following the currency union of 1837. Groschen were called Silbergroschen, especially in Prussia, to distinguish them from the pre-1821 coins of the same name.

The currency of the southern German states was the Gulden, also often referred to as Florin, which was based on the Gulden or Florin (abbreviated fl.) that originated in the Holy Roman Empire. The southern German states also introduced the Kreuzer (abbreviated x.) (1 Florin/Gulden = 60 Kreuzer). Throughout the German states, 1.34 Gulden/Florin was worth 1 Thaler. The system of currency in both the northern and southern German states was in place until German unification in 1871. In mid-century the exchange rate between British and German currencies was approximately 7 Th per £.

French currency is also occasionally mentioned in this volume. The franc was the national currency in France from 1795 until 1999 when France adopted the Euro. In the mid-nineteenth century the exchange rate between British and French currencies was approximately 25 francs per £.

TIMELINE OF JOHN TYNDALL'S LIFE

Year	Event
c. 1822	John Tyndall born at Leighlin Bridge, County Carlow, Ireland
c. 1836	Attends John Conwill's National School in Ballinabranna
1839	Begins employment as civil assistant with Ordnance Survey of Ireland in Carlow
1840	Becomes civil assistant in the Ordnance Survey Office at Youghal, Cork County
1841	First appearance in print (poem in *Carlow Sentinel* under pseudonym 'W[alter] S[nooks]')
1842	Joins English Ordnance Survey in Preston
1843	Leads written criticism of Ordnance Survey in *Liverpool Mercury*; dismissed November
1844	Home in Ireland, mostly unemployed until obtaining a position with the firm of Nevins and Lawton, Surveyors, of Manchester
1845–47	Works for Richard Carter, Surveyor, of Halifax
1845	Meets Thomas Archer Hirst in Halifax
1847	Begins teaching mathematics and surveying at Queenwood College, Hampshire
1848–50	At University of Marburg, working for PhD with Robert Bunsen and Friedrich Stegmann
1850	Returns to England: meets Michael Faraday and William Francis and attends his first British Association meeting before returning to Marburg
1851	Moves from Marburg to Berlin (April–June); then returns to Queenwood
1852	Elected Fellow of the Royal Society
1852	Friendship developed with Thomas Henry Huxley
1853	First lecture at Royal Institution (RI)
1853	Meets Herbert Spencer
1853	Appointed professor of natural philosophy at RI
1853	Awarded Royal Medal for magnetic work; declines when controversy arises over award
1857	Climbs Mont Blanc for first time with Hirst
1859	Demonstrates existence of greenhouse gases and climatic implications
1859	Joins Government School of Mines as professor of natural philosophy
1860	*The Glaciers of the Alps* published
1860	Meets John Lubbock and his wife, Ellen Lubbock

Year	Event
1861	First ascent of the Weisshorn
1861	Delivers his first Christmas Lectures at RI
1862	*Mountaineering in 1861* published
1862	Begins to champion Mayer's priority in discovery of conservation of energy
1863	*Heat Considered as a Mode of Motion* published
1864	Awarded Royal Society's Rumford Medal
1864	First meeting of X Club
1865	*On Radiation* published
1865	Engages in public controversy over the efficacy of prayer
1866	Succeeds Michael Faraday as scientific adviser to Trinity House
1867	Michael Faraday dies; Tyndall becomes superintendent of the RI
1867	*Sound: A Course of Eight Lectures* published
1868	Successfully climbs Matterhorn
1868	Discovers the cause of light scattering, to be known as 'Tyndall Effect'
1868	*Faraday as a Discoverer* published
1868	Lecture on 'Scientific Materialism' at the British Association
1868	Resigns from Royal School of Mines
1869	Joins Metaphysical Society
1870	'On the Scientific Use of the Imagination' lecture at British Association
1870	*Researches on Diamagnetism* published
1870	*Three Scientific Addresses* published
1870	*Researches on Diamagnetism and Magne-Crystallic Action* published
1871	Meets Louis Pasteur for first time while in Paris
1871	*Hours of Exercise in the Alps* published
1871	*Light and Electricity* published
1871	*Fragments of Science* published
1872	Debates over how to measure the efficacy of prayer, aka the 'Prayer-Gauge Debate'
1872	*Contributions to Molecular Physics in the Domain of Radiant Heat* published
1872	*The Forms of Water in Clouds and Rivers, Ice and Glaciers* published
1872–73	USA Lecture Tour
1873	*Six Lectures on Light* published
1874	As President of the BAAS, delivers the 'Belfast Address'
1876	Marries Louisa Charlotte Hamilton
1877	Develops 'Tyndallization' (discontinuous heating and cooling process to destroy heat-resistant spores)
1877	Tyndall and Louisa build summer cottage 'Alp Lusgen' at Bel Alp, northern side of Valais, above Brig
1877	*Fermentation and its Bearings on Phenomena of Disease* published
1879	*Fragments of Science*, which had gradually expanded, first published in two volumes
1881	*Essays on the Floating Matter of the Air* published
1885	Tyndall and Louisa build Hindhead retreat, Surrey Downs

Year	Event
1887	Resigns from RI
1890	Fight with Gladstone in *The Times* over Irish Home Rule
1892	Hirst dies
1892	*New Fragments* published
1893	Dies Haslemere, Surrey, accidentally poisoned

TIMELINE OF EVENTS IN JOHN TYNDALL'S LIFE SPECIFIC TO VOLUME 4

Month and year	Biographical details	Notable social, political, and cultural events
January 1853	Visits (1) Margaret Allen in Preston; saw her dying son	The UK annexed (20) Burma, ending the Second Anglo-Burmese War
	Tyndall's paper 'On Molecular Influences' was read (6) at the Royal Society	
	Attends Faraday's Friday evening lecture (21) at the RI 'Observations on the Magnetic Force'	
	Spent much time translating for the *Scientif. Mem.*	
February 1853	Short letter 'On the Temperatures of Conductors of Electrical Currents' published in the *Phil. Mag.*	
	Gives Friday evening lecture (11) at the RI 'On the Influence of Material Aggregation Upon the Manifestations of Force'	
	Was offered to give a course of four lectures at the LI (16)	
	Was offered to give a course of four lectures at the RI (19)	
	Corresponded with Huxley (22, 25) about his professional prospects	
March 1853	Gave a lecture (16) 'On Certain Points Connected with the Transmission of Heat through Bodies' at the LI	Pope Pius IX recovered Catholic hierarchy in the Netherlands
	Offered a professorship at the LI (18)	
April 1853	Began preparing for his course of lectures at the RI	First passenger rail opened in India (16) connecting Bori Bunder, Bombay to Thane

Month and year	Biographical details	Notable social, political, and cultural events
April 1853	Began researching for a paper on animal electricity for the *Medico-churgical Review* (26)	
May 1853	Gave first lecture (21) on 'Air and Water' at the RI Elected an honorary member of the Société Royale Hollandaise des Sciences (20) Offered the Professor of Natural Philosophy position at the RI (23) Gave second lecture (27) on 'Air and Water' at the RI	Argentina adopted (1) its constitution Great Industrial Exhibition held in Dublin (12 May–31 October) HMS *Resolute* from the west meets (18) the HMS *Investigator* from the east, establishing the existence of the Northwest Passage in the Arctic World's first public aquarium opened in the London Zoo
June 1853	Gave third lecture (3) on 'Air and Water' at the RI Gave fourth and final lecture (11) on 'Air and Water' at the RI Tyndall was informed about the possibility of a position opening up at the Government School of Mines	UK naval fleet traveled (20) to Besika Bay to defend Ottoman Empire against Russian threats The Gadsen Purchase (24) where US bought 29,670-square-mile (76,800 km^2) territory from Mexico, which today is in southern Arizona and southwestern New Mexico
July 1853	Tyndall was officially elected professor of natural philosophy at the Royal Institution (4) Last day as a master at Queenwood College (6) Offered a position at the Government School of Mines (6) Read Thomas Carlyle's *Discourse on the Nigger Question* (10) Traveled to Berlin (11) and visited du Bois-Reymond, Magnus, Clausius, and Poggendorff Repeated some of du Bois Reymond's experiments (21) Spent some time with du Bois-Reymond (24)	US Navy visited Japan (9) and requested trade rights (14)
August 1853	Remained in Berlin until 22 and experimented on the Trevelyan instrument and engaged in experiments in connection with Peltier's discovery of the generation of cold in the circuit	New Zealand acquired (12) self government Establishment (23) of the International Meteorological Organization in Brussels

Month and year	Biographical details	Notable social, political, and cultural events
August 1853	Finished article on animal electricity for the *Medico-churgical Review* After leaving Berlin (22), Tyndall traveled through the Harz Mountains before stopping in Marburg (29) Spent a few days in Marburg (29 August–September 1) visiting Knoblauch, Kohlrausch, Dove, and Wiedemann	
September 1853	Left Marburg (2) and arrived in Paris (4) via Brussels Visited Magnus at the College de France (5–9) and saw Richard Owen read a paper at the Academy of Sciences (5) Met up with Bence Jones in Paris (18–19) and travelled to Folkestone, UK (22–23) and then to Queenwood for six days (24–29) before heading to London (30)	Antoinette Blackwell became the first US woman ordained as a minister (15) France annexed (24) New Caledonia
October 1853	Spent much time at the RI preparing his lectures on 'Heat' and experimented with the electro-magnet Translated Helmholtz's paper for the *Scientif. Mem.* (22) Began writing (24) a critique of James Forbes's researches on vibrations that will be published in *Phil. Mag.* Translated Magnus's paper for the *Scientif. Mem* (28)	The Crimean War (4) broke out between the Ottoman Empire and Russia Maastricht-Aken railway in Netherlands opens (23) Lockout of Preston cotton mill workers (31 October–15 May 1854)
November 1853	Tyndall was informed (4) that he has been awarded the Royal Medal Tyndall met (10) with Henry De la Beche to discuss a position at the Government School of Mines Gassiot informed (15) Tyndall that some members of the RS were in opposition to him being awarded the Royal Medal	Richard C. Carrington determined (9) the solar rotation rate by watching low-latitude sunspots Battle of Sinop (30), Crimean War, in which the Russian fleet destroys the Turkish fleet

Month and year	Biographical details	Notable social, political, and cultural events
November 1853	Tyndall declined (16–18) the Royal Medal At Huxley's request, Tyndall's letters declining the Royal Medal were read (30) at a meeting of the RS	
December 1853	Tyndall engaged in a series of frustrating exchanges (1–5) with Henry De la Beche about taking up a position at the Government School of Mines Spent the day (11) with Captain Wynne and his family Received a proof (16) of his article on animal electricity Visited Queenwood College (17–20) Confronted Edmonson (18) about Mummery's purchase of a German doctorate Attended Queenwood College's Christmas Concert (19) Finally received (27) a clear offer from the Government School of Mines *Scientif. Mem.* was published	UK signed a protocol (3) with France, Austria, and Prussia beginning the Crimean War Dinner party is held (30) inside a life-size model of an Iguanodon created by Benjamin Waterhouse Hawkins and Richard Owen in South London
January 1854	Declined (1) the Government School of Mines position Became a 'conductor' of the *Phil. Mag.* Gave the first (17) of twelve lectures on 'Heat' at the RI Essay on animal electricity was published in the *Medico-churgical Review* 13 (1854), pp. 126–42 Dined with the Red Lions Club (19) and was toasted as the youngest lion Read a paper at the RS (26) Gave Friday evening lecture (27) at the RI 'On the Vibrations and Tones produced by the Contact of Bodies having different Temperatures'	Lincoln University, an African American college, chartered (1) in Oxford, Pennsylvania

Month and year	Biographical details	Notable social, political, and cultural events
February 1854	Delivered (2) first of six lectures at LI on 'Electricity and Magnetism' Before a large audience at the RI (3), Faraday introduced Tyndall to Herschel, a public recognition of Tyndall Wrote to Edmonson (7) demanding that he stop abetting Mummery's fraud at Queenwood Dined (16) with the Red Lions Club, sitting next to Huxley Disputed (25) with Jane Edmonson about Tyndall's treatment of her father	UK recognized (17) the independence of the Orange Free State UK sent ultimatum (27) to Russia to withdraw from two Ottoman provinces
March 1854	Asked (10) to contribute a section on 'Natural Philosophy' for Edward Hughes's series of school books (eventually published in *Hughes Reading Lessons*) Dined (16) with the Red Lions Club Gave Friday evening lecture (19) at the RI on 'Phenomena Connected with the Motion of Liquid'	Charles Miller patented (7) first US sewing machine to stitch buttonholes Royal Navy fleet sailed (11) under Vice Admiral Sir Charles Napier
April 1854	Twelfth and final lecture (4) on 'Heat' at the RI Visited Queenwood (6) Spent a week (7–14) with his American cousin, Hector Tyndale, who was visiting Gave first (25) of seven lectures on 'Ebullition, Combustion, and other Phenomena of Heat' at the RI	Dickens's *Hard Times* was first published (1) in serial form in *Household Words* San Salvador destroyed (16) by earthquake
May 1854	Burnt his fingers experimenting on phosphorous (6) Gave Friday evening lecture (19) at the RI 'On Some Phenomena Connected with the Motion of Liquids' Gave lecture (27) 'On the Importance of Physics' at the RI	

Month and year	Biographical details	Notable social, political, and cultural events
June 1854	Attended (12) a concert at Queenwood College Offered (23) a lectureship at St. Bartholomew's Hospital, which he declines Received (28) a £100 grant from the Royal Society	The Crystal Palace reopened (10) in Sydenham, South London The first class (10) of the US Naval Academy students graduated
July 1854	'On the Vibrations and Tones Produced by the Contact of Bodies Having Different Temperatures' was published in *Phil. Mag.* Meeting (12) of the Entomological Club Spent a week in Broughton Wrote first installment on 'Natural Philosophy' for *Hughes Reading Lessons*	
August 1854	Travelled (31 July–3 August) with Hector Tyndale to the Isle of Wright Spent many days translating and experimenting Visits (20) Queenwood Finished translating (22) Wertheim Corrected proofs (23) for *Hughes Reading Lessons* Experimented on the polarity of bismuth (26)	Oxford University Act of 1854 abolished the requirement for undergraduates to undergo religious tests Smith & Wesson patented (8) metal bullet cartridges Epidemic of cholera in London killed 10,000 (31 August–8 September) Battle of Petropavlovsk (30–31), Crimean War
September 1854	Began writing (2) a memoir 'On the nature of the Force by which Bodies are repelled from the Poles of a Magnet; to which is prefixed, an Account of some Experiments on Molecular Influences' Traveled (20) to Liverpool for the 24th meeting of the BAAS Dined with the Red Lions Club (25)	John Snow traced the source of the cholera, validating his theory that cholera is water-borne Battle of Alma (20), Crimean War, in which the British and French defeated the Russians Beginning of the Siege of Sevastopol (25 September–8 September 1855) that would eventually lead to Russia's defeat
October 1854	Trip to Liverpool and then Wales Stayed with Edward Frankland in Richmond Grove near Manchester on 15–16	Florence Nightingale and a group of volunteer nurses leave England (21) to care for injured British troops during the Crimean War

Month and year	Biographical details	Notable social, political, and cultural events
October 1854		As part of the Battle of Balaclava (25), the disastrous cavalry Charge of the Light Brigade occurred
November 1854	submitted memoir (2) to *Phil. Trans.* Decided that his next course of lectures at the RI will be on 'Electricity' Spent ten days with Hirst in Broughton in early November Death of Edward Forbes (18)	As part of the Crimean War, the British and French defeated force of 50,000 Russians at the Battle of Inkerman (5)
December 1854	Informed (10) that his recent memoir submitted to the *Phil. Trans.* was chosen for the annual Bakerian Lecture of the Royal Society Visited Queenwood (16–19) for the Christmas concert (18) Appointed (28) secretary to the Committee for Magnetism and Electricity for the 1855 Paris Exhibition	Pope Pius IX proclaims (8) the Immaculate Conception

The Correspondence of John Tyndall

1853

From Richard Carter 7 January 1853 0697

Halifax | 7 Jan^{y}, 1853

Dear Sir,

I have been solicited by a lady,[1] a friend of yours, to recommend a suitable Educational Establishment where she might place her son, a fine youth of 12 years—I at once thought of Queenwood College, having seen, a little while ago, that you had renewed your connection with it. Will you favour me with a little general information as to changes &c in order that the lady in question may be able to determine whether she will send her little charge in otherwise. I am very anxious to render her any assistance I can & you will therefore imagine how gratified I shall feel for such council, as in your judgment it w[oul]d be her intent to follow.

The father[2] of the little fellow, is an officer in India and the parent of a very intelligent & I think an admirable youth.

I cannot allow this opportunity to pass with congratulating you on the successful presentation of your scientific researches, resulting as I have been gratified to observe in a Fellowship of the Royal Society[3]—Such a distinction is alike honourable to your reputation; as it must be pleasing to your own mind, in the associations of noble champions of science with whom you have secured a position.

I had the pleasure too of receiving a few weeks ago the Inaugural dissertation[4] of our mutual friend Hirst—It was sent to me with 'the Authors kind regards' but unaccompanied by any note or other testimony which would enable me to make any recognition of the pleasing Event.

If you can give me any account of his whereabouts when you write I should feel obliged

Say also if during the vacation you do not intend to pay a visit to our industrious locality

I should be very glad to see you, & in order that I may have a little more of your Society than in your later visit to Halifax, I should be further glad to offer you a welcome to such hospitality as my humble domicile affords

Wishing you very happy returns in the New Year I remain Dear Sir | Very faithfully yours | Richd Carter | John Tyndall Esq. F.R.S.

RI MS JT/1/C/61

1. *a lady*: not identified.
2. *The father*: not identified.
3. *Fellowship of the Royal Society*: Tyndall was elected FRS on 3 June 1852 (see Journal, 3 June 1852, JT/2/13b/568). Also see letter 0633.
4. *Inaugural dissertation*: Hirst's (1852) doctoral thesis at the University of Marburg was entitled 'On Conjugate Diameters of the Triaxial Ellipsoid'.

To Thomas Archer Hirst 9 January 1853 0698

Queenwood. | 9th January. 1853

My dear Tom.

I wish I could write you a nice letter as a return for the pleasure which your letter of the 2nd of this month has given me.[1] But I cant write fluently for this reason—I am terribly costive and my brain is always cloudy when my bowels are not free.[2] I send you a draft for £30 for which you will receive cash on presenting it to Asher the bookseller. Before the end of January I will send you some more as I shall want you to pay for some crystals which Prof. Knoblauch has ordered for me.[3] A happy new year to you Tom also. On New year's night I was in Preston[4] and witnessed one of the saddest, solemnest, and most beautiful scenes that have yet crossed my experience: it was a little boy about three years old with a face full of beauty and radiant with intelligence, the son of a friend of mine, lying on his back dying.[5] You could scarcely believe it, he appeared to be whispered away by angels. This morning your note came to me enclosed in one from Pridie[6] containing some remarks on poor Booth's health which agree but too well with your own. Poor fellow he appears to have received his summons. Pridie I have learned to admire for his behaviour in this affair; in fact I have got quite an affection for him. I will send your letter to Bevington and thus you can tell your own story. I know Sylvester and will write to him or to Thomson on that subject. Beforehand however I wish you would give me a notion of what the nature of the thing is. I know Sylvester himself at the request of the British Association is drawing up a report of some new system of Mathematics.[7] The "Memoirs" have heretofore been in bad repute on account of irregularity, and its present Editors must endeavour to outlive this and gain it a better name.[8] Why Tom there was nothing personal that could be avoided in that letter regarding Adie. Really I think most people would have been more severe than I was. I quite agree with you that the less the scientific man is present in person the better, and I shall endeavour to profit by your remarks—you see that Adie is not yet convinced, and he has said something in his last remarks which I might be very severe upon if I wished—but you will see that he will hardly feel my remarks I shall rub him down so softly and tenderly. I had a letter this morning from Carter making some enquiries regarding Queenwood, as a lady friend of his wishes for his advice regarding an educational establishment for her son.[9] I was glad to hear from Richard,[10] his letter aroused images of old times, which rose before me like the tintings of pictures painted two centuries ago. He mentions having received your dissertation.[11] I chanced to drop in to the Royal Society on Thursday evening last, and heard a paper of my own read[12]—not having the slightest anticipation that such a thing was going to take place. The secretary[13] made some remarks regarding the absence of the author, and at the end of the paper a gentleman arose and made a kind of critical speech, closing it by a certain question which he said he should like to ask the author. The president[14] a minute before had learned that I was in the room, and he smilingly said so adding that the author would perhaps answer the question himself. I got up and replied. It was the first time I had opened my mouth in the Society and I was totally unprepared, but nevertheless what I said seemed to please them for when I sat down they gave me a round of applause. You see where they have put me in the programme of the Royal Institution.[15] It certainly is a position of honour, but it is a position of terror also. There I twinkle like the end of a burnt stick between two such orbs as Airy & Stokes.[16] Well boy in such a case the heart is half the matter and I trust that wont fail me. And now I will pause Tom. If ever I go to Berlin I must look more closely into Dirichlet. Do you *[ever]* see a young fellow named Wiedemann (a son-in-law of Mitscherlich[17]) On Thursday last I had a conversation with Faraday regarding a memoir of his.[18] I returned from *[London]* last night and have not yet had time to read the paper, but if it be a good one, as I believe it is, I will translate it for the Memoirs. I will stop now—may the Lord bless you Tom, and sustain you with corn and wine, and grant of the dew of heaven and of the fatness of the earth.[19] Good bye boy.

Your affectionate | John

Carter says "I had the pleasure of receiving a few weeks ago the Inaugural Dissertation of our mutual friend Hirst. It was sent to me "with the Author's kind regards", b[ein]g unaccompanied by any note or other testimony which would enable me to make any recognition of the pleasing event.

If you can give me any account of his whereabouts when you write I should feel obliged.

RI MS JT/1/T/558
RI MS JT/1/HTYP/224–26

---

1. *your letter*: letter missing.
2. *costive*: constipated (*OED*).
3. *crystals . . . Knoblauch has ordered for me*: Tyndall had been trying to purchase crystals from Knoblauch since July 1852 (see letters 0638, n. 9; 0650; 0651; 0743, n. 4; and 0748, n. 2).
4. *Preston*: a city in Lancashire and home of Tyndall's friends Robert and Margaret Allen.
5. *the son . . . dying*: Charles Edward Allen, son of Robert Copeland Allen and Margaret Allen. See letter 0699.
6. *Pridie*: W. R. (Roby) Pridie, a friend of Hirst from Halifax, and son of the Rev. James Pridie, an Independent Minister in Halifax.
7. *I know . . . system of Mathematics*: see J. J. Sylvester, 'On a Theory of the Syzygetic Relations of Two Rational Integral Functions, Comprising an Application to the Theory of Sturm's Functions, and That of the Greatest Algebraical Common Measure', *Phil. Trans.* 143 (1853): 407–548.
8. *Memoirs*: this is in reference to the *Scientific Memoirs*, a series devoted to publishing the translations of the most important papers that were published in foreign journals. In 1852, Tyndall was appointed an editor (see Volume 3), and at this point he was working on editing the 1853 edition (Tyndall and Francis [eds.], *Scientif. Mem.*). On 16 January, Tyndall remarked that the translation work required for the *Scientific Memoirs* weighed heavy on his mind and he wished that he had not encouraged the continuance of the project (Journal, JT/2/13b/595).
9. *I had . . . Carter*: see letter 0697.
10. *Richard*: Richard Carter.
11. *your dissertation*: see letter 0697, n. 4.
12. *a paper of my own read*: On 6 January, Tyndall's debut at the Royal Society was a paper read discussing heat transfer through various substances. J. Tyndall, 'On Molecular Influences. Part I. Transmission of Heat Through Organic Structures', *Phil. Trans.* 146 (1853): 217–31.
13. *secretary*: Samuel Hunter Christie (1784–1865), British mathematician.
14. *president*: William Parsons (1800–67), the third Earl of Rosse, Irish astronomer.
15. *programme of the Royal Institution*: Tyndall was scheduled to give a prominent Friday evening lecture at the RI on 11 February (see letter 0712, n. 3).
16. *Airy & Stokes*: Sir George Biddell Airy (1801–92), the Astronomer Royal, and the physicist George Gabriel Stokes.
17. *Mitscherlich*: Eilhard Mitscherlich (1794–1863), the German chemist responsible for discovering isomorphism and polymorphism in crystals.
18. *regarding a memoir of his*: Presumed to be Wiedemann's (1852) paper 'On the Motion of Liquids in a closed Galvanic Circuit', which was translated and included in Tyndall and Francis (eds.), *Scientif. Mem.*, pp. 232–56.
19. *sustain you . . . the earth*: Genesis 27:28.

From Margaret Allen 11 January 1853 0699

Stanley Terrace[1] | Jan 11th / 53.

My dear Tyndall

How shall I thank you sufficiently for your beautiful letter[2] the contents are indeed calculated to warm and console my heart. I had more than once thought of writing to acquaint you with the state of my poor darling,[3] being assured from the sympathy you exhibited while here that you would be obliged to *[come]* again. I thought that probably you had not returned to Queenwood and that it would be better for me to wait.

He is truly suffering but so patiently, he looks so sweet & confused, his beautiful eyes constantly turned upwards that I ask myself, can this be the last *[evening]*? Poor dear, since last Wednesday almost every day has brought with it a resumption of convulsive struggles which are painful to behold, indeed I have asked what has such innocence provoked that he should thus suffer? between these he remains perfectly motionless, and unconscious of all around, he has not be sensible since the Saturday you were here. I hope from this that he does not feel the acuteness of the pain which his body evidently suffers. For four days until yesterday he had not spoken, when I observed his lips moving and I heard him whisper with difficulty 'I must come, I must come', and tried to articulate 'Mama' but the sweet sound was not to be heard and none has since escaped but the gentle moan of distress.

We have thought him nearly gone when he has again *[revived]* it is really astonishing how he recovers from these attacks, but Nature clings apparently, with such tenacity to life, in the case of children. After finding your letter I pressed my lips upon his brow, you will not be displeased I became your proxy.

I feel I must apologize for my seeming want of attention, when you were in Preston, the happiness & surprize caused by your visit, together with the situation you found me in, made one forget not only these little civilities which are requisite, but my mind appeared almost shrouded in chaotic darkness so that I both forgot and could not remember, the many things I wished to impart to you. the whole seemed a dream & had it not been for the tangible proof of your brotherly affection I believe I should altogether have doubted having seen you. I propose devoting this kindness to a payment for my pet's likeness it had not been sent to me or else you should have seen it, it is exceedingly like him, though Sister Penny[4] says and I think it is not so handsome or so sweet looking as our dear Charley; she is a favourite sister. I had thought of gratifying a long desired intention but forego the pleasure rather than ask for assistance where I perhaps should receive a *[1 word illeg]*; now when I gaze upon that beautiful forehead, pleasing memories of *[1–2 words illeg]* will fill this poor heart. Ah too 'My heart's *[*oppressed*]*, my peace is o'er', but, in full hope that youth and nature will provide a relief, I do not apply the latter clause of the verse, although the first is ever present in my mind.[5] You will remember from whence the quotation is taken. I had more to say, something in regard of your letter to Robert,[6] but I must conclude ever believe me your obliged

and sincere friend | Maggie Allen.

There is a postscript! I sincerely thank you for the circulars,[7] I shall earnestly hope for your success in every sense of the word.

RI MS JT/1/A/69

---

1. *Stanley Terrace*: in Preston, Lancashire, England.
2. *your beautiful letter*: letter missing.
3. *my poor darling*: Allen's son, Charles Edward Allen, who was dying (see letter 0698).
4. *Sister Penny*: Allen's sister.
5. *My heart's . . . is o'er . . . in my mind*: quotation from Johan Wolfgang von Goethe's *Faustus: From the German of Goethe* (London: Thomas Boosey and Sons, 1821). The 'later clause' is 'I know no rest, No, nevermore'. (See also letter 0918, n. 7.)
6. *Robert*: Robert Copeland Allen, Margaret Allen's husband, and a friend of Tyndall's from his days with the Irish Ordnance Survey (see Volume 2). He was at the time in America doing surveying work for Messrs Bridgens & Moody Publishers of Maps & Lithographers. Last mention of him was in June 1851, when he was in Charleston (see letter 0492). See also letters 0779, n. 2 and 0865, n. 4.
7. *circulars*: not identified.

From Margaret Allen — 11 January 1853 — 0700

Tuesday Evening | Jan 11th/53

My dear Tyndall

The spirit of my beloved child[1] has just winged its flight to the realms of everlasting blessedness. At 20 minutes to eight this evening he departed. Excuse your friend is disconsolate.

Yours sincerely | Maggie Allen

His last words were 'Mama, Mama'

RI MS JT/1/A/70

---

1. *my beloved child*: Charles Edward Allen. See letter 0699.

From Margaret Allen 14 January 1853[1] 0701

Stanley Terrace | Jan. 14th./53.

My dear Tyndall

You will have received the sad news though not unexpected of our beloved darlings death.[2] I now send you the card and a lock of his hair through which his little fingers were often put both in childish glee, and patient endurance. He was interred this morning, laid by his little sister's[3] side in Trinity Church yard, sweetly do they both rest in the arms of the Great Destroyer,[4] and still more happily united in the Paradise of God. Ah! all my prospects for the future are darkened, I had hoped he would have proved a 'crown of rejoicing', even in this world; the budding vigour of his intellects gave me great reason to believe that I should one day see him a wise and good man, but fond Mother thy hopes, prayers, & anticipations so far as it regards this present sphere have been blighted, but he is gone to bloom and bring forth everlasting fruit in that Better Land where none of thy anxieties are required.[5] I do indeed feel my present circumstances keenly, they require an amount of moral and physical courage which I feel I have not in store, but I do try to take your advice and believe that in the end I shall see that these trials are for my benefit.

How shall I write to Robert?[6] how endeavour to fortify his mind? he will feel this loss deeply his dear boy; will you make some mention of your *[1 word illeg]* respecting Charley in your letter let that letter (I am speaking or rather writing to you, believing that you are my friend, in the fullest truest sense) be faithful, searching, but kind, if you think that it would in any way aid you when writing. I will send you Roberts last letter, it might give you an incite into his mind, and enable you to *[*direct*]* your *[*shafts*]* to the right *[*quarter*]*. I also propose this on the above considerations. You have declared yourself My Brother, shall I not have a sister's confidence in you at all times, three last kisses did I plant on that marble brow, beautiful serenely calm after such severe conflicts. Mamma desires her respects to you and hopes the next time you are near Preston, you will come and dine or *[*remain*]* with us, in which I heartily concur

Yours very sincerely | Maggie Allen

Dr. Tyndall

RI MS JT/1/A/71

---

1. *14 January 1853*: Although the handwritten letter is dated 14 January, parish records at Holy Trinity in Preston, Lancashire, England state that Charley was buried on 19 January

('Burials at Holy Trinity in the Town of Preston', http://www.lan-opc.org.uk/Preston/Preston/trinity/burials_1846–1855.html [accessed 13 August 2015]; Register: *Burials, 1816–1888*, p. 299, entry 2387; Source: LDS Film 1470818).

2. *death*: Charles Edward Allen had died on 11 January. See letter 0700.
3. *his little sister's*: Emily Ellen Allen, the daughter of Robert Copeland Allen and Margaret Allen, who died at the age of one year in early 1852 (Burials at Holy Trinity in the Town of Preston', http://www.lan-opc.org.uk/Preston/Preston/trinity/burials_1846–1855.html [accessed 2 February 2015]; Register: *Burials, 1816–1888*, p. 295, entry 2356; Source: LDS Film 1470818).
4. *the Great Destroyer*: death.
5. *'crown of rejoicing'*: 1 Thessalonians 2:19.
6. *Robert?*: see letter 0699, n. 6.

From Henry Bence Jones 14 January 1853 0702

My dear Sir

The proof[1] arrived safely. I am sorry to hear that I missed you when you were in London. Will you dine with me at a quarter before six on Friday next. We will go to Mr. Faraday's lecture afterwards.[2]

Believe me | Yours very truly | H Bence Jones

Jan, 14. 1853. | 30 Grosvenor Street

RI MS JT/1/J/37

1. *The proof*: most likely refers to the physics portion of Liebig's *Annual Report* that Tyndall was translating for Bence Jones, published as J. Liebig and H. Kopp, *Annual Report of the Progress of Chemistry and the Allied Sciences, Physics, Mineralogy, and Geology*, vol. 4, ed. A. W. Hofmann and H. Bence Jones (London: Walton and Maberly, 1853). See letters 0671, n. 1 and 0683, n. 1.
2. *Faraday's lecture*: at the RI on 'Observations on the Magnetic Force' (Journal, 21 January, JT/2/13b/596).

To Thomas Archer Hirst [14 or 15 January 1853][1] 0703

My dear Tom.

Push your visit to London[2] as far forward towards the 21st as you can. I want to be there on the 21st. Write to me however and I will meet you any day you name. I am preparing to meet the emergency stated in the enclosed programme.[3] I looked out upon the blue sky this morning as your letter[4] reached me. The sun shone and it was calm and beautiful—so it spread ages ago and so it will spread when you and I are gone. Poor Booth—he has been soon whispered away—I do not wish to die but I often wish for a closer communion with thoughts which would be natural to a dying bed. The flanks of the Calder[5] will soon be a closed book to Booth. Well when I do die I should like to die somewhat in his fashion; with a friend like you beside me—who knows boy. You may close my eyes also:—bid the little fellow farewell for me. He is but a few paces in advance of us. We are travellers to the same "bourne"[6] and mayhap on the shores of another sea, and clothed in other forms may look upon each other again. One thing is sure, however—all will be right at last.

Your affectionate | John

call to see Carter he mentioned you lately in a letter to me.[7]

RI MS JT/1/T/1010
RI MS JT/1/HTYP/226/1

---

1. *[14 or 15 January 1853]*: on 16 January, Tyndall noted in his journal having 'A morning or two ago received a letter from Tom; it was dated from Brussels'. On 17 January, he also noted having received a reply from 'Tom' arranging to meet in London (JT/2/13b/595). His reply to that letter, sent on 17 January, is letter 0705.
2. *your visit to London*: Tyndall reported meeting Hirst at Waterloo station in his journal entry for 18 January (JT/2/13b/595). See also letter 0705.
3. *enclosed program*: not identified.
4. *your letter*: letter missing (see n. 1).
5. *flanks of the Calder*: the banks of the River Calder near Halifax, Calderdale, West Yorkshire, England, where Booth lived.
6. *bourne*: a small stream or brook (*OED*).
7. *Carter . . . letter to me*: letter 0697. This postscript was written on the top right corner of the first page.

From Henry Bence Jones 17 January 1853 0704

My dear D[r] Tyndall

Mr Faradays lecture is next Friday the 21st.[1] I had asked D[r] Hofmann to dine with us but a few minutes since Sir James Clark[2] I sent to ask me to go there to talk to D[r] Playfair about future schemes. I told Sir J C that you and Hofmann were coming to me and he has asked me to introduce you to him on that evening so that instead of dining here if you are not so inclined we will go there. We will go together if you will call here at a quarter before six next Friday. I have had a letter from D[r] du Bois[3] this morning. I have been in correspondence with him regarding a manual of electricity like Fownes on Chemistry.[4] I thought he might think it worth his attention. He declines. Will you look at De la Rives book[5] & at Fownes and tell me whether such a work would be agreeable to you. The terms are 100 pounds each edition. It must be illustrated like Fownes.

If you say yes, you could see the publishers on Saturday morning, or on Friday before four oclock.

Believe me | Yours very truly | H Bence Jones

Jan[ry] 17. 1853.

RI MS JT/1/J/38
RI MS JT/2/13b/596

1. *Mr Faraday's lecture*: Bence Jones had invited Tyndall to attend the lecture. See letter 0702, n. 2.
2. *James Clark*: personal physician to Queen Victoria who encouraged Albert, the Prince Consort, to back the foundation of the Royal College of Chemistry and the appointment of Hofmann.
3. *du Bois*: Emil du Bois-Reymond.
4. *Fownes on Chemistry*: G. Fownes, *Manual of Elementary Chemistry, Theoretical and Practical* (London: Churchill, 1844).
5. *De la Rives book*: Auguste Arthur de la Rive's *Traité de l'électricité théorique et appliquée*, which was translated into English as *A Treatise on Electricity, in Theory and Practice* (London: Longman, 1853).

To Thomas Archer Hirst [17 January] 1853 0705

Monday[1]

My dear Tom

I start at 8 20 tomorrow morning and hope to be at the Waterloo Station[2] at 11 or a little after. If you are not there to meet me I would not insure your safety across the Channel for the gods will be sure to punish you.

John

RI MS JT/1/T/953
RI MS JT/1/HTYP/226

1. *Monday*: LT wrote 'Jan 18/53'. This is a mistake as the 'Monday' heading to this letter was 17 January. Tyndall noted in his journal entry of 17 January: 'wrote telling him [Tom] that I would start early in the morning' (JT/2/13b/595).

2. *Waterloo Station*: a major train station near the south bank of the Thames. Tyndall was meeting Hirst, who had come from Germany to visit his dying friend, Francis Booth (see letters 0698 and 0703).

From James B. Bevington[1] [20] January 1853[2] 0706

It is not mine but the good genius's which gave it to me to distribute.[3] [...][4] You must not flatter yourself that the cheque which I enclose has any reference to the friendship we may entertain for each other, it is simply for little Booth[5] for whom I have a tender regard, were any circumstance to occur that would render 5 times the sum of the slightest use to you it would not rest in my pocket for an hour.

I am my dear Tyndall | Yours faithfully | James B. Bevington

RI MS JT/2/13b/597
LT Typescript Only

---

1. *From James B. Bevington*: this is only a fragment of the original letter, which is no longer extant. Tyndall copied out parts of the letter in his journal on 22 January (JT/2/13b/597).
2. *[20] January 1853*: Tyndall noted receipt of Bevington's letter on 22 January (see n. 1).
3. *to distribute*: Tyndall noted in his journal, 'Found letters at home, one of which I will further notice: It was from Bevington enclosing £20 for Booth: I was puzzled what to do with it and at last determined to forward it to Booth' (22 January, JT/2/13b/597).
4. *[...]*: after quoting the previous sentence in his journal, Tyndall wrote that Bevington 'wants me to go and spend two or three days with him at Wandsworth, and concludes his letter with the following words' (22 January, JT/2/13b/597). Those words are the last sentence in the letter above.
5. *little Booth*: Francis Booth, a friend who was unwell. See letters 0698 and 0703.

To E. D. Steuart 24 January [1853][1] 0707

24th. Jan.

My dear Mrs Steuart,

I most heartily sympathise with you in your great loss.[2] Believe me that every thing that gives you sorrow gives it to me also.

I thought of not troubling you with the enclosed;[3] but on second thought decided to send it to you.

Yours ever | John Tyndall

RI MS JT/1/TYP/10/3336
LT Typescript Only

1. *24 January [1853]*: while the letter was dated 24 January it was likely written in 1853. Letter 0842 may be a reply to it.
2. *your great loss*: likely in reference to Elizabeth Steuart's husband, William Richard Steuart (1798–1852), who had been seriously ill (see letters 0586 and 0593) and died in 1852. See also letter 0842.
3. *enclosed*: not identified.

To Michael Faraday 26 January 1853 0708

Queenwood College | near Stockbridge Hants |
26th Jan. 1853.

Dear Prof. Faraday,

Will you allow me to introduce to your notice an experiment I made in Berlin,[1] which I thought very remarkable at the time, but which I have had no opportunity to follow up since. It appears to me to lead to the same conclusion, if it be not the same in principle as those you have recently made. —It proves that the law of decrease[2] for iron is different from that of a salt of the metal. Do you not consider it probable that a difference of this kind between water and bismuth is the cause of the greater repulsion of the latter at an increased distance. As the bismuth is carried to a greater distance it may in fact be regarded as immersed in a new fluid altogether, and if the difference of repulsion between this fluid and the bismuth be greater than the corresponding difference near the poles the facts observed would of course follow.

I confess I never entertained the thought of applying the experiments described on the accompanying leaf[3] as you have done, and though startled by the profound ingenuity of your argument on Friday night,[4] for I did not imagine that the conclusion could be at all approached by the route chosen, the old magnetic law[5] has still I confess a hold upon my convictions.

Believe me dear Sir | Most faithfully Yours | John Tyndall

APS Mss.Ms.C011.200
Faraday Correspondence, 4:2630

1. *Berlin*: in 1851, Tyndall had spent time in the laboratory of Heinrich Magnus.
2. *law of decrease*: refers to the inverse-square law, which states that the decreased intensity of a magnetic force is inversely proportional to the square of its distance from the source of said force. Both Faraday's and Tyndall's independent experiments found that this did not

uniformly apply for all magnetic bodies immersed in water. In some cases, as expected, at greater distance there was a decrease in repulsion, but in other cases there was an increase in the intensity of repulsion.

3. *the accompanying leaf*: not identified.
4. *Friday night*: Tyndall had attended Faraday's lecture in London on 21 January. He described it in detail in his journal (JT/2/13b/596–97).
5. *the old magnetic law*: the inverse-square law.

From T. C. Higginson 28 January 1853 0709

Camp *[*Cuttagh[1]*]* | Jan. 28th 1853

My dear Tyndall,

Am I not a negligent fellow not to have written to you since we last met in London. My only hopes of pardon are derived from the idea, that as I write now, you will understand that I have not forgotten you, no indeed my old friend I have not and many times wished you were with me to enjoy the many scenes I have visited since we last met. Your love of the picturesque would have been fully gratified in viewing the beautiful places in the United States—Niagara[2] &c. &c. —I have visited within the last two years, but here I am again returned to old Ireland and located in a bleak spot[3] of the <u>auld</u>[4] country I do not grumble as others do at the misery of such habitations as these little huts make, because I have been in much worse. Still I find it difficult to keep out old Boreas,[5] he will pop his cold nose in no matter what you can do. There is some probability of our going to England next year, so I hope to have another shake of your auld fist. I hope this will find you as well as could be wished by.

Your very sincere friend | T. C Higginson | 16th Rg$^{\underline{t}}$

RI MS JT/1/TYP/2/599
LT Typescript Only

---

1. *Cuttagh*: this appears to be a typographical error by Louisa Tyndall and should read, Curragh. During this period the British Army had a makeshift base on Curragh Plain in County Kildare, Ireland.
2. *Niagara*: three waterfalls that are in New York State on the border between the United States and Canada. Tyndall would visit Niagara in 1873, and gave a discourse on 'Some Observations on Niagara'.
3. *a bleak spot*: the Curragh is a flat, open plain with little cover.
4. *auld*: old (*OED*).
5. *Boreas*: the Greek god of the north wind.

From Michael Faraday 29 January 1853 0710

Royal Institution | 29 Jany 1853

My dear Sir

I have only at this moment been able to read your letter,[1] having laid all I could aside that I might write some simple plain notes of last F. evening:[2] and now that I take it up, I cannot venture to judge of your experiments[3] or draw conclusions from them not having seen the experiments or the apparatus employed. I never can decide upon a result until I have seen the experiment it depends upon made and then never until I have considered repeated & varied the experiment Pluckers conclusion[4] is one that I do not know of my own knowledge to be correct I mean the varying rate of change of the Paramagnetic[5] & diamagnetic form[6] of force though I have seen many experiments.[7] I know that some of the forces of his experiments are truly stated and think he may be right, but would not vouch for it or give it as my own decided opinion. There is no end to the cases in which I think it safest not to come to a conclusion least it should be erroneous

Ever Truly Yours | M. Faraday

Dr. J. Tyndall | &c &c &c

APS Mss.Ms.C011.200
Faraday Correspondence, 4:2632

---

1. *your letter*: see letter 0708.
2. *Last F. evening*: refers to Faraday's lecture at the RI on Friday, 21 January (see letter 0708 n. 4).
3. *your experiments*: Tyndall had previously written to Faraday regarding experiments conducted while in Berlin in which he noted a difference in intensity of repulsion of iron and salt of iron when immersed in water (Journal, 21 January, JT/2/13b/597).
4. *Pluckers conclusion*: this may likely refer to Plücker's conclusion that the magnetic and diamagnetic forces changed differently with respect to distance from the poles.
5. *Paramagnetic*: the form of induced magnetism whereby materials, usually elements, are attracted by an externally applied magnetic field and form internal magnetic fields in the direction of the applied magnetic field.
6. *diamagnetic form*: the form of induced magnetism whereby materials are repelled by an externally applied magnetic field and form internal magnetic fields in the opposite direction of the applied magnetic field.
7. *I mean . . . many experiments*: diamagnetism is 'an extremely weak and complex property of matter, easily overpowered by contamination with minute amounts of paramagnetic materials and dependent on the nature of the magnetic field in relation to the size and shape of substances' (Jackson, 'Tyndall and Diamagnetism', p. 438). Since 1847, there had been

ongoing experiments done by the likes of Faraday, Plücker, Wilhelm Weber, and Tyndall seeking to understand the diamagnetic properties of substances (Jackson, 'Tyndall and Diamagnetism').

To Editor[1] January 1853 0711

To the Editors of the Philosophical Magazine and Journal.[2]

GENTLEMEN,

I FEEL some difficulty in writing this note, and yet I think it requisite to say a closing word upon a subject which has already occupied too much of your space.[3] It is nearly ten years since Mr. Adie made his experiments: would not a careful repetition of them at present show more wisdom upon his part than a reliance upon results which have such a mass of scientific evidence against them? With regard to his theory of the metallic cross*,[4] I have one request to make of Mr. Adie; and that is, that he would kindly place his pencil on the point C, where he supposes the current to be generated, follow the course of the current, and get back to C, *without passing through the battery.*[5] Some oversight must, I imagine, have occurred in Mr. Adie's direction of the cross; as the matter stands at present, I frankly confess my inability to understand it.

JOHN TYNDALL. | Queenwood, Jan. 1853.

J. Tyndall, 'On the Temperatures of Conductors of Electrical Currents', *Phil. Mag.* 5:30 (1853), p. 147.

---

1. *Editor*: the editors in 1853 were David Brewster, Richard Taylor, Robert Kane, and William Francis.
2. *To the . . . Journal*: the letter was given the title 'On the Temperatures of Conductors of Electrical Currents' by the editors of the *Phil. Mag.* See Volume 3 for Tyndall's extensive contributions to the *Phil. Mag.*
3. *too much of your space*: Tyndall was referring to a series of letters published in the *Phil. Mag.* between Richard Adie and himself in fall of 1852 regarding experiments on the ability of electricity to cause the absorption of heat or reduction of the temperature of conductors. The original letter sparking the discussion was R. Adie, 'On Some Thermo-electrical Experiments', *Phil. Mag.* 3:17 (March 1852), pp. 185–87. See also letters 0663; 0670; 0672; and 0685.
4. *theory of the metallic cross*: this is described as the observed effect of heating or cooling at the cross-joint of two metal bars when electricity is applied to one end of each of the bars (often of different metals) and measured through the opposite ends of the same bars by a galvanometer (an instrument for detecting electric current). When the current passes in one direction a current is generated in the adjacent bar suggesting that the cross-joint

connected to the source of the current is cooled. However when the current is passed in another direction or varied in tension this leads to different temperature effects. The point at issue, for Adie, appears to be a lack of sufficient time allowed for the metallic cross to exhibit temperature change, thus not allowing Tyndall to adequately observe the cooling effect taking place (R. Adie, 'On the Temperatures of Conductors of Electrical Currents,' *Phil. Mag.* 5:29 [January 1853], pp. 46–49).

5. *pencil on the point C . . . battery*: refers to a diagram of the metallic cross included with Adie's letter. See n. 3. Adie acknowledged and responded to Tyndall's objections and lack of understanding in 'On the Electrical Currents induced in the Metallic Cross', *Phil. Mag.* 6:34 (July 1853), pp. 46–50.

To Michael Faraday 15 February 1853 0712

Queenwood, 15th, Feb, 1853.

Dear Prof Faraday

I never calculated on my expenses being paid, but am perfectly content to abide by your usual practice on such occasions—My travelling expenses are the only ones worth naming and they amounted to about thirty shillings—This sum I shall be equally contented to receive or not to receive.

When a man feels deeply on any point I believe as a general rule he will not talk much about it. I once heard the silence of Goethe[1] for a year during which he was in love with a girl in Frankfort[2] accounted for in this way, and I have always thought the hypothesis a probable one. Friday night[3] is to me what his sweetheart was to Goethe—I shall never forget it but cannot well write about it.[4] The memory of it I trust will serve to keep me more loyally to my task, to assure me when I doubt, to strengthen me when I falter and to establish my faith in a sentiment which I have met somewhere in your own writings, that the patient and conscientious investigator of natural truth is ever sure of his reward

Believe me dear Professor | Most sincerely Yours | John Tyndall

RI MS JT/TYP/12/4022
Faraday Correspondence, 4:2638

1. *Goethe*: Johann Wolfgang von Goethe (1749–1832), German writer and philosopher.
2. *Frankfort*: Frankfurt am Main, Germany.
3. *Friday night*: Tyndall's talk of the evening of 11 February, his first at the RI, entitled 'On the Influence of Material Aggregation Upon the Manifestations of Force' (*Roy. Inst. Proc.*, vol. 1, pp. 254–59). He conducted several experiments on diamagnetism.
4. *cannot well write about it*: Tyndall gives a summary of the lecture in letter 0713.

To Thomas Archer Hirst [14–]16 February 1853[1] 0713

Queenwood | 16th Feb. 1853 | J Tyndall.

My dear Tom.

You are a good boy for helping your old brother in his need. Your note was sent to me by Roby accompanied by a brief account of Booth's state[2] from Roby himself.[3] Poor fellow, I believe his sunset approaches quickly. I spent yesterday and part of the day previous with Bevington—He is one of the truest-hearted fellows I ever met. Yesterday while the good people were at church he and I were discharging snowballs at each other in the middle of Richmond Park.[4] We fought while we had breath. I went up to town on Thursday and on the same evening made sundry experiments before the Royal Society in Somerset House[5]—It passed off very cordially and I seem to have rivetted one or two friends, who first argued with me and afterwards gave in—But Friday night[6] was my night of trial when I was to appear before the fashionable audience of the Royal Institution—Well I had mastered my subject and made up my mind not to let the audience master me. At 9 o'C the lecture commenced—the Duke of Northumberland[7] was in the chair—many men of the first calibre were scattered all round; well I got my work done—pretty much to my own satisfaction and judging from the hurricane of applause which assailed my closing sentences, to the satisfaction of my audience also—When I sat down and while the tempest was raging, with bravo's mingling like thunderclaps through it, Faraday crossed the platform to where I sat and grasped me by the hand before the whole audience. The Duke of Northumberland came forward afterwards and did the same—I believe no fellow has ever the good fortune to be more petted and patted and overloaded with all manner of odoriferous utterances than I was. I was presented to this person and that person, in my bewilderment I did not know whom. Hands were ready to grasp mine at all corners and finally Faraday took me up stairs where his wife and niece[8] (a sweet intelligent girl) were at supper, and stuffed me with sherry and sandwiches.

One little incident had a good deal of originality about it—while talking from one to the other and endeavouring to have a word for each I often noticed a thin, tall, dark looking man standing perfectly silent and passive quite near me—evidently hearing what was said though appearing to have his attention abstracted. I once or twice made room for him to pass me and others but he shewed no disposition to move. Finally having got loose from a group of questioners I was turning away when my tall thin friend confronted me—I have waited he said until you were perfectly at liberty—I dont know you, you dont know me, but I must know you for I believe you are worth knowing—I am perfectly conversant with public assemblies both the house of Commons[9] and other assemblies—I have heard speakers of all ranks and degrees, but I never heard anything to equal your lecture tonight—now these are the words of a stranger' he says 'who knows you not and of whom you know nothing—I shall remember you, and you will perhaps bear me in mind—I am Whitbread the brewer!"[10]

I have now got home again and at the present moment there are about 40 little boys spread out before me—some at chess, some at novels, some writing, some chattering, and I can truly say that the glories of the Royal Institution had nothing so sweet in them as the feeling with which I look upon some of these lads—I actually love them Tom almost as well as you can ever love a sweetheart. A few nights ago I dreamt that one of them threw his arms about my neck and kissed, and now when my eye rests upon the lad's brow, and the rosy glimmer of his cheek I am almost tempted to enact my dream. I believe there are schemes concocting among men of influence in London to get me up there, but I shall weep to part with some of these lads—I think of them at night, and the thought of them is worth two extra blankets on a winter night—the longer I am with them the more I love them—the less of a schoolmaster do I become, and then with my scanty experience I have a power over them which the practical veterans around me cannot obtain. We *[joy]* over our science with a good humour and heartiness which resembles a musical chant, and the little fellows are never weary of working with me. I pelt them with snowballs and they pelt me—sometimes I have a try at shinty[11] with them—and I sometimes lay my old parchment cheek against theirs, soft as the down of peaches, and the little fellows do not shrink from me. But you will call me an old dotard[12] if I go on in this strain—and I am a dotard at times, till some shock comes to cause the iron within me to vibrate, and people find that I can be stern on occasion. At all events I am a first-rate egotist and have contrived to cram this sheet to saturation with my own personality. No matter you will be able to endure the dose and I am all the better for the discharge of so much vapour. I sent Bevington one or two of Booths letters—he has them still, safely preserved, and when he returns them I will send them to you. Bevington is delighted with them. He is softer than either you or I and his enjoyments probably more exquisite.

now goodbye. | your affectionate | John

<hereafter LT Transcript Only>

Tuesday[13]

Your translation[14] has just reached me, you have given me a great lift. I shall have very little trouble in polishing up and you have written it far more beautifully than I could have thought possible to your scrambling fingers. I dont want to tax you too much. If you could help Wiedemann at his translation it would perhaps forward me.[15] There is also a short paper by Clausius in No 11[16] of Poggendorff 1852 commencing at page 415 and ending at 426.[17] Of this I have given the 1st. page in the last no. of the Memoire and have therefore translated the said page, but the whole of the remaining portion is to be translated.[18] Clausius can write English, try and get him to do it and you can correct it. Next month will be quite early enough for me.

'A hemisphere' some writers use an but a is correct for the h is not silent, and it is only before a silent h that the a becomes an. I will have a copy of Magnus's paper[19] sent to you. Now I think I cannot afford you more time, we have frost and snow here at present. Good bye boy | John

I have had a note[20] this morning offering me a course of 4 lectures next month in London at 5 guineas a lecture. I have offered to give one, as I did not like to bind myself to 4.

RI MS JT/1/T/559
RI MS JT/1/HTYP/227–228

---

1. *[14–]16 February 1853*: it appears most likely that the letter was begun on 14 February (Monday), finished on 15 February (Tuesday), and sent on 16 February (Wednesday).
2. *Booth's state*: Francis Booth was dying (see letters 0698 and 0703).
3. *Roby himself*: see letter 0698, n. 6.
4. *Richmond Park*: large park in southwest London.
5. *Somerset House*: large building on The Strand in central London, which housed the Royal Society (among other institutions).
6. *Friday night*: Tyndall's Friday evening talk at the Royal Institution (see letter 0712, n. 3).
7. *Duke of Northumberland*: Algernon Percy (1810–99), fourth Duke of Northumberland and President of the Royal Institution, was the chairman for the evening of Tyndall's talk of 11 February.
8. *wife and niece*: Sarah Barnard (1800–79) and Faraday had no children of their own. Mrs Faraday's niece (Jane Barnard) lived with them at the Royal Institution for nearly twenty years.
9. *house of Commons*: the lower house of the British parliament.
10. *Whitbread the brewer*: likely Samuel Whitbread (1796–1879), who was interested in astronomy and meteorology, and became FRS in 1854. His family ran the Whitbread brewing company.
11. *Shinty*: Scottish team sport played with a stick and ball.
12. *dotard*: a stupid or silly person (*OED*).
13. *Tuesday*: the addendum to the letter appears only in the typescript, but Tyndall noted on 16 February that he had received a note from Bence Jones regarding the offer to conduct a series of lectures at the LI (Journal, 16 February, JT/2/13b/601); letter missing.
14. *your translation*: Hirst translated Magnus's paper for the *Scientif. Mem.* (Hirst, 'Journals', 5 February 1853). It was published as G. Magnus, 'On the Deviation of Projectiles; and on a remarkable Phaenomenon of Rotating Bodies', in Tyndall and Francis (eds.), *Scientif. Mem.*, pp. 210–13 (Hirst, 'Journals', 5 February 1853).

15. *Wiedemann . . . forward me*: Tyndall was working on translating foreign scientific works into English, including one by Wiedemann. The amount of time it was taking was of concern to him (see letter 0698, nn. 8 and 18).
16. *No 11*: refers to the November 1852 edition of *Poggend. Annal.*
17. *415 . . . at 426*: Clausius, 'Ueber die bei einem sationären elektrischen Strom in dem Leiter gethane Arbeit und erzeugte Wärme' appeared in *Poggend. Annal.* 87 (November 1852), pp. 415–26.
18. *translated*: the translation of Clausius's article (see n. 17) was entitled 'On the Work performed and the Heat generated in a Conductor by a Stationary Electric Current', in Tyndall and Francis (eds.), *Scientif. Mem.*, pp. 200–209.
19. *Magnus's paper*: see n. 14.
20. *note*: see n. 13.

From Michael Faraday 16 February 1853 0714

Royal Institution, | 16 Feb. 1853.

My dear Tyndall,

I do not know what post office near you would be convenient for a money order, so I send a cheque for the 1.10.0,[1] with many thanks on the part of the Institution, and I may truly say of many members who were at your discourse.[2] If the cheque is inconvenient to you, return it to me by post, and tell me what I should do.

Ever truly yours, | M. Faraday.

RI MS JT/TYP/12/4132
Faraday Correspondence, 4:2639

---

1. *1.10.0*: one pound, ten shillings (i.e., 30 shillings). Tyndall was receiving money in compensation for travel costs and expenses incurred for his inaugural lecture at the RI (see letter 0712).
2. *discourse*: see n. 1 and letter 0712, n. 3.

From Henry Bence Jones 17 February 1853 0715

Feb. 17. 1853.

Dear Dr. Tyndall,

I had not time to tell you more about the London Institution I will send your letter[1] to Mr Tite.[2]

The audience is much the same or not quite so good as the Royal

Institution. Playfair is lecturing there now and the vacancy offered[3] to you is caused by Dr Mantell's death.[4] Grove gave his insulation lectures[5] there.

I am stopped and must end.

Yours very truly | H. Bence Jones.

RI MS JT/1/TYP/2/671
LT Typescript Only

---

1. *your letter*: letter missing.
2. *Mr Tite*: William Tite (1798–1873), architect and antiquarian, was the honorary secretary of the London Institution from 1824 to 1869 (*ODNB*).
3. *vacancy offered*: Tyndall noted that Bence Jones was authorized to offer him the opportunity to give four lectures for a total of twenty guineas (see letter 0713, n. 13).
4. *Dr Mantell's death*: Gideon Mantell (1790–1852), a surgeon and geologist, who had been lecturing at the London Institution died on 10 November 1852.
5. *insulation lectures*: Grove gave a series of six lectures 'On the Correlation of Physical Forces' in November and December 1843, which were later published by the Institution (W. Grove, *On the Correlation of Physical Forces: being the Substance of a Course of Lectures delivered in the London Institution in the Year 1843* [London: Samuel Highley, 1846]). See M. L. Cooper and V. M. D. Hall, 'William Grove and the London Institution, 1841–1845', *Annals of Science* 39:3 (1982), pp. 229–54, on pp. 247–48.

From George Wynne 17 February 1853 0716

Whitehall[1] | 17 Feb[r] '53

My dear Tyndall

I heard of your success from my son[2] and nephew who I think were as much gratified by the applause you received as by the pleasure they derived from the lecture.[3] I wish I could have been present myself, but better luck next time—

I would be much inclined to offer you the advice to be very /weary/ in giving lectures except to decidedly scientific bodies and not to let yourself be brought forward as a popular lecturer, I should feel great doubts whether the Finsbury Institution[4] (I know nothing of its Constitution) would be a desirable plan to lecture in immediately after the Royal Institution but you will have an opportunity of consulting persons who can offer a much better opinion on the matter than I can. My <u>own</u> idea is that tho' popular lecturers are very interesting to those who hear them being kind of Royal highnesses of learning they do not advance those who give them inasmuch as they do not <u>generally</u> proceed from first class men, occasionally first class minds such as Faraday condescend to gratify the public by a popular exposition of their views and it is very right of them to do so & very kind, but thus they are established men and can afford to take liberties. I think a popular lecture may be given to a mechanics institute[5] or such like place when it *[2 words illeg]* a place of known */*distinction*/*—Tomorrow I can better come of the nature of the audience, I w[oul]d believe the *[1 word illeg]* early *[4 words illeg]* like to *[4 words illeg].*

Very truly yrs | Geo. Wynne

RI MS JT/1/W/95

---

1. *Whitehall*: area in the City of Westminster (London), which is known for being the location of many governmental offices.
2. *my son*: George Wynne (1804–90), together with his first wife Anne, had three sons: Henry Le Poer Wynne (b. 1836), Edward Toler Wynne (b. 1837), and Francis George Wynne (b. 1846). This likely refers to either Henry or Edward.
3. *the lecture*: Tyndall's lecture of 11 February at the RI (Tyndall discusses the lecture in letter 0713).
4. *Finsbury Institution*: refers to the LI at Finsbury Circus, where Tyndall had recently been offered the opportunity to lecture, which filled a vacancy left after the death of Gideon Mantell (see letters 0715 and 0713, n. 13).
5. *mechanics institute*: an educational institute created to provide technical education for working men.

From Henry Bence Jones 19 February 1853 0717

Dear D[r] Tyndall

Your letter[1] to me this morning leads me to write to you a few days sooner than I otherwise should have done regarding LI & RI.

Mr Gassiots letter[2] which you have received before this will make the first course clear. I need only say that if you want any apparatus if you ask Mr Faraday for the loan of it he will let you have it. It is a common request; and Newmans in Regent Street[3] will lend you any thing they have. Give the one lecture on Heat. It will tell. But regarding the future. It is highly probable that you will be offered a professorship at the London Institution with about £200 per annum.

Last night I shewed Mr Barlow how desirable it was that you should give a course at the R.I. <u>this year</u> so as to qualify you for the Professorship at the Royal Institution[4] & we settled that he & I should speak to Mr Faraday. Mr

Barlow being most willing that you should give a course of four lectures after Easter, about May. I need not tell you that Mr. Faraday & not Mr B or D^r^ G[5] influence the managers. For another week I shall be unable to tell Mr Faraday why it is most desirable that another candidate for the Professorship should if possible be obtained. I can now only say that it is not likely that the College of Chemistry Council will let Dr Hofmann[6] accept the appointment. I have little doubt that Mr Faraday will agree & be most anxious that you should lecture with a view to qualifying yourself for the Professorship at the Royal Institution. Whether you could also hold the London Institution place is very doubtful. It will be a matter for your consideration which place you would like best. The pay at the RI is 150 & 2/3 of lecture fees making perhaps 100 more.[7] The professorship at the RI is a Professorship of Chemistry. Lectures on Heat would do quite as well as lectures on Chemistry, but a course of chemical lectures in future years would not trouble you. I had some conversation with Mr Huxley about all these proposals last night & if you like to consult him, do so.[8] You shall hear from me again when I have anything to say

Believe me yrs most truly | H Bence Jones

Feb 19. 1853

RI MS JT/1/J/41

---

1. *your letter*: letter missing but presumably in response to letter 0715. In his journal, Tyndall discussed the various offers he received and his delight and ambivalence regarding which to accept or undertake (Journal, 20 and 22 February, JT/2/13b/601).
2. *Mr Gassiots letter*: letter missing but it was from John Peter Gassiot who conducted research into electricity and was an associate of William Grove. Gassiot was a manager of the LI and a contact of Tyndall's regarding the professorship on offer there.
3. *Newmans in Regent Street*: refers to a shop located at 122 Regent Street, London from 1827 to 1860, run by John Newman.
4. *Professorship at the Royal Institution*: in order to qualify for a professorship at the RI, it was necessary to have already given a course of lectures there. (See also letter 0721.)
5. *Mr B or Dr G*: Barlow and Gassiot.
6. *College of Chemistry Council . . . Dr Hofmann*: the Royal College of Chemistry was founded in 1845 on Oxford Street in London in order to teach practical chemistry. August Wilhem Hofmann was its first director.
7. *150 . . . 100 more*: the salary for the professorship at the RI was £150 per year plus two thirds of the lecture fees, amounting to about £100 additional income.
8. *Mr. Huxley . . . consult him, do so*: Tyndall consulted with Huxley about the positions in letters 0719 and 0721.

From Thomas Henry Huxley [21] February 1853[1] 0718

Do for heaven's sake alter the molecular condition of your mind so as to suspend the operation of that confounded modesty of yours and dont sell yourself too cheaply. You had better stay where you are than take a third-rate place in London, however tempting it may look. If you come to town you must be prepared to fly high or believe me you wont fly at all. [. . .][2] It seems probable that two openings will occur for you both at the Royal Institution and at the London Institution[3] and I know that the most influential people at the Museum of Economic Geology[4] would get you there if it could be managed

RI MS JT/2/13b/601
LT Typescript Only

---

1. *[21] February 1853*: in his 22 February journal entry Tyndall wrote: 'Yesterday had a letter from Huxley in which he raises a warning note' (JT/2/13b/601). Tyndall then copied out excerpts from a letter from Huxley. What is included here, therefore, is just a fragment from an original letter that is no longer extant.
2. *[. . .]*: After quoting three sentences from the letter in his journal, Tyndall wrote here 'In another place he says' and then quoted the last sentence included in the letter above (JT/2/13b/601).
3. *two openings . . . at the London Institution*: see letter 0721.
4. *Museum of Economic Geology*: this is in reference to the Museum of Practical Geology and the Government School of Mines and Science as applied to the Arts (often just referred to as the Government School of Mines), established in 1851 on Jermyn Street (Piccadilly), and which had developed from the Museum of Economic Geology.

To Thomas Henry Huxley 22 February 1853 0719

Queenwood | 22nd Feb. 1853.

My dear Huxley

Twice have you written to me[1] and each time an odour of sanctity (Gott vergebe mich![2]) has breathed through your letters—they have propped my faltering soul and added 50 per cent to my courage and my kindly feeling towards humanity in general. Now what would you have me to do? Speak plain and fear no consequences. Supposing the London Institution makes the offer[3] would you have me wait? What kind of a place is it? Is it on a safe footing? Of course the Royal[4] would be far more to my taste but it is a chair of chemistry and not of physics. I have worked hard at chemistry for some time but any man who has made it his particular study could gravel me upon it, especially on organic chemistry. I think I could lecture pleasantly on the subject but not profoundly without an amount of previous study which I am afraid would divert me from my professed career. Of course I should not object to lecturing on chemistry at intervals but my ordinary ground would be experimental physics. Remember what I want. I want the means of living and the time for working. Can you give me an idea of the duties to be performed at the R.I. Give me your full advice You may depend upon it it will have great weight with me.

believe me dear Huxley | ever Yours most faithfully | John Tyndall

RI MS JT/TYP/9/2858
LT Typescript Only

---

1. *written to me*: the first of these letters is missing. A fragment of the second one is letter 0718.
2. *Gott vergebe mich!*: German for 'God forgive me'.
3. *the offer*: see letter 0715, n. 3.
4. *the Royal*: the RI.

To John Barlow 23 February 1853 0720

Queenwood | 23rd Feb. 1853.

My dear Sir

Accept my sincerest thanks for the trouble you have given yourself: I shall be most happy to respond to the invitation of the Managers[1] of the Royal Institution.

To yourself or Professor Faraday I gladly leave the arrangement of the time, but allow me to make one remark: As you can readily conceive my duties here[2] occupy me a good deal and leave me but little time for the preparation of lectures. I should like if possible that some portions of the lectures at least should be original matter, and the preparation of this costs considerable time. If therefore your thoughts be already nearly balanced as to whether you will have me immediately after Easter[3] or 6 weeks afterwards, allow me to throw my petition into the scale of the latter decision. However I shall cheerfully abide by any arrangement you think proper to make.

My matter would I think range itself suitably under the title "On Air and Water".

believe me dear Sir | most faithfully yours | John Tyndall

RI MS AD/03/A/02

---

1. *the Managers*: from May 1852 to April 1853, the acting managers were: William Wilberforce Bird (1784–1857), William Thomas Brande (1788–1866), B. Bond Cabbell (1782/33–1874), Henry John Codrington (1808–77), J. Griffith Cole (1805–65), George Dodd (1808–81), Charles Fellows (1799–1860), Aaron Asher Goldsmid (1785–1860), William Robert Grove (1811–96), Lord Londesborough (1805–60), George Moore, Frederick Pollock (1783–1870), Lewis Powell John Webster, and C. Wheatstone (1802–75) (*Roy. Inst. Proc.*, vol. 1, p. 190). Bence Jones had suggested that Tyndall give a series of lectures at the RI so as to qualify himself for the available professorship (see letter 0717).
2. *my duties here*: at Queenwood College.
3. *Easter*: Easter Sunday was 27 March.

From Thomas Henry Huxley 25 February 1853 0721

6 Upper York Place | St. John's Wood.[1] | Feb. 25th. 1853.

My dear Tyndall,

Having rushed into more responsibility than I wotted of,[2] I have been ruminating and taking counsel what advice to give you. When I wrote I hardly knew what kind of work you had in your present office, but Francis[3] has since enlightened me. I thought you had more leisure. One thing is very clear—you must come out of that. Your Pegasus is quite out of place ploughing.[4] You are using yourself up in work that comes to nothing, and so far as I can see cannot be worse off.

Now what are your prospects? Why as I told you before you have made a <u>succès</u> here and must profit by it. The other night your name was mentioned at the Philosophical Club[5] (the most influential scientific body in London) with great praise. Gassiot, who has great influence, said in so many words 'you had made your fortune'—and I frankly tell you I believe so too if you can only get over the next three years. So you see that <u>quoad</u>[6] position, like Quintius Curtius[7] there is a 'fine opening' ready for you.[8] Only mind you don't spoil it by any of your horrid modesty.

So much for glory,—now for Economies. I have been trying to ferret out more nearly your chances of a post, and here are my results (which I need not tell you must be kept to yourself).

At the Museum in Jermyn St.,[9] Playfair, Forbes, Perry,[10] and I think Sir Henry,[11] would do anything to get you, and eliminate Hunt,[12] but so far as I can judge the probability of his going is so small, that it is not worth your while to reckon upon it. Nevertheless it may be comforting for you to know that in case of anything happening these men will help you tooth and nail. Cultivate Playfair when you have a chance—he is a good fellow, wishes you well, has great influence and will have more. Entre nous,[13] he has just got a new and important post under government.[14]

Next, the Royal Institution. This is where, as I told you, you ought to be—looking to Faraday's place. Have no scruple about your chemical knowledge; you won't be required to train a college of students in abstruse analyses; and if you were, a year's work would be quite enough to put you at ease. What they want, and what you have, are clear powers of exposition—so clear that people may think they understand even if they don't. That is the secret of Faraday's success, for not a tithe[15] of the people who go to hear him really understand him.

However I am afraid that a delay must occur before you can get placed at the Royal Institution, as you cannot hold the Professorship until you have given a course of lectures there, and it would seem that there is no room for you this year.[16] However I must try and learn more about this.

Under these circumstances the London Institution looks tempting. I have been talking over the matter with Forbes[17] whose advice I look upon as first-rate in all these things, and he is decidedly of the opinion that you should take the London Institution if it is offered to you. He says that lecturing there and lecturing at other Institutions and writing you could with certainty make more than you at present receive, and that you would have the command of a capital laboratory and plenty of time.

Then as to the position—of which I was doubtful—it appears that Grove has made it a good one.

It is of great importance to look to this point in London—to be unshackled by anything that may prevent you taking the highest places—and it was only my fear on this head that made me advise you to hesitate about the London Institution. More consideration leads me to say, take that if it will bring you up to London at once, so that you may hammer your reputation while it is hot.

However, consider all these things well, and don't be hasty. I will keep eyes and ears open and inform you accordingly. Write to me if there is anything you want done, supposing always there is nobody who will do it better—which is improbable.

Ever yours | T.H. Huxley

IC HP 8:9
RI MS JT/TYP/9/2859–2861
LT Typescript Only

1. *St. John's Wood*: area in northwest London.
2. *wotted of*: know of (*OED*).
3. *Francis*: William Francis.
4. *Your Pegasus . . . place ploughing*: Pegasus is a symbol of wisdom and fame. The allusion is that Tyndall (akin to the mythical winged horse, Pegasus) should aspire to greater things.
5. *Philosophical Club*: the Philosophical Club of the Royal Society was formed in 1847 and was a dining club that met before the monthly meetings of the Society.
6. *quoad*: from the Latin, with respect to (*OED*).
7. *Quintius Curtius*: Quintus Curtius Rufus, Roman historian from the first century BCE, wrote the only extant Roman historical text on Alexander the Great. He is not known to have written anything else, nor was his work ever referred to by any other ancient text.
8. *'fine opening' . . . you*: referring to the opportunity Tyndall has before him to make a name for himself within the scientific community, as Curtius did for Roman history.
9. *At the Museum in Jermyn St*: see letter 0718, n. 4.
10. *Perry*: likely refers to either James or Stephen Perry, brothers who invented a slit pen nib and owned a pen manufacturing company in Red Lion Square.
11. *Sir Henry*: Henry De la Beche.
12. *Hunt*: Robert Hunt (1807–87), a scientist involved in research on light and experimental physics, was at the time Professor of Mechanical Science at the Government School of Mines.
13. *Entre nous*: French, between ourselves.
14. *Playfair . . . new and important post under government*: Playfair had recently been made secretary for science in the Department of Science and Art.
15. *tithe*: a tenth part of (*OED*).
16. *no room for you this year*: note that Tyndall had already been offered (letter 0717) and agreed (letter 0720) to give a course of lectures at the RI.
17. *Forbes*: LT placed a cross here and noted 'Ed Forbes' in the margin.

To Thomas Archer Hirst 25 February 1853 0722

Queenwood | 25th. Feb. 1853.

My dear Tom

The image[1] of you sitting on the ice with a star beneath your stern-works convulsed all my frame with laughter. I suppose by this time you will have received my short account of the proceedings in London.[2] Nothing definite is settled yet. It is likely that two openings will occur one at the Royal Institution one at the London Institution. But as yet I have had no definite offer. I am to give four lectures at the Royal Institution in May and June,[3] and a lecture on the 16th of March at the London Institution.[4] I am so much occupied that I have but little time to prepare and I want you to assist me a little. I want a few notes regarding the difference of temperature between day and night in the desert of Sahara and also the difference between day and night in some forest regions in the same latitude as the desert. My object is to compare the influence of silica with that of *[regation]* upon climate. Now nobody can give such information better than Professor Dove and nobody I think will give it more readily. If he is acquainted with any peculiarities of the desert as regards temperature I should be thankful to be informed of them. The fact is quartz or pure silica, the prevalent mineral in Sahara is a better conductor of heat than many of the metals and this peculiarity must I think have great influence upon the climate of the desert. I imagine à priori that a maximum and minimum thermometer ought to shew a far wider range of temperature in the 24 hours in the desert than in a wooded region in the same latitude. Ask Prof. Dove does he not think so. Now I want you to send me this information immediately. I send you a cheque for 20 pounds as I shall want to draw upon you to pay for some crystals which I am getting out in Frankfort.[5]

Your affectionate | John

RI MS JT/1/T/560
RI MS JT/1/HTYP/229

1. *The image*: not identified.
2. *proceedings in London*: see letter 0713.
3. *lectures in May and June*: see letters 0717 and 0720.
4. *lecture . . . London Institution*: Tyndall noted his invitation to give this lecture in his journal (Journal, 20 February, JT/2/13b/601).
5. *some crystals . . . Frankfort*: Tyndall had ordered some crystals from Knoblauch, which he finally received in April (Journal, 26 April, JT/2/13b/603–4) (see letter 0748).

To William Francis 27 February [1853][1] 0723

Dear Francis—

Would you like a good fellow enclose the enclosed[2] with the magazines to the persons to whom they are addressed. You will be able no doubt to obtain Whitbread's[3] Christian name for me, to put it on the paper intended for him, and to have it posted to him for me. What do you think of the signs of the times there in London?[4]—I am a tranquil looker on waiting patiently for the evolution of the matter.

ever yours | J Tyndall
27th Feb.

RDS 27/5

1. *27 February [1853]*: while '27 Feb 1852' has been written in the upper right corner by another hand, Tyndall's comments regarding London (see n. 4) and Whitbread's (see n. 3) suggest that 1853 is more likely.
2. *the enclosed*: not identified.
3. *Whitbread's*: see letter 0713, n. 10.
4. *signs . . . London?*: refers to Tyndall's predicament regarding which direction to take his career, whether to the LI or the RI (see letters 0715; 0717; 0718; 0719; 0720; 0721; and 0722).

From Henry Bence Jones 28 February 1853 0724

Dear Dr Tyndall,

I am very glad you have chosen a subject[1] which will be so popular, but do not limit yourself in your treatment of it. Do you happen to know Schleiden's Plant.[2] It is worth your looking at in connection with your subject.

The politics of science have not advanced this week.

I have today received a long & violent answer by Matteucci to Mullers[3] abstract of Du Bois[4] work but he very unfairly writes it to me.

Speaking of Du Bois he says le génie irritable de ce savant[5] & elsewhere he calls him jeune Physiologiste[6] & he makes Muller out almost a liar because Muller said there was not only a want of clearness but sometimes of fairness in Matteucci's writings. I must send it to Du Bois if Matteucci has not sent him a copy for his instruction.

Believe me | Yours most truly | H. Bence Jones

Feb 28. 1853

RI MS JT/1/J/42
RI MS JT/1/TYP/672

1. *chosen a subject*: Tyndall proposed the subject 'On Air and Water' for his upcoming course of four lectures at the RI (see letters 0720 and 0726, n. 4).
2. *Schleiden's Plant*: Matthias Jakob Schleiden (1804–81), *Die Pflanze und ihr Leben* (1848), trans. Arthur Henfrey, *The Plant: A Biography in a series of lectures* (London: H. Bailliere, 1848).
3. *Mullers*: Johannes Peter Müller (1801–58), German comparative anatomist and physiologist under whom Emil du Bois-Reymond worked and studied (Finkelstein, Emil du Bois-Reymond, pp. 39–54). Du Bois-Reymond was undertaking research on animal electricity, and was critical of Matteucci's previous work on the subject.

4. *Du Bois*: Emil du Bois-Reymond
5. *le génie irritable de ce savant*: French, 'the terrible genius of this learned man'.
6. *jeune Physiologiste*: French, 'young physiologist'.

To William Francis 4 March [1853][1] 0725

4th Mar.

My dear Francis.

I intended to send you a note[2] but I really forget whether I did or not. I wanted to ask you to be kind enough to send me the younger Mr Brooke's[3] address—I want to send him two or three bits of magnetic substance which I promised he should have when I saw him last. I should also be obliged if you would purchase Herschel on Sound[4] for me—It is one of the reprints of the Encyclopedia Metropolitana[5]—price 5/- Have you ever seen Schleidner's plant?[6] I should like much to look into it or to purchase it if it is not too dear. I shall call upon you on the 16th and hope to see you in the evening at Finsbury Circus.[7]

ever yours | John Tyndall

RDS 27/6

---

1. *4 March [1853]*: the references to Tyndall's upcoming lecture at the LI (see n. 7) as well as to 'Schleidner's plant' (see n. 6) suggest that this letter was written in March 1853.
2. *a note*: possibly in reference to letter 0723. Note that he did not ask for the address in that letter.
3. *younger Mr Brooke's*: Refers to Charles Brooke, FRS (1804–79), son of Henry James Brooke (1771–1857), a well-known mineralogist. The younger Brooke was a surgeon and an inventor who wrote on mathematical and experimental physics.
4. *Herschel on Sound*: Herschel's treatise on Sound was initially published as part of the *Encyclopaedia Metropolitana* in 1830 and reprinted in 1845 (J. F. W. Herschel, 'Sound', in E. Smedley [ed.], *Encyclopaedia Metropolitana; or, Universal Dictionary of Knowledge, on an Original Plan Comprising the Twofold Advantages of A Philosophical and an Alphabetical Arrangement, with Appropriate Engravings. Second Dvision. Mixed Sciences.*, vol. 2 [London: Baldwin and Cradock, 1830], pp. 747–820).
5. *Encyclopedia Metropolitana*: initially edited by Samuel Taylor Coleridge and published in 1817–45, the *Encyclopedia Metropolitana* was thirty volumes in length and contained fifty-nine parts that discussed the arts and sciences.
6. *Schleidner's plant*: misspelling of Schleiden. See letter 0724, n. 2.
7. *Finsbury Circus*: the LI. Tyndall was to give a lecture on 16 March at the LI (see 0722, n. 4).

To Thomas Henry Huxley [6 March] 1853[1] 0726

Queenwood Sunday. 1853

My dear Huxley,

Many thanks for your valuable letter.[2] I write now to let you know that the disability you mention[3] no longer exists, as I have been invited to give a course of 4 lectures at the R.I.[4] <u>this year</u>. All these things appear to be guided by the gods—with perhaps a little help from my sublunary[5] friends. At all events it seems as if I had nothing to do but to resign myself to the kind offices of gods and men and that all goes well.

I am not at all impatient, and if I thought remaining in my present position[6] for a year or two would make matters better in the end I should be perfectly content to stop here. I have chosen a subject which will enable me to discharge a battery of chemistry upon them, and fortunately for me the lectures dont come off till towards the end of the Session[7] and I am thus allowed time for preparation. I lecture at the London[8] on the 16th.,[9] and am endeavouring to chasten the chaos of my brain into some kind of harmony. The subject is naturally a dry one and I must contrive to infuse a little sap into it. I find the letter bag is gone and I must run after it. Good bye

ever Yours faithfully | John Tyndall.

IC HP 8:11
RI MS JT/TYP/9/2858a
Typescript Only

---

1. *[6 March] 1853*: as this was a response to Huxley's letter of 25 February (see n. 2), Tyndall was to give the named lecture (see n. 9) on 16 March, and Huxley responded with a congratulatory message on 7 March (see letter 0729), this was most likely written on 6 March.
2. *your valuable letter*: see letter 0721.
3. *disability you mention*: Huxley mentioned in letter 0721 that there would be a necessary delay in Tyndall taking a position at the RI, as he would first be required to give a series of lectures in order to qualify for a professorship.
4. *R.I.*: See letters 0717 and 0720. Tyndall also noted the receipt of the offer from the RI in the entry to his journal on 22 February (JT/2/13b/601). In the typescript LT inserted a superscript X here and noted: 'On Air and Water'. R. I. May 21, 28, June 4, 11.
5. *sublunary*: terrestrial; of, in, or belonging to this world (*OED*).
6. *my present position*: at Queenwood College.
7. *the Session*: the teaching term.
8. *the London*: LI.

9. *on the 16th.*: First mentioned in letter 0722, n. 4. In the typescript LT inserted a superscript X here and noted: 1st London Institution lecture: March 16. 53 'On certain Points connected with the transmission of Heat through bodies.'

From Thomas Archer Hirst 6 March 1853 0727

Mittel Strasse No 5 | Berlin | March 6th / 53

My dear John,

Our last letters crossed[1] but mutually answered each other as well as if they had been purposely made to do so. The account[2] you gave of the proceedings at your lecture of course interested me as much as you. By the bye, I wish you enlighten me a little about the London Universities. Who are the professors of the Royal Institution, and what kind of students attend. What is the London Institution, is it synonymous with London University?[3] and who are the principal men at these places?

About a week ago Booth wrote to me and enclosed a letter from Bevington which you had sent him together with Bevington's £20.[4] Of course the little fellow asked me what he should do, and I answered him that he must receive it by all means, that granted I could find such givers as Bevington I would consent to his amassing a small fortune. He is a manly sensitive fellow, that Bevington, his letter made me cry, and laugh too. One might travel far before meeting his equal. So you are to warn me that he is a prosy[5] old fellow and has lost some of his teeth!! Tell him I promise faithfully never to look into his mouth. A strange notion to be sure, as if I should judge him as I would a horse! Another letter received also during the week has pleased me very much, it was from Jemmy Craven.[6] In his own small way Jemmy is a good natured lad (by the bye I forgot he is older than me, I always did forget that fact). He has the slightest trace of something better in him than he has ever yet shewn, just sufficient to make him uneasy at times and make him conscious that there is a little hollowness about his life. But in Jemmy this feeling will never grow intense enough to make him shake off the net of shams and conventionalisms he has been inveigled into. He will remain always a little sceptic, never troubled with inconvenient aspirations yet never entering with heart and soul into the trivialities he sees practised about him, and in which for the sake of not being eccentric he will take a shallow part. Poor Jemmy! He will never see the frustrated possibility he is—he was bred and born among shams, in his youth he cast many of his conventional wrappages[7] away, just sufficient to make him ever after a little unquiet—his life, to himself, in his best moments, will appear hollow, and he will die a respected citizen, he held up perhaps as a model to the generation which will follow. Poor Jemmy!

And now a word about your last letter.[8] You know what sort of a fellow Dove is, how difficult to catch, how busy, etc. etc. Well, I asked him your question,[9] he said at the moment he did not know and that information on the subject would be difficult to come at, but in the course of a few days he would see what he could find. Accordingly yesterday I went to him again, he had little more certain information to give. He said Bruce[10] had made a journey to Sahara & published certain facts. He says that perhaps he will be able to tell you something about the coast of Guinea. Moreover that the daily variation of temperature in the desert is well known to be extraordinarily great. Bruce says that camels have been killed by it. In a week's time Dove promises to tell you more. I have seen Wiedemann, he had not begun to translate his paper[11] yet but promised to set his wife to work immediately. Clausius will not undertake the translation of his, he has given me a copy of the original and in the course of a fortnight I have promised to do it for him. (perhaps less).

Some time ago I was coming away from a party with old Knoblauch, I saw that in the cold frosty night he breathed badly. I made him take my arm and in spite of his polite protestations I would see him home. It was well I did or he would not have got there, the poor fellow suffered greatly, could get no breath whatever. At last after great difficulty a sledge[12] came luckily past and I got him home, after opening about 10 locked doors in his house. It is very characteristic of the good hearted weak old man to see how he remembers that act of mine now that he is well. He has not the eye to see that for me there was any duty in the matter, in his harmless, selfish way he sees only a kindness rendered to him, which must be repaid. Since that night I have received tickets of admission to concerts, feasts, lectures, etc. etc. in abundance. Whenever I can I accept them, for Knoblauch is not the man to understand refusal. He expects his son and son's wife here soon. Prof. K. has got a call (as independent parsons say) to Halle[13]—I do not know whether or not he intends to accept it.

From Marburg there is a report that poor Wrightson has been making Miss Bücking an offer, and has been refused! Poor fellow, he used to make a confidant of me and I tried to discourage him as much as possible for I knew there was not the least hope for him (having heard her hint as much). He was blind, blind as a bat, he construed her good-natured manner towards him as an encouragement, not being able to see that her manner was just as 'encouraging' to everybody else. I have watched him pay her marked attentions, and I have seen plain enough how they were received, he could not see, but has gone blindly on annoying her until at last I hear the brother has shewn him the door.

Since writing the last I have been helping a dentist to pull Dickinson a tooth, he was glad to part with it and tells me now to remember him to you kindly.

Love to Debus. | Yours affectionately | T A Hirst.

Riess has just published an important book in two volumes, "Die Lehre von der reibungs Electricität".[14] My brother Bill[15] writes me this morning that he has just made me into a veritable 'Uncle Tom'.[16] My address now is 'Uncle Tom's Cabin'.[17]

RI MS JT/1/H/177
RI MS JT/1/HTYP/230–231

---

1. *last letters crossed*: while Hirst's letter is missing, it would have 'crossed' with Tyndall's letter dated [14–]16 February (letter 0713).
2. *The account*: letter 0713.
3. *London University*: founded in 1836, it was comprised of University College London and King's College. London University was not synonymous with the LI.
4. *About a week ago . . . Bevington's £20*: see letter 0706.
5. *prosy*: talking or behaving in a commonplace, dull, or tedious way (*OED*).
6. *Jemmy Craven*: James ('Jemmy', 'Jimmy') Craven.
7. *wrappages*: that which wraps or covers (*OED*).
8. *your last letter*: letter 0722.
9. *your question*: Tyndall requisitioned Hirst to ask Dove whether or not there would be a wider range of temperatures recorded over 24 hours in the desert than in a wooded region at the same latitude (see letter 0722).
10. *Bruce*: James Bruce (1730–94), Scottish explorer who claimed to have traced the source of the Nile River in North Africa and wrote a five-volume account of his travels and experiences in Africa, J. Bruce, *Travels to Discover the Source of the Nile, In the Years 1768, 1769, 1770, 1771, 1772 and 1773,* 5 vols (London: G. G. J. and J. Robinson, 1790).
11. *Wiedemann . . . paper*: see letter 0713, n. 15.
12. *sledge*: a carriage mounted on runners used for travelling over snow and ice (*OED*).
13. *Halle*: the University of Halle in Germany. In 1853, Knoblauch left Marburg for Halle (see letter 0823, n. 4).
14. *'Die Lehre von der reibungs Electricität'*: Riess's 'Doctrine of Static Electricity' was a treatise on his experiments seeking to understand static electricity. P. T. Riess, *Die Lehre von der Reibungselektricitüt,* 2 vols (Berlin: Hirschwald, 1853). It was not translated into English.
15. *Bill*: William (Bill) Aked Hirst (1828–?).
16. *veritable 'Uncle Tom'*: Bill's first child was born in 1853.
17. *'Uncle Tom's Cabin'*: Hirst was referencing the famous abolitionist novel by Harriet Beecher Stowe, *Uncle Tom's Cabin; or, Life Among the Lowly* (London: Thomas Bosworth, 1852). The novel was incredibly popular in Britain, selling a million and a half copies in England and its colonies between 1852 and 1853 (Claire Parfait, *The Publishing History of Uncle Tom's Cabin, 1852–2002* [Aldershot: Ashgate, 2007], p. 108).

From Robert Bunsen 6 March 1853 0728

Heidelberg | 6^{n} März 1853.

Theuerster Freund

Ich habe mich von ganzem Herzen über den schönen Erfolg Ihrer Londoner Vorlesungen gefreut, und schöpfe daraus die beruhigende Hoffnung, daß Ihnen im eigenen Vaterlande eine Zukunft bevorsteht, die alle bösen Reisegedanken nach Australien am besten verscheuchen wird.

Die Wahl Ihrer nächsten Luft- und Waßervorlesen ist glaube ich eine sehr glückliche, weil sich darüber am besten vor einem gemischten Publicum reden läßt. Ich gebe Ihnen nachstehend eine Disposition für die Waßervorlesungen wie ich sie etwa halten würde, woraus Sie sich dann, was Ihnen etwa für Ihr Publicum paßend scheint, entnehmen können.

I Chemische Beziehungen des Waßers

1. Qualitative Zusammensetzung: a Verbrennen von H in trocknem Sauerstoff.

b Verbrennen von O in H.

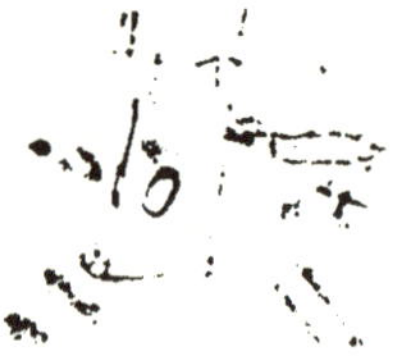

Man läßt zuerst das O in einem <u>nicht zu starken</u> Strahl ausströmen entzündet das H am Halse der Flasche und senkt sie langsam über die O-ausströmungsspitze a, wobei sich das O entzündet und in der H Atmosphäre fortbrennt. Einige Diskurse über die Flamme. Gelbe Farbe u Leuchte kommt vom verflüchtigten Natron des Glases—Temperatur der Flamme aus der Verbrennungswärme und der specifischen Wärme des Waßerdampfs berechnet (Explosion von Knallgas??)

2 Quantitative Zusammensetzung: Im Eudiometer mit <u>Quecksilber</u> ist in einer Vorlesung unausführbar. Am einfachsten durch einen geeigneten electrolytischen Apparat mit Thonzelle wobei man beide aus Entwicklungsröhren entweichende Gase in graduirten Röhren auffängt. Der Apparat ist so /eingerichtet/, daß die Gase nachdem sie in den Röhren aufgefangen, unter gleichem Waßerdruck stehen. —Wenden Sie ein Eudiometer an, so ist um ein <u>Zerspringen</u> zu verhüten darauf zu sehen, daß von dem <u>einen</u> Gase mindestens 2–3 mal soviel vorhanden ist, als zur vollständ. Verbrennung nöthig ist.

3. Chem. Eigenschaften: a Waßer als Säure (/Löschen/ von Kalk) b Waßer als Basis (waßerfreie flüßige in Waßer gegoßen) c Waßer als Krystallwaßer (Erstarren von im Krystallwaßer geschmolzenen $\genfrac{}{}{0pt}{}{Na_2}{\#}\}\ddot{P}h-24ag$) d, Verhalten gegen einfache u zusammenges. Körper. Zersetzung des # durch Kalium Natrium etc. Davy'sche Vulkanentheorie. Natürliche Waßerzersetzungen und Aufklärung über diese Davy'sche Theorie finden Sie in der beigesandten Abhandlung A von p 42 an u. folgende: Chemische Zerstörung der Gesteine.

II a Farbe des Waßers cf. beigesandte Abhandl B p. 44

b Anomalie der Ausdehnung-Explosion von Glasröhren in warmer #, *[2 words illegible]* Kältemischung.

Die Röhren müßen bei a durch Zusammenfallen vor der Lampe etwas verdickt und die Haarröhrchen sehr fein sein, damit der Raum von b bis c beim Zerschmelzen waßerleer sein kann. Aus dem Versuch abgeleitet: die zerstörenden mechanischen Wirkungen des Frostes auf Gesteine und Pflanzen. Wohlthätiger Einfluß des Gefrierens auf den Ackerboden (Pulverisirmethode der /Natur/) *[1 word illeg]* Kriechen der Gletscher etc. Große Warmecapacität des Waßers. Der Ocean als großes Wärmemagazin zur Regulirung der klimatischen Temperaturverhältniße. —Continentalklima u. oceanischer oder Küstenklima. Der Winter in Mailand (Milano Lombardei) ist strenger als der in Island etc. etc.

III Vereinte mechanische und chemische Wirkungen des Waßers. Vulkane und Erdbeben cf. Abhandl A p. 42. Metamorphose der Gesteine durch Waßereinwirkung (cf. Bischoffs chemische Geologie) Ich glaube daß Sie an dieser Stelle eine Reihe prächtiger Versuche über die Isländischen Geisirquellen anstellen können, die sehr schön und überraschend sind, und die sich mit dem in Zeichnung beiliegenden von Dr Bromeis (einem Schüler von mir) angegeben[en] Apparat leicht ausführen laßen.

a a ist ein Rohr von Zinkblech auf der Kohlenpfanne B. C ist eine zweite korbförmige Kohlenpfanne um die seitliche Erhitzung des Geisirrohrs nachzuahmen. D eine große Blechschaale das Geisirbecken repräsentirend: Die Röhre a a wird mit kochendem Waßer [gefüllt] (kochend um nicht zu lange auf die Erhitzung zu warten) und dann gefeuert. Sobald das Waßer zum Kochen kommt, beginnt das periodische Spiel des Geisirs, mit allen Eigenthumlichkeiten wie ich es in der Abhandlung B von p 24 an beschrieben habe. Verstopft man die Oeffnung k ganz locker mit einem Pfropf, so hat man nach einiger Zeit die Strokkereruption in ihrer ganzen Schönheit en Miniature. Durch verengte Aufsätze auf die Offnung kann man auch die Erscheinungen der übrigen Islandischen Springquellen nachahmen. Ueber diese müßen Sie Mackenzie's und des Mißionars Hendersons Reisen in Island nachlesen. Die Versuche sind die *[effect]*vollsten die man sehen kann, und der ganze Gegenstand ist sehr anziehend. Versäumen Sie ja nicht im Britischen Museum und in der Royal Society nachzuforschen, was dort von Solander's und Bank's Papieren über ihre Beobachtungen am Geisir noch etwa aufzutreiben ist. Solche ungedruckte Papiere sollen sich dort noch finden. Mir würde Auskunft darüber und Einsicht dieser Papieren von großem Werth sein. Ich lege Ihnen noch ein Buch bei was Ihnen als Quelle sehr erwünscht sein wird, bitte Sie aber mir die sämmtlichen Drucksachen nebst der Zeichnung des Geisirapparats später wieder *[zukommen]* zu lassen, da es die einzigen Exemplare sind die ich *<several words missing>* *[1 word illeg]* brauche.

Von Herzen | *<1 line missing>*

Heidelberg | 6th March 1853.

Dearest friend,

I was deeply pleased about the wonderful success of your London lectures[1] and draw from them the reassuring hope that a future lies in store for you in your own homeland that will best drive away all the bad thoughts about travelling to Australia.[2]

The choice of your next lectures on air and water[3] is, I believe, a very happy one because this is the best topic to talk about before a mixed audience. I am giving you in the following an outline for the lectures on water as I would more or less present them, from which you can then take what seems to you to be more or less suitable for your audience.

I Chemical relations of water

1. Qualitative composition: a Combustion of H in dry oxygen

b Combustion of O in H.

One first lets the O stream out in a not too strong jet, ignites the H at the neck of the bottle and lowers it slowly over the tip of the O stream *a*, whereby the O ignites and continues to burn in the H atmosphere. Some discussions about the flame. Yellow colour and glow come from the evaporated sodium bicarbonate of the glass—temperature of the flame calculated from the heat from the combustion and the specific heat of the steam (oxyhydrogen explosion??)

2 Quantitative composition: Cannot be carried out in the eudiometer[4] with mercury in a lecture. Simplest using a suitable electrolytic apparatus with a porous pot cell,[5] whereby one collects both gases in graduated tubes as they escape from delivery tubes.[6] The apparatus is [set up] in such a way that the gases are under equal water pressure after they are collected in the tubes. —If you use an eudiometer, make sure that, in order to avoid bursting, at least 2–3 times as much of the one gas is present than is necessary for complete combustion.

3. Chem. properties: a Water as acid ([slaking] of lime)[7] b Water as base

(water-free liquid, poured in water) c Water as crystallisation water (solidification of $^{Na_2}_{\#}\}\ddot{P}b-24ag$ melted in crystallisation water)[8] d, Reaction towards simple and compound bodies. Decomposition of # through potassium sodium etc. Davy's theory of volcanoes.[9] You will find the natural decompositions of water and an explanation of this theory of Davy's in the enclosed paper A[10] from p 42 on and the following: chemical destruction of rocks.

II a Colour of the water cf. enclosed paper B[11] p.44

b Anomaly of the expansion-explosion of glass tubes in warm #, [as in] cold mixture.

The tubes must be somewhat widened at *a* due to their converging before the lamp and the capillary tubes must be very fine, so that the space between *b* and *c* can be free of water during melting. Deducted from the experiment: the destructive mechanical effects of frost on rocks and plants. Beneficial influence of freezing on soil (pulverization method of *[*Nature*]*), *[1 word illeg]* creeping of glaciers, etc. Great thermal capacity of water. The ocean as large heat reservoir for the regulation of climatic temperature conditions. —Continental climate and oceanic or coastal climate. Winter in Milan (Milano Lombardy) is more severe than that in Iceland, etc. etc.

III Combined mechanical and chemical effects of water. Volcanoes and earthquakes cf paper A p. 42. Metamorphosis of rocks due to the actions of water (cf. Bischoff's chemical geology).[12] I believe that at this point you can do a series of splendid experiments about the Icelandic geyser springs[13] that are very nice and surprising and that can be carried out easily with the apparatus of Dr Bromeis[14] (a student of mine) depicted in the enclosed sketch.[15]

a a is a tube made of zinc plate on top of the coal pan B. C is a second, basket-shaped coal pan to simulate the lateral heating of the geyser's pipe. D is a large tin bowl representing the geyser's basin: Tube *a a* is [filled] with boiling water (boiling so as not to wait too long for it to heat up) and then fired. As soon as the water comes to the boil, the periodic play of the geyser begins, with all the peculiarities as I have described in paper B, from p 24 on. If one blocks the opening k quite loosely with a stopper, one has after some time the Strokkur eruption[16] in all its beauty in miniature. By placing constricting caps on the opening, one can also simulate the appearance of the other Icelandic springs. About the latter, you will have to look up Mackenzie's and the missionary Henderson's travels in Iceland.[17] The experiments are the most [effective] that one can see, and the whole subject is very appealing. Do not neglect to try and find out at the British Museum[18] and at the Royal Society what can be still found there from Solander's and Banks's papers[19] about their observations of the geyser. Such unpublished papers are still to be found there. Any information about them and a look at these papers would be of great value to me. I am enclosing one more book[20] for you which will be very welcome to you as a source, but I ask you to /return/ <u>all printed matter together with the drawing</u> of the geyser's apparatus to me later, as these are the only copies that I <*several words missing*> need.

Sincerely | <*1 line missing*>

RI MS JT/1/B/144

---

1. *your London lectures*: a likely reference to Tyndall's lecture at the RI on 11 February (see letter 0713).
2. *Australia*: in mid-1851, Tyndall had applied for two jobs: one in Toronto, Canada and one in Sydney, Australia. The position in Sydney was for a professorship in experimental philosophy at a new university that was being established there (see letters 0558; 0561; 0568; and 0587). There is no mention of apprehension regarding the prospect of moving abroad in Tyndall's journals despite what Bunsen is alluding to here (see, for instance, 4 November 1851, JT/2/13b/550).
3. *your . . . air and water*: Tyndall was invited to give four lectures at the RI (see letter 0717) after the success of his inaugural appearance (see n. 1). He accepted, proposing that the lectures be given under the title, 'On Air and Water' (see letter 0720). The lectures were scheduled for May and June (see letter 0722).
4. *eudiometer*: an instrument for testing the quantity of oxygen in the air (*OED*).
5. *electrolytic . . . porous pot cell*: the term 'electrolytic' refers to the nature of electrolysis, which is the process of passing a current through a liquid to produce a chemical reaction at the electrodes (*OED*). A porous pot cell is a battery-like apparatus constructed using unglazed pottery as a barrier between the anode and the cathode. The porous barrier allows for the passage of ions, while keeping the respective chemical solutions on either side of the barrier from mixing.
6. *both gases . . . delivery tubes*: 'both gases' refers to hydrogen and oxygen.
7. *slaking of lime*: the process of hydrating quicklime (calcium oxide), turning it into lime (calcium hydroxide) by adding water.
8. *solidification of* $\genfrac{}{}{0pt}{}{Na_2}{\#}\}\ddot{P}b-24ag$ *melted in crystallisation water*: the mathematical equation for explaining this melting process.
9. *Davy's theory of volcanoes*: Humphry Davy's theory asserted that eruption of volcanoes was caused by the heat generated in the chemical reaction between metal oxides and air and water below the earth's surface (H. Davy, 'On the Phaenomena of Volcanoes', *Phil. Trans.* 118 [January 1828], pp. 241–50).

10. *enclosed paper A*: not identified.
11. *enclosed paper B*: not identified.
12. *(cf. Bischoff's chemical geology)*: Karl Gustav Bischof (1792–1870), a German chemist and founder of chemical geology, applied chemical theory to explain the mechanisms behind geological systems.
13. *Icelandic geyser springs*: geysers are springs that intermittently shoot water and steam into the air with force. Geologically, they require the presence of a water supply that circulates, and a heat source. They are usually in close proximity to areas with volcanic activity, like Iceland, which is famous for its geysers and hot springs.
14. *apparatus of Dr Bromeis*: Bunsen developed an apparatus to recreate artificially the geyser in the lab, accredited here to Johann Konrad Bromeis, his former student. The apparatus consisted of a basin of water with a long tube inserted through the center. The tube would then be heated at the bottom and near the middle of the tube, resulting in a geyser-like eruption of water once the temperature and pressure were raised sufficiently (J. Le Conte, 'Geysers and How they are Explained', *The Popular Science Monthly* 12 [February 1878], pp. 407–17).
15. *enclosed sketch*: not identified.
16. *Strokkur eruption*: Strokkur, located east of Reykjavik, is one of Iceland's most famous fountain geysers. It erupts every 4–8 minutes with a height of 15–40 meters.
17. *Mackenzie's and the missionary Henderson's travels in Iceland*: George Steuart Mackenzie (1780–1848) was a Scottish mineralogist who travelled to Iceland in 1810 in order to explore its geology. His account of his travels was documented in *Travels in the Island of Iceland* (Edinburgh: Archibald Constable and Company, 1812). Ebenezer Henderson (1784–1858) was a Scottish missionary who travelled to Iceland in 1814–15 in order to distribute bibles. His account was entitled *Iceland: Or a Journal of a Residence In That Island, During the Years 1814 and 1815*, 2 vols (Edinburgh: Oliphant, Waugh and Innes, 1818).
18. *British Museum*: established at London in 1753 from the collections of the naturalist Hans Sloane (1660–1753). Exhibits showcased all aspects of human cultural history.
19. *Solander's and Banks's papers*: Joseph Banks (1743–1820), a naturalist and patron of the sciences, travelled to Iceland in 1772 with Daniel Charles Solander (1733–82), a Swedish naturalist who catalogued collections at the British Museum. Their observations were published in Uno Von Troil, *Letters on Iceland* (London: J. Robson, 1780).
20. *enclosing one more book*: not identified.

From Thomas Henry Huxley 7 March 1853 0729

6 Upper York Place | March 7th. 1853.

My dear Tyndall,

I congratulate you, and look upon your way now as clear. Stick to the Royal,[1] if you can.

I have been so busy lately that I have seen nobody and know nothing, so I have no news to send.

Ever yours faithfully | T. H. Huxley.

IC HP 8:12
RI MS JT/TYP/9/2862
Typescript Only

1. *the Royal*: the RI, where Tyndall had been asked to give a series of four lectures (see letter 0726, n. 4).

From Henry Bence Jones 8 March 1853 0730

Dear Dr Tyndall

Very many thanks for your very friendly letter.[1] I quite agree with you that no public answer at present should appear. I must hear what DuBois[2] thinks. I have told him that it is important to his interests here that nothing should be done rashly. I left out very much that was far more abusive of Matteucci than the passage which has angered him.[3]

If you have Müller's Bericht 1851 p778 you will see what I consider far more personal.[4]

I shall send the page to Matteucci as my answer.

Thanks for the offer of a notice in Phil. Mag. I think not at present. Nothing new here at present

Believe me | Yours most truly | H. Bence Jones
March 8. 1853

RI MS JT/1/J/43
RI MS JT/1/TYP/673

1. *friendly letter*: letter missing.
2. *DuBois*: Emil du Bois-Reymond.

3. *passage which has angered him*: Matteucci was upset by an abstract of du Bois-Reymond's work written by Müller. He had written to Bence Jones about this in late February (see letter 0724).
4. *Müller's Bericht 1851*: Johann Heinrich Jakob Müller, *Bericht über die neuesten Fortschritte der Physik* (Braunschweig: Vieweg 1851).

To William Francis 10 March [1853][1] 0731

My dear Francis

Should your thoughts turn towards Finsbury circus on Wednesday next[2] the enclosed card[3] will open your way to the theatre. The doors are thrown open at 7½oC.[4] and lecture commences at 8.

Be good enough to put Henfrey's address[5] on the enclosed[6] and have it posted for me.

I have got the translation of Magnus' paper from Berlin[7] and hope soon to have that of Wiedemann[8]

You did not send me the address of Mr Brook[9]

ever yours | J Tyndall

10th Mar.

If you would like another card or two let me know

RDS 27/7

---

1. *10 March [1853]*: although there is no year listed in the date of this letter, the mention of Magnus and Wiedemann also place the letter in March 1853.
2. *Finsbury circus on Wednesday next*: reference to Tyndall's upcoming 16 March lecture at the LI (see letters 0722, n. 4; 0725, n. 7; and 0726, n. 9), which was located on Finsbury Circus.
3. *enclosed card*: not identified. Presumably it is the ticket or invitation necessary to gain entry to his lecture on 16 March.
4. *7½oC.*: 7:30 pm.
5. *Henfrey's address*: Arthur Henfrey (1819–59), a botanist and lecturer at the medical school of St George's Hospital. Along with Huxley, Henfrey edited the *Natural History* volume of the *Scientif. Mem.* (A. Henfrey and T. H. Huxley [eds.], *Scientif. Mem.: Natural History* [London: Taylor and Francis, 1853]).
6. *the enclosed*: see n. 3.
7. *Magnus's paper from Berlin*: see letter 0713, n. 14.
8. *Wiedemann*: see letters 0698, n. 18; 0713, n. 15; and 0727, n. 11.
9. *Mr Brook*: Charles Brooke (see letter 0725, n. 3).

From Henry Bence Jones 11 March 1853 0732

Dear Dr Tyndall

You are quite at liberty to talk to Mr Gassiot as you would to me about the Royal Institution.[1] He also will in a few weeks be one of the managers[2] and he has every desire to serve you. I do not think it is impossible for you to hold both professorships.[3] The two would make up a tolerable income & not give you too many lectures. In all not 30 I think at most.

As to Dr Frankland's note,[4] I hope before June you will see our London atmosphere clearer.[5] If you could tell him that until after that time nothing can be done by you about Manchester[6] it will be well I think.

I received a letter from DuBois[7] yesterday rejoicing in your success. He is furious against Matteucci & asks for his letter[8] which went yesterday with an extract from your letter[9] as to the course you thought best for him which I also urged. At all risk he must not be abusive. It will injure him more than it will hurt Matteucci.

I hope to be at your lecture on Wednesday.[10] Mr Gassiot has the strongest desire to help you. I met him last Tuesday.

Ever yours truly | H. Bence Jones

March 11 1853

RI MS JT/1/J/44
RI MS JT/1/TYP/674

---

1. *Mr Gassiot . . . Royal Institution*: Gassiot was a manager at the LI who Tyndall relied on for career advice.
2. *few weeks . . . managers*: new managers of the RI were elected at the annual meeting of the RI at the beginning of May each year.
3. *both professorships*: Tyndall had been offered a professorship at the LI, as well as the opportunity to lecture at the RI in order to qualify him for a professorship there (see letters 0715 and 0717).
4. *Dr Frankland's note*: not identified.
5. *London atmosphere clearer*: likely refers to the confusing situation regarding Tyndall's career options and the hopes that decisions would be made by June.
6. *Manchester*: the reference to a note from Frankland in n. 4 suggests that Tyndall was possibly being considered for a position or series of lectures at the University of Manchester. Frankland, however, was chosen as the first Professor of Chemistry at the University of Manchester (Russell, *Edward Frankland,* p. 149).
7. *DuBois*: Emil du Bois-Reymond.
8. *his letter*: refers to a vitriolic letter Bence Jones received from Matteucci on 28 February (see letters 0724, n. 3 and 0730, n. 3).

9. *your letter*: perhaps the unidentified letter mentioned by Bence Jones in letter 0730.
10. *lecture on Wednesday*: Tyndall was scheduled to give a lecture at the LI on 16 March (see letter 0731, n. 2).

To Thomas Archer Hirst [11–14 March 1853][1] 0733

My dear Tom.

I wish I had time to tell you how your last letter[2] pleased me and how amid a wrecked body my mind gathered up her forces as she read it. The wreck is now repaired but I cant spare the time to talk about it

I must lean upon you for a little more help. I want that translation of Wiedemann's.[3] It wont do to leave it depending upon him. I know nothing myself of the Professorships at the Royal or London Institutions but shall know something on Wednesday[4]—Faraday you know is professor at the former. they are in no way connected with the London university.[5]

lie to tell an untruth is regular—lie to lie down is irregular—the former is present imperfect, I lie I lied, perfect future I have lied I shall lie. The latter is present imperfect I lie I lay, perfect 1st future I have lain I shall lie. 2nd fut. I shall have lain.

I am to give 4 lectures at the R.I. after Easter[6] and shall probably have to run over to Paris for some instruments. This is the reason why I want you to help me—the lectures are necessary to qualify me to become a candidate for the Professorship[7]—I wish you would ascertain and let me know very soon what the price of the Sirene modified by Dove[8] is?—I shall either have one from Berlin or Paris—I wish I could take a run over to you.

affectionately yours | John

RI MS JT/1/T/875
RI MS JT/1/HTYP/232

1. *[11–14 March 1853]*: Hirst wrote 'March 1853' at the top right corner of this letter. While LT noted that the letter was probably written on March 25–29, this cannot be right as Hirst responded to this letter on 18 March (see letter 0735). Thus this letter was most likely written 11–14 March.
2. *last letter*: letter 0727.
3. *translation of Wiedemann's*: see letter 0713, n. 15.
4. *shall know something on Wednesday*: Tyndall was due to lecture at the LI on 16 March (see letter 0731, n. 2).
5. *London university*: see letter 0727, n. 3.
6. *Easter*: see letter 0720, n. 3.
7. *Professorship*: at the RI (see letter 0717, n. 4).

8. *Sirene modified by Dove*: a siren is an acoustical instrument for producing musical tones and used in numbering the vibrations in any note (*OED*). In the 1840s, the eponymous, polyphonic Dove Siren was invented and it was used in the study of interference (R. T. Beyer, *Sounds of Our Times: Two Hundred Years of Acoustics* [New York: Springer-Verlag, 1999], p. 30). See also letter 0735.

To William Francis [17–21] March 1853[1] 0734

My dear Francis

I should feel greatly indebted to you if on receiving this you would send a boy to the Royal Society with the enclosed note[2] for Mr Weld,[3] to be opened by Mr White[4] if he is absent. It is to borrow Mackenzies travels in Iceland,[5] and when you receive it I should feel doubly indebted to you if you would send it to me by post or rail on the same day. I want to have it by Wednesday as I go to London on Thursday[6] and my object is to leave the book with a person *[here]* who is going to copy some of the drawing for me.

I think the paper of Senarmont Annales de Chim. [3] XXXIII. 391[7] would form a very suitable subject for the Scientific Memoirs.[8] If you would get it translated I should be *[happy]* to go halves on the expense.

ever yours | J Tyndall

RDS 27/28

---

1. *[17–21] March 1853*: this letter had to have been written after 6 March, as the recommendation to read Mackenzie was made by Bunsen in a letter dated 6 March (see n. 5). Additionally, Tyndall lectured in London on Wednesday, 16 March (see letter 0731, n. 2), so the Thursday he was travelling to London was definitely not 17 March. As it is unlikely that he received Bunsen's letter before 9 March and he was in London on 16 March, he must have written this after he returned to Queenwood from his lecture.
2. *enclosed note*: not identified.
3. *Mr Weld*: Charles Richard Weld (1813–69), assistant secretary of the Royal Society.
4. *Mr White*: Walter White (1811–93), who spent forty years working in the library of the Royal Society from 1844 until he retired in 1884.
5. *Mackenzie's travels in Iceland*: see letter 0728, n. 17.
6. *Thursday*: Tyndall was in London on Thursday, 24 March (Journal, 24 March, JT/2/13b/603).
7. *Senarmont... 391*: M. H. de Senarmont, 'Researches upon the Optical Characters of Double Refraction in Isomorphous Substancs', in Tyndall and Francis (eds.), *Scientif. Mem.*, pp. 257–93. The paper was initially published as 'Recheres sur les propeiétés optiques biréfrignetes des corps isomorphes', *Annal. Chim. et Phys.* 33 (1851), pp. 391–437.
8. *Scientific Memoirs*: see n. 7.

From Thomas Archer Hirst [18 March 1853][1] 0735

Mittel Stasse No 5 | Friday

Dear John,

Your few words[2] arrived yesterday. I have called on Wiedemann, he has nearly finished his paper,[3] in a few days he will bring it to me for perusal and after correcting I will send it.

I have called on Sauerback[4] the maker of Dove's Improved Sirene.[5]

The best kind with four tones and "Zähler Neck"[6] costs 32 Thalers. With one tone and Zähler, 25 Thalers, with 4 tones but without Zähler 26 Thlrs. Of course you must have the one at 32. Du Bois[7] says they are cheaper here than in Paris. In 3 weeks (at most) after ordering it it will be ready. I think you had better come and fetch it yourself. Clausius wishes the enclosed[8] to be inserted as note in the paper of his[9] I sent you a day or two ago.

Since I began writing this short note the postman brought me a letter from Roby[10] stating that poor Booth died on Tuesday last.[11]

Let me hear from you soon.

Yours affectionately, | T. A. Hirst.

Prof. Knoblauch is in Berlin.

RI MS JT/1/HTYP/233; 235–236
LT Typescript Only

---

1. *[18 March 1853]*: Hirst noted in his journal on 18 March that 'News came that on Tuesday last Booth died' ('Journals', 18 March 1853).
2. *your few words*: letter 0733.
3. *his paper*: see letter 0713, n. 15.
4. *Sauerback*: misspelling of Sauerbach.
5. *Dove's Improved Sirene*: see letter 0733, n. 8.
6. *Zähler Neck*: not identified.
7. *Du Bois*: Emil du Bois-Reymond.
8. *the enclosed*: not identified.
9. *paper of his*: see letter 0713, nn. 17 and 18. It is presumed that Hirst was forwarding the fully translated version of Clausius's paper that Tyndall had requested.
10. *Roby*: W. R. Pridie.
11. *poor Booth . . . Tuesday last*: see n. 1.

From Henry Bence Jones 21 March 1853 0736

Dear Dr Tyndall

Your letter[1] caused me some surprise. I thought they were more liberal in the city. I think by far the best plan would be to let the matter remain in suspense until June.[2]

The Session[3] is nearly over & there will be plenty of time between June & October to get ready for lectures. Let Mr. Gassiot know all you think about their offer[4] but say that until all the proposals are before you you are not competent to decide what is best for yourself & science. I have again heard a rumor of your receiving an offer from the School of Mines.[5] Some of the changes there are to be settled this week. I must stop.

Believe me | Yours most truly | H. Bence Jones | March 21 1853 | 30 Grosvenor St

March 21. 1853 | 30 Grosvenor St.

If you are in London & have time to come & dine here send or leave word with my servant at what hour you will come. I am generally in at six.

RI MS JT/1/J/45
RI MS JT/1/TYP/675

1. *Your letter*: letter missing.
2. *the matter remain in suspense until June*: the 'matter' is in reference to the offer of a professorship at the RI that Bence Jones suggests delaying until June (see Journal, 24 March, JT/2/13b/602).
3. *The Session*: teaching term.
4. *their offer*: the offer for a professorship at the LI. Tyndall recorded in his journal that 'Mr. Gassiot's proposal is that I should deliver a course of 6 lectures; give one evening lecture; attend the meetings of the Laboratory Committee once a month and have all the rest of my time free; it is a sin qua non that my scientific publications should be dated from the London Institution' (24 March, JT/2/13b/602–3).
5. *School of Mines*: see letter 0718, n. 4.

From Emil du Bois-Reymond 22–23 March 1853 0737

Berlin (vom ersten April an 5 Schweitzerstr. im Urban'schen Hause) | 22 März 1853

Verehrter Freund.

Es ist mir ein kleines Unglück geschehen. Ich habe mit vervollkomm[n]eteren Hülfsmitteln eine Wirkung nachweisbar gemacht die ich in der zu Belfast gelesenen Abhandlung gelaügnet hatte. Damit fällt ein Theil des in dieser Abhandlung befindlichen Raisonnements zusammen und die ganze Mittheilung büsst ihre Pointe ein. Wenn die Sache schon gedruckt, muss ich mich darin finden. Wenn es aber möglich wäre den Druck des Ganzen oder eines Theiles desselben zu verhindern, so würde mir diess natürlich aüsserst angenehm sein. Kennen Sie die Verhältnisse auf die es hierbei ankommt? Wenn es gestattet würde einen Theil zu streichen so würde sich ein Mittel finden nur das Richtige stehn zu lassen und das Falsche zu entfernen, obschon freilich die Abhandlung nur noch ein verstümmeltes Ganze sein würde. Es müsste nämlich dieselbe mit dem Satze enden: „The electric current, therefore, acts in the same way on both nerves and muscles viz., it decomposes the peripolar groups and forces the dipolar molecules into a certain arrangement different from the natural one." Damit wäre schon geholfen, am besten wär' es jedoch, wenn man noch, mit theilweiser Weglassung einiger folgenden Phrasenglieder, fortfahren könnte: In the nerves, this preternatural arrangement extends on either side of the portion submitted to the current, and even, though with decreasing regularity, over the whole length of the nerve; . . . (hier würde fortfallen: „But it vanishes as soon as the exciting current itself", was nämlich, wie ich jetzt gefunden habe, für die unmittelbar vom Strome betroffene Strecke falsch ist). . . . ; in the muscles, on the contrary, the preternatural arrangement does not extend to any sensible degree beyond the portion immediately acted upon by the exciting current Und damit müsste geschlossen werden. Da Sie Sekretar der physikalischen Section waren so sind Sie vielleicht im Stande mir diesen Dienst zu leisten. Aber ich vermuthe, es ist zu spät, gewiss hat Ihr pünktliches Council den Bericht schon gedruckt, oder wenn nicht, so wird es doch statutenwidrig sein, überreichte Mittheilungen zurückzunehmen.

Mit welcher Freude ich von Ihrem Erfolge an der Royal Institution gehört, brauche ich Ihnen nicht zu sagen. Auch Magnus hat den lebhaftesten Antheil daran genommen. Was Matteucci betrifft, so halt' ich es doch für passend, ihm einige ganz thatsächliche Bemerkungen mit aller Würde entgegenzuhalten. Ich bin dabei, so etwas zu schreiben, weiss aber noch nicht, auf welche Art es bekannt zu machen sei. Ein fliegendes Blatt drucken zu lassen geht nicht an weil man ja nicht weiss wem Matteucci das seine zugeschickt hat und es also doppelt aufdränglich scheinen würde wenn [man] das Blatt Leuten zuschickte die von Matteucci nichts bekommen haben. Ich würde Sie bitten mir eine Seite kleinen Druckes im Philosophical Magazine zu gewähren, aber werden Sie nachher auch im Stande sein Matteucci die Aufnahme fernerer Invectiven zu verweigern. Ich fürchte er wird sich darauf berufen dass ihm nach Brauch nun auch ein Raum in Ihrem Journal offenstehen müsse. Sie können freilich antworten, Sie sähen schon meinen Artikel als Erwiderung an und ich hätte im Journal erwiedern müssen weil ich nicht hätte wissen können, wem er Exemplare seiner Schmähschrift zugeschickt. Und dabei könnte die Sache ihr Bewenden haben. Auch wird das was ich sagen werde der Art sein, dass kein verständiger Mensch (ich rede nicht von einem Ehrenmann) sich einfallen lassen würde darauf zu antworten. Ich glaube also dass es zuletzt so am besten sein wird. Es sei denn dass Sie mein statement of facts zu dem ihrigen machen wollten. Bence Jones, den ich bedauere in einen Streit mit solchem literarischen Highwayman verwickelt zu haben, schreibt mir, Sie wollten im Magazine etwas sagen. Ich mache Sie zu diesem Behufe aufmerksam auf das Urtheil welches die mir feindliche, Matteucci freundliche Commission der Pariser Akademie über Matteucci's Arbeiten gefällt hat (Comptes Rendus, 15 Juillet 1850. t. xxxi. p. 28). Vielleicht können Sie es benützen.

Oh weh, eben fällt mir ein, dass mir Hirst gesagt hat, Sie reisten nach Paris zu Ostern. Dann bekommen Sie diesen Brief wohl erst nach der Rückreise.

Knoblauch ist hier, ich habe ihn noch nicht gesehen, hoffe ihn aber heut Abend bei Poggendorff zu sehen, wo ich auch Hirst sehen werde. Knoblauch soll den Ruf nach Halle angenommen haben.

Werden Sie in der letzten Woche des August in London sein? Diese denke ich mit meiner jungen Frau, die ich mir alsdann aus Ambleside geholt haben werde, in Bence Jones Hause zuzubringen, der mich mit wirklich einziger Freundlichkeit eingeladen hat. Leben Sie wohl.

Ihr | E du Bois-Reymond. | Gelegentlich meine Grüsse an Francis.

23 März 1853

Gestern Abend erfuhr ich dass Matteucci jetzt auch Berlin mit seinem Pamphlet überschwemmt hat. Dove, die Rose's, Poggendorff waren der Meinung ich solle ihm mit einer französischen Antwort auf's energischste und ausführlich entgegentreten. Ich bin sehr unschlüssig was zu thun ist. Das leichteste und, meiner Meinung nach, ehrenhafteste wäre den letzteren Rath zu befolgen. Denn eigentlich kommt es mir erbärmlich vor, in der Furcht, mir in England vielleicht Hrn Grove noch weniger freundlich zu machen als er schon ist, der literarischen Welt die Wahrheit und mein unumwundnes Urtheil wie ich es längst im Deutschen bekannt gemacht habe vorzuenthalten. Da Sie nach Paris gehen so haben Sie wohl Gelegenheit mir Nachricht zu verschaffen, woran übrigens wohl nicht zu zweifeln ist, ob auch die dortigen Gelehrten beehrt sind. Leben Sie nochmals wohl. *[*E du B-R*]*.

Gestern Abend empfingen auch Hirst und ich ihren Vortrag in der R. I. Die Schlussphrase ist vortrefflich, und mir aus dem Herzen gesprochen. Ich könnte Matteucci morden wenn es ihm gelingen sollte the greatest worker of the age an mir irre zu machen.

Berlin (from the first of April onwards 5 Schweitzerstr.
in Urban's house)[1] | 22 March 1853

Dear Friend

I have had a little misfortune. Using more improved means to assist me, I have made demonstrable an effect that I disavowed in the paper read at Belfast.[2] With this, part of the argument found in this paper collapses and the whole communication loses its point. If the thing has already been printed, I shall have to reconcile myself to it. If, however, it were possible to prevent the printing of the whole thing or part of it, then I would naturally be extremely grateful about this. Do you know the circumstances on which this would depend in this regard? If it were permitted to delete a part, then a way would be found to leave only what is correct and to remove what is wrong, although admittedly the paper would only be a mangled whole. It would have to end, namely, with the sentence: "The electric current, therefore, acts in the same way on both nerves and muscles viz. it decomposes the peripolar groups and forces the dipolar molecules into a certain arrangement different from the natural one." This would certainly help, but it would be best, however, if one could continue with the partial omission of some of the following phrases: In the nerves, this preternatural arrangement extends on either side of the portion submitted to the current, and even, though with decreasing regularity, over the whole length of the nerve; . . . (the following would be omitted here: "But it vanishes as soon as the exciting current itself", which, namely, as I have now found, is wrong for the section affected directly by the current) . . . ; in the muscles, on the contrary, the preternatural arrangement does not extend to any sensible degree beyond the portion immediately acted upon by the exciting current And here it would have to end.[3] As you were Secretary of the Physics Section[4] then perhaps you are in a position to do me this service. But I suspect it is too late, for undoubtedly your punctual Council[5] has certainly published the report[6] already, or if not, then it will nevertheless be contrary to the statutes to retract communications already submitted.

I do not need to say with what pleasure I have heard of your success at the Royal Institution.[7] Magnus also has taken the liveliest interest in it. As far as Matteucci is concerned, I do think it is fitting to point out, with all dignity, some quite factual remarks to him. I am in the process of writing something of the sort, but do not yet know in which way it should be published. It will not do to have a pamphlet printed because one does not know of course to whom Matteucci has sent his and it would therefore appear doubly obtrusive if [one] sent the leaflet to people who have got nothing from Matteucci. I would ask you to grant me a page of small print in the Philosophical Magazine, but will you afterwards also be in a position to refuse Matteucci the taking up of further invective. I fear that he will refer to the fact that, according to convention, a space will now have to stand open for him in your journal[8] also. You can, admittedly, respond that you already viewed my article as a reply and I had to reply in the journal because I could not have known to whom he had sent copies of his diatribe. And that could be the end of the whole affair.[9] Also, what I am going to I say may be of the sort that no reasonable man (I am not talking about a man of honour) would consider responding to. I think, then, that in the end it will be best like this. Unless you want to make my statement of facts your own. Bence Jones, whom I regret to have embroiled in a dispute with such a literary highwayman, writes to me that you want to say something in the magazine. To this end, I draw your attention to the judgment which the Commission[10]—hostile to me, kind to Matteucci—of the Paris Academy[11] has made about Matteucci's works (Comptes Rendus,[12] 15 July 1850. T. xxxi. p. 28). Perhaps you can use it.

Oh dear, it just occurs to me that Hirst has told me that you were travelling to Paris at Easter.[13] You probably won't get this letter then until after your return.

Knoblauch is here, I have not seen him yet, but hope to see him this evening at Poggendorff's,[14] where I shall also see Hirst. Knoblauch is said to have accepted the offer made to him of a professorship at Halle.

Are you going to be in London in the last week of August? I am thinking of spending it with my young wife,[15] whom I shall by then have fetched from Ambleside,[16] in Bence Jones's house, who has invited me with truly singular kindness. Fare well.

Your | E du Bois-Reymond | My greetings to Francis if you happen to see him.

23 March 1853

Yesterday evening I learned that Matteucci has now inundated Berlin with his pamphlet[17] too. Dove, the Roses,[18] Poggendorff were of the opinion that I should confront him in the most energetic terms and in full with a reply in French. I am very undecided as to what should be done. The easiest and, in my opinion, most honourable course would be to follow the latter advice. For it actually seems miserable to me to withhold from the literary world the truth and my frank judgement as I have long made it public in German, for fear of making Herr Grove even less friendly to me in England than he already is. As you are going to Paris, you will probably have the opportunity of obtaining news for me, about which incidentally there is probably no doubt, as to whether the scholars there have also been honoured. Fare well again. *[*E du B-R*]*.

Yesterday evening Hirst and I also received your lecture in the R. I.[19] The final phrase[20] is said exactly what I think. I could murder Matteucci if he should succeed in misleading the worker of the age about me.

RI MS JT/1/D/145

---

1. *Urban's house*: not identified.
2. *treatise of Belfast*: at the meeting of 22 September 1852, du Bois-Reymond gave a report entitled, 'On a New Effect produced on Muscles by the Electric Current' (*Brit. Assoc. Rep. 1852*, pp. 78–80).
3. *It would have to end*: it appears that the requested changes were not made, as the report does not end at the newly requested statement, nor are the requested statements omitted from the published report (see n. 2, p. 79, paragraph 5).
4. *Secretary of the Physics Section*: Tyndall was Secretary of Section A, the Mathematics and Physics Committee, along with William John Macquorn Rankine (1820–72), Rev. Robert. V. Dixon, and John Stevelly (1795–1868).
5. *your punctual Council*: refers to the Council governing the BAAS.
6. *the report*: after each annual meeting, the BAAS published a volume including reports from the various committees, as well as all the reports given by attendees at the annual meeting.
7. *Royal Institution*: du Bois-Reymond was likely referring to the success of Tyndall's lecture of 11 February (see letters 0712, n. 3 and 0713), as well as his success in securing an offer for a series of lectures at the RI (see letters 0717 and 0720); he was in correspondence with Bence Jones, who was intimately involved in Tyndall's progress.
8. *your journal*: the *Phil. Mag.*
9. *the affair*: du Bois-Reymond and Matteucci were in the midst of a very public disagreement regarding their respective perspectives on electrical activity in animals. Both men believed that their professional reputations were at risk in the dispute (see letters 0724; 0730; and 0732). See also G. Finkelstein, 'Matteucci and du Bois-Reymond: A Bitter Rivarly', *Archives Italiennes de Biologie* 149:4 (2011), pp. 29–37).
10. *the Commission*: the Committee of the Académie des Sciences comprised of François Magendie (1783–1855), Alexandre-Edmond Becquerel (1820–1891), César-Mansuète Despretz (1791–1863), Pierre François Olive Rayer (1793–1867), and Claude Pouillet (1790–1868).
11. *the Paris Academy*: the Paris Academy, or Académie des Sciences at the Institut de France, was founded in 1666 by Jean-Baptiste Colbert under Louis XIV in order to encourage scientific inquiry and research.
12. *Comptes Rendus*: the proceedings of the French Academy of Sciences (Paris Academy) published under the name *Comptes rendus hebdomadaires des séances de l'Académie des Sciences* from 1835 to 1965.
13. *Easter*: see letter 0720, n. 3.
14. *Poggendorff's*: Johann Christian Poggendorff's home in Berlin.
15. *my young wife*: fifteen years his junior, Jeannette Claude (1833–?) was du Bois-Reymond's second cousin. They met at a party in Berlin in 1851 and married in August 1853 at Ambleside following a long engagement (Finkelstein, *Emil du Bois-Reymond* p. 152).
16. *Ambleside*: Ambleside, Cumbria, located in northwest England, is a small town on Lake Windermere in the Lake District.

17. *his pamphlet*: *Lettre de Charles Matteucci à M. H. Bence Jones, F.R.S. &c. &c., éditeur d'une brochure intitulée* (Florence: Imprimerie Le Monnier 1853).
18. *the Roses*: the brothers Heinrich (1795–1864) and Gustav Rose (1798–1873) were professors of analytical chemistry and mineralogy, respectively, at the University of Berlin.
19. *your lecture in the R.I.*: Tyndall's first lecture at the RI on 11 February (see n. 7).
20. *the final phrase*: referring to Faraday, Tyndall concluded his lecture stating, 'I rejoice, Ladies and Gentlemen, in the opportunity here afforded me of offering my tribute to the greatest worker of the age, and of laying some of the blossoms of that prolific tree which he planted, at the feet of the great discoverer of diamagnetism' ('On the Influence of Material Aggregation upon the Manifestations of Force', *Roy. Inst. Proc.*, vol. 1, p. 259).

To Thomas Archer Hirst 2 April [1853] 0738

2nd April

My dear Tom

Would you glance at the Analyses of Mem I. & II.[1] and at the historic remarks between them?

I came home on Saturday after two days of wet weather found 5 packages of proofs[2] here waiting for me & was altogether so oppressed by the quantity of work weighing on my mind, that I resolved not to this till I had broken the heart of it.

Yours aff[ectionate]ly | John—

RI MS JT/1/T/887

---

1. *Mem I. & II.*: possibly a reference to Tyndall and Francis (eds.), *Scientif. Mem.*, parts of which Hirst was helping to translate (see letter 0713).
2. *proofs*: not identified.

To Thomas Archer Hirst [5–6 April 1853][1] 0739

My dear Tom

I received your note[2] this morning—I have now only time to request you to order the Sirene at 32 Th.[3] for me:—I hope Herr Sauerbach will prove himself a punctual man: there are many other things I should like to have and within the coming week I will write to you again.[4] Things are still in such an indefinite state that I hardly like to write to you about them. It is very probable that this is my last half year at Queenwood—: if Edmondson is wise his course is very clear—he ought to strive to get you in my place,[5] but to this bargain there are two parties and I dont know what you are going to say to the proposal. As yet however we are in the land of conjecture but a very few weeks will reveal all and shew us the path we ought to follow.

affectionately | John

RI MS JT/1/T/886
RI MS JT/1/HTYP/234

1. *[5–6 April 1853]*: as Hirst dated this letter as 'April 1853', it was possibly written sometime between 5 and 6 April, though this is an uncertain estimate.
2. *your note*: letter missing.
3. *Order . . . 32 Th.*: see letters 0733, n. 8 and 0735.
4. *write to you again*: likely letter 0743.
5. *in my place*: Tyndall is referring to his position at Queenwood, and is suggesting that Hirst take it over after his departure to either the RI or the LI (see letters 0742 and 0746, n. 8).

From Henry Bence Jones 7 April 1853 0740

Dear Dr Tyndall

I have written out my version[1] will you look at the last paragraph especially the last lines, may it pass?

If you do not like my introductory note will you put any other that pleases you in your name.

I have had no good opportunity of speaking to Mr Faraday about you.[2]

Frankland began to day.[3] Without effect.

He read the whole & only made one experi[ment]. Igniting common phosphorous by friction & shewing that red phosphorous would not ignite.[4] The matter was very good indeed & in the best possible time. Excepting perhaps a little too strong of Playfairism.[5]

I have written to Du Bois.[6] I was able to say a few words about him to Col. Sabine last night.

Ever yours truly | H. Bence Jones

April 7. 1853

RI MS JT/1/J/46
RI MS JT/1/TYP/676

1. *my version*: not identified.
2. *speaking to Faraday about you*: Bence Jones was to consult Faraday about whether or not it would be appropriate for Tyndall to accept simultaneous positions at both the LI and the RI (see letter 0741).

3. *Frankland began today*: Frankland had been asked to give a series of ten lectures on Technological Chemistry at the Royal Institution following Easter ('General Monthly Meeting, Monday, March 7', *Roy. Inst. Proc.,* vol. 1, p. 276).
4. *common phosophorous*: i.e., white phosphorus. Red phosphorus, a polymer, is formed by heating white phosphoros or exposing it to sunlight.
5. *Playfairism*: Playfair had a reputation amongs his contemporaries for wordiness, as well as pomposity. Bence Jones was therefore likely suggesting that Frankland was too wordy in his presentation.
6. *Du Bois*: Emil du Bois-Reymond.

From Henry Bence Jones 8 April 1853 0741

Dear Dr Tyndall

I saw Mr Faraday to day & had a long talk with him about you. He is most anxious you should come to the Institution & what he would like best would be that you should be left to your own good judgment & right feeling without any engagements & that you should receive enough from the institution to make you free of any other engagements. But here unfortunately is the first stop. The institution cannot give you three hundred a year[1] which is the least we think you could live on and it is not near but we could give you 200[2] for certain. I told him the offer of Mr Gassiott[3] & that you had not closed with it because you would not decide hastily on anything which might affect your wishes regarding the Royal Institution. I asked him about your holding Both Professorships. He did not altogether approve thinking one must interfere with the other. But saw no difficulty in your taking one professorship & one lectureship, Mr Brand[4] did. What he would like best would be that you should be Professor of RI & lecturer at London Institution but this is not exactly what Mr Gassiott wants. I may have some further conversation in a day or two after he has answered what are the best inducements we could offer to you. I think at present £150 a year & perhaps two thirds of the profits of the lectures is all we have; but I think I shall make out something more clear about this soon.

Tell me on what terms £.s.d you consider you could come. You being perfectly free to take lectures elsewhere or private practice or anything except what Mr Gassiott wants namely to devote yourself heart & soul or rather head & hand to the London Institution.

Your letter I should like to be able to read to Mr Faraday if I think fit.

Goodbye for the present | Yours most truly | H. Bence Jones

April 8 1853

RI MS JT/1/J/47
RI MS JT/1/TYP/677

1. *three hundred a year*: A salary of £300 per year.
2. *200 for certain*: A salary of £200 per year.
3. *Gassiott*: Gassiot, a manager at the LI, was involved in offering Tyndall a professorship there (see letter 0736, n. 4).
4. *Mr. Brand*: William Thomas Brande (1788–1866), a chemist and honorary professor and manager at the RI, was most noted for his *Manual of Chemistry*, which was first published in 1819.

From Thomas Archer Hirst [10 April 1853][1] 0742

by Cayley, Boole and Salmon.[2] Such experiences are not at first agreeable but certainly salutary. But no matter, there is plenty of time and plenty of subjects yet to treat.

And now a word about my willingness to accept the Queenwood Situation[3] if offered me It is a difficult question for me to answer, your advice would influence me most, for as usual it requires a decision that I feel to want to throw partly off my shoulders on to thine. It will interfere seriously with my plans I have at least 2 years unfinished work on the continent which I fear if left now unfinished would remain so. I do not think it would be advisable to break up my student life now, even if there were a prospect of resuming it, it will or rather can never again be so well timed and effectual as at present, to break into the present course of study I have set out for myself would be seriously to injure its success. On the other hand I can scarcely bring myself to refuse such an offer because on its side also are great and necessary advantages. Like a selfish rascal I feel as if I should most like to refuse it at present with the permission to accept it in two years or a year and a half. But that of course is out of all reason. In this letter therefore I will come to no decided conclusion upon the matter I will first hear what you have to say about it for you know full well my position and your advice placed in either scale will draw the balance.

I am anxious to hear something definite of your prospects.[4] I have heard whispers from Du Bois[5] which are almost too good to be true I wait to hear from you the real fact of the matter.

Thine affectionately | T. A Hirst

I have had a letter[6] from R. Carter which has pleased me very much. It was like a breath of wind from the olden time. Three full pages of large letter paper closely written, such eloquent verbosity, beautifully rounded paragraphs with high sounding words but alas little kernel. But it was ever so with Carter, and I have got half to like this habit of his. Before to day I have amused myself often by pruning his letters and seeing into how much space, what they actually contain can without injury be compressed. How much corn and how much chaff. I will give you a sentence excusing his delay in answering my note (this excuse occupies the first page nearly) hear how beautiful it reads. "And I must admit that my temporary silence has in some measure arisen from the anxiety I have felt to acknowledge more definitely than I could do in the language of an ordinary letter some of the sentiments to which you give expression—yielding as I confess they did an unexpected gratification to my own feelings in the agreeable retrospect they clearly convey of the past association between us" That definite acknowledgement he mentions is a sublime delusion I wish to goodness he had sent me the ordinary letter for I strive in vain to get much definiteness from this I suppose, extraordinary one. But nevertheless, Carter is a decent fellow for whom I have a kind of liking. His religious remarks are also in keeping with the whole man he generally shoves them into parenthesis as here. "The knowledge of science and of nature constitutes (next to the blessed truths of revealed Religion) the noblest and most estimable achievement to which the human mind and energies can be directed"*

P.S. What have you been doing in Manchester[7] again?

* How it would startle the little fellow to tell him his parenthesis (in his mouth) is a damned lie. It is as clear as day that the parenthesis is a more tacit admission a matter of ettiquette between R Carter and the blessed truths in question. He bows politely to Revealed Religion[8] but that is all.

RI MS JT/1/H/492
RI MS JT/1/HTYP/235–236

---

1. *[10 April 1853]*: the first part of this letter is missing. LT noted that this letter was written about 9 April 1853. It was most likely 10 April as Hirst noted in his journal that he wrote two short letters to Tyndall on 10 April (Hirst, 'Journals', 10 April 1853).
2. *Cayley, Boole and Salmon*: the mathematicians Arthur Cayley (1821–95), George Boole (1815–64), and George Salmon (1819–1904). Because the first page of the letter is missing, it is difficult to determine what Hirst was alluding to here.
3. *Queenwood Situation*: Hirst was referring to the vacancy that would be created at Queenwood assuming that Tyndall would come to terms with the RI or the LI (see letter 0739).
4. *definite about your prospects*: Tyndall referred to the 'indefinite state' of things regarding the possible positions at the LI and RI in letter 0739.
5. *Du Bois*: Emil du Bois-Reymond.
6. *I have had a letter*: letter missing.
7. *Manchester*: Tyndall travelled to Manchester for the Easter weekend and wrote Hirst from there (see letter 0733).
8. *Revealed Religion*: scriptural authority.

To Thomas Archer Hirst [10 April 1853][1] 0743

Queenwood Sunday

My dear Tom,

I have written off about the Cambridge and Mathematical Journal.[2] At the last meeting of the British Association Sylvester was requested to draw up a statement of the theory of determinants:[3] and the matter was then and there spoken of as something not yet published: so I think you must wait a little. Never mind the crystals:[4] I shall manage to get on without them, but I trust the fellow will not be sending them to me after my investigation has been closed and when they are no longer of service. I wish M. Wiedemann would contrive to say exactly what he means; one would then know how to act. The Memoirs[5] will be published on the 1st of next month and I have made arrangements for the printing of Wiedemann. Now I ought to have it by the 16th, but in case of hardship I might let it run on to the 20th or 21st—Can you let me have it by that time. If not let me know by return as I must then set about something else without delay. In fact I shall be placed in a great difficulty if the paper is not forthcoming at the date stated. I mentioned in ordering the Sirene[6] that I should like you to send me one or two of Seebeck's[7]—don't—they would make the parcel awkward and I can easily make them here. My going to Berlin to fetch the Sirene[8] is now out of the question. I hope Sauerbach is pushing it forward—and I hope you will manage to keep him to his word—a portion of one lecture depends wholly upon the Sirene and I want to become familiar with the use of it beforehand. I wish you would let me know in your next what is the price of the instrument drawn at page 371—2* in the 2nd vol. of Pouillet.[9] It is for the compression of water and Gases—I always thought this instrument was Orsteds[10] but Pouillet seems to say that it belongs to Magnus—will you decide this point for me? I bought Tennyson's poems[11] the other day—they are wine to the intellect. I am afraid in my haste I forgot to allude to DuBois intended marriage although he informed me of it in a previous note.[12] She must be a brave girl if she is able to love Carlyle.

I think I have now said all I wanted to say. but no—we dont say 2° 10' lat. west of Paris Tom. Latitude is reckoned north and south.

Ever yours affectionately | *[J Tyndall]*

*Used also to shew the compression of water

RI MS JT/1/T/884
RI MS JT/1/HTYP/238

1. *[10 April 1853]*: as this was written on 'Sunday', and refers to the siren requested in letter 0739, it was likely written on 10 April.
2. *Cambridge and Mathematical Journal*: *The Cambridge and Dublin Mathematical Journal* succeeded the *Cambridge Mathematical Journal* in 1846 and became the *Quarterly Journal of Pure and Applied Mathematics* in 1855.
3. *theory of determinants*: a determinant is a value calculated from a square matrix in linear algebra. Sylvester's report was to be presented in September at the Dublin meeting, but the inclusion of his 'Provisional Report on the Theory of Determinants' in the report of the meeting shows that he had not yet completed it (*Brit. Assoc. Rep. 1852*, p. 66).
4. *crystals*: Tyndall had requested crystals from Knoblauch who was acquiring them from Frankfurt (see letter 0698, n. 3). Knoblauch reported sending the crystals on 16 April (letter 0748, n. 2).
5. *The Memoirs*: see letter 0698, n. 8.
6. *I mentioned . . . Sirene*: while Tyndall had asked Hirst to order him a Sirene priced at 32 Thaler in letter 0739, n. 3, he did not mention 'Seebeck's' (see n. 7), suggesting that a letter is missing in this exchange.
7. *Seebeck's*: refers to apparatus used to demonstrate the Seebeck Effect, which is the generation of electricity from the temperature difference between two dissimilar electrical conductors. It was first identified by Thomas Johann Seebeck (1770–1831), a German physicist, as a thermomagnetic effect; it was further identified as a thermoelectric effect by the Danish physicist Hans Christian Ørsted (1777–1851).
8. *fetch the Sirene*: Hirst suggested that it would be faster for Tyndall to go to Berlin and get the Sirene himself in letter 0735.
9. *the 2nd vol. of Pouillet*: Likely refers to figure 185, which is mentioned in section 246 in Chapter 2, Book 2 of C. Poulliet, 'De l'existence du calorique rayonnant et de l'idée qu'on peut se frmer des rayons calorifique', *Élémens de Physique Expérimentale et de Météorologie* (Paris: Chez Béchet Jeune, 1827), p. 371.
10. *Orsteds*: see n. 7.
11. *Tennyson's poems*: Likely refers to A. Tennyson, *Poems* (London: Edward Moxon, 1833) or A. Tennyson, *Poems in Two Volumes,* 2 vols (London: Edward Moxton, 1842).
12. *DuBois intended marriage . . . a previous note*: see letter 0737, n. 15.

From Henry Bence Jones 11 April 1853 0744

Dear Dr Tyndall

I was glad to get your letter[1] tho' your answer did not quite tell me what I want to know. I had today another conversation with Mr Faraday; With the utmost desire that you should be at the R.I. It is not easy with many managers to get every thing done quickly.

He evidently does not like the two professorships & he will do every thing that he can do to make it possible for you to accept our professorship. If I knew what you could live in London for or (still better) on what terms you could accept our professorship, with liberty to take any course or courses of lectures elsewhere I think a proposal would soon be made to you.

All that would be asked for the largest stipend we could give would be one lecture a week during each division of the season. In all 19 lectures for this we agreed to pay 200gs[2] might? be got with any sum for materials apparatus diagrams &c. that you pleased in reason £20 or 30 perhaps. You would have the laboratory to yourself to work in & Mr Anderson or some other assistant?

I have proposed to Mr Faraday an alteration in the Professors payment namely that the fees (whole or part?) should go to the lecturer & that the paid sum should be smaller. This may or may not be; & certainly cannot be settled quickly.

I mentioned the sum of 200 a<u>nd</u> a portion of the fees for the 19 lectures & he thought we should have difficulty in getting it & we should certainly unless <u>he</u> (Mr F) took a portion of his fees which he does not do.

The whole apparatus in the Institution & in Newmans shop[3] is at the disposal of the Professors & any thing special is bought; Last year the Logeman exhibition magnet[4] for Mr Fs experiments. I have no doubt Mr Gassiot is very anxious you should take his London professorship[5] & it would not satisfy him that you should give a course of lectures only; at least I fear not. I will see what he says on Thursday, tho I feel it will not be what he wants.

Think over this question. On what terms can will you take our Professorship;[6] to give 19 lectures (Holding any other lectureship but not Professorship)??

As soon as Mr Faraday can make a reasonable proposal he will attack the managers.[7]

I asked Mr Faraday if there was no one treasurer &c to whom he could speak about the arrangement. He said no. It was a question between science & money which no one could settle & so I fear it troubles you also but I have no doubt he will think over the best arrangement & if you will do the same I think an agreement may be found

If not you must accept the London Professorship & give us a course of lectures occasionally.

I have no other news to tell you except that I had a conversation with Grove about Du Bois[8] & Matteucci[9] He told me Matteucci had done more than Du Bois. I asked him how he knew that, had he read his book.[10] No he could not read German. He had read the abstracts! & seen his experiments. I assured him 20 hours would not have enabled Du Bois to shew all he had done. This from Grove shows what some others also think or are led by him to think.

Ever yrs truly | H Bence Jones
April 11. 1853

RI MS JT/1/J/48
RI MS JT/1/TYP/678–679

1. *your letter*: letter missing.
2. *200gs*: 200 guineas.
3. *Newmans shop*: see letter 0717, n. 3.
4. *Logeman exhibition magnet*: Wilhelm Martin Logeman (1821–94) was a Dutch optician who displayed the three strongest magnets at the Great Exhibition of 1851. In 1850 he sent Faraday a magnet (see *Faraday Correspondence,* 4:2278).
5. *London professorship*: at the LI (see letters 0715 and 0717).
6. *our Professorship*: at the RI. It should be noted that 'can' was written above 'will' in this sentence.
7. *the managers*: the managers of the RI (see letter 0720, n. 1).
8. *Du Bois*: Emil du Bois-Reymond.
9. *Du Bois & Matteucci*: they were embroiled in a public conflict regarding their respective work on animal physiology (see letters 0724; 0730; 0732; and 0737).
10. *his book*: du Bois-Reymond's book on animal electricity *Untersuchungen Über Tierische Elektrizität*, 2 vols (Berlin: G. Reimer, 1848; 1849).

To William Francis [14 April 1853][1] 0745

My dear Francis.

I thought I had exercised my */*limit*/* of caution regarding the memoirs.[2] I mentioned a paper of Senarmont[3] several weeks ago as worthy of translation and you said you would look to it. I know that it together with the two I have already sent you[4] and Wiedemann[5] would be more than sufficient for the forthcoming number.[6] That I have not neglected Wiedemann the enclosed scraps of letters[7]—one of which reached me this morning[8] will prove, and notwithstanding the delay occasioned by the fellow himself I have full reliance that Hirst will send me the paper[9] in time for publication. If you will allot me my portion of the Scien. Mem.[10] I will engage to have my translation[11] in your hands at least half a month previous to publication—I think some arrangements ought to be made so that we may all pay for translations which our own engagements prevent us from executing. I should be very glad indeed if even the expenses of publication could be equally distributed over all our shoulders.

Bunsen's paper[12] I have never seen but I would take the man as a guarantee that anything he writes will be good. I have been reading many things from him lately and they are all first rate. Senarmont appears in the Annalen de chimie S. III. I. XXXIII.[13]

ever yours sincerely | John Tyndall

RDS 27/9

---

1. *[14 April 1853]*: there is an annotation from RDS that reads 'July ? 1852' in the upper right corner of the letter; however, due to its contents related to the *Scientif. Mem.* and issues regarding getting the translation from Wiedemann, it is clearly from early 1853. On 16 February (letter 0713, n. 15), Tyndall requested that Hirst aid Wiedemann in his translations so as to not hold them up. As of 6 March (letter 0727), Wiedemann had not yet begun the translation, but by late March (letter 0735, n. 3) Hirst made mention that the Wiedemann paper was nearly complete. On 14 April (letter 0746), Tyndall wrote to Hirst, responding to comments he had made in his last letter; he also mentions that he had just written earlier that day to enquire about the status of the *Scientif. Mem.* As such, it is likely that this is that letter.
2. *the memoirs*: see letter 0698, n. 8.
3. *a paper of Senarmont*: see letter 0734, n. 7.
4. *two I have already sent you*: as papers by Clausius, Magnus, Wiedemann, and de Senarmont were published in Part III of the *Scientif. Mem.*, Tyndall likely means the papers of Clausius and Magnus (see letter 0713, nn. 14, 17–18). He mentioned sending Magnus's manuscript to Francis in letter 0750.
5. *Wiedemann*: see letter 0698, n. 18.
6. *the forthcoming number*: see n. 2
7. *enclosed scraps of letters*: Tyndall had been working with Hirst in order to get the Wiedemann article translated (see letters 0733; 0735; 0743; 0746; and 0752). It is presumed that Tyndall enclosed portions of letters regarding this correspondence in order to bring Francis up to speed, including the missing portion of letter 0742 (see also letters 0713; 0733; and 0735, n. 3).
8. *Reached me this morning*: likely letter 0742.
9. *the paper*: see n. 5.
10. *my portion of the Scien. Mem.*: it is presumed that Tyndall was referring to the papers that he was responsible for translating for the publication.
11. *my translation*: it seems that Tyndall was trying to work out a system or schedule for completing their respective work required to produce the *Scientif. Mem.*
12. *Bunsen's paper*: presumably Bunsen's (1851) paper 'On the Processes which have taken place during the formation of the Volcanic Rocks of Iceland', which was translated and included in Tyndall and Francis (eds.), *Scientif. Mem.*, pp. 33–98.
13. *Annalen de chimie S. III. I. XXXIII*: The *Annales de Chimie et de Physique* was a French scientific periodical founded in 1815 and based in Paris (see n. 3).

To Thomas Archer Hirst 14 April 1853 0746

My dear Tom,

I hope you will be well when this reaches you—your remarks concerning Wiedemann[1] did not take me by surprise—they fulfil my own instinct regarding him. Right Tom[2] to tell him your mind upon the matter and let the fellow feel that he has knocked against a man. I dont know how the memoirs stand but have written off today[3] to enquire—I always calculated on Wiedemann and I fear Francis has done so too—If you have the half of it ready send it to me—it is perhaps as much as we shall require this time.

I cannot speak very definitely as to my prospects yet.[4] I believe I might have 200 a year,[5] an assistant laboratory and £50 yearly for experiments at the London Institution and should only have to deliver 7 lectures during the year. But Faraday I believe is anxious to have me at the Royal Institution—it was proposed to him that I should hold both professorships but he seems to think that one would interfere with the other but would have no objection to my holding a <u>Lectureship</u> at the London Institution in connexion with the professorship at the Royal.[6] I dont think they will be able to offer me more than the London and the lectures instead of 7 will be 19. However even this is not much: and I shall probably close with the Royal. I have weighed the matter of your coming here—it is a near balance. On the whole I think you had better come for a year or two—I know your studies are not finished and I should like you to take a run over to Paris[7] and spend a year or so there. I am commissioned to offer you £130 a year, they will make it 150 when the pupils number 70—we have now 60.[8] In making this arrangement £20 have been taken off the amount allowed to me and added to Debus's salary. he will have 100. You will have about 3 hours a day and two lectures a week—<u>no farmers</u>.[9] The practice will be of great value to you—it has been to me. If you could come some time in May all the better—it would help me amazingly and I should have the satisfaction of seeing everything settled before my departure. I write as if all were settled but there is nothing settled! Nevertheless I see no likelihood of my stopping here another half year. The Cambridge and Dublin Mathematical Journal[10] <u>you</u> can have at £4–4–0. The usual price is £4–16–0.

I received a similar letter[11] from Carter, and he sent me a paper yesterday. Hawkshaw[12] the engineer is proposed as Fellow of the Royal Society this year. There is only one t in the 1st syllable of etiquette[13]—nothing could express the exact relation of Carter to the 'blessed truths' better than what you have said.

I was much amused by Dickinson's scrap.[14] Why does not the little fellow pay some attention to his spelling—he will find a want in this respect like a millstone round his neck in the world. Your improvement has been surprising. Post time draws near and I must close—I hope you stir up Herr Sauerbach[15] now and then.

ever yours affectionately | John Tyndall
14th Apr. 1853

RI MS JT/1/T/885
RI MS JT/1/HTYP/237

---

1. *your remarks concerning Wiedemann*: letter missing or Hirst's remarks were in the missing portions of letter 0742.
2. *Right Tom*: Tyndall wants Hirst to write to Wiedemann regarding the translation of his paper for the *Scientif. Mem* (see also letter 0745, n. 7).
3. *the memoirs stand . . . written off today*: Tyndall sent a letter to Francis discussing the organization of the *Scientif. Mem.* (see letter 0745).
4. *I cannot . . . yet*: this is a direct response to Hirst's desire to know Tyndall's decision on the professorship offers he had received (see letter 0742, n. 4).
5. *200 a year*: £200 per year salary.
6. *But Faraday . . . at the Royal*: see letters 0741 and 0744.
7. *take a run over to Paris*: Hirst would visit Paris in May (see letter 0759). Hirst left for Paris on Friday, 22 April, his birthday (Hirst, 'Journals', 24 April 1853)
8. *I am commissioned . . . now 60*: Tyndall is formally offering Hirst a position at Queenwood, first mentioned in letter 0739. Hirst eventually accepted the offer.
9. *no farmers*: Hirst was referring to Tyndall's negative past experiences with farmers, as discussed in letters 0540, n. 3; 0597; 0639; and 0658, n. 3. In his journal entry of 12 December 1852, Tyndall mentioned that four students, who were studying to be farmers, were expelled for getting 14 or 15 boys drunk (JT/2/13b/592).
10. *Cambridge and Dublin Mathematical Journal*: see letter 0743, n. 2. Tyndall had enquired about the journal for Hirst.
11. *a similar letter*: Hirst recounts a letter from Carter in letter 0742.
12. *Hawkshaw*: John Hawkshaw (1811–91), a civil engineer who was elected FRS in June 1855.
13. *only one t . . . etiquette*: Tyndall was correcting a spelling mistake from the postscript of Hirst's letter (see letter 0742).
14. *Dickinson's scrap*: likely a reference to something written by Carl Dickinson on a 'scrap' of paper (now missing) that was enclosed in Hirst's last letter.
15. *Herr Sauerbach*: see letter 0735, n. 4.

From Henry Bence Jones[1] 15 April 1853 0747

April 15th 1853.

Dear Dr Tyndall

I have so much to say that I do not know how to begin. Thanks for your note[2] of this morning, I thought the best plan was to put all your letters before Mr Faraday. And the result of this and a conversation with Mr Barlow is this. That Mr Faraday will at the May meeting of the Managers <u>write</u> them a note advising them to appoint you Professor of Physics with 200[3] you giving, in two courses a lecture each week during the season and Mr Faraday implied he should also probably alternate years (if not all) ask you to give the juvenile course,[4] which is 6 lectures for £50. The other duties of a Professor we read over and I answered that there was nothing you would object to, (about care of instruments and doing what you were asked)

Mr F. has been thinking over the possibility of your having rooms[5] but he sees no way whilst he remains.

All this must be a matter of private arrangement until this May meeting[6] is over and then you can not be <u>elected</u> Professor, for by the laws you must be proposed in April, having given a course of lectures previously. So that the election must be put off until next year. The Managers however, or Mr Faraday's note, will determine the salary which will commence—when? This partly must depend on your convenience. Let me know your position in this respect. The first duty will be in Jan. 1854. and until you hear the result of the proposal to the Managers you ought not to give up your present duty. I say this not that I doubt Mr Faraday's omnipotence but that accidents will happen, tho' unforeseen. Mr Gassiott[7] I had only a few words with but you are free evidently and he has told you so.

Mr Barlow, I found quite agreeable this morning. And *[if]* you tell me about times I will see that Mr Faraday writes to you or you to him.

Yours ever | H. Bence Jones.

RI MS JT/1/HTYP/239
LT Typescript Only

---

1. *From Bence Jones*: LT noted: 'original *[put]* with Bence Jones'.
2. *your note*: letter missing.
3. *with 200*: see letter 0744.
4. *the juvenile course*: since 1825, the RI ran lectures during the school holiday between Christmas and into the New Year that were aimed at a 'juvenile auditory' in order to introduce

youths to a scientific subject through spectacular demonstrations ('Christmas Lectures Founded', at http://www.rigb.org/christmas-lectures/history [accessed 14 August 2015]).

5. *having rooms*: Faraday lived at the RI.
6. *May meeting*: see letter 0732, n. 2. The meeting was held on 2 May ('Annual Meeting', *Roy. Inst. Proc.*, vol. 1, p. 309).
7. *Mr Gassiott*: Tyndall was also negotiating with Gassiot about a position at the LI (see letter 0744, n. 5).

From Karl Hermann Knoblauch 16[–17] April 1853 0748

Dr John Tyndall | Queenwood college | near Stockbridge | Hampshire | England

with a little | box in black | under the same | address. | containing crystals | 24 Thaler Worth. | franco.[1]

Marburg April 16th 1853

My dearest friend.

I have just received the crystals from Mr Albert[2] and I am very happy to be able to send them to you in time before the 1st of May.

It has not been my fault that you must wait such a long time I have requested very often but always in vain. I hope you will be satisfied with them, though you will find them a little expensive.

I copy Mr Albert's bill:

1 cube of brasilian topaz[3]	6fl.	30 x
1 ——— green turmaline[4]	5	30
1 ——— white fluate of lime[5]	4	30
1 ——— beryl[6]	6	30
1 ——— dichroite[7]	4	
1 ——— brown rock crystal[8]	4	
1 ——— amethyst[9]	5	30
1 ——— arragonite[10]	4	30
	41fl. = 23 Thaler 13 [Sgr.][11]	

I shall send the original with the money to Mr Albert, to get the bill „quittirt"[12] and then shall send it to Mr Hirst according to your wish in your letter.[13]

I beg you, to give me notice whether the crystals have reached you in time. I am in a great hurry to send them immediately to the post-office. Do you wish still other crystals or instruments from Germany? I shall always be very happy to render you any service in my power.

I have marked the optical axes[14] on the cubes. —My serene[15] costs 25 Thaler. —The whole series of pipes, bellows etc (inside the serene) costs 100 Thaler, but this series is only the chief part; a more complete one (such as my own for instance) will cost about 115 Thaler.

I am very glad to hear of your great success, and that you have given up the Toronto- and Sidney-affair.[16]

I am just returned from Berlin where I have engaged myself for a professorship in Halle (the „ordinariat",[17] 1200 Thaler a year, 200 for instrument, and the directory of the public cabinet[18]). So I shall leave my good Marburg as you did, my Dear friend, and the little room, where we have worked together and which I leave with „Wehmuth".[19]—I shall read the last time in Marburg this summer.[20]

Pray, keep me your friendship and don't forget me, when your thoughts coming over to our lovely Marburg don't find me more in your working room. I hope we shall see each other again.

Can you perhaps send my a copy of my memoir[21] on the different transmissive power of crystals in different directions? I have not yet seen the translation in the scientific Memoirs.

M[rs] Knoblauch sends her kind regards[22] to the friend of her husband and wishes to have soon the opportunity (to make)[23] his acquaintance.

You must try to get married also! You will find it much better than to remain alone.

Excuse my bad letter with my great hurry. It is the last moment to send it off. Wishing you every success in your experiments

I am | most faithfully yours | Hermann Knoblauch.

Marburg Apr. 17th. I am very angry, that they have sent back my letter and the crystalls. They tell me that letter and box must be separated, and the letter sent with a French adress. So I do now.

I add a notice. You are asking in your letter,[24] from whom you can get the sticks for longitudinal vibrations.

You can get these as well as all acustic instruments for the purpose of lectures, experiments etc in Berlin from a man, whom I can fully recommend as a „gewißenhafter und sorgfältiger Arbeiter".[25]

Ferdinand Lange, Orgelbauer, Instrumentenmacher und academischer Künstler[26] in Berlin.

My acustic apparatus, bellows, pipes, sticks, plates, discs, bells etc are all from him

As you know them, you are able to judge yourself;

RI MS JT/1/K/19

1. *Dr Tyndall . . . franco*: the address and note regarding the contents of an accompanying box appeared on the back of the letter along with the wax seal.
2. *crystals from Mr Albert*: Tyndall first ordered the crystals in June or July of 1852 (see letters 0698, n. 3 and 0743, n. 4). Tyndall noted receiving the crystals in his journal on 26 April (JT/2/13b/403–4). Albert is Johann Wilhelm Albert of Frankfurt, who dealt in optical and physical objects (Central-Organ für den deutschen Handelsstand [ed.], *Sammlung der Deutschen Handels-Register*, vol. 1 [Köln: M. DuMont-Schauberg, 1862], p. 41) (see letter 0651, n. 5).
3. *brasilian topaz*: a deep yellow colored topaz gemstone.
4. *green tourmaline*: tourmalines are semiprecious stones found in shades of red, green, blue, and brown.
5. *white fluate of lime*: a phosphorescent, brittle crystal that can be produced by passing fluoric acid through limewater; it can also be found in veins in sandstone, limestone, and slate.
6. *beryl*: a mineral crystal, clear in its purest form, coloured blue, yellow, red, white, and green by impurities. The green form is commonly known as emerald.
7. *dichroite*: a synonym of the gem corderite or iolite, dichroite is glass-like in appearance, while often found in shades of blue or yellowish-grey. The term 'dichroite' means 'two-colored rock'.
8. *brown rock crystal*: brown rock crystal refers to brown quartz.
9. *amethyst*: a purple variety of quartz.
10. *arragonite*: a naturally occurring carbonate mineral formed through biological and physical processes, often in aqueous environments. Corals and stalactites are forms of aragonite.
11. *23 Thaler*: at an exchange rate of 52 guldens/pound of silver, 41 guldens = .79 pounds of silver.
12. *quittirt*: German term for 'receipted'.
13. *your letter*: letter missing.
14. *optical axes*: the optical axis refers to the direction which when light is transmitted along it, the internal structure is such that the light will pass through without being refracted. Crystals are classified as either having one (uniaxial) or two (biaxial) optical axes.
15. *serene*: misspelling of siren (see letter 0733, n. 8).
16. *the Toronto- and Sidney-affair*: Tyndall had hoped to receive a position in either Toronto or Sydney (see letter 0728, n. 2).
17. *ordinariat*: more correctly, 'Ordinariate'; German for 'full professorship', i.e., one with a chair.
18. *public cabinet*: public museum.
19. *Wehmuth*: German for 'sorrow'.
20. *I shall . . . summer*: this note appeared in the left-hand margin.
21. *my memoir*: Knoblauch's article in the *Scientific Memoirs* that Tyndall was working on (see letter 0698, n. 8): 'On the Dependence of Radiant Heat in its passage through Crystals upon the direction of transmission', in Tyndall and Francis (eds.), *Scientif. Mem.*, pp. 99–113.

22. *Mrs Knoblauch sends her kind regards*: this sentence (with the exception of the added correction 'of making') is written in another hand, probably Knoblauch's wife.
23. *(to make)*: 'of making' is written above '(to make)'.
24. *your letter*: letter missing.
25. *gewißenhafter und sorgfältiger Arbeiter*: German for 'conscientious and meticulous worker'.
26. *Orgelbauer, Instrumentenmacher und academischer Künstler*: German for 'organ builder, instrument maker, and academic artist'.

To William Francis [17 or 18 April 1853][1] 0749

My dear Francis,

I send you a most instructive little article[2] for next months magazine—it is from the piece of du Bois[3] although his name is not attached to it.[4] It is worth a [thick] load of such papers as Donovan's[5] and will prove a salutary corrective to his lucubrations.[6] The following note may appear at the foot of the first page:[7]

"In connexion with this lucid exposition of the views of Ohm[8] on a point of great importance and of some difficulty to certain of our readers, we refer to a report[9] upon the same subject Phil. Mag. Vol. 3. p.321." Ed.

most truly yours | John Tyndall

RDS 27/26

---

1. *[17 or 18 April 1853]*: the article that Tyndall refers to in this letter was published in May 1853 (see n. 2). As Tyndall mentioned in a subsequent letter (0750) that the article would reach Francis 'tomorrow', this letter would be dated just before or on the same date as that letter.
2. *little article*: 'On the Intensity of Electric Currents', *Phil. Mag.* 5:33 (May 1853), pp. 363–67.
3. *du Bois*: Emil du Bois-Reymond.
4. *name is not attached to it*: the article (see n. 2) was published anonymously.
5. *Donovan's*: Michael Donovan, Esq. (1788–?), an Irish physician and chemist and member of the Royal Irish Academy.
6. *lucubrations*: pedantic or over-elaborate meditations, often a product of nocturnal study (*OED)*. Refers to a series of articles by Donovan on electricity appearing in *Phil. Mag.* in 1852. See M. Donovan, 'On the supposed Identity of the Agent concerned in the Phenomena of ordinary Electricity, Voltaic Electricity, Electromagnetism, Magneto-electricity, and Thermo-electricity: On the Constitution of the Electric Field', *Phil. Mag.* 3:16 (February 1852), pp. 117–27; 3:17 (March 1852), pp. 198–213; 3:18 (April 1852), pp. 290–99; 3:19 (May 1852), pp. 335–47; 3:20 (June 1852), pp. 445–57; 4:22 (July 1852), pp. 33–42; 4:23 (August 1852), pp. 130–38; and 4:24 (September 1852), pp. 210–19.

7. *the following note . . . page*: Tyndall's editorial comment was not included in the article (see n. 2), but was instead replaced by a note from Bence Jones, who had originally requested the piece from du Bois-Reymond (Finkelstein, *Emil du Bois-Reymond* p. 143). See *Phil. Mag.* 5:33 (May 1853), on p. 363.
8. *Ohm*: Georg Simon Ohm (1789–1854), the German physicist and mathematician responsible for Ohm's Law, which characterizes the relationship between current and voltage when electricity is passed through a conductor.
9. *report*: J. Tyndall, 'Dr. Tyndall's Reports on the Progress of the Physical Sciences', *Phil. Mag.* 3:19 (May 1852), pp. 321–30. This report reviewed a series of experiments to verify Ohm's theory on voltaic current which was undertaken by Friedrich Kohlrausch and originally appeared in *Poggend. Annal.*

To William Francis [18 April 1853][1] 0750

Monday.

My dear Francis,

I send you back Magnus du Bois[2] will reach you tomorrow. I have sent it through Bence Jones lest his corrections and mine should clash. du Bois name is not to appear[3] as he does not wish it. I have been pondering upon your note[4] of the 15th where you say you are almost used up—you must not let this thought get possession of you. I know nothing in this life more necessary for a man than courage and there are inexhaustible sources of courage in the world. Our difficulties are great by our own sufferance in most cases. Shake them away man like dewdrops from the mane of the lion.[5] Dont let fear get possession of your heart—the old bible says that it is the worst counsellor.[6] Things are not so cloudy. *[The harm]* is chiefly within us and we magnify our afflictions—courage I say again! I can hardly conceive of a man being overwhelmed except when a Railway embankment or something of the kind falls upon him. You must not suffer yourself to be overwhelmed. It will never do—the darkest cloud is in motion and will soon vanish and let the sun appear once more. When I am in difficulty I get into a rage and find it often of service to me—it lifts me across the obstacle and places me at the other side. You must not faint. Swear valiantly that you will beat down your difficulties and they will instantly hide their diminished heads.

ever yours sinc[ere]ly | J Tyndall

RDS 27/27

---

1. *[18 April 1853]*: Tyndall enclosed the promised article by Emil du Bois-Reymond in letter 0749. He mentioned that it was to be included in next month's magazine and the article

appeared in the May edition of *Phil. Mag* (see letter 0749, n. 2). Tyndall noted here that he was responding to Francis's 'note of the 15th', which is presumably a response to Tyndall's letter of 14 April (see letter 0745). As this letter is dated 'Monday', 18 April is a reasonable estimate of the date.

2. *I send you back Magnus du Bois*: Tyndall is returning manuscripts from Heinrich Magnus and Emil du Bois-Reymond. Magnus's manuscript was presumably for the *Scientif. Mem.* that Tyndall was editing with Francis (see letter 0745, n. 4), while du Bois-Reymond's was for the *Phil. Mag.* (see letter 0749, n. 2) and was also being examined by Bence Jones (see letter 0754).
3. *not to appear*: see letter 0749, n. 4.
4. *your note*: letter missing.
5. *mane of the lion*: 'Fortunes dire weight, the patron's cold distain | Shook off like dew drops from the lion's mane'. James Boswell, *The Life of Samuel Johnson*, vol. 5 (London: John Murray, 1831), p. 453.
6. *worst counsellor*: not identified.

From Robert Bunsen 19 April 1853 0751

Theurerster Freund

Eine dienstliche Reise in den Schwarzwald, von der ich erst eben zurückgekehrt bin, hat mich verhindert Ihnen früher zu antworten. Was zunächst die Geisirformeln anbelangt, so ist ihre

Ableitung folgende: Die Aufgabe ist: Wie hoch ist die Dampfsäule H bei 100°c, welche aus einer Waßersäule h von der mittleren Temperatur t°c erhalten wird, wenn diese Waßersäule h aus der Tiefe des Geisirrohrs an die Oberfläche deßelben, also unter den Druck $0^{m}.76$ Quecksilber gelangt?

1 Litre #, wenn es aus der Tiefe des Geisirrohrs an die Oberfläche gelangt, wird von t° auf t°-100°c durch Verdampfung abgekühlt, oder was daßelbe ist, giebt t°-100° Wärmeeinheiten (Calories) w oder 637 Wärmeeinheiten verwandlen 1 Ltr Waßer in Dampf t°-100° verdampfen daher w:1=(t°-100°):x= $\frac{t-100}{w}$ Litres Waßer. Ist ferner das specif. Gewicht des Waßerdampfs bei 0° (das spec. Gew. des # als Einheit genommen) s, so liefern diese $\frac{t^{\circ}-100^{\circ}}{w}$ Litres Waßer $\frac{t^{\circ}-100^{\circ}}{ws}$ Litres Waßerdampf von 0° oder $\frac{(t-100)(1+0.00368t)}{ws}$ Litres Dampf von 100°. —h Litres Waßer oder eine h_1^{m}-hohe Waßersäule von beliebigem Querschnitt giebt daher $\frac{h(t-100^{\circ})(1+at)}{ws}$ Litres Waßerdampf oder eine $\frac{h(t-100^{\circ})(1+at)}{ws}$ Metres hohe Dampfsäule von 100° Temp /u. dem Querschnitt der verdampften W S/. —Daß die Adhäsion des # an den Geisirwänden bei der Eruption irgend eine erhebliche Rolle spielt, glaube ich nicht. Bei dem Strocker gelang es mir wenigstens nicht durch eine verschloßene mit Schwefeläther gefüllte, in der Tiefe nahe dem Waßer explodirende Flasche eine Eruption hervorzurufen,

was nothwendig hätte der Fall sein müßen, wenn die Temperatur des Waßers durch Adhäsion an irgend einer Stelle höher gewesen wäre als der Preßion der über einer solchen Stelle ruhenden Waßersäule, in Beziehung auf den Kochpunkt entspricht. Daß das Geisirphänomen bei den Dampfkeßelexplosionen eine Rolle spielen könnte, ist sicher.

Hat Dampf a und Waßer b im Keßel die 20 Atmosphären Druck entsprechende Temperatur 230°c und wird das Ventil c geöffnet, so kann zwar der Druck des sich im Waßer b nacherzeugenden Dampfes während der Ausströmung niemals den Druck von 20 Atmosph. wieder erreichen, wie langsam oder schnell oder plötzlich derselbe auch erzeugt werden mag. Allein es ist möglich, daß sich zu dem durch Dampfausströmung verminderten Druck noch ein Stoß von überwiegender Kraft hinzuaddirt, der unter Umständen hinreichen könnte, den Keßel zu sprengen. Nimmt man den Dampf in a hinweg, so ist das # in b um 130° überhitzt, es erfolgt Dampfbildung die das Waßer (wie im Geisir) gegen die obere Keßeldecke schleudert. Die Kraft dieses Stoßes addirt sich zu dem gleichzeitig in a regenerirten Dampfdruck. Macht man die Hypothese daß die Dampfbildung in überhitztem Waßer in einer verschwindend kleinen Zeit erfolgt, so muß die Kraft des Waßerstoßes gegen die obere Keßelwand mit der Schnelligkeit der Dampfentleerung in a wachsen, und unendlich groß werden, wenn die Dampfentleerung von a in einem unendlich kleinen Zeittheil vor sich ginge. Der Stoß müßte daher um so heftiger sein, je kleiner der Dampfraum a und je größer das Ventil ist. Entwickelt man 1 Cubiczoll aus nahe einer Flüßigkeitsschicht, so wallt die Flüßigkeit nur gelinde auf, entwickelt man dieselbe Menge (1 Cbz) Cl u N aus Chlorstickstoff unter einer Flüßigkeit, so wird diese mit ungeheurer, heftige Zerstörungen bewirkender Gewalt gegen die Zimmerdecke geschleudert, eben weil die Zeit, in welcher daßelbe Volumen Gas erzeugt wird, bei dem ersten Versuch eine lange, bei dem letzten aber eine verschwindend kleine ist.

Wie weit indeßen diese Bedingungen einer Geisirexplosion bei den Dampfkeßeln vorhanden sind läßt sich a priori gar nicht sagen. Es wird sich dies aber leicht experimentell entscheiden laßen, wenn man beobachtet ob beim Aufschlagen des Sicherheitsventils ein Waßerstrahl aus diesem hervorgeschleudert wird, und ob die Kraft mit der dies geschieht eine bedeutende ist. Findet ein solches Hervorschleudern nicht statt, so ist die ganze Erklärung [unzuläßig].

Mit der Bitte um freundliche Empfehlungen an Dr. Debus | von Herzen | Ihr | RW Bunsen

Heidelberg d 19 April 1853.

Dearest friend

A business trip to the Black Forest,[1] from which I have just returned, has prevented me from replying to you earlier.[2] As far as the geyser formulae are concerned first of all, they are deduced as follows: The question is: How high is the column of steam H at 100°C, which results from a column of water h of the average temperature t°C, when this column of water h, comes from the depths of the geyser's pipe to its surface, that is, at a pressure of 0.76m of mercury?

1 litre #, when it comes from the depths of the geyser's pipe to the surface, is cooled down from t° to t°-100°C through vaporisation, or, which is the same thing, it gives off t°-100° units of heat (calories) w, or 637 units of heat transform 1 10f water into steam t°-100°, and therefore vaporise: w:1= (t°-100°):x=$\frac{t-100}{w}$ litres of water. If, in addition, the specif. weight of steam at 0° (the spec. weight of # taken as a unit) is s, then these $\frac{t^\circ-100^\circ}{w}$ litres of water supply $\frac{t^\circ-100^\circ}{ws}$ litres of steam at 0° or $\frac{(t-100)(1+0.00368t)}{ws}$ litres of steam at 100°. —h litres of water or a column of water of height $h_1{}^m$ of any cross-section therefore produces then $\frac{h(t-100^\circ)(1+at)}{ws}$ litres of steam or a $\frac{h(t-100^\circ)(1+at)}{ws}$ metre high column of steam of 100° temp. *[*and the cross-section of the vaporised water col.*]*. —I do not believe that the adhesion of the # on the walls of the geyser plays any considerable role or other during eruption. With the Strokkur,[3] at least, I did not succeed in causing an eruption in the depths near the water using a sealed bottle filled with sulphur ether,[4] which would necessarily have had to have been the case if the temperature of the water had been higher at any place or other through adhesion than what corresponds to the pressure of a column of water over such a place, in relation to the boiling point. That the geyser phenomenon could play a role in steam-boiler explosions is certain.[5]

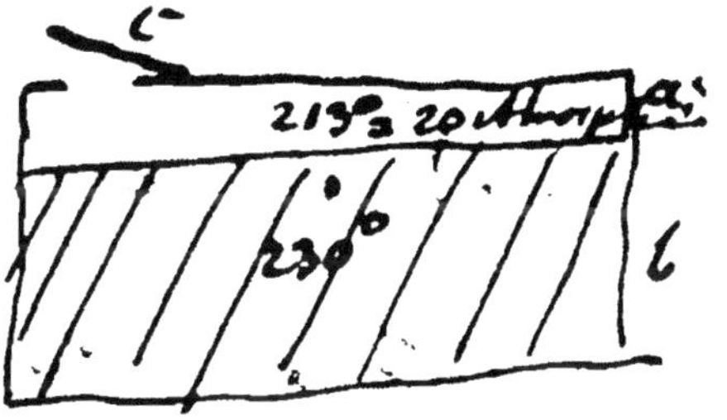

If steam a and water b in the boiler have the temperature of 230°C, corresponding to a pressure of 20 atmospheres, and if the outlet c is opened, then the pressure of the steam subsequently being generated in water b during emission will indeed never reach the pressure of 20 atmosph. again, no matter how slowly or quickly or suddenly it may be generated. However, it is possible that another burst of overwhelming force is added to the pressure reduced by emission of steam, which possibly could suffice to blow up the boiler. If one takes away the steam in a, then the # in b is overheated by 130° and steam forms, which hurls the water (as in the geyser) against the ceiling of the boiler. The force of this burst is added to the pressure which has regenerated at the same time in a. If one hypothesises that steam formation occurs in overheated water in a vanishingly short time, then the force of the burst of water against the boiler ceiling must increase with the speed of the discharge of steam in a, and become infinitely large if the discharge of steam from a should happen in an infinitely short fraction of time. The burst would therefore have to be all the more violent the smaller the area of steam a is and the larger the outlet is. If one produces 1 cubic inch from near a layer of liquid, then the liquid bubbles up only gently; if one produces the same quantity (1 cub in) of Cl and N from nitrogen chloride under a liquid, then this will be hurled against the ceiling with enormous force causing severe damage, precisely because the time in which the same volume of gas is generated is a long one in the first experiment, but a vanishingly short one in the second.

Just how far, however, these conditions for a geyser explosion are present in steam-boilers cannot be stated at all a priori. But this can be easily decided experimentally, though, if one observes whether, when opening the safety outlet, a jet of water is hurled up out of it, and whether the force with which this happens is a significant one. If such a hurling up outwards does not occur, then the whole explanation is [invalid].

With the request to give my kind compliments to Dr Debus | sincerely | Your | RW Bunsen

Heidelberg 19 April 1853.

RI MS JT/1/B/145

1. *Black Forest*: the Schwarzwald, or Black Forest, is a forested mountain range in southwestern Germany.
2. *replying to you earlier*: letter missing. Presumed to be a response to Tyndall's reply to letter 0728.
3. *Strokkur*: see letter 0728, n. 16.
4. *sulphur ether*: an ether in which the oxygen has been replaced by sulphur.
5. *That the geyser phenomenon . . . is certain*: on 24 March, Tyndall received a note from Wynne regarding some questions he had on boiler explosions. Tyndall went to see him about the matter (Journal, JT/2/13b/602). It is presumed that Tyndall had written to

Bunsen beforehand regarding his ideas, as he remarked in his journal on 26 April that 'Bunsen has replied to my question regarding the explosion of steam boilers' (Journal, JT/2/13b/604) (see n. 2).

To Thomas Archer Hirst [19 or 26 April 1853][1] 0752

Queenwood | Tuesday

My dear Tom.

Since my going to London[2] has been a topic of speculation and of conversation I have constantly looked through the entanglement of circumstances to the prospect of your coming here to replace me; but I said in my heart I will not write to Tom on the subject. I will not disturb his thoughts until I see my way clear, notwithstanding my caution. I am afraid I have written you a little too soon.[3] The enclosed note[4] will let you know the exact state of matters at present. But when I think on the matter I am not inclined to regret my writing to you; for even should the matter be deferred to till January 1854 I am strongly disposed to allow myself the next six months to brush myself up in Latin and many other matters and probably to spend the said 6 months in Berlin.[5] If however I put this notion into practice I should want you to let me have a little money. this if you were here I think you could manage to do—possibly I might not want it as I might be able to scrape a little together by writing but I like to prepare for possibilities. These are merely the speculations of the moment, but I almost think the gods offer me this breathing time for this express purpose of enabling me to make a little preparation for my future life. Wiedemann has placed me in great difficulty.[6] I find that it will be hardly possible to get the memoirs[7] out in time. I hope my last note[8] did not reach you before you had completed the translation,[9] for we want the <u>whole</u> of it. I am working hard at my lectures—On June the 3rd (a fashionable night) I am to give a lecture on some of the eruptive phenomena of Iceland[10] and hope to be able to fulfil whatever expectations may be entertained regarding me. Besides this I have a course of 4 to give commencing in May.[11] No more at present *[from]* | Your brain-stuffed brother | John

RI MS JT/1/T/896

---

1. *[19 or 26 April 1853]*: Hirst wrote 'May 1853' on this letter, likely reflecting when he received it in Paris (he travelled there on 22 April). As this letter was written on a 'Tuesday', and refers to letters 0746 and 0747, it was most likely written on 19 or 26 April.
2. *going to London*: Tyndall had been discussing his thoughts regarding his prospective employment options at both the LI and RI (see letters 0722; 0733; and 0746).
3. *too soon*: see letter 0746, n. 8.

4. *enclosed note*: most likely letter 0747.
5. *6 months in Berlin*: letter 0754 from Bence Jones nullified these plans.
6. *great difficulty*: he was taking too long to provide a translation (see letters 0713, n. 15; 0746; and 0735).
7. *the memoirs*: see letter 0698, n. 8.
8. *my last note*: see letter 0746.
9. *the translation*: see letter 0746.
10. *eruptive phenomena of Iceland*: J. Tyndall, 'The Eruptive Phenomena of Iceland', *Roy. Inst. Proc.,* vol. 1, pp. 329–35.
11. *in May*: see letters 0717 and 0720.

From William Hopkins 25 April 1853 0753

Cambridge | April 25 /53.

Dear Sir,

As you have been experimenting on the conductive powers of various substances,[1] I am desirous of knowing whether you could undertake the determination of the conductive powers of certain substances solidified under pressures varying from the atmospheric pressure to 6, or 8000 lbs. per inch. I am now determining the effect of similar pressures on the temperatures of fusion of a variety of substances. We have applied pressures to Spermaceti[2] and wax (the only substances we have yet tried) which have produced an increase of 15°Fah;[3] in the temperature of liquifaction,[4] and I have reason to believe that the conductive powers of these substances may possibly be very much increased when solidified under great pressure. Our experiments will also be continued with a variety of other substances: and what I want to know is whether you would undertake to determine the conductive powers of these substances, specimens of which I would supply you with, solidified under different amounts of pressure. I suggest this to you knowing that you have a delicate apparatus[5] ready for the work. The results will have a very important bearing on certain questions in terrestrial physics.

We have the means of forming cylinders of these compressed materials of about 7 or 8 inches long and of which the diameter of a transverse section is 3 inches. You could probably obtain from such cylinders specimens quite large enough for the purpose.

I should be glad to hear from you on this subject as early as may be convenient to you.[6] If you would like to undertake the experiments, as I hope you will, I will have some specimens immediately prepared for you.

I may add that should you have to incur any expenses in these experiments

I will with pleasure defray it. The Committee for the disposal of the Government grant have behaved very liberally to me respecting the experiments I am carrying on, on the temperatures of fusions.

Yours very grat[eful][ly] | W. Hopkins.

RI MS JT/1/TYP/2/603
LT Typescript Only

1. *Experimenting... substances*: it was possible that Hopkins was in attendance when Tyndall gave his lecture on 6 January at the RS (see letter 0698, n. 12).
2. *Spermaceti*: wax that forms in the head of sperm whales.
3. *15°Fah*: fifteen degrees Fahrenheit.
4. *liquifaction*: misspelling of liquefaction, the reduction to a liquid state (*OED*).
5. *a delicate apparatus*: this is the 'ducky little instrument' Tyndall invented for experiments on conduction (see letter 0638, n. 11).
6. *I should... to you*: Tyndall acknowledged receipt of this request in his journal entry of 26 April (JT/2/13b/603) (see letter 0862, n. 7).

From Henry Bence Jones 2 May 1853 0754

Many thanks for the paper[1]
Dear D[r] Tyndall

Today the new managers are elected[2] and Mr. Faraday has already told the old ones what he proposes regarding you & Mr. Barlow has written to the president[3] Duke of North[d] to tell him. So that at our first meeting[4] I hope something official will be sent to you. Tho' I think I told you the absolute appointment cannot take place until next year. Still the arrangement can be made & you can begin duties & Salary. The time when this will be, I think will be left to you but of this I am not sure. I heard from Du Bois[5] on Saturday he will come with Helmholtz to Hull[6] & will be in England until the middle of Sept I expect. I send you his last paper.[7] He is writing his reply to Matteucci in English & will send us proofs very soon.[8] An article from you will be better than a review;[9] but follow your own course. I am asked to give lectures at the College of Physicians[10] next year & possibly I may take this as my subject. At present I am writing out some clinical lectures which are to be published & this takes all my spare time leaving very little for work after my daily patients are ended.

I had a long talk with De la Rue[11] about the two Institutions.[12] I think to day is the question at the London about selling off. The ballot will be decided against the motion and next week I think you will have an offer from them better than we can give. I expect £300 &c &c. You shall hear as soon as anything is said at the Royal I.

Believe me, | Yours most truly | H. Bence Jones

May 2. 1853

RI MS JT/1/J/49
RI MS JT/1/TYP/680

---

1. *the paper*: a manuscript written by Emil du Bois-Reymond for the *Phil. Mag.* (see letter 0750, n. 2).
2. *new managers are elected*: managers of the RI (see letter 0732, n.2).
3. *the president*: see letter 0713, n. 7.
4. *our first meeting*: the first meeting between Faraday and the new managers of the RI is discussed in letter 0756.
5. *Du Bois*: Emil du Bois-Reymond.
6. *he will come to . . . Hull*: this is presumably a reference to the BAAS meeting of 1853 in Hull. Tyndall could not attend as he was in Paris at the time.
7. *last paper*: not identified.
8. *He is writing . . . soon*: du Bois-Reymond's reply to Matteucci was published later in May (E. du Bois-Reymond, *On Signor C. Matteucci's Letter to H. Bence Jones M.D., F.R.S. &c. Editor of An Abstract of Dr. du Bois-Reymond's Researches in Animal Electricity* [London: John Churchill, 1853]).
9. *better than a review*: this may be in reference to the review of du Bois-Reymond's experimental researches that Tyndall was considering writing at this time (Journal, 26 April, JT/2/13b/604). It would be published as J. Tyndall, 'Rise, Progress, and Present Condition of Animal Electricity', *British and Foreign Medico-chirurgical Review* 25 (1854), pp. 126–42.
10. *College of Physicians*: The Royal College of Physicians.
11. *De la Rue*: Warren de la Rue (1815–89), chemist and astronomer. He was for many years President of the London Institution.
12. *the two Institutions*: the RI and the LI.

From Michael Faraday 4 May 1853 0755

R. Institution, | 4 May, 1853.

My dear Tyndall,

Anderson has asked me about the enclosed tables.[1] My pencil notes will point out the points questioned. Will you settle them and let us have the tables as soon as may be, that he may be employing his spare portions of time on them.

Ever yours, | M. Faraday.

RI MS JT/TYP/12/4133
Faraday Correspondence, 4:2672

---

1. *enclosed tables*: Tyndall mentioned having received this letter in his journal, but makes no mention of the tables it included (Journal, 7 May, JT/2/13b/604).

From Henry Bence Jones 9 May 1853 0756

My dear Dr Tyndall,

The Managers met to day[1] and Mr Faraday stated that he wished to mention a subject in order that at our next meeting (this day fortnight) it might be placed on the minutes. He then made a statement & proposed you as Professor of Physics to receive £200 & to give 19 lectures and in order to shew us what we might & ought to do he read a statement of what the then managers had done for Davy, & he said he saw no reason why we should not do the same now. Davy was given £100 & rooms the first year. The second year £200 The third £300 The fourth £400 The fifth an extra hundred to go on excursion. The sixth it was repeated & he was allowed to take travelling an assistant & have his expenses paid. The seventh he had a multitude of other appointments After this Mr Faraday said we might do anything.

When he finished Mr Gassiot said he thought it right to say that the London Institution were going to make an offer to you. Hereupon it was unanimously agreed that Mr Faraday should make a proposition in writing at our next meeting on the 23rd.

I have little doubt that a scale sliding upward will result and you will be allowed any sum you please however great for experiments. We agreed to pay £28 for the illustration of Hofmann's lectures[2] & ten times as much if you want it would be given for original research.

Nothing could be better than Mr Faradays statement. It was every thing that could be wished and so I hope in a fortnights time all this will be settled.

I have nothing else new. Du Bois[3] tells me he has nearly finished his answer but I have already told you this.[4]

Believe me | Yours most truly | H. Bence Jones
May 9, 1853| 30 Grosvenor St.

RI MS JT/1/J/50
RI MS JT/1/TYP/681

1. *Managers met today*: the Managers of the RI.
2. *Hoffman's lectures*: throughout 1853 and 1854, 'A Course of Lectures on Organic Chemistry' delivered by Hofmann at the RI were published in the *Medical Times and Gazette,* beginning with the 5 February 1853 issue (*Medical Times and Gazette* 27 [1853], pp. 131–34). It is likely that they are referring to these lectures.
3. *Du Bois*: Emil du Bois-Reymond.
4. *already told you this*: see letter 0754, n. 8.

To Thomas Archer Hirst [10 May 1853][1] 0757

My dear Tom

I have been waiting until I should have something to say, otherwise I should have written to you before. I believe you are already in possession of the state of matters up to lately. I will now send a copy of a note[2] received from Bence Jones this morning which will let you know the precise position of things at present. Here it is—

"The managers met today and Mr Faraday stated that he wished to mention a subject in order that at our next meeting (this day fortnight) it might be placed on the minutes. He then made a statement and proposed you as Professor of Physics to receive £200 and to give 19 lectures. and in order to shew us what we might and ought to do he read a statement of what the then managers had done for Davy & he said he saw no reason why we should not do the same now. Davy was given £100 and rooms the first year, the second year £200, the third £300, the fourth £600, the fifth an extra hundred to go on excursion. The sixth it was repeated and he was allowed to take travelling an assistant and have his expenses paid. The seventh he had a multitude of other appointments, after this Mr Faraday said we might do anything.

When he finished Mr Gassiott said he thought it right to say that the London Institution was going to make an offer to you—Thereupon it was unanimously agreed that Mr Faraday should make a proposition in writing at our next meeting on the 23rd.

I have little doubt that a scale sliding upwards will result and you will be allowed any sum you please however great for experiments. We agreed to pay £28 for the illustration of Hoffman's lectures and ten times as much if you want it would be given for original research. Nothing could be better than Mr Faraday's statement. It was every thing that could be wished and so I hope in a fortnight's time all this will be settled."

May the gods grant your brother[3] strength to remain the same man in prosperity that he has been in adversity, if any thing an humbler man.

Stay in Paris as long as you can my dear boy. I may take a run over to you when my labours here are ended.

Ever yours affectionately | John Tyndall

RI MS JT/1/T/561
RI MS JT/1/HTYP/240

---

1. *[10 May 1853]*: Tyndall would likely have received letter 0755 on 10 May. While Hirst wrote '12th May 1853' at the top of the letter that presumably refers to the date he received it.
2. *a note*: see letter 0756.
3. *your brother*: Tyndall himself.

To Henry Bence Jones 11 May 1853 0758

Queenwood 11th, May, 1853

My dear Dr Jones

When I cast my eye back upon my lifetime I sometimes feel inclined to be superstitious; for surely if a special guardian had been appointed to watch over my interests they could not have been more carefully looked to. For my own part I have resigned myself to the current of events, my chief care being to get a little work done, and now I find that the reward has come of itself. I have been often blamed by my more experienced friends for not making better use of my acquaintances to push myself on in the world but I have made little effort in this way. I have held on to my task and left events to the ruler of them. There seem to be moral affinities in the world which settle all these matters according to their own sure laws, and I have found friends rising around me, slowly but healthily, friends who like yourself have managed my affairs far better than I could have managed them myself. Had I the choice of a destiny I dont think I would make any alteration in my present one, for I do not think that the external world could do any thing more for my gratification than has been done. I dont know any recognition which the world could accord equal in my estimation to that of Faraday—Could I get science out of my head and resign myself for an instant to pure feeling I would say—God bless him. If I told the Managers of the Royal Institution how little I care about their professorship[1]—how little it weighs with me at the present moment whether they appoint me or not—I am afraid they would discard me. Their hundreds can never purchase a pleasure equal to that which I have already experienced.

The paper of Du Bois[2] which you have sent me is a most interesting one, I have been thinking of sending a report of the investigation to the Phil Mag and shall probably do so when I have got the lectures off my hands.

Francis informs me in a note received this morning that there is "a gentle pitch into me" by Matteucci in the Comptes Rendus.[3] I dont know what the nature of the 'pitch' is—my molecular notions dont seem to please him. Very good, God, no doubt, will defend the right.

ever most faithfully
Yours | John Tyndall

RI MS JT/TYP/3/791
LT Typescript Only

---

1. *Royal Institution . . . professorship*: see letter 0755.
2. *The paper of Du Bois*: most likely Emil du Bois-Reymond's reply to Matteucci as discussed in letters 0754 and 0756.
3. *Comptes Rendus*: see letter 0737, n. 12. Refers to C. Matteucci, 'Sur l'influence de la chaleur, de la compression, de la form cristalline et de la composition chimique sur les phénomènes diamagnétiques', *Comptes Rendus Hebdomadaires des Séances de L'Académie des Sciences* 36 (1853), pp. 740–44.

From Thomas Archer Hirst [15 May 1853][1] 0759

Hotel Corneille, | Rue Corneille, | Paris, | Whit Sunday,[2]

My dear John,

Thy letter, or rather Bence Jones's,[3] pleased me very much. Thy prayer for strength to keep thyself humble is not unnecessary, for thy prosperity and the brilliant prospect before thee are sufficient to upset any resolution a whit weaker than thine. Nevertheless to me it causes no surprise, it seems to be but the natural order of things. I did actually prophecy it and that four years ago. It was in Perkinton's[4] back snuggery.[5] Tom himself, and Jemmy Craven were present, the room was full of smoke, we were talking about you. Tom called you a 'damned fool' and me another because I was going to follow in your foolish footsteps. I got earnest in my enthusiasm and offered to wager any amount that before 6 years were past we should see you acknowledged amongst the first men of your profession. Tom said you would be acknowledged as one of the greatest fools of the day, but at bottom he did not think so for he did not accept my wager. If he had, you see, he would have lost. I shall wait anxiously for their decision next fortnight, don't neglect to send me word.

We are quite settled here now,[6] we hear some splendid lectures at the College de France and Sorbonne—Chiefly mathematical of course. Lamé[7] on the mathematical theory of heat, an exceedingly able series of lectures of which I hope to make some use. I have a little project in view with respect to your investigation on the conductive power of wood, of which we will talk some time. Then Liouville[8] on Differential Equations, very elaborate but which I do not yet thoroughly understand. Then again Bertrand[9] on Analytical Mechanics, also of the very highest class. I hear also Dessain[10] on Physics (Light), very interesting but rather popular, the experiments please me best, they are excellent and on a grand scale. Regnault's lectures unfortunately are at the same time as Lamé's, so that I have not heard him yet. I shall do so however before I leave. Come and see us by all means in June if possible. I have to-day called on Verdet Professeur de Physique a l'Ecole Normale, Du Bois[11] introduced me to him, apparently a decent fellow. Has he done anything? We were speaking of you and Plücker. He is a thorough partisan of yours and impeaches Plücker with unfairness. I left two boxes of books etc. in Berlin to be forwarded to England, I want you to tell me two things: 1st Whether I can order them to be sent direct to Queenwood, or whether it is requisite that I should be present in London to allow the Custom House officer to examine them. 2nd; If the latter be the case where must I order them to be left in London until I come.

I want you to order the back numbers of the Cambridge and Dublin Journal[12] for me, and to get them to Queenwood. Are the Scientific Memoirs[13] published? Is Magnus's paper therein? If so send him a few separate copies. Did my translation[14] of Wiedemann's paper arrive in time?

An interesting letter or two has passed between me and Phillips which I will show you when you come. A week ago I wrote to him for the first time since seeing him in England and spoke rather plainly to him. The letter touched him rather sorely, so that to-day I have had an answer of four pages (!!) extraordinarily long for January,[15] and very characteristic. I have only room to give you here his concluding words. 'And now, good-bye, old Praise God Bare Bones! Old Puritan Long Shanks![16] Old Roundhead![17] Old Pharisee![18] Old friend! and believe me most | Affectionately yours, | Geo. Phillips.'

Tell Debus that his letter has been forwarded to me here and received. I learn from it that the Sirene[19] arrived safely. Has your lecture on the Geysers[20] come off yet? It is a new subject for you. How did you come at it?

Yours affectionately, | T. A. Hirst.

RI MS JT/1/HTYP/241–242
LT Typescript Only

1. *[15 May 1853]*: Hirst's reference to Whit Sunday (see n. 2) dates the letter to 15 May.
2. *Whit Sunday*: traditionally identified as the fiftieth day, or seventh Sunday after Easter.
3. *Thy letter, or rather Bence Jones's*: in his last letter (0757), Tyndall included an extract from a letter from Bence Jones (see letter 0758).
4. *Perkinton's*: Tom Perkinton.
5. *snuggery*: a snuggery is a 'cosy or comfortable room, esp. one of small size, into which a person retires for seclusion or quiet' (*OED*).
6. *settled here now*: Hirst was visiting Paris (see letter 0746, n. 7).
7. *Lamé*: Gabriel Lamé (1795–1870), French mathematician and the chair in mathematical physics at the Sorbonne, was most notable for the mathematical theory of elasticity.
8. *Liouville*: Joseph Liouville (1809–82), French mathematician and chair in mathematics at the Collège de France, is notable for his work on number theory.
9. *Bertrand*: Joseph Louis François Bertrand (1822–1900), French mathematician and professor at Collège de France, notable for his work on number, differential, and probability theories, as well as for his work on thermodynamics and economics.
10. *Dessain*: misspelling of Desains. Paul-Quentin Desains (1817–85), French physicist and chair of physics at the Sorbonne from 1853, notable for his work on radiant heat.
11. *Du Bois*: Emil du Bois-Reymond.
12. *Cambridge and Dublin Journal*: see letter 0746, n. 10.
13. *Scientific Memoirs*: see letter 0698, n. 8.
14. *my translation*: see letter 0745, n. 7.
15. *January*: George Searle Phillips often went by the pen name 'January'.
16. *Long Shanks*: a term applied to a long-legged person, most famously to Edward I of England (*OED*).
17. *Roundhead*: a person considered to be puritanical in character (*OED*).
18. *Pharisee*: 'A person of the spirit or character commonly attributed to the Pharisees in the New Testament; a legalist or formalist; a self-righteous person, a hypocrite' (*OED*).
19. *Sirene*: siren (see letters 0733; 0735; 0739; and 0743).
20. *lecture on the Geysers*: see letter 0752, n. 10.

To William Francis [15 May 1853][1] 0760

Sunday

My dear Francis.

I have read Carmichael[2] through. It is short, and this is its best recommendation. I only wish the magazine were in a position to deal as sparingly as it wished in such communications. There is Adamson spinning his yarn further in the last number[3] and promising to continue it. All these matters may be interesting in themselves but the Phil. Mag. is not their arena. Chowne I have also read,[4] and cannot recommend its publication in the magazine—such a thing would be best published as a detached work, and then the magazine if *[I]* saw anything in it might communicate such in the form of a review. From the abstract it is hard to form an idea of even the nature of the results—here and there an expression occurs which suggests a doubt as to the scientific qualifications of the writer. At all events I am afraid your readers would derive but little satisfaction from the perusal of the article

If you have any more papers don't hesitate to send them to me.

ever yours sincerely | John Tyndall

I will send you the ticket.[5] Does Daubeny mention the geysers in his work?[6] have you a *[dirty]* copy that you could permit me to look into?

I hope I shall soon have the pleasure of seeing Mr Taylor.[7]

StBPL T&F Author letters

1. *[15 May 1853]*: as this letter was written on a 'Sunday' and appears to precede letter 0761, a likely date is 15 May.
2. *I have read Carmichael*: refers to Robert Bell Booth Carmichael (1823–61), Irish mathematician and Fellow of Trinity College, Dublin. The manuscript is not identified.
3. *Adamson . . . last number*: not identified.
4. *Chowne . . . read*: although the paper is not identified, W. D. Chowne M.D. (n.d.) had conducted research on the movement of atmospheric air in tubes in 1847 and again in 1853, as outlined in a paper received by the Royal Society on 14 June 1855 ('Experimental Researches on the Movement of Atmospheric Air in Tubes,' *Roy. Soc. Proc.*, vol. 7, pp. 466–75).
5. *the ticket*: possibly admission to Tyndall's lecture on 21 May at the RI, although it is unclear if Francis attended, as Tyndall mentioned in letter 0761 that he did not see him while in London.
6. *his work*: this is most likely in reference to C. Daubeny, *Sketch of the Geology of North America Being the Substance of a Memoir Read Before the Ashmolean Society Nov. 26, 1853* (Oxford: T. Combe, 1839). (See also letter 0761, n. 5.)
7. *Mr Taylor*: Richard Taylor.

To William Francis [22 May 1853][1] 0761

Sunday

My dear Francis,

Here I am back again without having been able to see you. I reached town on Friday evening & heard Frankland lecture, worked until 20C. next day at the RI, had to run to my lodgings, dress and run back again, the last train starting at 50C. I had to drive straight to the station and lo! here I am once more. Whatever the audience may be kind enough to think I think rather poorly of the lecture myself.[2] *[Will]* you be kind enough to send me Daubeny[3] and Mr Gyde[4] will probably have a vol. of the Phil. Trans.[5] which might come along with it. I should gladly have remained in town Saturday night but the loss of Sunday would make me a bankrupt.

ever Sincerely yours | J Tyndall

RDS 27/13

1. *[22 May 1853]*: although 'Sept. 52' is written by another hand on the top of this letter, it was clearly written on 22 May 1853. On Friday, 20 May, Tyndall had travelled to London to attend Frankland's lecture on 'Observations, Economical and Sanatory, on the Employment of Chemical Light for artificial Illumination' (*Roy. Inst. Proc.*, vol. 1, pp. 319–25) (see also letter 0740, n. 3). The following day, Saturday, 21 May, Tyndall delivered a lecture at the RI, before returning to Queenwood that evening (Journal, 19 and 21 May, JT/2/13b/605).
2. *think . . . myself*: Tyndall also recorded in his journal that his 21 May lecture at the RI 'fell far below my own ideal but nevertheless happened to please the audience' (JT/2/13b/605).
3. *Daubeny*: Tyndall had inquired about Daubeny earlier (see letter 0760, n. 6).
4. *Gyde*: Charles Gyde was a clerk at Taylor and Francis (see Volume 3).
5. *Phil. Trans.*: perhaps refers to Daubeny's article in the *Phil. Trans.* on thermal springs ('On the Quantity and Quality of the Gases Disengaged from the Thermal Spring Which Supplies the King's Bath in the City of Bath', *Phil. Trans.* 124 [January 1834], pp. 1–13).

From Henry Bence Jones 23 May 1853 0762

My dear Dr Tyndall

The managers met today[1] and I am requested to communicate to you officially, that in consequence of a recommendation from Mr. Faraday the managers are desirous of proposing you for election as Professor of Natural Philosophy with £200 a year. Your duties are defined in the bye law which I have marked.[2]

I am requested to report to the managers at their next meeting[3] your answer to this proposal.

Will you in the course of the next few weeks send me a short & sweet letter[4] or empower me to state in answer to them what I said today you had answered to me. Do this as you please.

I think you can be elected in July[5] & the £200 will commence from that time whether you begin to work at the RI or not. You are free to <u>lecture</u> anywhere <u>but</u> only to be <u>Professor</u> of the R. Institution. The £200 may & will be increased in a year & there is no reason why after a time you should not have

£400 (or more) if you devote yourself to R.I. Davy had; & Mr. Faraday has quoted him as a precedent for this proceeding.

We had a very full meeting & all were for you & some for offering you more at first & all thought it poor pay.

Mr Gassiott spoke highly of you & of your letter[6] asking to be free to accept an offer from the RI You can see some day on the minutes what Mr. Faraday said.[7]

Your lecture of Saturday[8] was highly approved | I must end.

We do not meet until the end of June

Believe me | Yours very truly | H Bence Jones

May 23. 1853.

RI MS JT/1/J/51
RI MS JT/1/TYP/682

---

1. *Managers met today*: see letter 0756, n. 1. At the 23 May 1853 RI Managers' Meeting, Faraday recommended offering Tyndall a professorship there stating that Tyndall 'has been an original and successful investigator of different departments of physical science;—that he has written several papers on research highly acceptable to philosophers;—that he is a good German and French Scholar, having thus free access to the science of the Continent; and is one of the Editors of the 'Scientific Memoirs' (Minutes of the Managers' Meeting, RI MS AD/02/B/02/A/11/13–14).
2. *I have marked*: document missing.
3. *next meeting*: the next Managers Meeting was 6 June (Minutes of the Managers' Meeting, RI MS AD/02/B/02/A/11/15–21).
4. *sweet letter*: letter 0765.
5. *elected in July*: 'General Monthly Meeting, Monday, July 4', *Roy. Inst. Proc.,* vol. 1, pp. 339–40.
6. *your letter*: letter missing.
7. *what Mr. Faraday said*: see n. 1.
8. *Your lecture of Saturday*: Tyndall lectured at the RI on 21 May (see letter 0761, nn. 1 and 2).

To William Francis [29 May 1853][1] 0763

Sunday

My dear Francis

I am rejoiced to learn that you have got over your difficulties[2] successfully. I wish you were here, the sun is aloft and Debus and myself are going to have a ramble[3]—I wish you were along with us. The country is splendid. The corn-crops *[1 word illeg]* velvet, the trees sprouting and the birds singing—God knows I want a few hours rest. The lining of my scull is creeping and simmering, but the fresh air will set me all right again. I am working hard at my lectures.[4] Thank you for the bit of intelligence from Sabine.[5] The circular[6] will do admirably. I think your remarks regarding the stamps &c are quite to the point—Let it be addressed to you. Hirst is in Paris. I had a note from him this morning[7]—he will probably come to fill my place here.[8]—On second thought I would not put stamps on the circulars. The man who would let this be an obstacle is not worth hearing from. But you will be the best judge yourself. I have just got a splendid work on the Beugungs Erscheinungen[9] by Schwerd. though published in 1835 I believe it is not yet known in England. It would be admirable for the Scien. Mem.[10] were it not so large—it contains about 160 pages small quarto, and a number of plates. It would be well if one of us would write to Challis[11] or Baden Powell and ask him whether the work or any portion of it has appeared in English—I think not.

ever yours truly | John Tyndall

RDS 27/31

1. *[29 May or 5 June 1853]*: this letter was probably written on 29 May as it was written on a 'Sunday' and refers to lectures that Tyndall was working on (see n. 4). Moreover, Hirst mentioned in his journal entry of 29 May that 'It is now certain that in August I shall commence my life as Tutor there [Queenwood]' (Hirst, 'Journals', 29 May 1853), which suggests that he and Tyndall were corresponding about this around that time (see n. 8).
2. *your difficulties*: not identified.
3. *a ramble*: a recreational walk or wander without a definite route (*OED*).
4. *my lectures*: presumably his 'Air and Water' lectures, the final two of which were delivered on 4 and 11 June (see letter 0717).
5. *intelligence from Sabine*: not identified.
6. *The circular*: not identified.
7. *a note from him this morning*: letter missing.
8. *fill my place here*: Tyndall had previously suggested that Hirst could be his replacement at Queenwood College (see letter 0746, n. 8).
9. *Beugungs Erscheinungen*: Schwerd's *Die Beugungserscheinungen aus den Fundamentalgesetzen der Undulationstheorie analytisch entwickelt und in Bildern daegestellt* (Mannheim: Schwan und Goetz, 1835) was a treatise on diffraction and wave theory.
10. *Scien. Mem.*: see letter 0698, n. 8.
11. *Challis*: James Challis (1803–82), English physicist and astronomer who was a supporter of the wave theory of light.

To Thomas Archer Hirst [5 June 1853][1] 0764

My dear Tom

Since last I wrote to you I have lectured twice[2] in London. The first lecture fell far below my own ideal,[3] nevertheless I was informed almost officially that it was highly approved. Yesterday I came on again, I was in better health and although my materials were of a less important character still I threw fire and freedom enough into the matter to please everybody myself included. Indeed the hour was one of enjoyment to me. I was crowded with questioners after the lecture: and some of my kind friends said that I had transcended myself. Faraday and an aristocratic looking man pushed through the throng and came up to me. The latter was Sir Henry Holland.[4] He has been in Iceland and has some beautiful paintings of the Geyser[5] which he wants to shew me so he engaged me to breakfast with him on Friday morning. 'You live in town? do you not' he asked—'no no said Faraday 'but he will do so very soon.' A professor of the Museum of Economic Geology*[6] called me aside and said that all the professors there were working to have me, and that now a post was open for me. I said that I was all but engaged to the Royal Institution—You can hold both was his reply. but this cannot be. They wanted to have me at the London Institution also: but although I am at liberty to lecture where I please I am to be professor at the Royal Institution only.[7] I shall be guided altogether by Faraday—I have received[8] the official expression of the wish of the managers to elect me as Professor at £200 a year. This it is said is but the commencement. A full meeting was unanimous in my favour, and some were for offering me a higher salary at first. At the end of June the seal will be set to the proceedings. I hope you will note every good experiment you see. The geyser came off on Friday.[9]

Magnus is published.[10] Wiedemann arrived luckily.[11] I used the Sirene[12] yesterday. Sauerwald has done his duty, it is a beautiful instrument. You can send your box direct to Queenwood, but I think it must be accompanied by a declaration of what it contains. I know M. Verdet very well, he is connected with the Annales de Chimie.[13] My compliments to him should you see him. This is a swift letter Tom and I have not time to moralize on passing events—but I could do so. I hope I shall be able to digest everything into nutriment; as yet there is no fear of your brother, he looks upon it all with a very philosophical eye.

Affectionately | John

* a government Institution

RI MS JT/1/T/562
RI MS JT/1/HTYP/243

---

1. *[5 June 1853]*: while another hand has written '28th May. 1853' at the top of this letter, the references to two lectures in this letter (see n. 2), suggest that it was written on 5 June.
2. *lectured twice*: likely refers to his 3 June lecture on geysers (see letter 0752, n. 10) and his 4 June lecture on 'Air and Water' (see letter 0726, n. 4).
3. *below my own ideal*: in his journal entry for 6 July, which summarized a month's worth of events, Tyndall said that 'the geyser lecture . . . fell far below my own hopes' (JT/2/13b/607).
4. *Sir Henry Holland*: 1st Baronet and physician, Holland (1788–1873) was known for his travel writings. He would become President of the RI in 1865.
5. *the Geyser*: the Great Geyser in southwestern Iceland. Holland had extensively travelled the country.
6. *Museum of Economic Geology*: see letters 0718, n. 4 and 0721, n. 9.
7. *professor of the Royal Institution only*: this is stipulated in letter 0762.
8. *have received*: see n. 7.
9. *geyser came off on Friday*: see n. 2.
10. *Magnus is published*: see letter 0713, n. 14.
11. *Wiedemann arrived luckily*: see letters 0745, n. 7 and 0759, n. 14.
12. *Sirene*: see letter 0759, n. 19.
13. *Annales de Chimie*: see letter 0745, n. 13.

To Henry Bence Jones[1] 5 June 1853 0765

Queenwood | 5th June, 1853

My Dear Dr. Jones,

I have to acknowledge the receipt of your note[2] of the 28th of May,[3] stating that in consequence of the recommendation of Professor Faraday, the Managers of the Royal Institution were desirous of proposing me for election as Professor of Natural Philosophy in the Institution.

I will not dwell upon the gratification which this proof of the confidence of the Managers; and more especially of him[4] whom I have long regarded as my chieftain and exemplar in scientific matters, has afforded me. —but will proceed at once to state as clearly and as briefly as I can, the reflections which your communication has suggested

On one point in the Bye Laws to which you have drawn my attention[5] I would beg to make a remark. It is stated to be one of the duties of the Professors, 'to superintend all experiments ordered by the Committee of Managers.' Now mind has its affinities as well as matter, and upon them the successful progress of scientific investigation must in a great measure depend. These affinities are not always under the control of the investigator himself, and much less under external direction. The poems of our Laureates which are destined for immortality, are rarely those which are made to order—their spontaneous utterance is best; and I think freedom of choice ought, if possible, to be granted to the man of science. It would, I think, paralyze his best force to have his field of enquiry determined for him before hand. To be brief:—I would undertake to give a Lecture every week during the season—19 in all—but beyond this I trust the Managers will grant me freedom of action.

On Thursday afternoon you mentioned to me a subject which I think may with propriety be introduced here. What you then said, and what I have learned from other quarters, make it very probable that a Professorship at the Museum of Practical Geology[6] will be offered to me. I have been swayed variously by this communication. —but to one simple principle I have remained fixed, and that is to act in an upright honourable manner towards the Royal Institution; not merely in a manner which I might deem upright; but in a manner which the men who have given me such proofs of their kindness and confidence would deem so. My determination throughout has been to adhere to that rule of action, which Professor Faraday might consider right, —not grudgingly or of necessity, but cheerfully and dutifully. —From this resolve I have never for an instant swerved, nor shall I. —no matter how tempting my prospects may appear in other quarters. —This premised, let me express my feelings here.

The Museum of Practical Geology has many attractions. It is a Government Institution. It has high hopes and aims; and is probably the herald of a new state of things in England. If it could be arranged I would gladly connect myself with it; but if not, I am content. I have the utmost confidence that the Managers of the Royal Institution will demand nothing of me but what they have a right to demand; and by their decision, whatever that may be, I will cheerfully abide.

Believe me ever most truly yours, | John Tyndall

Minutes of the Managers' Meeting, RI MS AD/02/B/02/A/11/17–18
RI MS JT/2/13b/606–7
Clerk's Handwriting and LT Typescript Only

---

1. *Henry Bence Jones*: there are two versions of this letter: the above transcript from the RI's Minutes of the Managers' Meeting, and a typescript from Tyndall's journal entry of 26 May. The Managers' Minutes have been followed here, as the journal copy is, as the entry states, 'considerably different'. While the content is roughly the same, in the journal entry Tyndall wondered if it would be possible 'to have a nook [of the laboratory] partitioned off

for me where I might quietly work whenever the laboratory' was otherwise in use. Moreover, in the final paragraph of the draft letter he also mentioned some concern about the fact that there were 'two classes' of people that attend RI lectures, namely men of science and the general public. Because of this Tyndall admitted that 'I am sometimes haunted by the fear that if I go deeply into a subject I shall not be understood by some while if I treat it popularly others may deem me superficial. When weak in health as I was yesterday for example this difficulty hampers me and almost deprives me of the power of utterance' (26 May, JT/2/13b/606–7).

2. *your note*: see letter 0762.
3. *28th of May*: this is either a transcription error or Tyndall misread '28' when the letter was actually dated '23'. The letter could not have been dated 28 May because Tyndall noted that he received it the 'previous day' in his journal entry for 26 May and had transcribed an approximate copy of his own response; see n. 1.
4. *him*: Faraday.
5. *Bye Laws . . . attention*: see letter 0762, n. 2.
6. *Museum of Practical Geology*: See letter 0764, n. 6.

From Henry Bence Jones 6 June 1853 0766

Dear Dr Tyndall

I interpreted your letter[1] with the conversation we had at the Royal Society to mean that I might give your assent to the proposal of the managers.[2]

I read your letter & Mr. Faraday told the Managers that the bye law was never acted upon & that it was most desirable to leave you free.[3] I thought the best plan would be to place your letter on the minutes so that you & the Managers may appeal to it & be guided by it.

You were recommended by the Managers for election and on the 6th of July[4] you will be elected & may do what you please till January. I am not sure of the wording regarding the salary £200 but it begins from the day of election.

The Managers did not enter into the question of whether you could simultaneously hold the Jermyn St. Professorship.[5] We had already decided that you were free to lecture anywhere but not to be professor elsewhere. So that this point was not spoken of. If you do not find that we make the Institution all you wish then you must try the Museum.

I gave a positive assent to your accepting our Professorship now & I spoke strongly on your letter being on the minutes and it is there.[6]

I hope you will never have cause to regret this step

I am sorry to hear of your head aches. Rest. Rest is my prescription.

Ever yours most truly | H Bence Jones

June 6. 1853

RI MS JT/1/J/52
RI MS JT/1/TYP/683

1. *your letter*: see letter 0765.
2. *the managers*: of the RI (see letter 0762).
3. *bye law . . . free*: see letters 0762, n. 2 and 0765, n. 5.
4. *Election . . . on the 6th of July*: Tyndall was actually elected on 4 July (see letter 0762, n. 5).
5. *Jermyn St. Professorship*: see letters 0764, n. 6 and 0721, n. 9.
6. *it is there*: letter 0765.

To Thomas Archer Hirst [12 June 1853][1] 0767

My dear Tom.

My work in London is over but no sooner am I free from it than I am overtaken by the Scientific Memoirs.[2] It is a terrible pull upon me and I am now plunging in the midst of an immense Memoir on Diffraction by Schwerd.[3] My election at the Royal Institution takes place on the 6th of July,[4] when the salary also commences. I start in 8 or 10 days for Berlin—I did think of going through paris but I almost think it best for my head and purse to keep my eye steadily on my journey's end and to endeavour to reach it by the quickest route possible. What did it cost you from Berlin to Paris? Answer this question by return as it may have some influence upon my decision. I now want to ask you a business question or two. A new edition of our prospectus here[5] is required of course my name sets and yours rises in its stead. How would you like your name to appear—Simply Dr. Thos. Hirst will look naked. I propose therefore that you state it thus—'Dr. Thomas Hirst of the Universities of Marburg and Berlin'—you might I think add Paris also. Give me your answer to this immediately. I hope you will bear in mind all the experiments that you have seen. From a note[6] I received this morning I learn that the directors of the Crystal Palace[7] are likely to have a geyser constructed in consequence of a remark made in my lecture at the R.I.[8] The drawings have already been submitted to Sir Joseph Paxton[9]—It will be a splendid affair if properly carried out. My experiments succeeded well—If I dont go through Paris I suppose I shant see you for the next 6 months—If you are here on the 1st of August it will be early enough. Lodgings &c you will arrange about when you come. Be definite in all your statements to Mr E[10] and compel him to definiteness also—you will thus get on well together.

Yours affectionately | John.

RI MS JT/1/T/563
RI MS JT/1/HTYP/244

1. *[12 June 1853]*: Hirst wrote '16th June 1853' at the top of the letter but this likely refers to the day that he received it. As Tyndall said that his 'work in London was over' this likely referred to the end of his course of lectures on 'Air and Water' that finished on 11 June. It follows that Tyndall would have written Hirst the next day as he often wrote to Hirst on Sundays.
2. *Scientific Memoirs*: see letter 0698, n. 8.
3. *Schwerd*: see letters 0763, n. 9 and 0768, n. 1. Schwerd was neither translated into English nor published in any of the editions of *Taylor, Scientif. Mem.*
4. *My election . . . 6th of July*: Bence Jones told Tyndall that his election would take place on 6 July, but it actually occurred on the 4th (see letter 0766, n. 4).
5. *our prospectus here*: at Queenwood College where Hirst was to take up a position with Tyndall's departure (see letter 0746, n. 8).
6. *a note*: letter missing.
7. *Crystal Palace*: a cast-iron and plate-glass building erected in Hyde Park as part of the Great Exhibition of 1851. It was relocated to Sydenham Hill in 1853 where it was used for other exhibitions and events.
8. *my lecture at the R.I.*: see letter 0752, n. 10.
9. *Joseph Paxton*: Sir Joseph Paxton (1803–65), an English architect, and designer of the Crystal Palace for the Great Exhibition of 1851. He also aided in its relocation at Sydenham in 1853.
10. *Mr E.*: George Edmondson.

To George Gabriel Stokes 13 June 1853 0768

My dear Sir

I have before me a work of high repute in Germany entitled Die Beugungs Erscheinungen aus den Fundamentalgesetzen der Undulationstheorie analytisch entwickelt by F.M. Schwerd.[1] I do not think it has yet been introduced into England—It is a most valuable work and if a translation has not yet appeared I would make arrangements for its publication in the next number of the Scientific Memoirs.[2] Would you kindly inform me[3] whether you have ever seen a translation of the work?

Believe me | most faithfully yours | John Tyndall
Queenwood | near Stockbridge | 13th June 1853

Cambridge University, Stokes Correspondence, Letter 7656

1. *Die . . . Schwerd*: see letter 0763, n. 9.
2. *Scientific Memoirs*: though note that Tyndall determined to withdraw himself as an editor of future volumes of the *Scientif. Mem.* (see letter 0771).
3. *kindly inform me*: reply is missing.

To Thomas Archer Hirst [19 June 1853][1] 0769

My dear Tom.

You may make your mind easy as to my going to Paris.[2] I shall undoubtedly go there. Mean time cast about for a lodging for me. Perhaps you can manage it in the Hotel Corneille.[3] Possibly once in Paris I may stop there for several weeks, but dont mention this as there may be people there who would expect me to call upon them and I dont intend to make calls until towards the end of my sojourn. I shall probably remain here for the next three weeks—before the middle of July I shall certainly be at your side. How long does the session last? I should certainly like to hear some of the lectures as it is a capital mode of improving in the language. Debus started yesterday to London. Tonight at 10 o'C[lock] he starts for Hamburg[4] and thence to Cassel.[5] Wrightson has sent me his dissertation.[6] I hope he may only get out of the discussion as cleverly as he has entered it. For some time I have been very feeble both in brain and body, but this morning and for the last day or two I feel a pulse of returning vigour so that when I come to Paris I shall be as well able to thrash you as ever I was. I may ask you to help me at the next no. of the Scientific Memoirs[7] but this one I shall complete myself. You have had your own share of trouble with them for the present. The A shall be attended to.[8] I trust you will have a full house next half year. Prospects were never better at Queenwood. The farm class is extinct.[9] It was a pest and you may congratulate yourself that you will have nothing to do with it. I hope you will keep the house full. Indeed if it fails it will be purely for want of good management. and now I think I have satisfied you and this being done I am satisfied myself. Goodbye

Your affectionate | John

RI MS JT/1/T/564
RI MS JT/1/HTYP/245

1. *21 June 1853*: Hirst wrote '21st June 1853' at the top of the letter, but that was likely the date he received it. It was presumably written a few days before, possibly 19 June, which was a Sunday. LT noted as well that 'Tyndall was still at Queenwood at the time of writing and only left there on 6 July.'
2. *my going to Paris*: Tyndall previously stated that he might not make it to Paris (see letter 0767).
3. *Hotel Corneille*: the Hôtel Corneille was a Parisian hotel located in the Latin Quarter of the city.
4. *Hamburg*: a city located on the river Elbe in northern Germany.

5. *Cassel*: Kassel, Germany, a town south of Hamburg, where Debus had studied under Robert Bunsen at the Higher Vocational School.
6. *Wrightson . . . dissertation*: Wrightson's dissertation was entitled, 'Ueber das Atomgewicht und die Constitution der Alkohole' (Marburg: Joh. Aug. Koch, 1853).
7. *Scientific Memoirs*: see letter 0698, n. 8.
8. *The A shall be attended to*: as in Thomas 'A.' Hirst. Tyndall had asked Hirst to respond 'immediately' to a query about how his name was to appear in the prospectus for Queenwood College (see letter 0767).
9. *The farm class is extinct*: see letter 0746, n. 9.

To Thomas Archer Hirst [21 June 1853][1] 0770

My dear Tom

I find myself compelled partly against my will to accept your offer of assistance[2] with regard to the Scientific Memoirs. I had 60 pages translated and hoped my labour was drawing to a close but now find that the said 60 pages cannot be made use of as the cost of engraving the plates belonging to them would be enormous.[3] I must therefore begin anew. Would you like a good fellow lay hold of the Annales du Chemie et de physique for October 1852. The first article is by M. Viard[4] on the 'Role Electro-chemique de l'Oxygene' translate it for me.[5] I am going to attach some German papers.[6] You need not copy the tables as these may be printed at once from the original. After the present part is complete I must make some new arrangement with regard to the Memoirs. They are certainly of great value but the quarterly draw upon my time is more than I can stand.[7]

Your affectionate | John

RI MS JT/1/T/565

---

1. *[21 June 1853]*: Hirst wrote '23rd June 1853' at the top of this letter, which was likely the day he received it suggesting that it might have been written a few days before that.
2. *offer of assistance*: letter missing.
3. *I had . . . enormous*: he is referring to his work on translating Schwerd (see letters 0768, n. 1 and 0771).
4. *M. Viard*: Henri-Stanislas Viard (1821–58) was a French professor of physics at the Faculté des Sciences de Grenoble, in Grenoble, France from 1850 until 1855, when he then moved to chair the faculty of science at Montpellier ('Notice sur M. Henri Viard', *Académie des Sciences et Lettres de Montpellier. Mémoires de la Section des Sciences*, vol. 4 [Montpellier: Boehm et Fils, 1858–60], pp. 131–32).
5. *translate it for me*: M. Viard, 'Du role électrochimique de l'oxygene', *Annal. Chim. et Phys.* 3:36 (October 1852), pp. 129–55.

6. *some German papers*: not identified.
7. *more than I can stand*: in a letter to Francis, dated 22 June, Tyndall suggested that he will not edit any more editions of the *Scientific Memoirs* (letter 0771).

To William Francis 22 June 1853 0771

22nd June 1853

My dear Francis

I am heartily tired of Reuben[1] and his set. He ought to have finished with it long ago. Or he ought not to have published anything about it until he had made up his mind as to what he ought to say. No reader of the Phil. Mag. will be glad on opening it to see a long article from Reuben upon this subject. Even had he important results to communicate I would recommend him to let them rest for a couple of years for really people have enough of the steam jet[2] for the present. In the regard to the sounds I hardly know what to say. I cannot recommend the publication of the paper.[3] There is nothing in them in common with the musical tones which de la Rive and others have found to be produced by magnetism and I am inclined to think that they present neither difficulty nor interest in a scientific point of view.

Reuben I fear does not pursue science for the mere love of it. He likes to see himself in print and he talks sometimes about things which he had far better let alone. It would be refreshing to see him fixing his eye steadily on one object and prosecuting that and to avoid thrusting in remarks here and there which to my notion appears intended to shew his readers how much he has read. I dislike this mode of writing exceedingly.

Previous to deciding on translating Schwerd[4] I had spoken to Baden Powell[5] and written to Stokes[6] and was informed by both that they never had met an English translation of the work.[7] This of the very highest repute in Germany and I sometimes meet a remark, in the Giessen Jahres Bericht[8] for instance that Schwerds investigation appears not to be well known in England—under these circumstances I thought the work a valuable contribution to English science and ignorant as I was of the cost of engraving, the plates did not strike me as an obstacle. If a sufficient number of subscribers could be found to pay for the translation it would be worth while to publish the work in an isolated form; however for the present I suppose it must drop[9] as I should feel reluctant to cut away the plates. I was anxious to set off to Paris and hoped that my work was nearly ended but I must endeavour to summon up strength to commence it anew.

This circumstance induces me to mention a subject at present which I intended to defer till the present part of the Memoirs[10] should be completed. In our conversation about the Memoirs we spoke if you remember of giving them a year's trial.[11] The present number[12] will complete the year. These papers are certainly of great value to English science and hence any man interested in its progress might be glad to engage in them, but it is a dreadful draw upon the time of a man like myself. Had I not private investigations in hand the work would be agreeable enough but I cannot repress the conviction that I should render science a greater service in working myself than in merely translating the work of others—translators are easily found and to such the work ought to be entrusted. This quarterly draw upon my time is like a millstone round my neck. Now I should be sorry to see the Memoirs discontinued and I should be most willing to recommend papers for translation and to inspect the translations but to continue furnishing such a large quota as I have hitherto done is more than I can promise. I look with a certain dismay at the mass of work now before me. I shall set about something immediately but after the present number some new arrangement must be contrived.

I think it best that perfect frankness should subsist between us and therefore I have written exactly as I have felt. The Memoirs at present are a dead weight upon me, dividing my thoughts and damping whatever little energies I possess. The idea of having to commence again may perhaps have some influence on my feelings at present but their pressure has been felt for a long time and I must contrive to get it lightened.

ever yours sincerely | John Tyndall

Those Reports[13] that used to appear in the Magazine are to my mind of the highest value. They require the brain of a man of Science and they furnish worthy occupation for him—but from these this translating which might be done by hundreds cuts me away effectually.

Have you got the Bulletin de la Classe Physics mathematique a l'academie de St. Petersburg?[14] In vol. IX p. 289[15] begins a memoir which I think would make a capital paper for the Memoirs. A translation of it appears in Krönigs journal vol. III. page 377.[16] but this I suppose you do not receive.

RDS 27/33

---

1. *Reuben*: Reuben Phillips (nd) was an English physicist who worked mostly on atmospheric electricity ('Preface', *On Atmospheric Electricity* [London: Robert Hardwicke, 1863], pp. i–iii). Tyndall was not impressed by Phillips (see, for instance, letter 0569, n. 6).
2. *steam jet*: see R. Phillips 'On the Electrical Condition of the Atmosphere', *Phil. Mag.* 4:23 (August 1852), pp. 126–29; R. Clausius 'On the Colours of a Jet of Steam and of the Atmosphere,' *Phil. Mag.* 4:27 (December 1852), pp. 416–17; and Phillips, 'On the Colours of a Jet of Steam and of the Atmosphere', *Phil. Mag.* 5:29 (January 1853), pp. 28–30.
3. *the paper*: not identified but see n. 2.

4. *translating Schwerd*: see letters 0767 and 0768.
5. *spoken to Baden Powell*: in a previous letter to Francis (letter 0763) Tyndall mentioned that he would write to Baden Powell to ask if Schwerd's manuscript had appeared in English translation.
6. *written to Stokes*: see letter 0768. Note that Stokes's reply is missing.
7. *the work*: see letters 0763, n. 9 and 0768, n. 1.
8. *Giessen Jahres Bericht*: refers to the annual report coming out of Giessen, a town 20 miles southwest of Marburg, Germany (H. Buff, E. Dieffenbach, C. Ettling, F. Knapp, H. Will, and F. Zamminer [eds.], *Jahresbericht über die Fortschritte der reinen, pharmaceutischen und technischen Chemie, Physik, Mineralogie und Geologie*, 8 vols [Giessen: J. Ricker, 1849–57], vols 1850 52).
9. *must drop*: see letter 0768, n. 1.
10. *Memoirs*: see letter 0698, n. 8.
11. *year's trial*: see letter 0472.
12. *present number*: Tyndall and Francis (eds.), *Scientif. Mem.*, vol. 1, part 4.
13. *Those Reports*: These are the 'Reports on the progress of the physical sciences' that Tyndall wrote for the *Phil. Mag.* beginning in 1851. See Volume 3, especially the 'Introduction'.
14. *Bulletin . . . St. Petersburg?*: The *Bulletin de la Class Physico-mathématique de l'Académie Impériale des Sciences de Saint-Pétersbourg* was a scientific journal published by the Imperial Academy of Sciences in St. Petersburg from 1843 to 1859.
15. *In vol. IX p. 289*: M. Jacobi, 'Sur la théorie des machines électromagnétiques', *Bulletin de la Class Physico-mathématique de l'Académie Impériale des Sciences de Saint–Pétersbourg* 9:211/212 (1851), pp. 289–310. Moritz Hermann von Jacobi (1801–74), German engineer and physicist after whom Jacobi's Law is named; he was the brother of mathematician Carl Gustav Jacob Jacobi.
16. *A translation . . . page 377*: It was translated from French to German (M. Jacobi, 'Ueber die Theorie der elecktromanetischen Maschinen', *Journal für Physik und Physikalische Chemie des Auslandes in Vollstaendign Uebersetzungen* 4:31 [1851], pp. 377–408).

To William Francis 26 June 1853 0772

<u>26th June 1853.</u>

My dear Francis

I feel truly grateful to you for your kind and considerate letter[1]—I was afraid you might deem me churlish but you have on the contrary felt as my friend and attributed to me nothing of the kind. With regard to the coming part[2] make your mind easy I will get all ready myself I send you a sheet and a half[3] by this post and hope by the end of the week to have most of the stuff hewn into shape. Your motives in continuing the memoirs were mine, and even now I regret that the thing is not able to support itself. I suppose most of our English men of Science can read the originals and hence do not support translations. Some of them I fear will regret the stopping of the publication and none perhaps more than Faraday—but I think we have done our duty in giving the public a year's publication for nothing—You I fear have lost something. Of course I cannot hear the bit of good news you communicate without a certain elation of heart. But my chief hopes and pleasure lie far in the future—if my health only hold out I trust to be able to justify the proposition which for the present I must regard as being partly prompted by the kindness of my friends. Excuse this muddled letter—my head has been muddled for some time lately however I begin to feel a pulse of returning thought and with a little quiet I hope soon to be myself again. ever yours truly | John Tyndall

a capital paper, that of Brunner[4]—I see Bunsen's finger throughout.

RDS 27/30

1. *your kind and considerate letter*: letter missing.
2. *the coming part*: the fourth and final part of Tyndall and Francis (eds.), *Scientif. Mem.*, which would be published later that year. (See letter 0698, n. 8.)
3. *a sheet and a half*: not identified.
4. *that of Brunner*: Carl Emanuel Brunner (1796–1867), Swiss pharmacist and Professor of Chemistry and Pharmacy at the University of Bern (F. Szabadváry, *History of Analytical Chemistry* [London: Pergamon Press, 1966], p. 294, n. 30). As Brunner wrote a large number of papers, it is difficult to determine which one Tyndall is referring to here ('Brunner, Carl', *RS Catalogue,* vol. I, pp. 679–81).

To Thomas Archer Hirst 26 June 1853 0773

Queenwood 26th June. 1853

Bother! Bother! Bother! Tom my muddled pate[1] is hardly able to fix upon a line of procedure. One thing is certain however, in England I must remain until after the 6th of July, perhaps three or 4 days longer—On the whole the best plan is as follows. I will see you in England, but dont keep me waiting I will wait till the 10th but dont push me much beyond it. I will go straight to Berlin and visit Paris[2] on my return. —I dont want to spend a week with you in England—a day will suffice to pour into you a concentrated extract of my experiences in Queenwood which you will find a salutary moral medicine and useful for your future guidance.[3] I want to talk to you both about principal[4] and boys. Wont you accompany /Stevens/[5] to Cambridge and see Stokes & Caley,[6] you must see Sylvester. Thomson is in Malta! I would accompany him everywhere if I were you—it will be a useful introduction for yourself to many people. Now the thing I hope is settled—I will see you either in London or here. If you can manage to be in London on the 10th I will be there & you may see me off.

Yours as usual | John

I hope you will get that paper[7] translated for me.

RI MS JT/1/T/566

1. *pate*: head *(OED)*.
2. *straight to Berlin and visit Paris*: Tyndall had been trying to organize a trip to Berlin and possibly to Paris in order to visit Hirst (see letters 0767 and 0769).
3. *Experiences in Queenwood . . . future guidance*: Hirst was taking up Tyndall's position at Queenwood College (see letter 0746, n. 8).
4. *principal*: the principal of Queenwood College was George Edmondson.
5. *[Stevens]*: not identified.
6. *Caley*: a possible misspelling of Cayley (see letter 0742, n. 2).
7. *that paper*: see letter 0770, n. 5.

To Alexander Cumming 27 June 1853[1] 0774

Queenwood 27th June 1853.

My dear Alexander

This is a wild evening at Queenwood; the wind is blowing, the windows shaking and the branches of the shrubs outside are tossed hither and thither. The wheat field at the opposite side of the road is furrowed with waves which roll up the side of the hill to the avenue at the top. It is a beautiful thing this wave motion and strange how the simple motion of the ears of corn to and fro can produce such an extraordinary progressive motion of the surface, just as the simple up and down motion of the particles of water can produce the waves of the sea.

For some time back I have been troubled with a headache which always attacks me when I commence thinking, it has been brought on I believe by thinking too much and I must endeavour to drive it away by thinking little—or at least if I think at all it must be on something pleasant. To day I have not attempted to do anything and this evening an occupation suggested itself to me which, as I find a pleasure in it, instead of making my head worse will make it better—the occupation is that of writing to you.

Prompted by this selfish motive—the simple motive of pleasing myself and making my headache better I take up my pen. But why in the name of fortune do I write to you? Why not take up my journal and write there to my heart's content and thus spare myself the penny stamp.[2] To all these questions I must answer 'I dont know! Perhaps somebody might account for it by saying 'oh you like Alexander perhaps' but this does not solve the riddle, for if any body asked me 'why do you like him?' I should still be forced to answer 'I dont know'. I dont know how it is that sometimes even at night and when it would be all the better for my head if I were fast asleep the thought of one or two particular boys should present itself to my mind, and that I should experience underneath my chest where anatomists say the heart lies, a certain feeling of warmth like that experienced by cold feet on a winter night when a bottle of hot water is placed against them. I am profoundly ignorant of the cause of all this, but I am very thankful that such feelings are granted to me, as a cold heart is far more dreary and uncomfortable than cold feet, bad as they are. But you will wonder to find me writing in this way, and perhaps you thought when I spoke of waves that I was going to preach science to you, but here you are mistaken. Indeed the feeling that was uppermost in my mind when the thought of writing to you occurred to me was that which a person experiences on separating from a friend for a long time—perhaps for ever—and who is unwilling to depart without saying goodbye. I dont know whether I shall see you again, or whether I must not content myself for the future with these midnight visits, when with the eye of my mind I shall look into your face and feel the bottle of water shedding its warmth underneath my ribs. I shall not be here when you come back, I shall be in Berlin, but even then I shall think of you and in my own silent way pray God to bless you. In Berlin I shall remain until the end of the year and afterwards my vocation will be that of Professor of Natural Philosophy in the Royal Institution[3] Albermarle St. London. Thus you see I am to teach you no more, perhaps see you no more unless you cheer me sometimes by a visit to London. I am to be succeeded by a gentleman named Hirst[4]—a friend of my own who has spent three years at the Universities of Marburg and Berlin and is now in Paris. I think you heard him lecture once here.[5] He is a very able young man, indeed I have rarely met one with so powerful a mind and the chief point is that his heart is as good as his head. I hope you will like each other—I hope he will love you ten times better than I do, and if he can contrive to do this you will be well beloved indeed. In writing thus far I have never mentioned George,[6] but I have thought of you both all through. Mr. Edmondson tells me that you are to bring two little brothers back with you, —how I should like to see them—but I shall be far away when you return. I hope you will take care of your brothers, and what is the best possible way of taking care of them? not by giving them good advice, though this may be well, but by showing them a worthy example. In my view all else is worthless, no matter what our tongues may utter it is our characters that really speak. A man's nature radiates through him—a good boy—which

I define to be an upright, truthful, loving, brave boy; works by a kind of magnetism instead of by words. But I must draw to a close. I must fancy my arms around your neck and that I am giving you a last kiss (what an old woman of a fellow I am to talk thus!). I know that you are ignorant of most of the vices and follies of the world—ignorant of many practices which are destructive even to boys—keep clear of them my son. Keep your heart and thoughts above them. Recollect the first temptation to sin of any kind is always the most easily strangled—strangle the first my boy and preserve your freedom. I hang upon these last lines, the thoughts within me are like the sound of dying music, but I must end them and bid you good bye. Do your duty my son—that's the grand maxim, and if too many duties present themselves do the nearest and you will gain insight by the effort. And now I throw my arms round you—press you closely to my heart—give you the kiss before alluded to and say once more good bye.

Your affectionate friend | John Tyndall.[7]

RI MS JT/1/TYP/1/1/255–256
LT Typescript Only

---

1. *27 June 1853*: LT noted: 'The very beginning of the correspondence and of the long friendship' | 'This was to | Alick Cumming | who was at school at Queenwood—27th June 1853'| and 'Alick must have been 13 or 14'.
2. *penny stamp*: i.e., the cost of postage.
3. *Professor of Natural Philosophy at the Royal Institution*: see letter 0762.
4. *succeeded by a gentleman named Hirst*: see letter 0746, n. 8.
5. *lecture once here*: see Volume 3.
6. *George*: George S. Cumming.
7. *John Tyndall*: LT notes: To the left of Tyndall's name appears, '[To | Master Alexander Cumming | (care of) James Coxe Esqu | 9 Atholl Crescent | Edinburgh.]'. Below Tyndall's name appears, '[Next */comes/* | Alex Cumming's letter]'. LT often copied out the address of the letter as it appeared on the envelope, as she has done here.

From Thomas Archer Hirst 30 June 1853 0775

Hotel Corneille | Paris | Thursday June 30th / 53

My dear John

This is the last day of June, I have as yet seen nothing of Prof. Steiner although I told him that I should not remain longer than the first of July in Paris. I have told you before what an eccentric man he is, and I am not in the least surprised at his not coming. I shall give him 3 more days to save himself, and then if he does not make his appearance I shall be off. You say in your last letter,[1] from which I could see very well that you are terribly fagged[2] in mind and body, that one day will be sufficient for us to be together. I beg to say it will by no means satisfy me, and that I don't intend being cheated. For that reason unless I am prevented by Steiner, I shall be at Queenwood on Tuesday next I shall leave Havre[3] for Southampton on Monday evening. It strikes me I can be of some use to you next week not only in the work which is at present pressing you but also in scolding you a little, at any rate I shall see when I come.

The Memoir by Viard[4] I have already translated I shall bring it with me or in case of delay I shall send it and tell you at the same time where I shall expect to meet you.

The lectures here are just about finishing. A hearty welcome awaits you in Berlin I know and after the fagging work you have gone through I hope you will submit yourself like a wise foreseeing man to the beneficial influence it can have on you. For the last two months I have discussed to myself all sorts of projects, fears, and hopes with respect to yourself, mostly vague ones requiring first of all to have more basis in fact. It was for this reason that I said a week of your company was necessary for me, and I will have it in spite of you. It will take so long to deliver myself of what I have to say, and for us both to know each other exactly, that is to what phase of existence we have arrived at and what modifications a year's time has effected in us both. In short, John, a week's close contact with thee has become a periodical necessity for me. I want to have somebody to whom I can talk all myself, whatever of significant and insignificant there may be in me. I want to get quit of it to thee, refuse me at thy peril!!

As ever | T A Hirst

P.S. Inasmuch as where one is best know we may expect most favour, I should advise Mr Edmondson to advertise in the Bristol Times, and Halifax Guardian, and perhaps also in the Leeds Mercury and Wakefield—(Guardian)? (I think).

RI MS JT/1/H/178
RI MS JT/1/HTYP/247

1. *last letter*: see letter 0773.
2. *fagged*: excessively fatigued (*OED*).
3. *Harve*: the French port of Le Havre.
4. *The Memoir by Viard*: see letter 0770, n. 5 and 0773, n. 7.

To William Francis [27 June–3 July 1853][1] 0776

My dear Francis,

I have just received the enclosed.

[...][2]

I find my head-ache still persistent. I dare I say I shall not be well until I get free from every thing. With reluctance therefore I must ask you to get Bunsen's paper translated.[3] Debus tells me it is a very important one and he does not think it *[1 word illeg]* out. I have Hirst at work on one in the Annals[4] and I am engaged at one of Plucker's myself on the mathematical theory of diamagnetism.[5] Were I not so very shaky in the very region which I require for this work I should be sorry to give you the least trouble with this concluding number.[6]

ever truly yours | John Tyndall

RDS 27/32

---

1. *[27 June–3 July 1853]*: 'July 1853' was handwritten in pencil in the upper right corner. Tyndall noted receipt of the enclosed letter (0776encl) 'a few days ago' in his journal entry of 6 July, but he had not yet received it at the time of his previous entry on 26 June (Journal, JT/2/13b/607–8).
2. *[...]*: Tyndall had copied letter 0776encl.
3. *Bunsen's paper translated*: see letter 0745, n. 12.
4. *the Annals*: Hirst was working on this translation (see letter 0775, n. 4).
5. *Plucker's ... diamagnetism*: see letter 0710, n. 4. Jackson states that it is unclear as to which paper this is (Jackson, 'Tyndall and Diamagnetism', p. 27).
6. *concluding number*: the last instalment of Tyndall and Francis (eds.), *Scientif. Mem.* (see letter 0771, n. 12).

From J. G. S. van Breda[1] 20 June 1853 0776encl

Harfleur le 20 juin 1853

Monsieur!

La Société Royale Hollandaise des Sciences a voulu donner un témoignage solennel du haut prix, qu'elle attache à vos travaux Scientifiques. Elle ose croire, que le moyen, qu'elle a employé pour remplir ce buts, ne vous rend pas tout à fait désagréable.

Vous avez été nommé le 21 mai dernier jour de sa cent et unième réunion générale annuelle, à l'unanimité des voix, Son *[1 word illeg]* honoraire.

Permettez Monsieur, qu'en vous faisait cette communication officielle, je vous témoigne en même temps combien il m'est agréable d'entrer aussi en relation scientifique avec un savant dont les écrits font depuis longtemps l'objet de mes études.

J'aurai l'honneur de vous envoyer dans peu le programme de la Société pour 1853, où vous trouverez peut être quelques questions que vous intéresseriez.

J'ai l'honneur d'être avec le plus haute considération

le secrétaire perpétuel de | la Société Hollandaise | des Sciences.

J. G. S van Breda

Harfleur,[2] June 20 1853

Sir!

The Royal Holland Society of Sciences wanted to give a solemn testimony of the high price that it attaches to your scientific work. The society dares to believe that the means it used to fulfil this purpose does not make you uncomfortable.

You were nominated on the 21st of last May, on the day of its hundred and first general annual meeting, by unanimous votes, its *[1 word illeg]* honorary.

Allow me Sir, by making this official communication to you, I testify to you at the same time how much it pleases me to enter into a scientific relationship with a savant whose works were the object of my studies for a long time.

I will have the honour to send you soon the program of the Society *[from]* 1853,[3] where perhaps you will find some questions that will be of interest to you.

I have the honour of giving you my highest consideration, the perpetual secretary of the Holland Society of Sciences.

J. G. S van Breda

RDS 27/32

---

1. *J. G. S. van Breda*: J. G. S. van Breda (1788–1867) was extraordinary professor of zoology and geology at Leiden in the Netherlands. He also researched in the area of physics, particularly on magnetism and electricity. (A. S. Breure and J. G. de Brujin, *Leven en werken van J.G.S. van Breda: (1788–1867)* [Haarlem: Hollnsche Maatschappij der Wetsenschappen & Groningen, 1979]).
2. *Harfleur*: A township in the Haute-Normandie region of northern France.
3. *le programme . . . 1853*: from 1773 to 1875, Royal Holland Society of Sciences published an annual program ('Programmes de la Société Hollandaise des Sciences de Harlem de l'année 1773 à 1875', *Société Hollandaise des Sciences à Harlem. Notice Historique, Liste des Protecteurs, Présidents, Secrétaires, Directeurs et Membres Résidants et Étrangers et Liste des Publications de la Société Depuis sa Fondation en 1752* [Harlem: De Erven Loosjes, 1876], p. vii).

From May Anne Coxe [1–3 July 1853][1] 0777

Burnt Island[2] | July 31st[3]

My dear Sir

I feel that I—as an utter stranger, ought to apologize for thus intruding upon you—but I really cannot resist the desire I feel, to thank you most warmly—most gratefully—for the exceeding kindness you have shown my boys.[4]

Words fail me to express the feelings with which I read your letter to Alick[5]—but you may understand them when I tell you that I had not read far before "tears—big tears" filled my eyes—nor had I concluded without those tears coursing each other down my cheeks. There was pure unmixed Pride and Pleasure that my boy should have made such a friend of such a Man—but along with these, was the deep regret, that he was no longer to be under the influence, of such precept and such example.

I have however the conviction, that you must look upon him as no common boy, or you never would have written him such a letter—and I earnestly pray that he may so conduct himself, as to make you satisfied that your kindness was not ill bestowed. I do hope you will not be displeased with my having seen your letter—but I was so surprised and gratified by the fact of your writing to him at all, that I insisted upon Alick's letting me see the letter—need I say—indeed I cannot—how much that gratification was enhanced by its perusal?

I have not ventured to tell him that I am writing to you but I hope you will pardon the liberty and allow for the feeling which impels me to write.

Most cordially do I hope your health may ere long be thoroughly re-established.

Pray do not think of troubling to notice this—but allow me to subscribe myself with much respect and in all sincerity

Yours most gratefully | and obliged | May Anne Coxe

RI MS JT/2/6/d
RI MS JT/1/TYP/1/258

---

1. *[1–3 July 1853]*: Coxe dated this letter 31 July but Tyndall noted receipt of this letter along with the letter from Alexander Cumming (see letter 0778) in his journal entry dated 6 July (Journal, JT/2/13b/608). It is therefore likely dated between 1 and 3 July.
2. *Burnt Island*: Burntisland, a town in Fife, Scotland on the Firth of Forth estuary.
3. *July 31st*: see n. 1.
4. *my boys*: Alick (Alexander Cumming) and his brothers (see letter 0774).
5. *your letter to Alick*: see letter 0774.

From Alexander Cumming [1–3 July 1853][1] 0778

N°. 5 Lammerlaws | Burnt Island[2]

My dear D^r^. Tyndall

I received your very kind letter[3] yesterday morning and I was much gratified to see that you did not disdain to write to a little boy like me. I prize your letter very highly, and I shall always keep it beside me, and years hence when we shall both be pursuing different courses in life, I shall find the greatest pleasure in reading it, and in profiting by the good instruction it contains. I am sure my Father[4] would be greatly delighted to hear of the interest you take in George[5] and me, for it is no ordinary thing for a man of your mind and abilities to trouble yourself with writing such a kind letter to me. I am sorry you are going to leave Queenwood for I shall never have such a kind and indulgent master again. Although I think I shall like D^r^. Hurst very much. Whenever I am in London and when I have an opportunity I shall not forget to visit you in London. I hope I shall never forget the good advices that your letter contains and that when I grow up a man I shall not deceive your expectations. George joins with me in sending you, our kindest and warmest love, Hoping to see you soon, I am ever

Your sincere friend | Alex^r^. Cumming.

RI MS JT/2/6/c
RI MS JT/1/TYP/1/257

1. *[1–3 July 1853]*: as this letter was sent along with letter 0777, it follows that the two letters were written on the same day (see letter 0777, n. 1).
2. *Burnt Island*: see letter 0777, n. 2.
3. *kind letter*: letter 0774.
4. *my Father*: presumably May Anne Coxe's brother, William Fullarton Cumming (1804–92) who graduated in medicine from Edinburgh University in 1827 and was then employed by the Bengal Medical Establishment (see letter 1087, n. 5).
5. *George*: George S. Cumming, his brother.

From Margaret Allen 4 July 1853 0779

Sunday Eveng. July 4th /53.

Dear Tyndall

I feel your kind consideration very deeply & my sympathies are also awakened concerning the present state of your health. I am well able to feel for you; I have been suffering almost to distraction with my head, but have found considerable relief from the application of Leeches,[1] from one anxiety and another my mind has been reduced almost to a state of inability yet I am still working resolutely.

You will think I have taken your advice, when you learn that I have not yet seen Robert,[2] prudential motives could alone influence such a state of circumstance. it was also Roberts proposition that I should remain here until he procu*[res]* a situation. I received a note[3] from him a few days ago, with the information that he had rec[d] a letter from Southampton directing him to proceed to Londonderry,[4] and report there to Capt. Leach,[5] having been appointed by Leut. Col. Hall,[6] that he had to correspond with Major *[*Tanster*]*[7] as to Southampton, but that I should hear very shortly from him again. I am now expecting the letter any post. I suppose the appointment will not be very remunerative but it will be a beginning—I am afraid there is not much chance of rising in the Profession under the Government but I may judge wrongly. I hope he will be actively employed and I wish we were somewhat settled. You will not be surprized when I tell you I have much to *[*undergo*]*, but I am not fearful regarding the future, tho' some may think I have everything to dread. My burden is not made an iota lighter by the actions of them by whom I am *[*surrounded*]*, instead of busying my mind with something like hope or even a good wish, I have to combat all kinds of gloomy assertions, visions of future wrath, even to the want of a friendly look of recognition on the part of those who ought to possess a natural affinity if even the milk of human kindness is scarce in the bread.[8] I am writing plainly for when I write to you I feel I address a beloved Brother and between true friends there can be no misunderstanding, nor any need for apologies. I have learned & passed through much during the past three years; have *[1 word illeg]* solid worth depreciated, and vanity and outward show flattened, but disappointment does not always *[*engraft*]* the lesson which it is intended it should teach upon the mind, nor will people be benefited even by bitter experience. But I shall not sacrifice my happiness at the altar of mere expediency nor to gratify the ambition nor please them whim of some of my would-be-friends. If I leave home under such circumstances as alluded to, I trust a strong mind will hear me conquer through, even though sent *[*Hagar*]*-like[9] into the wilderness. Thank Heaven my desires are written comforts, I want no luxuries, at least those things which many esteem such; I can purchase some at a cheaper rate, and of a more enduring quality; there is that which it is not in the power of the human to wrench from our grasp. If ever I have to become practical, I do not fear the result; and if I leave this house with a tear it will not be for my position, but that our poor Nature is become so selfish, and is so much belied.

I need not write any more on this subject, I do not wish to be cowardly, and trust that a proper consideration will mark all my proceedings. If you can write to me, my best friend, ere you leave for the Fatherland[10] do so, I shall often think of you and trust you will employ the greatest part of the time you are about for the purpose of recruiting your shattered health; do not study and make your self even worse; do take some little ease, or I shall not be pleased with you.

I have sent you a little memento which I believe you will not deem unworthy of your acceptance, a blade of Grass from the grave of William Wordsworth[11] such is the only epitaph on a plain black headstone in the churchyard of Grasmere,[12] no date, nor anything to indicate the great man lies here. I restrained that the tear at the grave of him, whose musings have often cheered the gloomy hour, and kneeling on the sacred spot I vowed my life should be more natural, and thus more holy for the Future. I actually walked 30 miles on that day, it was a time of perfect enjoyment which I shall never forget.

If I hear nothing very particular from Robert I shall probably be in Blackburn[13] on Thursday, or visit to M^r^. Heywoods[14] for a few days, perhaps you will address for me there. *[1 word illeg]* I hope to hear something of you when away, I trust you will return fully restored, more vigorous in mind, and body, Ever believe me—

Yours sincerely | Mrs Allen

RI MS JT/1/A/72

1. *application of Leeches*: leech therapy, the application of leeches to an area where one wants to relieve excess blood, was a medical treatment stemming from Hippocrates and used through to the mid-nineteenth century; in this case, it was likely used to relieve the pressure of headaches (P. J. Koehler and C. J. Boes, 'A History of Non-Drug Treatment in Headache, Particularly Migrane', *Brain: A Journal of Neurology* 133 [2010], pp. 2489–500).
2. *Robert*: Allen's husband, Robert Copeland Allen (see letter 0699, n. 6.), had a drinking problem and had been sent to America to reform himself (see letter 0492, n. 1; 0865, n. 4; and 0975, n. 19). Margaret had been living apart from her husband for some time.
3. *a note*: not identified.
4. *Londonderry*: City in Northern Ireland and an important port for emigrants embarking for North America.
5. *Capt. [Leach]*: George A. Leach R.E. Not further identified.
6. *Leut. Col. Hall*: Lieutenant-Colonel Hall R.E., superintendant of the Ordnance Survey in Ireland.
7. *Major [Tanster]*: not identified.
8. *the milk … bread*: a reference to a line from Shakespeare's *MacBeth*, the 'milk o' the' human kindness', refers to a sense of kindness or benevolence (I.v.15–18). Allen was suggesting that she does not receive much compassion or support from those around her.

9. *[Hagar]-like*: Hagar, a handmaiden, was sent into the wilderness by Abram (Abraham), the father of her son, Ishmael, and husband of her mistress, Sarai, as a punishment for Ishmael mocking their son Isaac (Genesis 21:9–13).
10. *the Fatherland*: Tyndall was about to embark on a trip to Berlin (see letter 0773).
11. *William Wordsworth*: William Wordsworth (1779–1850) was an English Romantic poet.
12. *Grasmere*: Wordsworth was buried at St. Oswald's, Grasmere in Cumbria, England after his 1850 death to pleurisy.
13. *Blackburn*: Blackburn, Lancashire, England.
14. *M^r^. Heywoods*: not identified.

From Emil Dubois-Reymond 4 July 1853 0780

Berlin, Neuenberger Str. 5 | Montag 4. Juli 1853

Lieber Freund,

Ich bin ein Sünder und Bitte um Gnade. Ihren Brief den Sie mir durch Hirst zukommen liessen habe ich damals erhalten, er wurde Hirst nach Paris nachgeschickt und trat von dort zum zweiten Mal den Weg nach Berlin an. Da er aber nichts enthielt was mich zu einem augenblicklichen Antwort aufforderte und ich damals durch Matteucci, wie jetzt leider durch eine Gedächtnissrede auf Paul Erman, schrecklich in Anspruch genommen war, so verschob ich es von Tag zu Tag Ihnen zu schreiben. Auch Hirst bin ich einen Brief schuldig. The Germans are not punctual, you know—passt also leider auch auf mich. —Nun zum Geschäft. Es thut mir erstaunlich leid Sie nicht hier zu sehen um mit Ihnen über ihren thierische elektrischen Artikel zu reden. Überdiess werden Sie nicht einmal Magnus hier treffen. Überhaupt Niemand, und ich halte es für das Gerathenste wenn Sie hier die Studien machen und ich Ihnen das Material dazu liefern soll dass es in meiner Wohnung bleibt wo Sie durch die Person die die Wohnung hütet, unter Vorzeigung Ihrer Karte, die Sache bekommen werden. Gehen Sie auf dem Hof drei Treppen hinauf, nicht vorn, wo Alles verrammelt sein wird. Sie können sich sogar in meiner Wohnung so lange aufhalten als Ihnen beliebt, wenn es Ihnen bequemer ist als die Sache mitzunehmen. Der französischen Artikel nach der Sie fragen ist ohne Zweifel der von Gavarret, Annales de Chimie et de Physique, 3me Serie T. XXV. p. 58 (1849). Sie finden aber dieselbe Geschichte sehr viel ausführlicher und auf genaueres Quellenstudium gegründet im 1. Bde. meiner Untersuchungen, denn Gavarret hat nur zufällig angeregt durch die neuen Gesamtherausgabe der Werke Galvani's jener Artikel verfasst ohne Volta's Schriften und die Deutschen Sachen selber zu lesen. Am nützlichsten glaube ich kann Ihnen nächst dem Abstract ein Aufsatz von Helmholtz sein den Bence Jones auch besitzt, in welchem er meine Ergebnisse für das grössere Publicum populär darstellt. Der Grund weshalb dieser Aufsatz im Dunkel geblieben ist ist der dass Helmholtz und ich selber nie davon sprechen. Es ist nämlich Helmholtz dabei ein kleines Unglück begegnet, er hat etwas gänzlich missverstanden und zwar einen Punkt von solcher Wichtigkeit, dass dadurch die Arbeit ganz entwerte wird. Indessen lässt sich der Fehler vermeiden und mit Beachtung dasselben kann Ihnen der Aufsatz, wie ich glaube aussserordentlich nützlich sein. Der Fehler besteht darin dass Helmholtz meinte, ich hätte gefunden, unversehrte Muskeln wirkten nicht elektromotorisch, sondern es müsste immer erst eine Schicht von natürlichem Querschnitt auf irgend eine Art erwirksam wurden. Nun ist diess nur richtig für Muskeln [von Thieren, d.h. Fröschen] die 24h lang oder länger auf 0° erkältet worden sind. Sonst findet man die Muskeln stets mit natürlichen Querschnitt wirksam, wenn auch aus Gründen, die ich vollständig aufgedeckt haben, nicht so stark als mit künstlichem Querschnitt; und wenn die Muskeln Thiere angehören die erfroren sind kommt es vor dass der natürlichen Querschnitt positiv statt negativ gegen den Längschnitt sich verhielt. (Abstract Chap. XV)

Es freut mich von Herzen dass Sie mit meiner Brochure gegen Matteucci zufrieden sind, aber ich fürchte Sie täuschen sich wenn sie glauben dass ich der Sieger bleibe werde. Sie kennen die Nichtswürdigkeit des Mannes nicht ganz. Glauben Sie denn dass Matteucci (ein Mensch der in Florenz so verhasst und verachtet ist wie in Berlin) nicht sehr gut Alles weiss was ich gegen ihn geschrieben habe? Er kennt aber sein Publicum noch besser. Er weiss sehr gut dass diese Brochure in zwei Jahren vergessen ist und dann bringt er, ein Spiel was er nun seit einer langen Reihe von Jahren ohne zu ermüden fortsetzt, genau dasselbe wieder vor wofür ich ihm jetzt abgestraft habe. Ich würde auch, für meinen Theil, das Ding nicht geschrieben haben: aber ich war es Bence Jones schuldig. Ich hoffe indes dass es Faraday die Tagen öffnen wird, und diess würde für mich eine grosse Entschädigung für die unsägliche Mühe sein, die mich das zusammensuchen aller der Beweisstellen und das Anordnen des Materials und nicht zu vergessen das Mässigen meiner Wuth, dann am liebsten hätte ich mich mit dem Spitzbuben *[2 words illeg]* hat.

Es ist Postzeit und ich schliesse diesen hastigen Brief mit dem Dank für Ihre Theilnahme an dem mich bevorstehenden Glucke wodurch, wie Sie leicht denken können, mein Sinn noch fester an Ihre herrliche grüne Insel geknüpft wird—und mit dem Wunsch dass es sich doch noch fügen möge dass wir uns irgendwo treffen. Bence Jones weiss allezeit um meine Pläne. Sie sind jetzt folgende: Anfang August gehe ich nach Ambleside, heirathe am 22. Aug und verlasse England etwa vierzehn Tage darauf; diese vierzehn Tage denk′ ich grössentheils in London und Folkstone zu verbringen.

Ihr, | E. DuBois-Reymond

Berlin | Neuenberger Str. 5 | Monday | 4 July 1853

Dear friend,

I am a sinner and plead for mercy. I received the letter[1] that you sent via Hirst, it was sent on to Hirst in Paris and took the route a second time back to Berlin. Since it did not contain anything that demanded an immediate answer and I was terribly occupied dealing with Matteucci,[2] much as I am now with the memoriam to Paul Erman,[3] I put off writing to you from day to day. I owe Hirst a letter as well. The Germans are not punctual, you know—this unfortunately also applies to me. —Now to business. I am very sorry not to see you here[4] to talk to you about your article on animal electricity.[5] Moreover, you will not find Magnus here either. Absolutely no one, and I think it most advisable that if you want do the studies here and I am to supply you with the material for them, that it is kept in my lodgings, where you can get the things when you show your card to the person who is looking after the apartment. Go up three stairs in the courtyard, not in front, where everything will be barricaded. You can stay in my apartment for as long as you would like, if it is easier for you than taking the things with you. The French article[6] that you inquired about is doubtless the one from Gavarret,[7] Annales de Chimie et de Physique, 3me Serie T. XXV. p. 58 (1849). In the 1. vol. of my experiments[8] you will find a much more extensive version of the same story based on a more precise study of the sources, since Gavarret was motivated to compose that essay by chance by the new complete edition of Galvani's works[9] without having read Volta's writings[10] or the German pieces[11] himself. What would be most useful to you, I believe, after the abstract, is an essay from Helmholtz[12] that Bence Jones also has in his possession, in which he gives a popular rendering of my results for a broad public. The reason why this essay remained unknown is because Helmholtz and I never talk about it. Namely, Helmholtz had the misfortune of making a small mistake, he completely misunderstood something in regard to a point of such importance that the entire paper became worthless. It is however possible to get around this mistake and, taking it into consideration, I believe that you will find the essay to be extremely useful. The mistake is such that Helmholtz thought I had found that intact muscles were not electro-motorically active but that first a layer with a natural cross-section had to somehow be activated. Now this is only correct for muscles from animals, that is, frogs, that have been cooled to 0° for 24h or longer. Otherwise, one finds that muscles with a natural cross-section are always effectual, even if, for reasons that I have thoroughly clarified, they are not as strongly so as those with an artificial cross-section; and when the muscles come from animals that were frozen, it can happen that the natural cross-section reacts positively instead of negatively to the longitudinal section. (Abstract Chap. XV).[13]

I am heartily pleased that you are satisfied with my brochure against Matteucci,[14] but I fear that you are fooling yourself, if you believe that I will remain victorious. You do not know the entire baseness of this man. Do you believe that Matteucci (a man who is as hated in Florence as he is in Berlin), does not very well know everything that I have written against him? But he knows his public even better. He knows very well that this brochure will be forgotten in two years time and then he will repeat a game that he has been playing untiringly for years, and do again exactly the same thing that I am pummelling him for now. For my part, I would have the thing unwritten, but I owed it to Bence Jones.[15] I hope in the meantime that Faraday will open it, and this would be a great compensation to me for the indescribable troubles I had searching for all the citations and sorting the material, not to mention curbing my rage, since I would have dearly loved to *[2 word illeg.]* the scamp.

It is time for the mail and I close this hasty letter with thanks for your interest in my coming happiness, which will, as I am sure you can easily imagine, bind my thoughts even more strongly to your wonderful green island[16] and with the hope that we will somehow find it possible to meet somewhere. Bence Jones keeps abreast of my plans. They are currently as follows: In early August, I will travel to Ambleside,[17] marry on 22 Aug[18] and leave England around fourteen days later; I intend to spend these fourteen days largely in London and Folkstone.[19]

Yours, | Emil Du Bois-Reymond

RI MS JT/1/D/146

1. *the letter*: letter missing.
2. *dealing with Matteucci*: see letters 0744, n. 9 and 0754, n. 8.
3. *memoriam to Paul Erma*: for the July 1853 meeting of the Austrian Academy of Sciences, du Bois-Reymond presented a memoriam of German physicist Paul Erman (1764–1851) (E. du Bois-Reymond, 'Gedächtnifsrede auf Paul Erman', *Physikalische Abhandlungen der Königlichen Akademie der Wissenschaften zu Berlin. Aus dem Jahre 1853* [Berlin: Druckereider Königlichen Akademie der Wissenschaften, 1854], pp. 1–27).
4. *not to see you here*: in Berlin.
5. *your article on Animal Electricity*: Tyndall was working on a review of the status of past and current research on animal electricity, which would be published in January 1854 (see letter 0754, n. 9).
6. *The French article*: M. Gavarret, 'Étude sur les Recherches Électro–Physiologiques de Galvani', *Annal. Chim. et Phys.* 3:25 (January 1849), pp. 58–82.
7. *Gavarret*: Louis Denis/Dominique Jules Gavarret (1809–90) was a French physician, notable for his advocation of the statistical method in medicine.
8. *In . . . my experiments*: see letter 0744, n. 10.

9. *Galvani's works*: likely referred to the publication of all of Galvani's writings in Italian, both previously published and unpublished, entitled *Opere Edite ed Inedite del Professore Luigi Galvani Raccolte e Pubblicate per Cuba dell'Accaemia delle Sciencze dell'Instituto di Bologna* (Bologna: Emidio dall'Olmo, 1841).
10. *Volta's writings*: Alessandro Volta (1745–1827), an Italian phyicist and chemist, long disagreed with Galvani regarding the nature of animal electricity and invented the first electric battery out of experiments to understand it. He published heavily on the topic from 1797 to 1817 ('Volta, Alessandro', *RS Catalogue*, vol. 6, pp. 204–5).
11. *German pieces*: not identified.
12. *essay from Helmholtz*: not identified. It is unknown if the article was published. If it was it would be listed in the Royal Society's Catalogue ('Helmholtz, H.', *RS Catalogue*, vol. 3, p. 270–71).
13. *(Abstract. Chap. XV)*: see n. 8.
14. *my brochure against Matteucci*: see letter 0754, n. 8.
15. *owed it to Bence Jones*: Bence Jones was involved in the dispute between Matteucci and du Bois-Reymond, as Matteucci's letter against du Bois-Reymond was initially sent to Bence Jones (see letter 0724).
16. *your wonderful green island*: he was referring to England.
17. *Ambleside*: see letter 0737, n. 16.
18. *marry on 22 Aug*: see letter 0737, n. 15.
19. *Folkstone*: Folkestone is a port town in Kent, England where du Bois-Reymond presumably stayed with Bence Jones, who had a house there.

From George S. Cumming 5 July 1853 0781

5 Lammerlaws | Burntisland[1] | July 5th. 1853.

My Dear Dr Tyndall

I hope you are quite well. We are enjoying our holidays very much. Alex received your kind letter[2] and we were very well pleased with it. Aunt Coxe[3] said that we may be very thankful that we have such a kind friend as you in the world for to write us such a nice long and kind letter and we both agree with her. Alick would have written a much longer letter had not the post being going away. There is to be a fare[4] here on the on next Tues–day the 12th of July and there is to be a great number of games &c played there. I hope your headache is quite well. I am sorry that you intend leaving Queenwood.[5] Our cousin Elma Fraser[6] came here to spend her holidays on Saturday last. Our cousin Grace Fraser[7] sister of Elma is going to be married on the beginning of August to a young gentleman named Thomas Gillian Rose Innes[8] and we are going to a week more than the holidays for to see to her marriage. This is a very fine bathing and fishing place and we bathe every good day. It has been very bad weather the last few days and yesterday was the worst of them all. Grace has been receiving a very great many preasents. An old watchmaker named Wighton[9] sent her a very beautiful broach and when Uncle[10] went to Edinburgh we were very sorry to hear that he was very ill with Collery.[11] Table moving[12] is getting very common now and also the sprit rapping[13] and it is said that in nearly every party in Paris they do it and Professor Pharody has put it in the Tines.[14] Alick and all the rest join in love with me to you

I remain | yours sincerely | G. Cumming.

RI MS JT/1/HTYP/271
LT Typescript Only

---

1. *Burntisland*: see letter 0777, n. 2.
2. *your kind letter*: see letter 0774.
3. *Aunt Coxe*: refers to May Anne Coxe (née Cumming).
4. *fare*: misspelling of fair.
5. *leaving Queenwood*: see n. 2.
6. *Elma Fraser*: not identified.
7. *Grace Fraser*: Grace Louisa Fraser.
8. *Thomas Gillian Rose Innes*: Thomas Gilyean Rose Innes (1799–?) would become the deputy lieutenant of Banffshire in 1854 (W. Anderson, 'Innes', *The Scottish Nation; or the Surnames, Families, Literature, Honours, and Biographical History of the People of Scotland.* vol. 2 [Edinburgh: A. Fullerton & Co., 1867], p. 536). They were married on 5 August 1853 at Saint Cuthberts in Edinburgh, Scotland ('Scotland, Marriages, 1561–1910', database, *FamilySearch* [https://familysearch.org/ark:/61903/1:1:XY9P-GDH (accessed 27 August 2015)], Thomas Gilyean Rose Innes and Grace Louisa Fraser, 5 August 1853; citing Saint Cuthberts, Edinburgh, Midlothian, Scotland; FHL microfilm 1,066,767, 103,072).
9. *Wighton*: likely William Wighton of Edinburgh or one of his family members.
10. *Uncle*: refers to James Coxe.
11. *Collery*: likely refers to cholera.
12. *Table moving*: 'the moving of a table by supposed spiritual agency, esp. during a séance (*OED*).
13. *spirit rapping*: the communication from or with spirits by raps or knockings (*OED*).
14. *Professor Pharody has put it in the Tines*: Faraday had written a letter to the editor on séance phenomena that appeared in *The Times* (M. Faraday, 'Table–turning', *The Times*, 30 June 1853, p. 8). Tyndall referenced this article in a letter to Faraday (see letter 0783, n. 5).

To John Barlow 5 July 1853 0782

Queenwood, 5th July 1853

Sir,

I am deeply sensible of the honour done to me by the Members of the Royal Institution in choosing me as their Professor.[1] Had I indeed the control and choice of the events of my life I could not have guided them to a result more in accordance with my own feelings. Knowing as I do the rank which the Royal Institution holds in the scientific thought of Europe, I might well be excused a feeling of pride at the present moment. But I trust such feelings, as far as they exist, may manifest themselves in the endeavour to contribute something not unworthy of the present renown of your Institution. In scientific matters this is my highest hope at present, and this shall constitute my aim in future.

I have the honour to be to be, Sir, your most obed[ien]t servant | John Tyndall

Rev. John Barlow, V.P.S.[2] | Secretary, Royal Institution.

Minutes of the Managers' Meeting, RI MS AD/02/B/02/A/11/31
Clerk's Handwriting Only

---

1. *choosing me as their Professor*: see letters 0762 and 0766.
2. *V. P. S.*: Vice President Secretary.

To Michael Faraday [9 July 1853][1] 0783

London, Saturday

Dear Prof Faraday

I regretted on reaching London that I had started two days too late to find you here.[2] I should gladly have had a little conversation with you regarding lectures and concerning a few instruments which I think we shall find useful. On the latter point however I will write to you again from Berlin.[3]

It struck me while looking over the apparatus[4] that a course on heat might be rendered interesting. If you have no objection therefore I would propose a course of 12 on this subject. Should it interfere at all with your arrangements I shall be most happy to choose something else.

Your table-turning article[5] I read with great relish—every body has read it. It has worked wonderful change in the minds of many. I believe you will find the ladies most impenitent upon the subject.

Should you desire to say any thing to me soon a letter addressed to the Post office Berlin will find me. As soon as I arrive however I will write to you[6] and send you my address.

I hope you will find your vacation pleasant and profitable and that you will return to town with renewed vigour. My head is still a little wary which you will probably have already inferred, but I am getting stronger daily and shall soon be in good working order.

My kind respects to Mrs Faraday | Most sincerely Yours | John Tyndall

RI MS JT/TYP/12/4018
Faraday Correspondence, 4:2699

---

1. *[9 July 1853]*: this would be the first Saturday after Tyndall's arrival in London (see n. 2).
2. *too late to find you here*: Tyndall recorded in his journal on 10 July that he went to the Royal Institution on Friday 8 July only to find that Faraday had travelled to North Wales (JT/2/13b/609).
3. *from Berlin*: Tyndall was about to embark on a trip to Berlin (see letter 0773, n. 2).
4. *apparatus*: not identified.
5. *Your table-turning article*: see letter 0781, n. 14.
6. *I will write to you*: letter 0788.

To Thomas Archer Hirst [13 July 1853][1] 0784

Wednesday night[2]

My dear Tom

As I came along I thought I would dot down the impressions and incidents of the journey, and had this thought not occurred to me thus it would not occur to me now. Every man has his hours of strength and clearness and he ought to know how to carry out the resolves then made through hours of weakness and muddiness. During the latter he ought to ask no questions, suggest no doubts, entertain no fears, as the obedient serving man of himself in higher moments. He may thus preserve his life from extreme and ridiculous oscillations, at one moment crying like Dean Swift[3]

> Here I go up, up, up!
> and at the next
> here I go down, down, downy![4]

I hate this undulatory life—its general level must be cast high for otherwise every Wellenberg[5] is infallibly compensated by a Wellenthal[6] the average height of existence being uninfluenced by either. I should like to have seen you before leaving London—the thing would have been more human though perhaps less firm. On the whole I think I prefer the latter to the former; but no I am an egotist I like to be firm myself but to see others human. I rose on monday with the intention of leaving that day, a little bird whispered me that I should succeed. I went to the Prussian Consul but could get no pass, to Mr Taylor's bankers[7] but could get no letter. I went to the Royal Institution wrote a letter stating who I was, took it to the foreign office in Downing St,[8] but the attendants assured me that no notice would be taken of it. The Board of Trade[9] was at hand I went in to enquire for Capt. Wynne and found that he was in Dublin. The thought occurred to me of going straight to the President of the Board m^r^ Cardwell.[10] I went, saw him, told him my perplexity.[11] Well said he I'll tell you how you perplex me. As a man of science I know Mr Tyndall's standing, but how am I to know that you are Mr Tyndall. You must be introduced to me. I went to Marlborough House,[12] saw Playfair got a note certifying that I was really myself, went back with it, found all doors open for me, and got my passport. On leaving Mr Cardwell he said to me "I hope when you come back that you will let me know you better" Started from London bridge at 6.10: reached Dover, walked about and dropped into a cosy little coffee room where I built a brick into my mortal edifice and smoked a cigar, at 12 oC[lock] loosed from the quay. travelled 2nd class and 2nd cabin—I stooped over into the latter but the smell created a million associations of sea sickness and I retreated. Spread my rug on deck and lay down upon it—my cat skin coat[13] being on. Slept there with the gray clouds for my curtains, during part of the time it rained upon me. From Ostende[14] the following morning. had to travel first class for which I swore, internally, if not audibly. There was no second class and I had to pay 16 francs supplement. At 6 in the evening got a view of the Rhine, stretching up and down under the calm sunset like a long memory. at 8 oC[lock] started from Cologne. the twilight advanced and the landscape darkened duller and duller, more and more monotonous, the eye wandered over it and at length for want of variety got numbed and listless, a dreamy dull consciousness crept over me. The lamps however were lighted and on the opposite seat was a lady,[15] to look at whom I turned from the leaden landscape outside. She at length became sleepy, endeavoured to put herself in a comfortable position, spread a little cushion on the seat, laid her white handkerchief upon the cushion, and her glossy black hair upon the white handkerchief as poor John Keats[16] has it, it resembled

> "the feathers of crow
> falling on a wreath of snow"[17]

Her face was beautiful as she lay there, the curves were drawn very tenderly;

over her countenance there was a freshness, not a rose pink tint laid merely upon the surface, but a flush coming from a depth underneath and resembling the eastern sky before day-dawn—By the way Tom Clausius has proved that this morning-red is due to the transmission of the light through the steam bladders in the atmosphere[18]—I thought I should like to kiss her and then I analysed this thought and endeavoured to get a clear view of where the angelic ends in these matters and the demonic commences. I compared the desire of kissing her with that of kissing a sweet little boy and I think I noticed the following distinction I should like to kiss the sweet little boy's cheek, but to kiss her lips!—perhaps the sole reason of this is that little boys are not over careful to preserve their lips clean and dry. I said that the lady was beautiful. I dont think she was divine—I wish we could get a divine girl—I wish we could get something to worship—out of pure selfish motives I wish it—I like to worship but it is so much more easy when the object of your worship takes an incarnate form, is it not possible among these millions of women to find one divine one? hang the baggages[19] I believe not. There is something of the pig in them all, and we must not cast the pearl of worship before a pig.[20] I will make an abstraction from them take all their good qualities, project them against yonder heaven and worship the image thus made. But the image at best is second hand—would that we could get one first hand. It would be splendid to have the soul interpreted by the beauty of a woman's face. But the pigs snout shews itself in the long run on the most beautiful that I have seen. The day dawned upon my weary eyes; the sun ascended *[early]* and I leaned out of the carriage window to let the wind enter through my hyacinthine locks[21] and cool the very skin of my head. On we went wearily wearily—a handsome lady in the carriage but all mere clay well moulded[22]—her voice rather hurts my tympanum.[23] Some of these women have very sweet voices, some of them set one's teeth on edge. I like the low musical voice whose syllables explode like my bubbles—softly and melodiously. At 2 o'c[lock] we reached Berlin the time between Cologne and Berlin being 18 hours—I am now well, sitting at Topfers[24] in 6 minutes I shall be in bed & I hope the gods intend to grant me a sound sleep.

Tomorrow or next day Schneider will send off your box. I have altered the direction putting Dunbridge[25] instead of Stockbridge.

I hope you won't forget to send the £15 to M[rs] Stewart.

Stewart Lodge | Leighlin Bridge[26] | Ireland

Send it by halves and let the last be a few days after the first.

RI MS JT/1/T/907
RI MS JT/1/HTYP/248–49

1. *[13 July 1853]*: Tyndall left London on the evening of Monday, 11 July and recounted journeying for two days in his journal; he arrived in Berlin at 2pm on 13 July (Journal, 11–13 July, JT/2/13b/610–11).
2. *Berlin Wednesday night*: in his journal entry for 13 July Tyndall wrote that he 'Reached Berlin at 2 o'clock and drove straight to Toepfer's' (JT/2/13b/611).
3. *Dean Swift*: Jonathan Swift (1667–1745) was an Irish writer famous for his book *Gulliver's Travels*. He was also dean of St. Patrick's Cathedral in Dublin, 1713–45.
4. *Here I go . . . downy!*: a line from *The Nurses Song*, a poem sometimes attributed to Swift ('The Nurses Song', in J. Struthers [ed.], *The Harp of Caledonia: A Collection of Songs, Ancient and Modern, (Cheifly Scottish.) with An Essay on Scottish Song Writers*, 2 vols [Glasgow: Khull, Blackie, & Co., 1821], vol. 1, pp. 354–55).
5. *Wellenberg*: German for 'wave mountain'.
6. *Wellenthal*: German for 'wave valley'.
7. *M[r] Taylor's bankers*: likely the bank used by Richard Taylor.
8. *foreign office in Downing St.*: located, in 1853, at 12 Downing Street in London, the Foreign Office is a governmental department involved in worldwide affairs, including consular affairs.
9. *Board of Trade*: the Lords Commissioners of Trade and Plantations, commonly referred to as the 'Board of Trade' until it was formalized in 1861, was located in Whitehall, London.
10. *President . . . M[r] Cardwell*: Edward Cardwell (1813–86), First Viscount Cardwell, a prominent politician, was president of the Board of Trade 1852–55.
11. *my perplexity*: in 1853, English citizens required either a passport issued by the Prussian minister in London or a referral from a Prussian consular official for passage into Prussia ('Prussian Passport', *Handbook for Travellers on the Continent; Being a Guide to Holland, Belgium, Prussia, Northern Germany, and the Rhine from Holland to Switzerland*, 9th edn. [London: John Murray, 1853], p. xx).
12. *Marlborough House*: located in The Mall, London, it was the new location of the Metropolitan School of Practical Art ('First Report of the Department of Practical Art', *The Sessional Papers Printed by Order of The House of Lords, or Presented by Royal Comman, in the Session 1852–53. Arranged in Volumes* [London: George E. Eyre and William Spottiswoode, 1853], vol. 79, Reports, Papers, &c., pp. 1–45).
13. *cat skin coat*: cat skin, a soft and light material usually taken from Lynx and Fitch (German pole-cats). In 1851, 55,000 Lynx skins and 65,091 Fitch skins were imported to Britain ('Fur', in J. Smith Homans [ed.], *Cyclopedia of Commerce and Commercial Navigation* [New York: Harper & Brothers, 1858], pp. 764–74, esp. table on p. 767).
14. *Ostende*: a coastal city in Belgium with a direct ferry connection to Dover, England, as well as a train station.
15. *a lady*: not identified.
16. *John Keats*: John Keats (1795–1821), an English Romantic poet.

17. *'the feathers . . . snow'*: line from Keats's poem, 'To —' (11–12).
18. *Clausius . . . atmosphere*: see R. Clausius, 'On the Colours of a Jet Stream and of the Atmosphere', *The Edinburgh New Philosophical Journal* 54 (October 1853), pp. 166–68.
19. *hang the baggages*: 'a worthless good-for nothing woman' (*OED*); also a reference from Shakespeare's *Romeo and Juliet* (III.v.8).
20. *cast the pearl . . . pig*: an allusion to Matthew 7:6, 'Give not that which is holy unto the dogs, neither cast ye pearls before swine, lest they trample them under their feet, and turn again and rend you'.
21. *hyacinthine lockes*: line from John Milton's (1608–74) *Paradise Lost* (IV.ii.301–2).
22. *clay well moulded*: a reference to the myth of Pandora from Ancient Greek mythology. Pandora, the first woman, was molded out of clay by Hephaestus on Zeus's order and sent to humans as a punishment for Prometheus's theft of fire.
23. *tympanum*: the eardrum (*OED*).
24. *Topfers*: Töpfer's was a hotel in Berlin (see letter 0479).
25. *Dunbridge*: the railway station close to Queenwood.
26. *Leighlin Bridge*: Leighlin Bridge (now spelled Leighlinbridge), the small town in County Carlow, Ireland, where Tyndall was born (see Volume 1). See also letter 0953, which includes a description by Hirst during his visit there in July 1854.

To Heinrich Debus 19 July 1883 0785

Universitäts Straße No. 4 | July 19th 1853.

Lieber Heinrich

Ich bin Dir einen langen Brief schuldig aber Du mußt froh sein wenn Du einen kleinen kriegst. Ich bin ein großer Schuldner in dieser Beziehung, seit ich Berlin erreichte bin ich so hin und her bewegt worden daß es mir kaum möglich ist, meine Gedanke[n] zu fesseln und einen Brief zu schreiben. Montag vor acht Tage[n] verließ ich London, Tom *[half]* ich in Queenwood ein paar Tage vorher und wir wohnten einige Zeit in London zusammen. Ich machte die ganze Reise von London bis Berlin ohne mich aufzuhalten—ungeheur müde war ich, und schmutzig wie ein Neger. Die Professoren werden bald alle wegreisen und dann werde ich ganz allein sein. Sie haben mich sehr freundlich aufgenommen—fast jeden Abend bin ich in Gesellschaft Heute habe ich ein paar Stunde[n] Gearbeitet mit du Bois Multiplicator und fand daß ich die Nadel nach belieben in bewegung setzen konnte. ich lese etwas jeden Tag. bestelle Apparate und benutze jede Gelegenheit um meine Kentniss zu vermehren. Ich werde meine immer starker und hoffe bald meine ganze Kraft wieder zu erhalten. ich habe ziemlich viel *[Bairisches Bier]* getrunken und finde dass es mir sehr wohl thut. Du wirst bald nach Queenwood zurüch kehren. hoffentlich wird dieser Brief nich[t] zu spät sein. Einige versuche werde ich hier machen. und habe die Apparate dazu schon bestellt. Ob etwas daraus kommt weiß die Götter allein. Kurz vor meine[r] Abreise habe ich einen Brief von Haarlem bekommen. *[worin]* stand daß ich als Mitglied der Konig. Gesellsc. der Wiss. zu Holland gewahlt war. Schreibe wen[n] du Zeit hast, und sage mir alles was du gesehen hast. Wenn du nach Marburg gehst, ein Gruß an Knoblauch.

Dein John Tyndall

Universitäts Straße No. 4 | July 19th 1853.

Dear Heinrich

I owe you a long letter but you will have to be glad it you get a short one. I am a great debtor in this regard, and since I reached Berlin,[1] I have been moved back and forth so much that it is hardly possible for me to get a hold of my thoughts and write a letter. I left London eight days ago on Monday. I *[helped]* Tom in Queenwood for a couple of days beforehand,[2] and we stayed together in London[3] for some time. I did the whole trip from London to Berlin without stopping—I was awfully tired, and dirty as a nigger. The professors will all be departing soon and then I shall be entirely alone. They have received me very kindly—I am in company almost every evening. I worked with du Bois's multiplier[4] for a couple of hours today and found that I could make the needle move any way I wanted. I read something every day, order apparatus and take every opportunity to increase my knowledge. I'm getting stronger and stronger, and hope to get all my strength back again soon. I have drunk quite a lot of *[Bavarian beer]* and find that it does me a lot of good. You will be returning to Queenwood soon. Hopefully, this letter will not be too late. I shall do some experiments here, and have already ordered the apparatus for them. Whether something will come of it, the gods alone know. Shortly before my departure, I got a letter[5] from Haarlem[6] *[in which]* it said that I was elected a member of the Roy. Soc. of Sci. in Holland.[7] Write when you have time, and tell me everything that you have seen. If you are going to Marburg, greetings to Knoblauch.

Your John Tyndall

RI MS JT/1/T/257

---

1. *reached Berlin*: Tyndall had arrived in Berlin on 13 July (see letter 0784).
2. *I helped Tom . . . beforehand*: Hirst had arrived at Queenwood on 5 July and he and Tyndall went to London together on 6 July (Hirst, 'Journals', 8 and 9 July 1853).
3. *stayed together in London*: Tyndall recounts his time in London with Hirst in his journal (Journal, 10 July, JT/2/13b/609–10).
4. *du Bois's multiplier*: Emil du Bois-Reymond was known for making modifications to a Schweigger multiplier, an early galvanometer, in order for it to function as he needed for his experiments (Finkelstein, *Emil du Bois-Reymond*, pp. 60, 179).

5. *I got a letter*: letter 0776encl. Tyndall noted his receipt of this letter and that he found these accolades encouraging, but, significantly, not as meaningful as the letter he received from Coxe (letter 0777) (Journal, 6 July, JT/2/13b/608).
6. *Haarlem*: The capital city of North Holland province in the Netherlands
7. *Roy. Soc. of Sci. in Holland*: La Société Royale Hollandaise des Sciences (see letter 0776encl).

To William Francis [19 July–31 July 1853][1] 0786

My dear Francis

It just strikes me that I should like to write a line to you although I have but a few minutes at present to accomplish it. I seize the thought as it comes for there is no knowing this hot weather if I put it off when it will return to me. I am comfortably fixed as far as lodging is concerned in

Universitäts Strasse No. 4

2 Treppen Hoch Rechts.[2]

I have seen most of our friends since I came and am at present making a few experiments in Magnus's laboratory.[3] The result although unbedeutend[4] are new and not altogether without interest. I was at Pöggendorffs[5] a few evenings ago. Mrs. P.[6] is evidently a great friend of Miss Francis[7]—the [beete] *[1 word illeg]* and *[*sauces*]* which were placed before me were staked by Miss Francis when she was in Berlin. Before I leave I hope to be able to send you something for the magazine[8] though as before hinted at this hot weather is sufficient to squeeze every particle of energy out of a man. Could you manage to have a copy of that note to Plücker's paper[9] sent to de la Rive with the magazine? I should feel greatly obliged if you would do this and reserve a copy or two for myself. Among the things I am scraping up a few materials for my lectures. I shall soon be alone here. all the professors are going away. If Magnus goes to Paris it will be a great temptation to me to go along with him as he knows every body there already. Don't be angry with me for sending you such a flat stale and unprofitable note. Kind remembrance to all at Richmond[10] to *[1 word illeg]* & *[1word illeg]*. & believe

*[*that I*]* sincerely yours | J Tyndall

StBPL T&F Author letters

---

1. *19 July–12 August 1853*: Similar to the text in this letter, in letters to Debus and Hirst dated 19 July and 23 July, respectively, Tyndall mentioned that he will soon be alone, as all the professors will be leaving (see letters 0785 and 0787).
2. *2 Treppen Hoch Rechts*: German, 'second floor, right'.

3. *Magnus's laboratory*: this was a private laboratory so likely located in Magnus's house.
4. *unbedeutened*: German, 'miniscule' or 'insignificant'.
5. *Pöggendorffs*: Johann Christian Poggendorff's home in Berlin.
6. *Mrs. P.*: Poggendorff's wife, Charlotte Eleanor Poggendorff (1805–65).
7. *Miss Francis*: Rachel Francis (1820–1913), Richard Taylor's daughter, who had accompanied William Francis to Berlin in October 1839 (Brock and Meadows, *Lamp of Learning*, p. 116).
8. *the magazine*: the *Phil. Mag.*
9. *that note to Plücker's paper*: see letter 0776, n. 5. Not otherwise identified.
10. *Richmond*: Richmond, Kingston upon Thames, a town southwest of London until 1890, when it became a suburban borough, called Richmond upon Thames. It was the home of the Taylor/Francis family.

To Thomas Archer Hirst 23 July [1853] 0787

Berlin. | Universität's Strasse No. 4. | 23rd. July

My dear Tom,

I find your friend Schneider makes some difficulty about sending a *[Mechesel[1]]* to London and I have ordered some instruments which I should like to be able to pay for when they reach me. Would you then like a good fellow send me the 1st halves of two £10 notes. Du kannst zwei Tage abwarten und dann die andere zwei Hälfte nachschicken.[2] If you are in Queenwood Wright will hand you over the balance of my account. Just let me know to what extent I may draw upon you. My orders for instruments must be guided by your answer. Whatever it may be these orders will not be heavy: I shall pay you very soon after my return. Dont forget to send the £15 to the place mentioned. I shall soon be alone here. The professors are all going away. Steiner is at some bathing place. Dirichlet I have seen but not yet spoken to. I shall probably call upon him to day or tomorrow—My lodgings here are very snug, and the people exceedingly attentive—10 thalers a month is what I pay. I saw several places but none pleased me so well as this. Let me know your prospects at Queenwood when you write—or rather your numbers. I hope you will have a full house. Remember me kindly to all. I wish all well and if Queenwood do not flourish I shall regret it. If it fail it will be a standing reproach to every body connected with it—You have a capital problem before you there boy. See that you solve it well—

ever yours affectionately | John.

RI MS JT/1/T/911
RI MS JT/1/HTYP/250

1. *Mechesel*: not identified.
2. *Du kannst . . . nachschicken*: translates from German as 'You can wait for two days and then forward the other two halves'.

To Michael Faraday 24 July 1853 0788

Berlin | Universitals Strasse No 4 | July 24th, 1853

Dear Professor Faraday

In commencing to write this letter I am violating to some extent an article of my own creed, that no man is at liberty to write or speak without some necessity. I do not clearly see the necessity at present, I have no distinct purpose in view, but still I write. Perhaps such writing is of a meteorological character depending on the state of the weather, and indeed there is a calmness and a beauty in this morning which might suggest an occupation of the kind. I have been here for the last 9 or 10 days and employ myself in extracting as much information as possible from my friends. A doubt of the value of such information is however often present with me: its value is chiefly conversational, it fills a gap in a lecture sometimes, but beyond this I believe it is very unprofitable. It tends to weaken the fixity of the mental axis and to keep it vibrating amid the multitude of facts. I sometimes think that education may be too liberal, that our books and our acquaintances may weaken our alliance with nature and thus do more harm than good. People often speculate on the possible achievements of an original thinker who has had to struggle with difficulty, had such difficulty not existed, forgetting that the private effort enforced by his circumstances may be the very thing to which his development is due. The tendency of our so called advantages is sometimes simply to dilute a man. Society is the enemy of work, and here I see a danger which lowers upon the foreground of my own future. I must circumscribe myself in London: must set my face against visiting if I would get any work done. I know that some who are now very friendly with me will dislike peculiarity, but I must trust to time for a true verdict. Here however my course is comparatively clear. You have already hewn your way through this jungle and I have nothing to do but follow your steps.

I have been repeating Du Bois[1] experiments thinking possibly that they might be made use of on some Friday evening.[2] Another subject out of which a Friday evening might be manufactured is the old one of Trevelyan's experiment.[3] Since Forbes's papers[4] I believe nothing has appeared on the subject in England and these papers cast a doubt upon your explanation of the matter.[5] Forbes seems to consider it essential that different metals must be used, I have had an instrument made since I came here and obtained it with a distinct tone yesterday, which continued a quarter of an hour, from copper *on copper*.

The artifice by which it is obtained is suggested by your explanation. I will try other metals this week and am not without hope of obtaining the action with quartz: at all events I am getting a crystal cut with the view of trying the experiments.

Another point to which I have turned my attention is the generation of cold at a bismuth and antimony joint.[6] Up to the present time the experiment has only succeeded when either bismuth or antimony is in the circuit. It is a most suggestive experiment. I have ordered an apparatus with which I hope to generalize the fact and which may lead to something else.

I have ordered a few instruments which will be required next year in London: but have restricted the order to a sum which will be no means render me bankrupt provided you do not consider the instruments necessary for the institution. There is a most convenient form of the galvanometer[7] used here: it consists of a magnetized polished steel disk with coils. I think it would be very useful both for private experiment and for the lectures—the deflection of the disk might be rendered very evident by the motion of a ray of light reflected from it. It would cost about three pounds. If you had time I should feel thankful if you would let me know whether I might devote 20 or 25 pounds to the purchase of apparatus.

Believe me dear Sir | Most faithfully Yours | John Tyndall

RI MS JT/TYP/12/4019
Faraday Correspondence, 4:2704

---

1. *Du Bois*: Emil du Bois-Reymond.
2. *Friday evening*: lectures were held at the Royal Institution on Friday evenings and attendance was open to members and guests.
3. *Trevelyan's experiment*: Arthur Trevelyan (1802–78), a Fellow of the Edinburgh Geological Society, discovered a phenomenon whereby sounds are produced by heated metals during the cooling process. He presented his experiments on the phenomenon to the Royal Society of Edinburgh on 17 January and 21 March 1831 (A. Trevelyan, 'Notice regarding some Experiments on the Vibration of Heated Metals', *Trans. Roy. Soc. Edin.* 12:1 [1834], pp. 137–46). Tyndall conducted similar experiments for a Friday evening lecture on 27 January 1854 (see letter 0876, n. 14).
4. *Forbes's papers*: see J. D. Forbes, 'Experimental Researches regarding certain Vibrations which take place between Metallic Masses having different Temperatures', *Trans. Roy. Soc. Edin.* 12:2 (1845), pp. 429–61.
5. *your explanation of the matter*: Faraday argued that the vibrations in the heated metal produce sound due to rapid expansion and contraction (M. Faraday, 'On a peculiar class of Acoustical Figures; and on certain Forms assumed by groups of particles upon vibrating elastic Surfaces', *Phil. Trans.* 121 [May 1831], pp. 299–318).
6. *generation . . . antimony joint*: see the correspondence of Richard Adie (1810–81) and

Tyndall in the *Phil. Mag.* (see letter 0711, n. 3). See also letters 0663; 0670; 0672; 0685; and 0693.

7. *galvanometer*: see letter 0711, n. 4.

From Michael Faraday 25 July 1853 0789

Royal Institution, | 25 July, 1853.

My dear Tyndall

I have received your letter[1] without date, so do not know whether you have been expecting to hear from me; I rather waited to learn your address, but by Dr Bence Jones' advice intend to send this to our friend Du Bois-Reymond's house. I really have nothing to say, but to convey my kindest remembrances and wishes to your and my friends at Berlin, and to set your mind free in every point however slight. I think your proposed course here[2] will be excellent, and know very well that I shall enjoy it, and believe that all others will do so with me. I thought I saw Du Bois-Reymond's face in the Railway the other day, on the Northern road, but it must have been a mistake. Give my heartiest remembrances to him. Also to M.M. Magnus, Rose,[3] and such other friends of mine as you may meet. In thought I send my homage and respects to Humboldt,[4] (and through him even to the King,[5] who has honoured me with the Order of Merit[6]), but I refrain from troubling him. Though very much his junior in years, I feel the burden of formalities, and so think I ought to spare him and such as him. Sometimes I fear I may carry this too far, and that it may assume the appearance of indifference on my part; but I hope that will not be the case and that I am favourably interpreted, especially in the case of the man whom I most deeply venerate.

I was with Dr. Percy[7] on Saturday night, who tells me they have been building an enormous geyser at Woolwich[8]—not on purpose, however, but accidentally. The fact is that they have been building a new boiler which, as far as I understand it, consists of two tall concentric iron cylinders, having water between them and fire in the centre. Percy compares it to the geyser, and thinks of it with some degree of apprehension.

I wish I could transport myself suddenly to Berlin, and be with you in some of the Laboratories and Workshops. I should luxuriate in some of the manufactories of philosophical instruments, but—but such wishes remain long after the ability or fitness to satisfy them has passed away. A single day of such would set my head a ringing, and I should have to run away.

Ever, my dear Tyndall, | Yours very truly, | M. Faraday

RI MS JT/TYP/12/4134
Faraday Correspondence, 4:2706

1. *your letter*: letter 0783.
2. *your proposed course here*: in letter 0783 Tyndall proposed a course of twelve on the subject of 'Heat'.
3. *Rose*: either Heinrich or his brother Gustav (see letter 0737, n. 18).
4. *Humboldt*: Friederich Wilhelm Heinrich Alexander von Humboldt (1769–1859), German explorer and naturalist, most known for his exploration of Latin America with an approach that encompassed many scientific disciplines.
5. *the King*: Friedrich Wilhelm IV (1795–1861) was the reigning monarch of Prussia, 1840–61.
6. *the Order of Merit*: the Orden Pour le Mérite für Wissenschaften und Künste (Sciences and Arts) was founded by the King (see n. 5) on the advice of Humboldt (see n. 4) and was awarded in recognition of extraordinary personal achievement. The award inducted the recipient as a lifetime member of the Order. Faraday received the honour in 1842.
7. *Dr. Percy*: presumably John Percy (1817–89), who was a lecturer on metallurgy at the Government School of Mines.
8. *Woolwich*: Woolwich a town in the parish of Woolwich St. Mary located on the south bank of the Thames, Kent, England. At the time it was a major center for several military facilities.

To Michael Faraday 28 July [1853] 0790

July 28th.

Dear Professor Faraday

DuBois[1] starts for England tomorrow and will carry this to Hull.[2] He came to me the evening before last with a countenance shining with pleasure, 'You would not guess what I have got' he exclaimed. I paused—one hypothesis alone seemed opened to me—'You have got a letter from Ambleside'[3] I replied. 'No' said he 'I have got a letter for you'—He opened his pocket book and handed me a letter bearing a superscription well known to both of us—the letter[4] in fact was yours.

Would that my wish and the wish of many able, honest and hospitable men added to my own could be effected in transporting you at a safe speed to Berlin. From one end of Germany to the other there is but one feeling towards you—a feeling which I believe it has never before been the lot of a man of science to excite: it will be a pity if you cannot afford time for a single visit to a country which I believe above all others pays you a noble reverence—You have visited France and Italy—establish the equilibrium of things by a visit to Germany. Whenever you make up your mind I offer my services as a guide. Mrs Faraday may rest assured that I will take great care of you—my greatest difficulty would be to preserve you from the loyalty of your clansmen, but even this I would undertake to manage.

I mentioned in my last letter[5] that I intended to try Trevelyan's experiment[6] with a new metallic body. I was led to choose quartz from the extraordinary conductive power which my last experiment proved it to possess. With the first crystal I had cut a permanent oscillation was obtained but no tone. With a second crystal cut somewhat differently I have obtained a distinct tone, sufficiently loud I think to be heard throughout the theatre when everything is perfectly quiet. I have failed to obtain vibrations with a piece of glass cut similarly to the quartz, this might have been anticipated. On the whole I think the matter might be worked up into a suitable Friday evening's lecture.[7]

I cannot resist the temptation to send you a note[8] which I received from a little friend of mine just before leaving England—a beautiful childlike little boy[9] about 10 years old. He was with me at Queenwood. His cousin Elma and cousin Grace[10] are strangers to me but he tells me all about them with the most perfect good faith that it will interest me. I should not venture to send you the note had not somebody told me that you could afford to play with little boys.

Believe me dear Prof Faraday | Most truly Yours | John Tyndall
Universitäts Strasse 4 | Berlin

RI MS JT/TYP/12/4023
Faraday Correspondence, 4:2707

1. *DuBois*: Emil du Bois-Reymond.
2. *Hull*: see letter 0754, n. 6.
3. *Ambleside*: see letter 0737, n. 16.
4. *the letter*: see letter 0789.
5. *my last letter*: letter 0788.
6. *Trevelyan's experiment*: see letter 0788, n. 3.
7. *Friday evening's lecture*: see letter 0788, n. 2.
8. *a note*: see letter 0781.
9. *little boy*: George S. Cumming, younger brother of Alexander.
10. *cousin Elma and cousin Grace*: see letter 0781, nn. 6–7.

From Thomas Archer Hirst 31 July 1853 0791

Queenwood College—| Sunday Morning. | July 31st 1853

My dear John.

On arriving here on Friday last I found two letters[1] from you. The first[2] interested me of course most, wherein you describe your journey and your fair travelling companion.[3] I rather smiled at your demanding a divine girl, and felt inclined to ask what you would do with her. For purposes of worship your imaginary girl projected against Heaven's clear blue is the proper article, you have no right to demand a more substantial one. For actual companionship and use your divine girl would be use<u>less</u>. She cannot keep house or mend stockings &&c above all your divine girls are not often good breeders. Excuse my very material answer to thy poetical letter for I do it intentionally. Your aspirations are strung up too high, or at least whilst tolerating them as aspirations and enjoying them as such I find it necessary to recall thee to the actual how to raise it as much as possible, what to expect from it and what to tolerate in it. In plain words I think you must alter somewhat your 'standpoints' when judging our veritable 'womankind'

You talk about this 'Pigs snout' (the term annoys me very much) 'There is something of the Pig in them all.' 'But the pigs snout shews itself in the long run in the most beautiful that I have seen' These are hard words John. and when I read them I accused thee of foul libel, of blasphemy. And then I set to work finding out thy meaning and why, when, and where thou hast seen all this. I cannot corroborate it. I protest against it, for the sake of one or two I have known, who far transcend myself in purity I protest against it. You are curiously constituted John and thy constitution has been a subject of many hours thought for me and though I have returned scores of times from such thought with a feeling of just pride and encouragement, yet there is always a trace of dissatisfaction which has never yet been able properly to give an account of itself but will do some day. Rousseau[4] somewhere says that the task of expression was often a source of great pain to him he scarcely ever sat down to write without feeling a strange disquiet, a turmoil within himself which held him by its fascination but yet pained him.[5] In this I often sympathise with him and never more than when I approach such fickle subjects—as the present. Oh there is a weight of something I feel certain could be shaken off, that presses on me and yet that I cannot throw off. I have nothing but patience to help me and the hope that like Christian in the Pilgrims progress[6] I shall throw down my load some day. Your strength and self command has very often been of mighty service to me—but in closer communion with thee I have felt with sadness that I could not did not possess thee altogether. I think the sadness perhaps arose from the thought that thy path through the world must be solitary that even I could not accompany thee every where, and I could not answer myself why. You like to be firm yourself but to see others human.[7] I have noticed it scores of times and with some sort of dissatisfaction. I have longed to be more human towards thee and could not and I believe solemnly that thou art to blame for it. You are not human enough for one that solves part of my difficulty. You have an ideal before you which no one can dispute you follow bravely and firmly, but you would absorb too much into it, it leads you into other failings which I perhaps think equal to those you would free yourself from. There is some allowance to be made I know for our different natures and that as long as ever this difference exists this want of total agreement will remain. But I have said enough we have both our own riddle to solve to a great extent this riddle for us both is the same but it has its modifications for each of us.

In the matter of marriage and 'womankind' from which we started, I believe you ought and must somewhat alter your point of view. You hesitate to ally yourself to one who might impede you in following this same ideal of yours, you wish to find some one who would rather encourage you in it & assist you. With regard to the latter you will be in some measure disappointed. Take care that here you do not deceive yourself. Take care that you do not shut yourself (I will not say from comforts or pleasures for you would spurn them) from experiences & sources of discipline equal to those you seek alone. Consider whether or not by deviating from your more solitary path you could not find one in company with some kind soul whom you could cheer and exalt & be thereby yourself highly benefited. Could you not thus humanize yourself without compromise. It is a problem for you to solve a problem you ought to solve, but one which in my humble opinion you are not exactly in the right way to solve at present.

After this long & in a measure futile attempt to say what I set out to say, I will pass on to more business like matters. I have received £35 from Mr E.,[8] the remaining £8 (about) yet due to you will be paid when asked for. Herewith I send the first halves of four £5 notes the second halves I will send in a day or two.[9] By this post I shall send the first halves of three £5 notes to Mrs Stewart.[10] I have at present only £20 at your disposal which with your own £8 will make a sum of £28 to expend immediately on apparatus. On October first I shall have another £20 which I can send you if required. I have arrived here under very disadvantageous circumstances for I am suffering very severely from Tic-douloureux,[11] it makes me almost frantic sometimes and has robbed me entirely of 3 nights sleep. The good folks here would nurse me well if I would let them. The other night Mr E. recommended a warm bath, I consented, at which he was pleased, saying 'I hope Thomas then wilt not take John Tyndall's example but will let us occasionally have the pleasure of nursing thee.' I replied 'I can make no promises, this time your remedy and mine happen to coincide, and for that reason I have so readily submitted.' Well the little fellow was very officious and at the end was disappointed, he wanted to help me to take a warm bath, he had it up in my bed room, poured in his hot and cold water and with a thermometer he succeeded in producing the orthodox temperature. I waited quietly by without speaking & shewed no symptoms of undressing, he then left me telling me to sit a quarter of an hour in it compared his watch with mine and said that at the end of that time he would return; I undressed then, sat in the bath ten minutes, jumped out, rubbed myself, and jumped into bed. True to his time he put his nose in at the door and uttered a cry of disappointment to see all finished, on his arm he had an immense number of clothes and flannels which he said he had brought to lap me in and to help me to rub myself. I thank him, chuckled & bid him good night. Next morning & ever since I have pretended to be better to avoid all further trouble. In reality however I am no better, and many a time have I to sit in agony opposite him at table, to keep talking and looking pleasant. I console myself however that it is no worse for thus suffering silently, nay perhaps the very discipline may be beneficial. Very few boys have arrived yet, and work will not begin until next week. All the tutors except Debus have returned. Edmondson expects 70 boys. Wright expects 65 at most, and will be satisfied with 60. As I expected, a delay has taken place with respect to my lodgings. No cottage can be found for the Gardener.[12] I know very well that if I were not daily to remind Mr E. that he would contrive very soon to forget all about it. I intend however to be very firm about the matter.

Remember me kindly to all Berlin Friends especially Dirichlet. Get to know from Schneider if you can what vessel he has sent my boxes by and to what port it will sail. I shall have to employ an agent at the port of arrival otherwise they will not send them on to Queenwood. Write soon.

Thine affectionately | T. A. Hirst

RI MS JT/1/H/180
RI MS JT/1/HTYP/251–253

---

1. *two letters*: letters 0784 and 0787.
2. *the first*: letter 0784.
3. *fair travelling companion*: see letter 0784, n. 15.
4. *Rousseau*: Jean-Jacques Rousseau (1712–78) was a Genevan philosopher of the Enlightenment noted for his many works on political and moral philosophy, as well as his autobiographical writings.
5. *somewhere says . . . pained him*: not identified.
6. *like Christian in the Pilgrims progress*: Christian, the protagonist in John Bunyan's (1628–88) seventeenth-century allegorical text, *The Pilgrim's Progress from This World to That Which Is to Come; Delivered under the Similitude of a Dream* (London: Nath. Ponder, 1678).
7. *see others human*: in letter 0784 Tyndall wrote that 'I like to be firm myself but to see others human'.
8. *received £35 from Mr E.*: Tyndall had asked Hirst to get the balance of his account from the school and send it to him (see letter 0787).
9. *Herewith I send . . . two*: see letter 0787.
10. *I shall . . . Mrs Stewart*: see letter 0784, postscript.

11. *Tic-Douloureux*: also known as Trigeminal neuralgia, which is characterized by severe pain in the face and spasm on one side of the face.
12. *the Gardener*: likely the gardener taking care of the grounds at Queenwood.

From Michael Faraday 1 August 1853 0792

Royal Institution, | 1 August, 1853.

My dear Tyndall,

Our letters have crossed,[1] and so I must teaze you again for a little moment, to say that I am sure you will do right in purchasing the apparatus[2] you think of; and also I am sure you will make heat most interesting to us.[3] Your letter[4] seems to imply that you are a little discouraged by the infinity of objects about you: but you are taking a rest now, and there is no reason why your mind should not in its quiescent state take the hue of every subject that comes near you; even though it do yield up that of the going, for that of the coming, subject. When you brace up your mind and settle with its undivided powers upon one subject, there is no fear that it will be any way deficient, or short of its ordinary vigorous tone. My kindest remembrances to all our joint friends.

Believe me to be | Ever truly yours, | M. Faraday.

RI MS JT/TYP/12/4135
Faraday Correspondence 4: 2711

1. *Our letters have crossed*: letters 0788 and 0789. Faraday responded to letter 0783 with letter 0789. Tyndall had already sent letter 0788, which Faraday had not yet received before composing letter 0789. This is Faraday's response to letter 0788.
2. *the apparatus*: see letter 0788. Tyndall wished to purchase multiple instruments, including a galvanometer (see letter 0711, n. 4).
3. *make heat most interesting to us*: Tyndall had earlier proposed that he teach a course of twelve lectures on the subject of 'Heat' (see letters 0783 and 0789, n. 2).
4. *Your letter*: letter 0788.

To Thomas Archer Hirst 9 August 1853 0793

Universitats Straße 4 | 9th Aug. 1853.

My dear Tom.

Your letter[1] has come—your two letters[2] and with them the notes[3] all safe. Your first was a hard one[4] Tom the only hard one you ever penned to me. I thought first of all I would write a reply in poetry and prove to you by the melody of my song and without any attempt at logic that you had in some measure wronged me. I walked towards the Linden[5] and saw the orange clouds sailing over the Brandenburger Thor[6]—the sun was tumbling down his last steep arch and *[smiting]* with gold the foliage of the Thier garten.[7] I was full of my idea—could I only have given it utterance I should have smelted you into tears. I composed a verse or two but at length thought that plain prose might answer. You protest against me—against my blasphemy—you protest meanwhile against your own notion of me which for the first time is a dislocated one. I hope you will ever continue to protest against this same notion—you protested against a misanthropist—I am not one. To your chivalrous defence of woman I say bravo with all my heart. Its only defect consists in its being misplaced. When I spoke against them I judged them from a superhuman ground and I thought you would see this. In stating their defects I merely expressed a necessity of my own soul. Turn to your Tennyson, to that Enone,[8] and see the description of Pallas[9] given there. I wanted a girl like Pallas—quite for other purposes I confess than sewing and washing and breeding. I never read that passage without a glow and a thrill which I would respectfully urge is quite as <u>useful</u> in its way as your conception of a good breeder. I pleaded for an object of worship and you hand me over a quality which a woman enjoys in common with a cow. There is a greater compliment in my censure than in your praise. Why did I connect such thoughts at all with a woman simply because she comes nearest to the thing I sought. This I conceive is praise sufficient for her and higher than any contained in your defence. You are too swift in your judgment of me. From the fact that circumstances compelled me to be solitary you infer that nature intended me to be solitary. Open your heart my boy and put another construction upon the position of your friend. I have cheerfully accepted my lot and made myself a throne in it and shall certainly continue my course as long as I believe it right. Put me in mind of it when we meet and I will tell you a little story. It may dissipate this mirage which you have contrived to get between you and the truth. When I think of my early feelings—feelings which though kept back are still unblasted—of the poetry I have written to bright eyes, of the friends who loved me and whom I loved—and then hear him whom I cherished on higher grounds than any of them uttering his dictum that nature intended me to be solitary the utterance is to me that of an unknown tongue. You say I must change my 'Stand punckt'[10]—you must change yours if you would judge correctly of your friend. You say you would like to be human towards me. I dont want you to be. If you had been less anxious about it you would perhaps have succeeded better. I never sought your affection, if it comes to me as a stone seeks the earth it is welcome, dont strive to love me—you cant do it. My relation to you has been as clear and independent as ever was the relation of man to man. You say you noticed with dissatisfaction my desire to be strong and to see others human. You noticed a feeling which if ever it gave itself utterance never to my knowledge manifested itself towards you. But I have harped too long on this old string—let me lift my heart above it. If you catch me writing such a letter again you may conclude your friend has become a dotard. Perhaps I should not have written it at all had I not just taken a glass of rum and water for the sake of my bowels which are very hard—but as it is I will let it go.[11]

<hereafter LT Transcript Only>

I have seen Schneider. Your box was sent off in July and reached London before the end of last month. It went over Hamburg.[12] One word about money matters. I thought you were rich and therefore did not bestow a thought on the matter till your letter[13] reached me. That £28 will go for instruments. I should like to spend a month in Paris. The other £20 will take me clean off the Continent, perhaps more, but then I shall require a little money in London. I dont know whether the payments of the R.I. are yearly or half yearly. If half yearly I shall need no money from you after the October £20, if yearly I shall. Can you advance it? I should like an answer to this soon, as upon it will depend my arrangements. The instruments I purchase are not at my own cost. The money will be refunded. All I want you to do is to be my banker pro-tempore[14]—if you can, well; if you cant also well—I will square my measures accordingly.

Your affectionate | John.

Send me a dozen postage stamps in your next letter.

RI MS JT/1/T/567
RI MS JT/1/HTYP/254–255

---

1. *Your letter*: letter 0791.
2. *your two letters*: see n. 1; second letter missing.
3. *the notes*: Tyndall had asked Hirst to send him 'the 1st halves of two £10 notes' (see letter 0787). In response, Hirst sent two £5 notes with letter 0791. Hirst must have sent the other two £5 notes with the missing letter.
4. *Your first was a hard one*: letter 0791. This letter represented Hirst's first real admonishment of Tyndall (excepting letter 0667), suggesting that at this point Hirst was becoming a mentor in his own right. It should be noted as well that letter 0791 was Hirst's first letter to Tyndall from Queenwood, his new station in life.
5. *the Linden*: Unter den Linden, a well-known avenue in Berlin named for the linden (lime) trees that grew along the street.
6. *the Brandenburger Thor*: the Brandenburg Gate, located at the western side of Unter den Linden. Built between 1788 and 1791, the Gate was commissioned by King Frederick William II (1744–97) as a sign of peace and was designed by Carl Gotthard Langhans (1732–1808).
7. *the Thier garten*: a garden in Berlin located by Unter den Linden and the Brandenburger Thor.

8. *Enone*: first published in 1832, Tennyson's 'Œnone', is based on the story of Œnone and Paris from Greek mythology, specifically the Judgment of Paris, which resulted in the Trojan War. In the poem, three goddesses tempt Paris. Rejecting offers from the goddesses Pallas and Herè, Paris accepts Aphrodite's offer of the most beautiful woman in the world, thus deserting his lover, Œnone (A. A. Markley, 'Tennyson', in Norman Vance and Jennifer Wallace [eds.], *The Oxford History of Classical Reception in English Literature*, 5 vols [Oxford: Oxford University Press, 2015], vol. 4, pp. 539–57, on p. 549). See also letter 0743, n. 11.
9. *Pallas*: one of the goddesses in Tennyson's poem, 'Œnone'. Pallas offers Paris 'Self-reverence, self-knowledge, self-control' (p. 142), but Paris rejects this offer (see n. 10). See also letter 0743, n. 11.
10. *Stand punckt*: German for 'standpoint'. Tyndall was responding to Hirst's comment in letter 0791: 'I think you must alter somewhat your "standpoints" when judging our veritable "womankind"'.
11. *I will let it go*: as the remainder of the handwritten letter is missing, it is LT Typescript Only from this point on.
12. *Hamburg*: see letter 0769, n. 4.
13. *your letter*: letter 0791. In this letter, Hirst told Tyndall '<u>I</u> have at present only £20 at your disposal'.
14. *protempore*: 'for the time being' (*OED*).

From Thomas Archer Hirst 11 August 1853 0794

Queenwood College. | Aug^st^ <u>11th</u> 1853

My dear John.

My first question is, have you received the £20 I sent 10 days ago?[1] We have all got well at work and things are in an exceedingly promising condition, there are already upwards of 60 boys here and some have yet to come. To night I give my 3rd Lecture,[2] the first two[3] have satisfied myself and I think every body else. My chief object in writing these hurried lines to day is the following. A

Mrs Graham[4]

formerly sent a boy to Queenwood called Newton,[5] through Playfair's interest, she has to day written again to ask sundry questions previous to sending another of her sons.[6] She intends sending him to Queenwood <u>because</u> (she says) she expects you are here still. When she finds you are not here she will of course hesitate until she knows something about the fellow who has taken your place. Mr Edmondson asked me to answer the letter as a matter of delicacy as well as satisfaction to Mrs Graham it would be best if you would tell Mrs G. <u>your</u> opinion of your successor for she will of course attach most weight to it. Will you do this by return of post?

Edmondson writes today to answer other questions and will of course have to tell her that I have taken your place, no time must be lost therefore. He will also mention the probability of her hearing soon from you.

Hoping your visit to Berlin is proving satisfactory both in a physical and intellectual point of view believe me as ever

Thine affectionately | TA Hirst

The four Cummings[7] have arrived, and are already under my eye. They are fine little fellows all. Yesterday afternoon I was playing cricket with them like any other youngster. My cottage is getting beautified rapidly in a few days I shall live there.

Instead of the above address M[r] E tells me you must write to | M[rs] Graham | 19 Queen's Road | Regents Park[8]

RI MS JT/1/H/181

---

1. *have you . . . 10 days ago?*: see letter 0793, n. 3.
2. *3rd Lecture*: not identified (see n. 3).
3. *the first two*: Hirst's first lecture at Queenwood was delivered on 4 August; he noted in his journal that it 'was a sketch of the subject of Physics, and at the conclusion a word of encouragement to boys and masters' (Hirst, 'Journals', 7 August 1853). In *The Queenwood Observer* of 31 August, this first lecture (which was misdated to 8 August) was noted to be 'an introduction to a course on Natural Philosophy'. After this first lecture, Hirst 'chiefly treated of the general and specific properties of matter, such as its divisibility, compressibility, expansibility, elasticity, density, inertia, malleability, ductility, &c.' in the remainder of August ('Lectures', in P. Grubb and G. Case [eds.], *Queenwood Observer* 5 [31 August 1853], pp. 37–44, on p. 37). Hirst's second lecture is not further identified.
4. *M[rs] Graham*: below her name, Hirst included Mrs Graham's address as 7 West Cliff Gardens, Folkstone—Kent. He crossed this out, however.
5. *Newton*: one of Mrs Graham's sons. Not further identified.
6. *another of her sons*: not identified.
7. *The four Cummings*: brothers who all attend Queenwood College. Three of the brothers are Alexander, George, and Charles Cumming. The fourth appears to be W. Cumming, who gave a paper on the porcupine on 3 December at a meeting of the Mutual Improvement Society at Queenwood (P. Grubb, G. Case, and H. Smithers [eds.], *Queenwood Observer* 9 [15 December 1853], pp. 69–72, on p. 69). Tyndall was fond of the brothers and remained in correspondence with the family after his departure from Queenwood (see letters 0774; 0777; 0778; and 0781).
8. *Regents Park*: a park located in northwest London, England.

To Thomas Archer Hirst 12 August 1853 0795

Berlin 12th. Aug. 1853.

My dear Tom.

I have discussed with Magnus once or twice the feasibility of my going to Paris—

He is going—and I think on the whole it will be most prudent to visit it while he is there. He knows everybody—all the instrument makers &c. So that I cannot hope to reap so much profit from a visit if I go there alone. I shall therefore leave Berlin in 7 or 8 days, so that I may allow myself a week or so on the Harz Mountains[1] breathing the pure fresh air—this I find is very needful for me. I am sometimes discouraged by this head of mine and feel half reconciled to throw this weary weed of a body into its trench and let it rot at once. Sometimes I am strong enough but a days effort of thinking is sufficient to bring me down and to set my head in confusion—the mountain air I think is the only medicine for me, and the effect his journey has had upon Debus, coupled with my own remembrance of what it has done for me heretofore induces me to hope much from it.

> "There is freshness in the mountain air
> And life that bloated ease can never hope to share"[2]

Now I want you to write to me immediately and to let me know Dickinson's address. I want the little fellow to look out for a lodging for me. Perhaps you or Haas would be able to name a hotel where I might remain until I should fix upon a lodging. Dickinson I know will be of great assistance to me. I shall write to you before leaving Berlin and request you to send me some money so that it may meet me at Dickinsons by the time I arrive. But I trust you will send me an immediate reply to this.

I wrote to you a day or two ago but handed the letter[3] to Prof. Helmholtz who is going to England but who wont reach London until Wednesday next. I wrote to Heinrich[4] at the same time—today I received his last letter[5] and a pleasant cheering one it is—It is more than I deserve from him. But Heinrich is kind hearted and now that I am far away from him will look leniently upon the faults and defects of his friend. He knows me well as well almost as you do. indeed were I to judge from his last letter alone he knows me better. I had a letter[6] from Dr. Pritchard[7] of Clapham[8] a few days ago trying to engage me to give his boys[9] two lectures a week—Two sons[10] of Sir John Herschell[11] & one of the Astronomer Royal[12] would be among my audience. I declined it for the present—I want to get some work done and will shut out every thing calculated to divert me from this object.

Faraday writes to me with extreme kindness[13]—it is a pleasure to be connected with him—He is a man one can trust—He is not only a man of science but a man of heart, honesty and truth, which between you and me Tom are better than all science. I hope you wont be offended with the shocking letter[14] I sent you by Helmholtz. Good bye.

Affectionately | John

When you receive this send the first halves of two five pound notes to Marburg wait a few days as before and send the second halves directed to Knoblauch for me so that I may have them on my arrival or better still send me the 1st halves here and the 2nd can go to Marburg.

RI MS JT/1/T/568
RI MS JT/1/HTYP/258–258a

---

1. *Harz Mountains*: mountains located in central Germany.
2. *There is freshness . . . hope to share*: lines from the narrative poem Lord Byron (1788–1824), *Childe Harold's Pilgrimage: A Romaunt* (London: John Murray, 1812), p. 23, I.xxx.349–50.
3. *the letter*: letter 0793.
4. *Heinrich*: Heinrich Debus. Letter missing.
5. *his last letter*: letter missing.
6. *a letter*: letter missing.
7. *Dr. Pritchard*: Charles Pritchard (1808–93), an astronomer and the headmaster of Clapham Grammar School (1834–62), which he founded. In 1849, he joined the Royal Astronomical Society, later becoming the society's president in 1866 (*ODNB*).
8. *Clapham*: Pritchard, who advocated for educational reform and the importance of science education, founded the Clapham Grammar School to put his ideas into practice (*ODNB*; see n. 8).
9. *his boys*: students at Clapham Grammar School (see n. 8).
10. *Two sons*: most likely Sir John Herschel's youngest two sons, Alexander Stewart Herschel (1836–1903) and John Herschel (1837–1921).
11. *Sir John Herschell*: misspelling of Sir John Frederick Herschel.
12. *the Astronomer Royal*: Sir George Biddell Airy (see letter 0698, n. 16) was head of the Royal Observatory Greenwich from 1835 until 1881. Hubert Airy (1838–1903), one of his sons, would have been the correct age to attend grammar school (*ODNB*).
13. *Faraday . . . extreme kindness*: letter 0792.
14. *the shocking letter*: letter 0793.

To Thomas Archer Hirst 13 August [1853] 0796

Universitäts Strasse 4. | 13th Aug.

My dear Tom

I wrote to you yesterday[1] requesting you to send me £10. one half to Berlin and the other half to Marburg. I made this arrangement intending to leave some instruments I had ordered[2] unpaid for: but on second thought I will clear up all my accounts in Berlin before I leave it. Cancel the above arrangement therefore and be kind enough to do as follows. Get those eight pounds from Mr Edmondson[3] add 12 of your own to them and send me the £20 exactly as you did before, by two successive posts.[4] I should be obliged to you if you would do it promptly, as I want to get away from Berlin as soon as I possibly can.

Magnus is off to Paris today.

Yours affectionately | John

Put a clean stamp upon the enclosed[5] and send it to post for me.

Would Mr Edmondson wish me to purchase some diagrams for him? It is on this subject that I am writing to Playfair. If he has 65 boys I would advise him to say yes, if not, no.

RI MS JT/1/T/569
RI MS JT/1/HTYP/260

1. *I wrote . . . yesterday*: see letter 0795, postscript.
2. *some instruments I had ordered*: see letter 0792, n. 2 (see also letters 0788 and 0793).
3. *eight pounds from Mr Edmondson*: see letter 0791, n. 8.
4. *send me . . . successive posts*: see letter 0793, n. 3.
5. *the enclosed*: a letter to Playfair (letter missing), the content of which is noted in the postscript.

To Thomas Archer Hirst 16 August 1853 0797

Berlin 16th. Aug. 1853.

My dear Tom

Hand the accompanying note[1] to Mr Edmondson it will answer his purpose with Mrs Graham, and if any thing similar should occur in future he may if he thinks it worth the trouble of copying make what use he pleases of it. I am gladdened by the tone of your last letter,[2] it smells of contentment. I am glad that you are not above playing with the little fellows. Indeed I knew you were not, but some men Debus for instance though he is an affectionate soul cannot manage to play well with children. I hope you will keep your heart warm and soft towards them. Little irritations may and will occur from time to time and you will hear the complaints of less loving souls but dont let this have any effect upon you—Overpower them by pure good feeling. You will probably do this better than I did it so I need say no more about it. By this time you are aware that the £20 have reached me and that I want a second £20.[3] In about a week I shall be moving—I purpose spending a few days on the Harz[4] If you could expand Mr Edmondson so as to enlarge his feelings towards the boys you would accomplish a great feat. Could you inspire him with the wish to please them rather than to please himself it would be a great thing—could you make him feel that the reward of his labour at Queenwood must in a great measure be the success of his labour, and not exactly the enjoyments which his income enables him to procure, you would do a great thing. You have a lever in the idea which he now evidently entertains and that is to leave Queenwood to his son.[5] Two years ago he had no thought of this kind but now he has. Queenwood was then dying (33 boys) but it is now prosperous and you can work upon his fatherly feeling to induce him to take an active part in its advancement. If you find him sensual tell him so. If you find him doing anything calculated to lessen his power either among boys or teachers tell him so. If you find him talking too much—which is a fearful peril—tell him so. If his feelings and the interests of Queenwood ever clash dont spare the former. The effect on them will be transient on Queenwood perhaps permanent. This morning for the first time I had a plunge into the Spree.[6] the proprietor[7] at first was hardly willing to let me in as he was not sure that I could swim. He made me half afraid. It's so long since I swam, I thought it possible that I might have forgotten the way: but in I went and the water bent beneath me like an elastic spring and I rode the Spree like a water Nimrod.[8]

Good bye | As ever yours | John

RI MS JT/1/T/570
RI MS JT/1/HTYP/261

1. *the accompanying note*: letter missing. Hirst had asked Tyndall to write a note to Mrs. Graham, who wished to send another one of her sons to Queenwood due to her high regard for Tyndall, in order to vouch for his successor (see letter 0794).
2. *your last letter*: letter 0794.
3. *By this time . . . a second £20*: see letter 0796, n. 4.
4. *the Harz*: the Harz Mountains (see letter 0795, n. 1).

5. *his son*: George William Edmondson.
6. *the Spree*: the Spree River, which flows through northeastern Germany and cuts through Berlin.
7. *the proprietor*: not identified.
8. *Nimrod*: a skilled hunter, or a person who enjoys hunting. The word also functions as a biblical allusion to Nimrod the hunter and great grandson of Noah; see Genesis 10:8–9 (*OED*).

From Thomas Archer Hirst 16 [August] 1853[1] 0798

Queenwood College. | July 16th 1853

My dear John.

I had somewhat anxiously expected the letter[2] from you which arrived only this morning. It is just as I expected you have not used Berlin medicinally and I am well pleased to see you have the courage to leave it and try to strengthen yourself.[3] If you did not do so it would be a neglect of duty for which no scientific discovery could, in my eyes, compensate. The letter which is to offend me[4] has not yet arrived. I defy it to do so however. For I never consider myself answerable for my past letters, they were true at the moment when written, that I can answer for, but to day I should never write the same. I may not understand you as you say, but if so it is in points of some intricacy that with others would never come into discussion. One thing satisfies me I understand a vast deal of thee & thank God for it, if I misunderstand the rest, I will patiently submit to a thrashing for it.

If there be a duty for you to fulfil at present it is to get strong and healthy again, and keep from you all discouraging notions of your body's weakness. Your body would with justice at all times reply that it has been badly used. I don't care to tell you that Health is better than all Honours though I believe it to be, but I may tell you, that sickness will react upon your work fearfully, and as a matter of pure prudence your duty is to give yourself a free fling and relaxation. Du willst es thun nicht wahr![5] First of all then Dickinson is not in Paris, he is here in England on a visit. If you stop a month in Paris you had better go at once to the Hotel Corneille[6] close by the Odeon,[7] it is now vacation and they are sure to have room, you will find it reasonable and convenient. If you do not like that you might stop a few days at the Hotel de Normandie,[8] Rue St Honorè[9] until you could meet with lodgings to suit you. Near the Odeon I can tell you a place where you may get a good plain dinner at a reasonable cost. It is the corner house[10] facing into the Streets Mr Le Prince[11] and Dubois.[12] From the Odeon you pass down the Rue Voltaire[13] until it joins the Rue de Mr Le Prince and you will see the corner house immediately opposite you. If you see Verdet and Ebersheim[14] remember me kindly to both, tell the former I was much disappointed at not seeing him before I left.

I have heard from duBois[15] inviting me to step up to London and see him and his wife. I think I shall.

There are some 65 or 66 boys here now and still more coming, we are beginning to find it difficult to accommodate all at meal times. I have some fine lads to teach and daily I love my task more for daily I feel my influence upon them increases. This I know you will be pleased to hear, and I hope we shall not only retain all we have got but next half year increase our number, we will see.

Enclosed you will find the £15[16] it will go somewhat further than £10 and save the trouble of sending again for I should like you to run about and spend it all. The dinner bell rings. Good bye

Thine affectionately | TA Hirst

There are 3 letters[17] here for you. Two are unmistakeably circulars which I shall keep. The third may be something else I shall enclose it with the next half notes.

RI MS JT/1/H/179

1. *16 [August] 1853*: while Hirst misdated this letter as July, it was clearly written in August as it directly responds to Tyndall's comment that he will soon leave Berlin and references Tyndall's warning that Hirst is to receive a 'shocking letter' (letter 0795, n. 14; also see letter 0793).
2. *the letter*: letter 0795.
3. *well pleased . . . strengthen yourself*: Tyndall was planning to retreat to the Harz Mountains for some fresh air (see letter 0795, n. 1).
4. *The letter . . . offend me*: letter 0793.
5. *Du willst es thun nicht wahr!*: a German statement, serving as both a request and an order, translates as 'You will do that, won't you?' It refers to Hirst's insistence that Tyndall get some rest.
6. *Hotel Corneille*: see letter 0769, n. 3.
7. *the Odeon*: L'Odéon, one of France's national theatres, located in Paris. Designed by the architects Marie-Joseph Peyre (1730–85) and Charles de Wailly (1730–98), the theatre opened in 1782 and was originally built to house the Comédie-Française (F. H. Londré, *The History of World Theatre: From the English Restoration to the Present* [New York: Continuum, 1999], pp. 117–19).
8. *Hotel de Normandie*: a well-known hotel in Paris.
9. *Rue St Honorè*: Rue Saint Honoré, a street in Paris located near the Seine River.
10. *the corner house*: not identified.
11. *M^r Le Prince*: Rue Monsieur le Prince, a street in Paris located near the Odéon Theatre.
12. *Dubois*: Rue Antoine Dubois, a street in Paris connected to Rue Monsieur le Prince.

13. *Rue Voltaire*: Rue de Voltaire, a street in Paris that connected the Odéon Theatre to Rue Monsieur le Prince.
14. *Ebersheim*: not identified.
15. *duBois*: Emil du Bois-Reymond.
16. *Enclosed . . . find the £15*: although Tyndall only requested £10 to be sent in two portions of £5 (see letter 0795), Hirst sent him £15 all at once. At this point, Hirst had not yet received letter 0796, which was a request from Tyndall to instead send £20 through two successive posts.
17. *3 letters*: not identified.

From Thomas Archer Hirst 17 August 1853 0799

Queenwood College. | Augst 17th/53

My dear John,

Yesterday I sent you the 1st halves of three £5 notes;[1] today having received your second letter[2] I send the remaining halves of the above, and with them I risk the safety of a whole £5 note, making in all £20 which I place to our account. I have received your balance £8. 1. 2 just now from Mr Wright, and therefore Mr. E.[3] and you are straight. I have yet a £10 note at your disposal but after that no more until Octr 1st when you can have about £20 and all this without a moments inconvenience. Should you require more however for instruments I can easily provide you with it. Mr Edmondson says that as we have now more than 65 boys you may buy the diagrams of which you speak,[4] and if they are such as to be of immediate service Schneider will send me them directly with other books he has for me. Mrs Graham and her son[5] are expected tomorrow. I hope before she comes she will have received your letter[6] about which I wrote you a week ago.

Playfair's letter[7] with a 'clean' stamp will be sent by this day's post but before it reaches him I am afraid the Post office authorities will dirty the stamp.

As this letter is risky I shall be anxious until I hear of its safe arrival.

And now go and get strong, with my blessing.

T.A. Hirst.

RI MS JT/1/H/182
RI MS JT/1/HTYP/262

---

1. *Yesterday I sent . . . £5 notes*: these notes were sent with letter 0798.
2. *your second letter*: letter 0796, in which Tyndall had requested that Hirst send him more money (see n. 4).
3. *Mr. E*: George Edmondson.

4. *the diagrams of which you speak*: Tyndall was wondering whether Edmondson wanted him to purchase more diagrams for Queenwood (see letter 0796).
5. *her son*: not identified. See letter 0794.
6. *your letter*: see letter 0797, n. 1. Hirst had requested (see letter 0794) a letter from Tyndall to be sent to Mrs. Graham, which Tyndall enclosed with letter 0797.
7. *Playfair's letter*: see letter 0796, n. 5, in which Tyndall enclosed a note he had written to Playfair. He had requested Hirst to send it for him.

To Thomas Archer Hirst [17, 19, 20 August 1853][1] 0800

My dear Tom.

I have worked pretty hard and pretty successfully today. I have had my supper, it is half past 9 oC. and I am afraid to work after supper lest I should petrify my bowels, so instead of working I take up my pen to play and amuse myself by writing to you. You are aware that one of the little jobs I had set before myself was an article on du Bois researches for one of the quarterly reviews:[2]—The writing of such an article is a horrible undertaking for a conscientious man. You have to read an immense lot and to digest it all and try to understand it if possible better than the writers. I have waded through these two mortal volumes of du Bois,[3] and through sundry papers of Matteucci's[4] spurred on by an anxiety to do no man an injustice. When I look at the volumes that have been written upon this subject, on the amount of real talent that has been expended upon it, an indescribable mixture of the solemn and the ludicrous fills my heart, I could sometimes sigh, sometimes swear, sometimes burst with laughter. If du Bois knew what was passing in my mind I believe he would think it doing God service to assassinate me. I came to Berlin a believer in animal electricity. I have looked at it, pondered it, I think mastered it, and my connection is that this thing about which du Bois and Matteucci are ready to crack each other's sculls,[5] on which Helmholtz and fifty others have edified the world is a pure non-entity, the child of misconception, the offspring of genius and ambition, the result of too little and of too much talent of too deep an insight and too shallow an insight, the result of a longing after nature's secrets, unchecked by an humble wish to accept her admonitions whatever they might be. In short I believe it to be the battling of ten thousand conflicting forces which belong to the greatest and to the littlest of men. I have never I think seen real talent so blindfolded as here. I have never before witnessed such an example of the lengths of error to which a single misconception can lead a man. If a man be once false to nature he appears to be afflicted with a judicial blindness in punishment of his crime. The sin against the holy ghost[6] was something of this nature. A man to be a true physicist must be capable of renunciations, tho' strong in will as an oak he must bend like a reed[7] before the revelations of that power whose laws he would investigate. But you will think that I am raving and I am willing that on the present you should believe with the crowd against me. I prophesy however that the fabric of Animal Electricity as it at present stands will share the fate of solomon's temple.[8] The ploughshare will pass over it. I began a believer, I end an infidel; I believed in authority, I doubt on convictions—I wonder if a man gazed steadfastly at the law of gravitation[9] would it stand his gaze? It is now a quarter past 10 oC and I will go to bed.

17th. Aug. 53. John

19th Aug. 1853

I must have been drunk when I wrote the above. there is a good deal of truth in it but it is so wildly expressed. dont mention it to mortal man. Your sweet letter[10] has just reached me Tom. It makes me like you better than ever. I never knew a truly noble fellow who could not love children and make himself at home amongst them. Its the best test I know of a real healthy and expansive soul. If I hear a teacher complaining of his profession I hear the words of a defective man. I knew you would like it boy. Something told me you would, *[perhaps]* it is all an arrangement of the Gods to make your pleasure and Queenwoods interests thus harmonize. I shall often come down to see you, when choked with the soot of London and bewildered by Science I will take a run down to you and cleanse my lungs and my affections at the same time. I will not neglect your advice as regards my health; it would be sheer madness. I know I shall soon get strong when I break away from all temptations to work. I have had most strange experiences of late. What Doctors call nervous attacks—Just imagine me getting out of bed to block up my bedroom door with chairs lest I should get up in my sleep and jump out of a window some 30 feet high. Yesterday I worked hard and eat nothing up to 2 or 3 oC. I felt a strange headache, and in the afternoon got on the top of an omnibus drove to Charlottenburg[11] and walked home again. The head ache stuck to me for a long time but at length it departed. It is now Friday. I shall leave on Sunday or Monday, travel to Halberstadt[12] and thence over the Harz[13] to Cassel.[14]

Do take a run up and see du Bois.[15] He is a profound fellow, an intellect as keen and cutting as the sabre of Saladin[16] and as eloquent as an orator of the forum.[17] Altogether a wonderful fellow, and the most instructive man that I have met. There is a great gulf between him and me on many points; but this does not prevent me from recognizing the manly proportions of his mind. He is a child of this world; prepared to battle in it like a man and keenly alive to its rewards and its inflections. In my eyes he would be nobler if he could realize at times a feeling of superiority to wordly chances, but this I believe he never does. He is a splendid animal—his bowels are always free and his head clear. What his feelings will be if he finds his animal electricity in peril I cannot imagine—in fact I dont like to imagine it. I wish I could undertake an investigation with him. I wish he could bring himself to look at least at the possibility of his being mistaken and to meet the question at all hazards. But when I look at these two terrible volumes[18] the sacrifice strikes me as an immense one. I have no doubt that he will be able to maintain his ground against all gainsayers, that is, nobody will be able to prove him absolutely wrong, but I do believe that a sufficient cause will be assigned for all his phenomena without ever calling in the aid of animal electricity. I trust this subject will not haunt me on the Harz. dont mention the matter at all to du Bois, only remember me to him most kindly.

65 or 66 boys and more to come! that is a trumpet note of triumph—I know you will keep them boy—I know that Queenwood will be filled—I said 12 months ago to a man in Salisbury[19] the time will come when people will think themselves happy in finding a place for their children at Queenwood—It has almost come already. God speed you and I need not tell you the best way to make him 'Speed' that is to help him.

'our valours are our best gods'[20]

night—

Since writing the last 4 pages I have been out. The nerves were at work and I took a cigar in my pocket and walked towards the Thier Garten.[21] I saw a number of fire-men pushing on in a certain direction as if they were in earnest about something. Shortly afterwards a heavy engine rattled by me. I saw at a distance the smoke of the fire they were going to quench, but walked listlessly on—I paused, and said mentally—here is something that will keep your brain from eating itself for one half hour. I went to the fire—a number of men were already upon the roof of the high house smashing down the flags, at intervals the heavy roll of fresh engines sounded like the rumble of cannon upon the pavement—the clouds of smoke sometimes enveloped the workmen—sometimes drew aside and let me see the brave fellows at work. I had often met them with their slouched leather hats and polished pickaxes and thought the latter more for ornament than use but here it was brought into play in good earnest, the spray of the hose split the smoke and showers were thrown upon the burning mass—man at length conquered and fire yielded. I turned towards the Thier Garten once more, lighted my cigar and at a distance observed illumination—It was Kroll's concert[22] in the public garden. I stalked towards it for the nerves had begun to work again; paid my 5 silbergroschen[23] and went in, had a glass of Weiss bier,[24] heard the music and enjoyed it. The moving crowd, the flicker of the lamps, the melody, the bright eyes and cheeks like the flush of roses, all had a medicinal effect—I went to see some officers shoot at a target—various instruments meant for proving strength were placed around: I tried my tragkraft[25] and my Shlagkraft[26] and found myself in each case above the average—so that whatever may be the state of the nerves the muscles have not yet lost their force. I returned very deeply impressed with the philosophy of folly. I felt improved and now trust the gods will grant me a good night's rest. | 19th Aug 1853.

You defend woman against me,[27] my dear boy if you knew my feelings towards a fair sweet face you would not deem this necessary. Man is a religious animal, a social animal a chivalrous animal, a loving animal. He is all in turn and if he were not all but one he would no longer be man but something else. You ought to know better than to take the affirmation of one of his moods for the totality of your friend John. Why man I caught myself singing 'love, young dream'[28] all unconsciously this morning. Sometimes a woman is a toy to me, sometimes a goddess, but I must stop or the nerves will begin once more.

10 oC. good night.

Remember me to Dr. Fox.[29] If he wants the model of an ear let him write to me to the care of Knoblauch Marburg—I shall be there in a week.

Saturday.

No historian ought to undertake to write the life of Chatham[30] without having first had an attack of the nerves. What solution can a fellow offer who has not made this experiment. How can the strong man's tears and womanly weakness be accounted for? There is a great deal of nonsense in the world through talented people writing about what they themselves have never experienced. I plunged three times into the Spree[31] today and bought a knapsack. —I leave it to you to connect the facts.

I gave your letter[32] to Schneider—The money has all come safe. It is plenty for the present. The diagrams[33] are not here but in Paris. I shall be off tomorrow or next day—The last letter[34] containing the notes has not been 10 minutes in my possession so you cant accuse me of delay.

ever your old friend | John.

RI MS JT/1/T/914
RI MS JT/1/HTYP/263

1. *[17, 19, 20 August 1853]*: Tyndall wrote this letter over three days, as noted within the text of the letter.
2. *article on du Bois . . . quarterly reviews*: Tyndall's review of Emil du Bois-Reymond's work on animal electricity would be published in January 1854 (see letters 0754, n. 9 and 0780, n. 5).
3. *two mortal volumes of du Bois*: most likely E. du Bois-Reymond, *Untersuchungen über Tierische Electricität*, 2 vols (Berlin: G. Reimer, 1848–49). Volume 1 was published in 1848; the first half of volume 2 was published in 1849. Du Bois-Reymond did not complete the second half of volume two until 1860 (Finkelstein, *Emil du Bois-Reymond* p. 89).

4. *papers of Matteucci's*: for a list of Matteucci's publications see 'Matteucci, Carlo', *RS Catalogue*, vol. 4, pp. 285–93.
5. *this thing . . . each other's sculls*: Matteucci and du Bois-Reymond were engaged in a heated public conflict over their views on animal electricity (see letters 0724; 0730; 0732; 0737; 0744; 0754; and 0780).
6. *The sin . . . holy ghost*: blasphemy, or the willful denial of divine truth (Mark 3:28–30).
7. *Tho' strong . . . like a reed*: an allusion to 'The Oak and the Reed' in *Aesop's Fables*, which teaches that it may be better to bend like a reed than stand sturdy like an oak in the face of powerful external circumstances (S. Croxall, *The Fables of Aesop; With Instructive Applications* [Halifax: William Milner, 1843], p. 74–75).
8. *the fate of Solomon's Temple*: according to the Old Testament, the Temple was built by King Solomon during the tenth century BCE, and then destroyed during the sixth century BCE when the Babylonians under Nebuchadnezzar besieged Jerusalem and burned it down.
9. *law of gravitation*: Newton's law of universal gravitation, which states that 'the attractive force of bodies varies directly as their masses and inversely as the square of the distance between them' (*OED*)
10. *Your sweet letter*: letter 0798.
11. *Charlottenburg*: a locality of Berlin, home to the famous Charlottenburg Palace.
12. *Halberstadt*: a town in Germany located near the Harz Mountains.
13. *the Harz*: the Harz Mountains (see letter 0795, n. 1).
14. *Cassel*: a city in central Germany located on the Fulda River.
15. *see du Bois*: Emil du Bois-Reymond had invited Hirst to visit him in London (see letter 0798, n. 15). Du Bois-Reymond had travelled to England to be married (see letter 0780, n. 18).
16. *the sabre of Saladin*: the sabre belonging to the Muslim hero Saladin (1138–93), who unified religious factions during a period of Muslim disunity, and captured Jerusalem during the Crusades. The sabre was found in Saladin's tomb, and has received much popular and scholarly attention (N. Chase, 'Saladin', in F. N. Magill [ed.], *Dictionary of World Biography: The Middle Ages* [New York: Routledge, 1998], vol. 2, pp. 809–11).
17. *the forum*: likely an allusion to the Roman Forum, a site of various activities such as public speeches, elections, communal feasting, and religious sacrifices in Ancient Rome (D. Watkin, *The Roman Forum* [Cambridge: Harvard University Press, 2009], pp. 11–16).
18. *these two terrible volumes*: see n. 3.
19. *a man in Salisbury*: possibly James Bevington (see letters 0681 and 0691; see also Journal, 19 November 1852, JT/2/13b/589).
20. *'our valours are our best gods'*: line spoken by Caratach in John Fletcher's 1613 play, *Bonduca: A Tragedy*.
21. *Thier Garten*: see letter 0793, n. 7.
22. *Kroll's concert*: a concert held by the Kroll Opera House in Berlin. Otherwise not identified.
23. *silbergroschen*: see the 'Note on Money'.
24. *Weiss bier*: German, 'white beer', brewed from wheat (Weizen).

25. *tragkraft*: German, 'capacity' (weightlifting).
26. *Shlagkraft*: misspelling of Schlagkraft; German, 'clout' (punching or pushing).
27. *you defend woman against me*: see letter 0791 in which Hirst suggested that Tyndall needed to alter his views when it comes 'to judging our veritable "womankind."' (See also letter 0784.)
28. *love, young dream*: 'Love's Young Dream', an 1811 song by the Irish poet and songwriter, Thomas Moore (1779–1852).
29. *D^r. Fox*: Luther Owen Fox (1812–79), the surgeon in Broughton who attended to the pupils and staff at Queenwood College. Tyndall often visited and dined with him.
30. *Chatham*: likely William Pitt (1708–78), the 1st Earl of Chatham; Pitt suffered from nervous and psychological disorders (*ODNB*).
31. *the Spree*: see letter 0797, n. 6.
32. *your letter*: letter 0799.
33. *The diagrams*: see letters 0796 and 0799, n. 4.
34. *The last letter*: see n. 32.

To Rudolf Clausius 21 August 1853 0801

Berlin 21st Aug. 1853

Dear Dr. Clausius

I am very much obliged to you for your kind, prompt and useful letter.[1] I am not without a hope that the instrument[2] may be so modified as to come very near to the conditions you mention, however my present object is simply to detect qualitative differences and when I come to quantitative matters I hope to be able either to change it or to invent something else. Helmholtz was here sometime ago we had a pleasant walk together through Zoological Garden.[3] I wish you had been here as I would very much like to hear you converse together über die Erhaltung der Kraft[4] I think he seems inclined to adopt Thomson's notion with regard to heat developed by friction when steam rushes out of a high pressure boiler.[5] You and Rankine[6] have found that if steam be permitted to exhaust a portion will be precipitated.[7] Helmholtz seems to think the heat developed by friction sufficient to prevent this precipitation and thus keep the steam dry.[8] He does not see exactly how you can make out a loss of heat here, as stated in your paper.[9] He also if I understand him aright thinks that it is not die Bewegungs große welche constant bleibt während die Lebendigekraft sich ändert sondern umgekehrt.[10] I wished very much that you were here as I know I should have derived great profit from listening to you converse on the subject.

Tomorrow I leave Berlin my purpose making a foot journey through the Harz.[11] This is necessary for my health which is not at all so strong as I should like it to be.[12] In about a week I hope to be at Marburg and I proceed thence to Paris. Magnus is already there. Indeed his being there has caused me to leave Berlin sooner than I otherwise should have done. He knows all the instrument makers &c. and it will therefore be pleasant to be there along with him.

I shall always be very glad to hear from you whenever you can spare time to favour me with a letter.

Believe me | most sincerely yours | John Tyndall

RI MS JT/1/T/170

---

1. *your kind . . . useful letter*: letter missing.
2. *the instrument*: not identified.
3. *Zoological Garden*: the Berlin Zoological Garden, first opened in 1844 and located in the Tiergarten.
4. *über die Erhaltung der Kraft*: German, 'on the conservation of force'. Also the title of Helmoltz's *Über die Erhaltung der Kraft: eine physikalische Abhandlung, vorgetragen in der Sitzung der physikalischen Gesellschaft zu Berlin am 23sten Juli 1847* (Berlin: Druck und Verlag von G. Reimer, 1847).
5. *Thomson's notion . . . high pressure boiler*: see W. Thomson, 'On a remarkable property of Steam connected with the Theory of the Steam-Engine', *Phil. Mag.* 37:251 (November 1850), pp. 386–89. In this letter originally written to James P. Joule (1818–89), Thomson claimed that steam passing through a high-pressure boiler heats up through friction, thus supporting Joule's mechanical theory of heat. Tyndall had an interest in steam boilers, as seen in his discussion in March and April 1854 with Robert Bunsen (see letter 0751, n. 5).
6. *Rankine*: William John Macquorn Rankine (1846–1922), a Scottish physicist and civil engineer notable for his work on thermodynamics, who was elected to the Royal Society of London in June 1853.
7. *You and Rankine . . . be precipitated*: see R. Clausius, 'On the Moving Force of Heat, and the Laws regarding the Nature of Heat itself which are deducible therefrom', *Phil. Mag.* 2:8 (July 1851), pp. 1–21; and W. Rankine, 'On the Mechanical Action of Heat, especially in Gases and Vapours', *Trans. Roy. Soc. Edin.* 20:1 (1853), pp. 147–90. This lecture was given in 1850, but was not published until 1853.
8. *Helmholtz . . . steam dry*: not identified.
9. *your paper*: see R. Clausius, 'On the Deportment of Vapour during its Expansion under different circumstances', *Phil. Mag.* 1:5 (May 1851), pp. 398–405.
10. *die Bewegungs große . . . sondern umgekehrt*: German, 'the momentum which remains constant while the kinetic energy changes but rather the other way round'.
11. *the Harz*: the Harz Mountains (see letter 0795, n. 1).
12. *my health . . . to be*: Tyndall was experiencing poor health, and recently had nervous attacks (see letter 0800).

To Thomas Archer Hirst [8 September 1853][1] 0802

Cite d'Autin N°. 3 | I believe it is Thursday

My dear Tom

Reached Paris on Sunday night[2] and from that time to the present, cannot say that I have had a tranquil breathing hour. I have just purchased some pens and paper and the first use I make of them is to write to you. I am tired, for I have been walking all day and it is now nearly 50C. therefore dont expect a long letter, but if you expect any thing expect a stupid letter. I hoped to hear from you in Marburg but no letter came. I spent 3 all pleasant days there just I was about to disappear Dove made his appearance and this induced me to extend my stay a day longer than I intended. Knoblauch has got a little son,[3] he made his debut in this world two days before I reached Marburg.[4] I made a gallant Journey through the Harz.[5] on the evening of the second day[6] I met a Konigsberg student[7] who proposed that we should travel together—after a little reflection I made up my mind to try him for a day. I found him good natured, talkative and on the whole found that I could travel with him without any trouble either to body or mind—we fuelled on harmoniously for 5 days and parted at the top of the Brocken.[8] The journey was of immense service to me—it restored my former self, and at the end of it I found that I could walk my companion off his legs. I am now engaged running about Paris making the most of Magnus while he remains here.[9] We have made many instructive visits together. I can conceive nothing more absurd and deplorable than condemnation to a land where you are a stranger to the language. I read French as easily if not more easily than German but the speaking of it has been a terrible infliction on me, if I commenced a sentence in French I was sure to end it in German. But I am getting better, my chains are getting loose. It is infamous to be so confined at my age, a child can wait and have patience, but I find that to succeed I must become as in other cases a little child here also. I have bought a play and I will go to the theatre every night till I can understand every word that is spoken on the stage.[10] Verdet I have seen, but next week I promise myself more of his company. I shall probably remain here for a month and then go to England travelling probably over Southampton. If you can afford me a bed I will stop a day or two. Kind greetings to Heinrich[11] "no more at present"[12] from

Your affectionate John

Seal the enclosed,[13] put a stamp on it and post it.

RI MS JT/1/T/921

1. *[8 September 1853]*: Tyndall arrived at Paris on the evening of Sunday, 4 September (Journal, JT/2/13b/622). The Thursday following his arrival was 8 September; Tyndall mentioned his purchase of a play and his intent to 'convert it into a lesson' (Journal, 8 September, JT/2/13b/623).
2. *Reached Paris . . . Sunday night*: see n. 1. After leaving Berlin (see letters 0784 and 0786, n. 1), Tyndall travelled through the Harz Mountains and stopped in Marburg before arriving in Paris (see Journal, 22 August–4 September, JT/2/13b/615–22).
3. *a little son*: not identified.
4. *two days . . . reached Marburg*: Tyndall reached Marburg on 29 August (Journal, JT/2/13b/620), which means Knoblauch's son was born on 27 August 1853.
5. *the Harz*: the Harz Mountains (see letter 0795, n. 1). Tyndall was experiencing health problems and believed a journey through the Harz would be beneficial; he discusses his journey heavily in his journal (Journal, JT/2/13b/615–20).
6. *on the . . . second day*: see Journal, 23 August, JT/2/13b/616.
7. *a Konigsberg student*: a student at the University of Königsberg, Germany. Not further identified.
8. *the Brocken*: the highest peak of the Harz Mountains; see n. 6.
9. *running . . . he remains here*: Tyndall visited Paris to see Magnus (see letters 0786; 0795; and 0801).
10. *I have . . . the stage*: see n. 1.
11. *Heinrich*: Heinrich Debus.
12. *no more at present*: not identified.
13. *Seal the enclosed*: not identified.

To Thomas Archer Hirst [14 and 16 September 1853][1] 0803

3 Cite D'Autin Wednesday

My dear Tom

At this moment you are preparing for Salisbury.[2] I wish you a strong voice and sufficient obstinacy of heart to get through your task manfully. I felt a little angry with Edmondson for asking you to engage in such a thing so soon, but this feeling vanished in the reflection that come what may you will be the gainer, even should you fail to satisfy yourself. I have derived profit from my experience in the same line though these experiences were not always agreeable. I regard the matter as an experiment upon yourself and experiments you know are not always successful; but this should not deter us from making them, for without experiments we should be without results. Even should you fail, which I confess I do not fear, you must put the fact behind you. It must not damp you and discourage you but whatever dissatisfaction you may feel must be converted into fuel for future and victorious efforts.

My stay in Paris will not be long. I will cut it shorter than I at first intended.

Magnus is off to Berlin this morning.[3] I will start on Monday or Tuesday. Many of the Savants[4] are absent but I have also seen a good many. I have met your friend Verdet twice and have agreed to breakfast with him on Friday morning. I see Wertheim very often. Magnus he and I dined together at Saint Germain[5] on Sunday. It was a glorious day and the look out from the terrace was magnificent. The sun flooded the landscape with gold and kissed tenderly the topmost branches of the chesnut as he sank below the horizon. Monday evening I spent at Soleils[6] and saw some splendid experiments. I made the acquaintance of Pasteur Moigno[7] and others there. I have been through the Cabinet of the College de France[8] with Regnault & Lenarmont.[9] today I dine with Despretz[10] at a hotel[11] in the city. If I remained I might make many more acquaintances, but I have got almost as much as I can at present digest. So some time next week I hope to see you and Heinrich[12] and take a look at Queenwood in its present vigor. At first I was almost bewildered by the magnificence of the apparatus in the various cabinets, but I soon mastered this feeling and fell back upon my old conviction that the man himself is the principal thing. My health still continues good and I long to turn it to some account. You have seen Dubois.[13] I should not be surprised to find that I have made him my enemy. I gave him high praise in that article that I wrote in Berlin;[14] but my friendship for him could not of course induce me to commit myself so far as to defend his views.[15] At the end of the article I expressed a wish that the matter might be placed upon surer foundations than it possesses at present,[16] and I should not wonder if the doubt I expressed excited the hostility of Dubois.

I should like to have been present at the reading of Brewster's paper,[17] but feel very tranquil about it. There is something sweet and pacifying in the conviction that a man stands upon the truth of nature. This is a rock against which the efforts of man will break in vain. Matteucci, I have been told, has attacked me here and Moigno has defended me. I know nothing further of the matter than what Moigno himself told me, and I do not trouble myself about it. These wrestlings must be left to the decision of time and I have patience enough to await this decision. I made an attempt today to find some diagrams for Queenwood[18] but failed. I know there are such but it is difficult to find where. It is good and wholesome at times to plunge into the world and meet your fellows face to face. I have done so, and now I long to withdraw myself and listen alone to the whisperings of nature, for this is the grand fountain of human power. I must try and shut out the world for a time and get some work done. I am afraid however my mission for some time will be rather destructive than productive. this animal electricity falls in with my experiments upon molecular influences[19] and I do not think it will stand the test *[5 words illeg]*. If it does I shall be its loyal defender, for my desire is even in the demolition of error is to arrive at truth.

Friday

I spent a pleasant evening with Despretz, we dined alone had a small glass of burnt brandy at a coffee house[20] and roamed through the streets for an hour or two. On my return home I found a note from Bence Jones[21] waiting for me, he had arrived at Paris. I found him at his hotel. He is gone to Vichy[22] to inspect the baths and will be back tonight. He wants me to go to Folkstone[23] and spend a week with him there. the beautifulness of the place and the sea bathing are both temptations; but my strongest inducement is the fact of his having repeatedly asked me to his house in London, whereas all my visits there have been flying ones. I fear he will conclude that there is some private reason which prevent me from accepting his invitations. I shall however endeavour to get to Queenwood. I wish you would get that large box of mine nailed up for me. and if you think it too unwieldly get it cut in two. do this for me like a good fellow.

Heinrich's warm hearted letter [24] is in my pocket. I will soon either write to him or stand before him face to face.

as ever | Your affectionate | John

Seal the enclosed[25] and send it to the post for me.

RI MS JT/1/T/922
RI MS JT/1/HTYP/267–268

---

1. *[14 and 16 September] 1853*: LT dated this letter to 14 September. However, part of it was written on a Friday, likely 16 September. Tyndall did not make any journal entries after 9 September until his last day in Paris (Journal, JT/2/13b/623). In this letter, he mentions that he will leave Paris on Monday or Tuesday, likely 20 or 21 September. As letter 0802 was written on a Thursday, likely 9 September, and details his arrival in Paris, it is unlikely that this letter was also written at the same time, dating it to the following week.
2. *preparing for Salisbury*: Hirst gave a lecture at Salisbury on Electricity and Magnetism on 14 September (see Hirst, 'Journals,' 11 and 18 September 1853).
3. *Magnus . . . Berlin this morning*: Tyndall likely decided to leave Paris early because Magnus was leaving for Berlin (see letter 0802, n. 9).
4. *Savants*: learned or scholarly individuals highly distinguished in their fields (*OED*).
5. *Saint Germain*: Saint-Germain-des-Prés, a neighbourhood in Paris known for attracting intellectuals.
6. *Soleils*: the workshop of Henri Soleil (d. 1879), son of French optical instrument maker Jean-Baptiste François Soleil (1798–1878). Soleil was a producer of scientific and optical instruments, and published a number of articles on crystal optics during the 1850s and 60s (P. Brenni, '19th-Century French Scientific Instrument Makers XIII: Soleil, Duboscq, and Their Successors', *Bulletin of the Scientific Instrument Society* 51 [1996], pp. 7–16).
7. *Pasteur Moigno*: François-Napoléon-Marie Moigno.

8. *College de France*: a well-known institution of research and higher education located in Paris, France.
9. *Lenarmont*: most likely M de Lenarmont, who performed experiments on conduction (Jackson, 'Tyndall and Diamagnetism', p. 23, n. 154).
10. *Despretz*: César-Mansuète Despretz (1791–1863), a physicist and chemist notable for his work on thermodynamics, heat transfer, and fluids (J. Wisniak, 'César-Mansuéte Despretz', *Revista CENIC Ciencias Químicas* 40:1 [2009], pp. 47–55).
11. *a hotel*: not identified.
12. *Heinrich*: Heinrich Debus.
13. *Dubois*: Emil du Bois-Reymond.
14. *that article . . . in Berlin*: see letters 0754, n. 9; 0780, n. 5; and 0800, n. 2.
15. *my friendship . . . defend his views*: Tyndall discussed his stance on the dispute between du Bois-Reymond and Matteucci in his journal entry of 21 August (Journal, JT/2/13b/613–14) (see also letter 0800).
16. *At the end . . . at present*: near the end of his article on animal electricity, Tyndall noted the 'extreme caution necessary to the successful prosecution of researches upon this subject' (J. Tyndall, 'Rise, Progress, and Present Condition of Animal Electricity', p. 107; see letter 0754, n. 9 for full reference).
17. *Brewster's paper*: not identified. It is possible the paper was read but not published.
18. *diagrams for Queenwood*: see letters 0796; 0799, n. 4; and 0800, n. 33.
19. *my experiments . . . molecular influences*: see J. Tyndall, 'On Molecular Influences, Part I: Transmission of Heat through Organic Structures', *Phil. Mag.* 6:37 (1853), pp. 121–38; and J. Tyndall, 'On the nature of the force by which bodies are repelled from the poles of a magnet; to which is prefixed, an account of some experiments on molecular influences', *Phil. Mag.* 10:66 (1855), pp. 257–90.
20. *a coffee house*: not identified.
21. *a note from Bence Jones*: letter missing.
22. *Vichy*: a city in central France, known for its thermal springs.
23. *Folkstone*: see letter 0780, n. 19. Rather than spending a week at Folkestone, Tyndall noted in his journal entry of 30 September that he only spent two days (JT/2/13b/624).
24. *Heinrich's warm hearted letter*: letter missing.
25. *the enclosed*: not identified.

To Thomas Archer Hirst [3 October 1853][1] 0804

London Monday.

My dear Tom.

The best plan will be to send my boxes to the Royal Institution Albermarle St. put a rope round the red one as it appears to be rather shakey. I am almost resolved on taking Francis's old lodgings[2]—they are the most airy that I have met and the only objection is that they are too distant from the Institution this perhaps has its advantages also. at all events I will try the lodgings for a couple of months and if I find them inconvenient I can move. After my arrival in London I regretted that I did not take Mr Edmondson's advice and remain at Queenwood until my cold had disappeared[3]—the damp fog brought it back again but it is now gone and I feel strong and active, able and anxious to get to work once more. I hope Heinrich is restored[4] tell him to write to me and let me know how he is. Send me the boxes as quickly as you can. Saturday I heard a lecture[5] at the Museum in Jermyn St.[6] by Prof. Forbes. he is a choice spirit. Some called the lecture dreamy, but they did not know the practical utility of such dreams. The dream is not to be despised which holds a man loyally to his task and keeps the pitch of his mind above that of 'the money loving herd'[7] I wished to be with you yesterday. Sunday is dreary in London but I lightened the weight of it by translating a little in the afternoon and thus earning the cup of coffee and cigar which I allowed myself after dinner. This is a fine cheerful day. They are striving to get up a statue of Newton at Grantham.[8] Mr Wright will of course subscribe. I have got a circular—I send you the Leader[9] tho there is nothing in it

as ever yours | John.

RI MS JT/1/T/571
RI MS JT/1/HTYP/269

1. *[3 October 1853]*: Although Hirst dated this letter 'Oct 2 1853', Tyndall wrote the letter on a Monday, dating it to 3 October.
2. *Francis's old lodgings*: 9 Chadwell St Myddleton Square Islington (see letter 0806, n. 2), located approximately 2.5 miles Northeast of the RI. Myddleton Square was a residential area in London built up in the 1820s ('Amwell Street and Myddleton Square area', in Philip Temple [ed.], *Survey of London: Volume 47, Northern Clerkenwell and Pentonville*, 47 vols. [London: London County Council, 2008], vol. 47, p. 192–216, at http://www.british-history.ac.uk/survey-london/v0147/pp192–216 [accessed 14 February 2016]).
3. *After my arrival ... had disappeared*: Tyndall left Queenwood for London on 29 September, having visited the school for six days. He made note of the cold he brought to London in his Journal (30 September, JT/2/13b/624).
4. *I hope Heinrich is restored*: Heinrich Debus had suddenly fallen ill at a dinner that Tyndall and Hirst had attended on 28 September (see Journal, 30 September, JT/2/13b/624). Reflecting on Tyndall's response to Debus's illness, Hirst admiringly stated that 'In such emergencies John shews wonderful presence of mind, so quiet and so determined' (Hirst, 'Journals', 29 August 1853).
5. *a lecture*: likely Edward Forbes's lecture given on 1 October (E. Forbes, *On the Educational Uses of Museums (Being the Introductory Lecture of the Session 1853–1854)* [London: Longman for HMSO, 1853]).

6. *the Museum in Jermyn St.*: see letter 0721, n. 9.
7. *the money loving herd*: from the line, 'Before the money-loving herd' (ii.206) in R. W. Emerson, 'Woodnotes', in *Poems* (Boston: James Munroe and Company, 1847), pp. 67–93, on p. 80.
8. *statue of Newton at Grantham*: in 1858, the sculptor William Theed (1804–91) built a statue of the famous natural philosopher Isaac Newton (1642–1727). The statue was erected in the town of Grantham, England where Newton went to school.
9. *the Leader*: a newspaper published weekly from 1850 to 1860 in London, England.

From François-Napoléon-Marie Moigno 3 October 1853 0805

Mon cher Mr Tyndal

Soyez assez bon pour /consulter/ un Docteur de mes amis qui va en Angleterre recommandant les excellents effets du cuivre comme préservatif du cholera. Il vous demandera de le mettre en rapport avec un des /mécaniciens/ de l'institution royale pour la /le confection/ des armatures metalliques.

Je /suis/ dans les sentiments de la considération la plus affectueuse et la plus distinguée.

Votre très humble serviteur | 3 8bre 1853 | L'abbé F. Moigno

21 Albermale Street | Monsieur | Monsieur Tyndall | Professeur à l'institution royale | à Londres

My dear Mr Tyndal

Would you be good enough to /consult/ a Doctor[1] of my friends, who is going to England recommending the excellent effects of copper as a prophylactic for cholera.[2] He will ask you to /get him in touch/ with one of the /technicians/ of the Royal Institution for the /fabrication/ of metallic armatures.[3]

Please accept my most affectionate and highest regards.

Your humble servant | 3 October 1853 | Father F. Moigno

21 Albermale Street[4] | Sir | Mr Tyndall | Professor at the Royal Institution | in London

RI MS JT/1/M/102

---

1. *a Doctor*: Tyndall reported meeting a 'Dr. Burg', recommended by Abbé Moigno, at the RI on 13 October (Journal, JT/2/13b/627). Jean Antoine Victor Burq (1822–84), a Parisian physician, was a pioneer of 'metallotherapy,' or the use of metals to heal disease (A. Crabtree, 'Burq, Victor', *Animal Magnetism, Early Hypnotism, and Physical Research 1766–1925: An Annotated Bibliography* [White Plains: Kraus International, 1988], pp. 658, 1016, 1054).
2. *effects of copper... cholera*: Tyndall mentioned reading a long report in the 'Siecle' by 'Burg'

on the curative properties of copper in cases of cholera; he noted that they are 'foolish and to a certain extent contemptible' (Journal, JT/2/13b/627). While the exact report is not identified, Burq wrote a text in 1853 on the curative properties of metal in the treatment of cholera (V. Burq, *Choléra, Préservation et Traitement par le Métaux* [Paris: G. Ballière, 1853]).

3. *metallic armatures*: 'a core of laminated iron wound round by coils of insulated copper wire in which electrical current is generated when it rotates in a magnetic field' (*OED*).
4. *21 Albermale St*: misspelling of Albemarle Street.

To Thomas Archer Hirst 5 October 1853 0806

London 5th. Oct. 1853.

My dear Tom

Well I am at length fixed I never took a wider range in search of lodgings in my life. I was in a strait between two. Francis was leaving his lodgings at Islington, they are very comfortable and the people thoroughly honest. But it was distant—full three quarters of an hour's walk from the Institution. This caused me to hesitate long, but at length balancing matters, patently I resolved on taking his lodgings notwithstanding the distance. Here I now sit more comfortable than the Queen[1] and very thankful to have such a block of a fellow as you to write to and discharge all my gossip. My lodging is 9 Chadwell St Myddleton Square Islington.[2] It is very quiet and very airy and not so expensive as places further west. Today I saw Faraday for the first time since my return—He came into the laboratory as I was there this afternoon. He had been doing something up stairs and his hands were black but that did not diminish the kindness of their pressure. We soon plunged into conversation, talked about my health and his health, about my troubles, about my future work. I had sent him a letter which little George Cumming had written to me[3] (I now send it to you for it is worth reading) today I found it in an envelope directed to me. I said that I ought to beg his pardon for sending it—Tyndall he replied I am glad to see such feelings within you Science itself is not the principal thing, we are men and we ought to cherish human feelings. He spoke of his being compelled to relinquish work, that his head would bear it no longer. I referred to the Catholic doctrine of supererogation,[4] and said that any thing he did at present would be a work of supererogation for that he had already done enough. As we spoke I felt my serenity increasing, and I could look calmly upon his silver hair and the masculine brow beneath it. It is a sweet privilege to have such a man to reverence, so great and yet so childlike. It answers to a want in my own nature and I thank the Gods for giving me the capacity to understand such a man. Alluding to his giving up work he said "Well Tyndall I am glad you are here to take my place" or some thing to that effect. He was about going several times but some remark would occur to hold him. I am sure he said on parting that if M[rs]. Faraday knew you were here she would desire her kind remembrances—I know she would—thus he left me. An odour remained about my heart and I spent an additional half hour very pleasantly among the dusty bottles.

And now I have told you the days experience. Goodbye. Love to Heins.[5] Kiss little Charles Cumming for me if he lets you, and give Tommy Grieve a little hug. Little Darton[6] is also a great friend of mine, but if I begin to commemorate I must go through the whole lot.

Goodbye again Tom | Your affectionate | John

RI MS JT/1/T/572
RI MS JT/1/HTYP/270–271

1. *the Queen*: Alexandrina Victoria (1819–1901), Queen of the United Kingdom of Great Britain and Ireland from 1837 to 1901 (*ODNB*).
2. *9 Chadwell St Myddleton Square Islington*: see letter 0804, n. 2.
3. *a letter . . . written to me*: see letter 0781. Tyndall sent Cumming's note to Faraday with letter 0790.
4. *the Catholic doctrine of supererogation*: 'the performance of good works beyond what God commands or requires, held to contribute to a store of merit which the Church may dispense to others to make up for their deficiencies' (*OED*).
5. *Heins*: an endearing nickname Tyndall used for Heinrich Debus.
6. *Little Darton*: presumably a student at Queenwood. Otherwise not identified.

From Thomas Archer Hirst [6–7] October 1853[1] 0807

nobody but you would have sent it to 'Pharody'[2] and no Professor except 'Pharody' would have known how to read it.

<Remainder of letter missing>

RI MS JT/2/13b/626
LT Typescript Only

1. *[6–7] October 1853*: this response to letter 0806, dated 5 October, was received by Tyndall on 8 October, which was noted in his journal (Journal, JT/2/13b/626). Hirst must have written it on either 6 or 7 October.
2. *'Pharody'*: Michael Faraday. Hirst was playing on George Cumming's misspelling of Faraday's name in a letter to Tyndall (see letter 0781, n. 14). Tyndall had circulated Cumming's letter to both Faraday and Hirst (see letters 0790, n. 9 and 0806, n. 3).

To Thomas Archer Hirst [12 or 13 October 1853][1] 0808

My dear Tom.

I think it better to send this[2] to you than to return it direct to Lordon[3] as if you do not agree with the few changes I have made you may strike them out. Do you think that the statement in the second paragraph 'In the material universe &c' will not be miscomprehended—are not the planetary motions examples of that persistency to which you deny an existence. If you limit your statement to this world it is perfectly true. and in your sense which is somewhat latent it is true of the planets—also for they may be looked as incessant labourers.* You use too many commas. What would you think of striking out the bit between the braces in page 4—The two 'its' coming together but referring evidently to two different things is a construction which ought to be avoided as much as possible. You will I think see the justice of the small correction in page 5—'more general properties &c' would imply that you had been speaking of <u>properties</u> whereas you have been speaking of Kingdoms. — The word conformed[4] refers almost exclusive to a chemical combination. an electric spark is not necessary you know—any spark will do.

I like the lecture exceedingly. Its suitability for the occasion appears to me as perfect as could be desired. Edmondson has shewn much wisdom in having it published and would have shewn more had he seen the wisdom of this without being prompted to it.

If you have that £20 beside you I wish you would send it to me.

ever yours | John

*you mean to say perhaps that there is no such thing as rest. this is universally true.[5]

RI MS JT/1/T/926
RI MS JT/1/HTYP/272

1. *[12 or 13 October 1853]*: LT dated this letter to either 12 or 13 October. On 12 October, Tyndall mentioned correcting a portion of Hirst's lecture, making it likely that he also wrote the letter on this date (Journal, JT/2/13b/627).
2. *this*: the first part of Hirst's introductory lecture at Queenwood (see letter 0794, n. 3), which he was working on for publication. The entire lecture would be published as *Introductory Lecture to the Course on Natural Philosophy*: *At Queenwood College, Hampshire* (Romsey: C. L. Lordan, 1853).
3. *Lordon*: Christopher Legge Lordan, a publisher and writer located in Romsey, England. It was Lordan who sent Tyndall the first part of Hirst's lecture (Journal, 12 October, JT/2/13b/627).

4. *conformed*: in chemistry, conformation refers to 'Each of the different three-dimensional structures that may be adopted by a particular molecule and can interconvert freely, esp. by rotations about single bonds' (*OED*).
5. *no such thing as rest . . . universally true*: while it is unclear to what exactly Tyndall is referring, it may refer to William Thomson's assertion that all particles are in vibratory motion (W. Thomson, 'On the Dynamical Theory of Heat, with numerical results deduced from Mr. Joule's equivalent of a Thermal Unit, and M. Regnault's *Observations on Steam*', *Phil. Mag.* 22 [1852], pp. 8–21). Alternatively, it may be an evocation of Leibniz's view that there is no absolute motion, contrasting Newton's view of space implying that there can be absolute motion.

To Thomas Archer Hirst 15 October 1853 0809

15th. Oct. 1853.

My dear Tom.

The cheque[1] has just reached me and I am very much obliged to you. I am just going out to meet Playfair at Marlborough House.[2] He wrote yesterday[3] requesting me to call upon him. I have no doubt it is something connected with the 'School of Science'[4] that he wants to talk to me about—The remainder of your lecture[5] has not yet reached me although I expected it today. I am preparing an Introductory Lecture[6] and will send it to you as soon as it is ready—I want to have the benefit of your scrutiny which I trust will be as severe as your tender heart will permit of. I shall probably send you a £10 note next week as I have little doubt of being able to get the draft cashed without loss. Make my respects to Mr Best when you see him appropos—Have you read that lecture of the Dean of Hereford[7] at the Society of Arts.[8] He sent me a copy before I went to Germany. It is valuable both on account of the information it contains but more so to you and me as affording a view of a sturdy practical educator who is not frightened by what would deter many—Edmondson has a copy.

Goodbye. Remember me to John[9] when he comes. | Yours affectionately | John

RI MS JT/1/T/573
RI MS JT/1/HTYP/273

1. *The cheque*: likely the £20 Tyndall requested from Hirst in letter 0808.
2. *Marlborough House*: see letter 0784, n. 12.
3. *He wrote yesterday*: letter missing.
4. *School of Science*: likely refers to Playfair's attempt to launch and co-ordinate a national

scheme of scientific schooling while he was Secretary of Science, which never came to fruition (*ODNB*). Tyndall noted in his journal on 17 October that Playfair had requested he write a treatise on physics to be used in schools. He also noted that there may be a post for him in the School of Science in the future (Journal, JT/2/13b/628). On 20 October he noted that he had written to Playfair to decline his offer of involvement (Journal, JT/2/13b/629).

5. *The remainder of your lecture*: Tyndall was awaiting the rest of Hirst's lecture from the publisher, Christopher Legge Lordan (see letter 0808, nn. 2 and 3).
6. *I am preparing . . . Lecture*: the first of twelve lectures Tyndall gave at the RI as part of a course on Heat in 1854 (see letter 0789, n. 2). The lecture took place on 17 January 1854.
7. *lecture of the Dean of Hereford*: the lecture was published as R. Dawes, *Schools and Other Similar Institutions, for the Industrial Classes. Remarks on The Importance of Giving Them, as Far as Possible, a Self-Supporting Character, and the Means of Doing So. A Paper Read before the Society of Arts, April 27th, 1853* (London: Groombridge and Sons, 1853). The educationalist Richard Dawes (1783–1867) was the Dean of Hereford from 1850 until his death, and is notable for his contributions to developing applied science for elementary education (*ODNB*).
8. *the Society of Arts*: shortened title for The Royal Society for the Encouragement of Arts, Manufactures, and Commerce. A British organization located in London, the Society was founded in 1754 by William Shipley (1715–1805) to 'embolden enterprise, to enlarge science, to refine Art, to improve our Manufactures, and extend our Commerce' (W. Shipley qtd. in D. G. C. Allan, *William Shipley: Founder of the Royal Society of Arts* [London: Hutchinson, 1968], p. 43).
9. *John*: Hirst's brother, John Henry Hirst (1826–82), who worked as an architect in Bristol.

To Thomas Archer Hirst 17 October 1853 0810

17th Oct. 1853

My dear Tom

I return you a proof[1] which I received this morning. I was about making some alterations in the portion referring to electricity but it will I think be better to tell you my thoughts and allow you to make the necessary alterations yourself. You have Riess on Electricity[2] have you not? Look out for the chapter on the back stroke[3] which Lord Mahon[4] discovered and explained. This back stroke it was which caused the convulsions of the frog in Galvani's first experiment.[5] The frog was first electrified by induction, the—E drawn towards the conductor and the + E repelled or vise versa. The student placed his scapel upon the frog—the—E flowed through it or through him to the earth, but when the conductor was discharged the natural state of the frog was restored and the sudden recomposition of the fluids caused the convulsion. Of course you need not say all this but I think you can with advantage be a little more precise than you have been.

Next the account of the lady's sickness[6] is in all probability a fiction to which currency has been given by Arago.[7] The truth seems to be that she was the first who saw the contraction of the dead animal and communicated the result to Galvani. In DuBois Book there is an Italian Sonnett[8] celebrating her praises for being the discoverer.

At the end of page 14 you draw a parallel between magnetism and the 'positive and negative electricities'—You never mentioned posit & neg. el. up to this point—would it not make it clearer if you stated the law of electric action[9] at its proper place in page 13.

I have now said all I intended to say—with these trivial exceptions the lecture has given me the greatest satisfaction—There is a serenity—a deep tranquility in its sentences—and withal a flush of life and hope which will commend it to everybody competent to judge of truly good writing.

Are you correct regarding the ignorance of our forefathers mentioned in page 13. They had thunder & lightning in those days had they not? Beware of keeping your glance on one sphere of electric action and founding a general statement upon that.

I should like to give a few copies of it among my great acquaintance here, for it is fit to be read by the first man of the land.[10] but you have almost taken this power away from me by mentioning me as you have done. I would propose the omission of the last paragraph. The thing will end beautifully with the word 'instructing.'

Yours affectionately | John

RI MS JT/1/T/574
RI MS JT/1/HTYP/274

---

1. *a proof*: likely the second part of Hirst's introductory lecture at Queenwood. See letters 0808, nn. 2 and 3 and 0809, n. 5.
2. *Riess on Electricity*: see letter 0727, n. 14.
3. *the chapter on the back stroke*: refers to vol. 2, ch. 1 of Riess's work, titled 'Der Rückschlag und die Seitenentladung', or 'The Recoil and the Side Discharge'.
4. *Lord Mahon*: Charles Stanhope, or Charles Mahon, 3rd Earl Stanhope (1753–1816), a politician and inventor. He was a fellow of the Royal Society and notable for his work on electricity (*ODNB*). According to Stanhope, the 'back stroke' (which he termed the 'returning stroke') was a sudden discharge of electricity between oppositely charged clouds

and earth, which allowed 'fatal Effects' to occur 'at a *vast Distance* from the Place where the Lightning falls' (C. Mahon, *Principles of Electricity: Containing Divers New Theorems and Experiments, Together with An Analysis of the Superior Advantages of High and Pointed Conductors* [London: P. Elmsly, 1779]).

5. *Galvani's first experiment*: refers to physician and scientist Luigi Galvani's (1737–98) chance viewing 'of contractions at a distance, when he observed the movements of a frog's leg coincident with a spark extracted from the electrical machine and with no material connection between the animal and the instrument' (M. Piccolino and M. Bresadola, *Shocking Frogs: Galvani, Volta, and the Electric Origins of Neuroscience*, trans. N. Wade [Oxford: Oxford University Press, 2013], p. 71). Galvani discussed this experiment in his work on animal electricity, which was first published in Italian as *De Viribus Electricitatis in Motu Musculari Commentarius* (Bologna: Accademia delle Scienze, 1791).
6. *the account of the lady's sickness*: when discussing Galvani's first experiment in his biography of Alessandro Volta (1745–1827), an Italian physicist and chemist, Arago (see n. 7) wrote that the frogs were originally prescribed to make broth as a remedy for Madame Galvani's illness (Arago, 'Alexandre Volta', *Ouvres de François Arago*, 2nd ed., 17 vols [Paris: Librairie des Sciences Naturelles, 1865], vol. 1, pp. 187–240, on p. 212).
7. *Arago*: Dominique François Jean Arago (1786–1853), a French physicist notable for his work on magnetism and optics ('Arago, Domique François Jean', *The Encyclopedia Britannica: A Dictionary of Arts, Sciences, Literature and General Information*, ed. H. Chisholm, 11th edn., 29 vols [Cambridge: Cambridge University Press, 1910], vol. 2, pp. 312–13, on p. 313).
8. *Italian Sonnett*: entitled, 'Sonett auf die Entdeckung des Galvanismus', in E. du Bois-Reymond, *Untersuchungen Über Thierische Elektricität* (Berlin: G. Reimer, 1848), p. 39.
9. *the law of electric action*: 'The attraction or repulsion between two quantities, q and q^1, of electricity supposed concentrated at two points at a distance d apart, is $q\,q^1/d^2$, and is in the line joining the two points'. The law was determined by the physicist Charles-Augustin de Coulomb (1736–1806) and the chemist and physicist Henry Cavendish (1731–1810). (R. Walmsley, *The Electric Current: How Produced and How Used* [London: Cassell and Company, 1894], p. 395.)
10. *first man of the land*: perhaps a reference to Faraday, otherwise not identified.

To Thomas Archer Hirst [19 October 1853][1] 0811

My dear Tom.

I send you my introductory lecture.[2] I want you to scarify it and where it is muddy to clarify it. Dont spare it boy. Tell me what you conceive to be its faults and I will adopt your notions just so far as they suit me. The corrections I have made in an awkward manner. You will have to open the sheet out to see the corrections referred to by the letter (a1 (b) &c in the text. You may keep it

as long as you like—take time and give it a good looking over. You might ask Heinrichs[3] opinion where you think it necessary.
affectionately yours | John

RI MS JT/1/T/575

1. *[19 October 1853]*: on 19 October Tyndall's journal stated, 'Finished my introductory lecture this morning and sent it off to Tom' (Journal, JT/2/13b/628).
2. *my introductory lecture*: see letter 0809, n. 6.
3. *Heinrichs*: Heinrich Debus.

To Thomas Archer Hirst [25 October 1853][1] 0812

My dear Tom.

The idea is a nice one. nevertheless you cannot I think say broadly that we are unable by the mere act of the will to change our state of rest. For the change of form you admit is in point of fact a change of rest. and I believe this changing of form is possible even when you have no fulcrum. This I see is your meaning. You require therefore simply to change the sentence 'no mere effort of the will can change our state of rest'—The parts of the body may move with reference to each other, but not the whole body with reference to external objects. I would restrict the phrase. 'where we see an inanimate body in motion' to 'where we see an animal walking.'

I would leave out the last short paragraph. The subject is so deep that the last allusion only creates a haze and will leave no clear impression. —end the matter with the word 'Motion'—

I could shew you almost the same series of thoughts written down in my Journal, and suggested by watching the movement of the horses in a post wagon[2] during a fine starlight night between Brunswick[3] and Göttingen.

Send me Pouillet Müller.[4] The news of Queenwood is gladdening,[5] glorious in fact. Go on and prosper.
your affectionate | John
I will go down to your Xmas concert if I can

RI MS JT/1/T/927
RI MS JT/1/HTYP/276

1. *[25 October 1853]*: the contents of this letter correspond to Tyndall's journal entry for 25 October, in which he recorded the receipt of a letter from Hirst noting the large number of boys at Queenwood (Journal, JT/2/13b/630). Furthermore, LT notes: 'Art[icle].

on Progressive Motion in Q[ueen]wood Observer written Oct 23, inserted Nov 9—see H[irst's]. J[ourna]l VII. This letter w[oul]d probly be about | 25th [October]'. In his journal entry of 23 October, Hirst recorded the writing of this article; it also contains the only known copy of the article (Hirst, 'Journals', 23 October 1853).

2. *the same series of thoughts . . . horses in a post wagon*: Tyndall was referring to his journal entry of 27 August, which states, 'A clear cold night, something spectral in the motion of the horses. Something very beautiful in the loyalty with which they breasted the road at the command of man. Watched the motion of the limbs, the hind-leg first drawn up by the contraction of the muscles, the hoof placed on the ground and the muscles then stretched to throw the animal forward. Thought of the nerves, their influence is wonderful, if we snap one of those strings which of itself will not bear a pound weight we deprive this great machine of the power of action. How is it that they are made the channels of the will? The fracture of a single thread in a factory is sufficient to stop the works, the finger of an engineer can set a locomotive in motion it merely loosens agencies which far transcend the power applied to set them free. Does something similar occur in the case of the nerves?' (Journal, JT/2/13b/620).
3. *Brunswick*: the English name for the city of Braunschweig, located in northern Germany.
4. *Pouillet Müller*: M. Pouillet and J. Müller, *Pouillet's Lehrbuch der Physik und Meteorologie, für deutsch verhältnisse frei bearbeitet von Dr. Joh. Müller*, 2 vols (Braunschweig: Vieweg, 1842).
5. *The news of Queenwood is gladdening*: see n. 1.

To Thomas Archer Hirst [28 October 1853][1] 0813

My dear Tom.

I send you the 1st half of a £10 note and will send you the 2nd as soon as I hear from you.

Since you told me that you are careless about your accounts[2] I have felt anxious lest you should permit our matters[3] to get into a state of complication. I hope you will avoid this by entering everything as it occurs.

Yours affection | John

RI MS JT/1/T/576

1. *[28 October 1853]*: Hirst wrote 'Octr 28 1853' at the top of this letter but that may refer to the date he received it.
2. *you told me . . . accounts*: not identified.
3. *our matters*: Tyndall had been borrowing money from Hirst since late 1849, when Hirst discovered his inheritance upon the death of his mother, Hannah Oates Hirst (1804–49) (see letter 0382).

To Thomas Archer Hirst 30 October [1853] 0814

Sunday 30th Oct.

My dear Tom.

I send you the 2nd half of the £10 note[1] and along with it a copy of the Leader.[2] I have been much pleased by many of the articles, that on Cobden at Barnsley[3] pleased me particularly. The paper certainly sustains the pitch onto which its first notes were thrown. But if I commence in this way I shall probably write you a long letter, which I by no means intend for you dont merit it—your letters are always so scanty! True they meet the material wants of the time but this is not all that I require of you—and I think I shall begin to act in the same way towards you just to see how you would relish it—You have protested before now when I did so, but the tables are turned and you are the offender. In the last page of your lecture[4] first change the word 'to' into 'in' in the phrase 'to our generation.' Was Galvani a believer in Animal Electricity before he discovered it? and if not could it have influenced him in the way you say?

Goodbye | affectionately | John

I am now going to allow myself a glorious dinner. roast beef potatoes and cauliflower, —perhaps an apple tart afterwards, and likely enough a glass of sherry to wash it down! My diurnal dinner is a bun and a pint of stout! I get no vegetables—the odour of the latter reaches up from Anderton's[5] although it is two miles distant.

Just care as little about your Andover defeat[6] as I do! if you do it wont keep you from sleeping—A thing of the kind spreads round a man's life like a vapour. If he is a fool he lives in it now scalds the eyes of his soul by the smoke.

I have been engaged during the last week in making utter smash of an investigation of a very eminent man.[7] but my task has been not all negation, for I have expanded the field of enquiry far beyond the limits which he ascribed to it—The subject the *[press]*, is not important. It will make a capital Friday even[in]g lecture.[8] Faraday was present when I made some of my experiments yesterday. He almost danced with pleasure. *[Sweet]* noble fellow, each day makes me like him more.

RI MS JT/1/T/932

1. *2nd half of the £10 note*: Tyndall sent the first half in letter 0813.
2. *the Leader*: see letter 0804, n. 9. The issue Tyndall included was *The Leader*, 29 October 1853.
3. *Cobden at Barnsley*: 'Public Education: Mr. Cobden and Sir Robert Peel', *The Leader*, 29

October 1853, which summarized a speech the politician and manufacturer Richard Cobden (1804–65) gave on education (*ODNB*).

4. *your lecture*: probably Hirst's introductory lecture at Queenwood. See letters 0808 and 0810.
5. *Anderton's*: possibly Anderton's Hotel and Tavern, located at 164 Fleet Street, London.
6. *your Andover defeat*: reference to the lecture Hirst delivered on 26 October in Andover, which he described in his journal as 'not at all to my satisfaction' (Hirst, 'Journals', 30 October 1853).
7. *an investigation of a very eminent man*: refers to J. D. Forbes, 'Experimental Researches regarding certain Vibrations which take place between Metallic Masses having different Temperatures', *Phil. Mag.* 4:21 (1834), pp. 15–28 and 182–94. In his journal entry of 25 October Tyndall noted, 'Forbes has found bismuth completely inert either as a rocker or bearer in Treveleyan's experiment. I produced distinct and enduring tones with it today' (Journal, JT/2/13b/629). On 26 October, he stated 'Forbes's experiments must have been coarsely made' (Journal, JT/2/13b/630).
8. *Friday even[in]g lecture*: Tyndall would indeed make this subject the focus of a Friday evening lecture (see letter 0876, n. 14).

From William Francis[1] 4 November 1853 0815

Viele Grüsse von W.F. Du hast den Medal bekommen.

Many greetings from W.F.[2] You have got the medal.[3]

RI MS JT/2/13b/631
LT Typescript Only

1. *From William Francis*: Tyndall copied this note in his journal, stating that he 'Found a card from Francis on my return, he had come to congratulate me' (4 November, JT/2/13b/631).
2. *W.F.*: William Francis.
3. *the medal*: the Royal Medal, one of the most prestigious awards granted by the Royal Society; between 1850 and 1965, two medals were presented each year, one in the physical sciences and the other in the biological sciences (Jackson, 'Tyndall and the Royal Medal', p. 151). Tyndall learned on 4 November that he was 'the elected candidate for the Royal Medal' (Journal, JT/2/13b/631).

To Thomas Archer Hirst [5 November 1853][1] 0816

My dear Tom

Thank you for the Observer.[2] It is excellent. Conducted as it is it must be productive of nothing but good to Queenwood. Those translations[3] are also a very good idea as they are a certificate that the boys learn what their teachers profess to teach them. The account of the mutual improvement society,[4] short as it is opened out a very pleasant view of your condition. On the whole the little paper has given me high gratification.

I sent Haas a semihumourous scrap[5] but he has rejected it. Were I in his place I should have rejected yours[6] and put in my own.

Side by side with your Observer came to me the notification that the Council of the Royal Society[7] had awarded me the medal.[8] Hofman[9] and myself were competitors (I did not know this) and the Council had great difficulty in deciding I think the best thing I could do would be take a little prussic acid[10] and kill myself quietly here while the halo is still around my head. People would then say 'if he had lived &c.'—'Poor fellow he went off just at the dawning of his day' I should thus get a hold on posterity no doubt

Yours affectionately | John.

I expect Simpson in a few minutes. I think you ought to come up and see Simpson yourself. You will have time tomorrow? You can come at half-fare.

I think I will go have a cup of tea with him tomorrow evg. as I really am anxious to see your daughter[11]

Such fellows as you whom I have satisfied so much ought to get up a penny subscription for me | Nicht wahr? Helmholtz ist hübsch[12]

RI MS JT/1/T/933
RI MS JT/1/HTYP/279

---

1. *[5 November 1853]*: this letter was dated by LT. Her dating is probably accurate, as Tyndall noted his visit with Simpson on 6 November (Journal, JT/2/13b/632). Furthermore, Hirst replied to the contents of this letter the following day (see letter 0819).
2. *the Observer*: *The Queenwood Observer*, a monthly periodical running from 1853 to 1860, produced and edited by students at Queenwood College. This likely refers to Tyndall's receipt of an article by Hirst that appeared in the October 1853 issue, which Tyndall had been helping him to edit (see letter 0812, n. 1).
3. *Those translations*: not identified.
4. *account of the mutual improvement society*: proceedings of the Queenwood Mutual Improvement Society's weekly meetings were frequently published in *The Queenwood Observer*. The society was, generally speaking, voluntary and devoted to the improvement of knowledge and skills.

5. *a semihumourous scrap*: possibly a letter Tyndall wrote to Charles Cumming, one of the boys at Queenwood College (see letter 0819, n. 15).
6. *yours*: not identified.
7. *the Council of the Royal Society*: the governing body of the Royal Society, composed of elected fellows. At the deciding meeting for the Royal Medals on 3 November there were 13 Council members present: Thomas Bell (1792–1880), James Booth (1806–78), Benjamin Brodie (1783–1862), Samuel Hunter Christie (1784–1865), John Peter Gassiot (1797–1877), Joseph Hooker (1817–1911), Henry Bence Jones (1813–73), George Newport (1803–54), Edward Sabine (1788–1883), William Sharpey (1802–80), William Smyth (1788–1865), William Spence (1782–1860), and Nathaniel Wallich (1785–1854) (Jackson, 'Tyndall and the Royal Medal', p. 156).
8. *the medal*: see letter 0815, n. 3.
9. *Hofman*: misspelling of August Wilhelm Hofmann.
10. *prussic acid*: hydrogen cyanide (HCN), a poisonous chemical compound.
11. *your daughter*: Tyndall was referring to Anna Martin, who would become Hirst's wife in December 1854. Hirst often referred to Martin, his future wife, as his daughter (see letter 0825, n. 8), and she called him 'father' (see letter 0634, n. 8). In his journal entry for 6 November, Tyndall noted seeing Maxwell Simpson, Mary Simpson, and Anna Martin. He 'rather liked the latter though she scarcely spoke' (Journal, JT/2/13b/632). Anna Martin and Mary Simpson (née Martin) were sisters.
12. *Nicht wahr? Helmholtz ist hübsch*: German, 'Is it not so? Helmholz is lovely'.

From Heinrich Debus 6 November 1853 0817

Queenwood College. | Stockbridge Hants | 6 Novbr 53.

Mein lieber John!

Dem Verdienste seine Ehre. Ich halte die Verleihung von Medaillen und Diplomen von Regierungen und wissenschaftlichen Gesellschaften als Anerkennung für tüchtige, gute Leistungen als eine vortreffliche Methode um ausgezeichnete Leute aufzumuntern und zu belohnen. Natürlich nicht die Person, nicht Intrigue, nicht Protection, sondern Gelehrsamkeit, Moral, und Fähigkeit müssen die leitenden Momente bei der Austheilung solcher Ehrenbezeuchungen sein.

Die angenehme Nachricht in Deinem Brief hat mich nicht überrascht. Du hast die Medaille verdient. Wenn daher meine Gratulation Dir angenehm sein sollte so hast Du sie von ganzem Herzen. Ich habe mich immer gefreut über Dein wissenschaftliches Streben, und noch mehr über die Resultate welche dasselbe bekleideten, aber besonders angenehm ist es mir und Dir muss es eine grosse Genugthuung sein dass die erste wissenschaftliche Gesellschaft dieses Landes Dein Wirken als ein erfolgreiches anerkennt und belohnt.

Aber es geht nicht immer so in dieser Welt. Heute morgen erzählte uns Mummery beim Frühstück dass er ein German Doctor-Diplom in London gekauft habe. Es thut mir leid dass eine deutsche Universität existirt welche mit wissenschaftlichen Ehren Handel treibt und an den Meistbietenden verkauft. Es ist eine Schmach dass Männer welche auf den Höhenpunct der Intelligenz stehen wollen welch ihrem Volke ein Leitstern in allen guten Dingen abgeben sollten dass solche Männer in den gewöhnlichen Schmutz hinunterkriegen und den einzigen Lohn welchen der Student für seine Arbeit erndten kann, das öffentliche Zeugniss der Intelligenz und Bildung für klingende Münze verkaufen. Wer mehr getadelt werden muss, der Mann der eine solche Sache kauft oder die Universität welche solchen Handel treibt weis ich nicht.

Dein, | H Debus

Queenwood College | Stockbridge, Hants | 6 Novbr 53.

My dear John!

For service its honour. I consider the conferring of medals and diplomas by governments and scientific societies as recognition for competent, good achievements to be an excellent method of encouraging and rewarding distinguished people. Of course, not the person, not intrigue, not patronage, but erudition, morals, and ability have to be the guiding factors when handing out such tributes.

The pleasant news in your letter[1] did not surprise me. You deserved the medal.[2] If, therefore, my congratulations should be agreeable to you, then you have them with all my heart. I have always taken pleasure in your scientific striving, and even more so in the results which attend it, but it is particularly agreeable to me, and it must be a great satisfaction to you, that the foremost scientific society in this country[3] is recognising your work as successful and rewarding it.

But it is not always like this in this world. This morning Mummery told us over breakfast that he has bought a German Doctoral Diploma in London.[4] I am sorry that a German university[5] exists which does trade with academic honours and sells to the highest bidder. It is a disgrace that men who want to stand on the pinnacle of intelligence, and who should be a guiding star in all good things to their people, that such men crawl down into the ordinary muck and sell for common coin the only reward which a student can reap for his work, the public testimony of intelligence and education. Who needs to be censured more, the man who buys such a thing or the university which does such a trade, I do not know.

Your | H Debus

RI MS JT/1/D/13

1. *your letter*: letter missing.
2. *the medal*: see letter 0815, n. 3.
3. *the foremost scientific society in this country*: the Royal Society of London.
4. *Mummery . . . bought a German Doctoral Diploma in London*: John Mummery, a teacher at Queenwood College, bought a German doctorate, much to the displeasure of Debus, Hirst, and Tyndall (see letters 0833, n. 10; 0836, n. 7; 0874, n. 2; 0876, n. 6; 0879, n. 5; 0881; 0882, n. 14; 0889; 0890; 0893; and 0910, n. 13).
5. *a German university*: not identified.

From Tommy W. Grieve 6 November 1853 0818

Queenwood College | November 6th. 1853.

My dear Dr. Tyndall,

The bonfire is going to be tomorrow, it is smaller this time than it was last time. There are seven new boys come since you have been here last[1] so we have seventy six boys, the room crammed full. One morning we had a very nice game at foot-ball about a fortnight ago Mrs Edmondson[2] invited us to take tea with her and that night we had the Panorama of London[3] and that night Tommy[4] and I sang a song and its name was Old Lang Sine[5] that is a Scotch song. Every Sunday the first class in singing go down into the lecture room from 3—4 and I am one of them.

Flight[6] sends his kind love to you and so do I

I remain | Your affectionate Friend | T. Grieve | Good bye.

RI MS JT/1/TYP/1/459
LT Typescript Only

1. *since you have been here last*: Tyndall last visited Queenwood in September (see letter 0804, n. 3).
2. *Mrs Edmondson*: Anne Edmondson (née Singleton), wife of George Edmondson. Anne greatly supported her husband's work as Headmaster of Queenwood.
3. *the Panorama of London*: possibly Thomas Horner's (1785–1844) Panorama of London, completed in 1829 and housed in the Colosseum at Regent's park until 1854.
4. *Tommy*: presumably another student at Queenwood. Not further identified.
5. *Old Lang Sine*: misspelling of 'Auld Lang Syne', a 1788 poem by the Scottish poet Robert Burns (1759–96).
6. *Flight*: Walter Flight (1841–85), a student at Queenwood during the mid-1850s. Flight eventually became a mineralogist at the British Museum, and was notable for his work on meteorites (*ODNB*).

From Thomas Archer Hirst 6 November 1853 0819

Queenwood College. | Nov 6th 1853

My dear John.

Your letter yesterday morning[1] caused such elevation of spirits that at the moment I accepted the invitation at the close, and told Mr Edmondson I was going up to London. In 5 minutes time I should have been on my road but my eyes happened to fall on a large amount of work that ought to be dispatched that day and without allowing myself time to argue the matter I seized my pen instead of my hat. I hope you went to Simpson's's[2] to tea for then you will have seen for yourself what sort of good folks they are. What a good natured little daughter[3] I have and what a beautiful little girl Simpson eldest[4] is. But about the Medal[5] I wish you had been a little more explicit, is it the Copley Medal?[6] How many of your medals are presented yearly, and what does the memorial[7] say which accompanies it? I am greatly pleased about it for it shews me once more that you have some staunch friends. The gift is graceful and encouraging in every respect. The effect it had on you was curious and to the donors I am sure unexpected. Were you to take Prussic Acid[8] I am afraid posterity would give you a far different reputation from the one you anticipate. Some (the Tom Perkintons[9]) would call you a clever fellow but a great ass or a 'damned fool' Others would say that like a mushroom you had grown up in 24 hours to reputation and not being able to sustain it you had quitted the scene in a rather cowardly way. No, to preserve the halo about you you must die a natural death, and this is out of your power to retard or accelerate. it is true you might accomplish death without poison by overwork but that would not be natural, and posterity would then say that in a few weeks you had compressed the work of as many years and had exploded with a puff. But to return to our old fundamental principles it is wisest not to care a straw about posterity and medals To go on your own way quietly and without misgivings, so that your life's labour shall not be a forced overgrown part but a solid proportionate whole. I write this under the conviction that you have anticipated it and that it is unnecessary. And in passing I may just remark that you have to thank such convictions for so many short letters. It is strange but yet true that for full & interesting correspondence there should exist less unanimity of thought and opinion than in our case This very unanimity destroys to a great extent the reflectional part, and limits interest to matters of news and fact which are far more concisely told and in my case at any rate far more scanty. If you were at Queenwood instead of Debus there would be much less spoken between us but much more understood. Ginty was here a week ago. He says you are growing morose, your a clever fellow he knows but 'I'm damned' says he 'If I know what's matter with him. His notions about money and the practicalities of life are in my opinion absurd in the highest degree' I enjoyed Ginty's company for a day very much. That vigour and acuteness of intellect he possesses is truly refreshing in these twaddling[10] days. It is true his ideal of Life differs materially from mine, indeed has very little in common with it.

Ginty requires an atmosphere of publicity and strife just as a fish requires water He must have something to conquer and he must measure his strength against other men's weakness. He can appreciate no work performed for its own sake or for discipline in his own study & without intended publicity before Chancery Court[11] or Manchester Athenaeum.[12] In his own words Mans object in the world is to improve the material condition of his fellow men and of himself. Such men are of immense value but there must be men above them whose will they shall unwittingly execute—

I was almost entirely a listener during his graphic talk, only once we clashed about Carlyle and Emerson and to my surprise he gave way rather than defended his position. He admires Emerson more than Carlyle, nay he believes the latter is but an imitator of the former! His great mistake arises from his want of sympathy with what constitutes the highest merit of both. He judges them as mere book writers. Emerson gives him sound deep glances into things gives him a solid hint or two and above all does not interfere with his own theory of action. Carlyle on the contrary possessing his own vigour incites him to combat, and he finds in Carlyle's action nothing practical nothing tangible or in accordance with his theory. Lacking this its deep earnestness is lost on him. Engaged in his own whirling combats he fails to see in Carlyles 'grumblings' the most radical & certain of all remedies.

Tomorrow we celebrate our 5th of Novr[13] Queenwood is in a glorious condition, there is but one drawback and that is Coombes[14] he is a perfect fool. To prevent rows we have had to take work away from him and we are now almost inclined to dismiss him and share his classes between us.

Haas was far too scrupulous about your letter to Charley Cumming,[15] he would not publish it, but he read it to the Mutual Improvement Society[16] suppressing the name of the boy to whom it was addressed. All recognized your writing and foolishly Edmondson set them at guessing to whom it was addressed. After a short silence George Case[17] gave it as his opinion that <u>it was addressed to me</u>

Haas is a splendid fellow but we differ much in opinion, of which more another time. You need not send me the Leader[18] any more I see it every week now. Did you notice that sound letter of Palmerston's?[19] Tell me the nature of those experiments which militate against those of some eminent man.[20]

Yours affectionately | <u>T.A. Hirst</u>.

RI MS JT/1/H/183
RI MS JT/1/HTYP/280–281

---

1. *Your letter yesterday morning*: letter 0816.
2. *Simpson's*: the home of Maxwell and Mary Simpson.
3. *little daughter*: Anna Martin (see letter 0816, n. 11).
4. *Simpson eldest*: one of Maxwell and Mary Simpson's six children (*ODNB*). Not further identified.
5. *the Medal*: Tyndall had recently been notified that he received the Royal Medal (see letter 0815, n. 3).
6. *Copley Medal*: the most prestigious award given by the Royal Society, first established in 1731 and awarded for distinction and achievement in the sciences (M. Yakup Bektas and M. Crosland, 'The Copley Medal: The Establishment of a Reward System in the Royal Society, 1731–1839', *Notes and Records of the Royal Society* 46:1 [1992], pp. 43–76).
7. *the memorial*: not identified.
8. *Prussic Acid*: see letter 0816, n. 10.
9. *the Tom Perkintons*: not identified.
10. *twaddling*: 'trifling, unimportant' (*OED*).
11. *Chancery Court*: 'The court of the Lord Chancellor of England, the highest court of judicature next to the House of Lords' (*OED*).
12. *Manchester Athenaeum*: a society founded in 1835 to support the cultivation of knowledge and the arts in England.
13. *5th of Novr.*: Guy Fawkes Day, an annual celebration in Britain commemorating the failure of Fawkes's Gunpowder Plot of 1605, which was a plan to blow up the Houses of Parliament while the King, Lords, and Commons were assembled there (*OED*).
14. *Coombes*: Edwin Coombes, a teacher at Queenwood. Not further identified.
15. *your letter to Charley Cumming*: letter missing (see letter 0816, n. 5).
16. *the Mutual Improvement Society*: see letter 0816, n. 4. The Mutual Improvement Society at Queenwood consisted of both pupils and staff, and strived to maintain egalitarianism between the youth and adults at meetings (Brock, 'Queenwood', p. 17).
17. *George Case*: a student at Queenwood.
18. *the Leader*: see letter 0814, n. 2.
19. *that sound letter of Palmerston's*: Hirst was referring to a letter written by Henry John Temple (1784–1865), third Viscount Palmerston, who was Home Secretary between 1852 and 1855, and would become Prime Minister in 1859 (*ODNB*); see 'Lord Palmerston on Prayer and Fasting', *The Leader*, 29 October 1853, p. 1034.
20. *those experiments . . . some eminent man*: see letter 0814, n. 7.

From Heinrich Gustav Magnus 7[–8] November 1853[1] 0820

Berlin | 7 Novemb | 1853

Lieber Tyndall

Gleich nach Empfang Ihres Briefes ging ich zu Sauerwald, der mich versicherte die Instrumente schon am 8 Octob an Asher übergeben zu haben. Am folgenden Tage war ich auch bei Asher der mir in seinem Buche zeigte, daß er sie am 12ten Octob nach Hamburg gesandt habe um an Nutt in London befördert zu werden. Er konnte nicht begreifen weßhalb Sie die Instrumente noch nicht erhalten, meinte aber, daß sie jetzt wohl unzweifelhaft in Ihren Händen wären. Ich hoffe dies auch, sollte es aber nicht der Fall sein so veranlassen Sie doch Mr. Nutt nach Hamburg zu schreiben und Erkundigung einzuziehn, und schreiben Sie mir gleichzeitig, dann werde ich Asher auch veranlassen nach Hamburg zu schreiben.

Es ist nicht ganz leicht Ihre Handschrift zu lesen, ich bin deßhalb auch nicht sicher ob das Leidenfrost'sche Experiment Ihnen von Dr. Franz hier gezeigt worden. Doch habe ich diesem geschrieben, und ihn gebeten mir die Flüssigkeiten zu nennen, hoffentlich erhalte ich die Antwort bis morgen, bevor dieser Brief abgeht und nenne Ihnen dann die Flüssigkeiten. Uebrigens hat Marchand eine ziemlich ausführliche Arbeit über das Leidenfrostsche Experiment bei Anwendung von 2 Flüssigkeiten gemacht, doch kannte man damals den Stern noch nicht. Auch wenn ich die Arbeit nicht gleich finden; vielleicht ist sie im XIX Bd. seines Journal's unter Lampensäure enthalten.

Sie schreiben von meiner Abhandlung: 'Contrary to my expectation the chemical paper pleased me better than the physical one.' Wenn ich dies richtig verstehe, so hat ihnen das physicalische nicht gefallen. Mir schien dies auch schon in Paris schon, bitte sagen Sie mir was Ihnen nicht gefallen. Mir will das physicalische viel besser erscheinen, aber ich weiß daß die Darstellung etwas kurz ist.

Ich bin jetzt mit den Flüssigkeiten beschäftigt und habe einige recht hübsche Resultate erhalten. Herr Weber ist dabei sehr brauchbar.

Was Sie mir von Faraday schreiben freut mich sehr, ich habe ihn, als wir in Birmingham zusammen waren, ausserordentlich lieb gewonnen, verehrt habe ich ihn immer, aber es ist so selten daß Leute, die so eminent in der Wissenschaft sind, einen so liebenswürdigen Charakter haben; nur Berzelius war ähnlich, aber nicht so frisch wie Faraday. Sagen Sie ihm doch recht viel herzliches und freundliches von mir, und ebenso auch my best remembrances to Missis F.

Dubois werde ich so bald ich ihn sehe berichten daß man einen Brief in London von ihm erwartet. Er hat jetzt einen starken Schnupfen. Wie hat Ihnen sein Letter on Signor Matteucci gefallen? Mir ganz gut.

Heut traf ich Dove der sehr erfreut ist über die Auszeichnung die ihm durch die Copland Medal zu Theil geworden.

Meinem Bruder werde ich Ihre Grüsse anrichten, wir correspondiren jetzt häufig, da er mir die Instrumente die ich bestellt habe bezahlt. &. &. &.

Mir scheint es am besten wenn Bence Jones Sohn Engländer und mein Paul Deutscher bleibt. Wenn aber B.J. seinen Sohn auf einige Zeit hierher schicken will, so könnte er mit Paul zusammen aufwachsen und in die Schule gehn, und dafür würde dann später Paul mit ihm nach England gehn können. Grüßen Sie Bence Jones viel mals so wie Alle die sich meiner erinnern und schreiben Sie recht bald wieder

Ihrem | Ihnen von Herzen zugethanen | G Magnus.

Berlin | 7 November | 1853

Dear Tyndall

Immediately after receiving your letter,[2] I went to Sauerwald who assured me that he had handed the instruments[3] over to Asher on October 8th. On the following day, I also saw Asher who showed me in his book that he had sent them to Hamburg[4] on October 12th to be conveyed to Nutt[5] in London. He could not understand why you had not yet received the instruments but he was of the opinion that they must undoubtedly be in your hands now. I hope so as well, but should this not be the case, have Mr Nutt write to Hamburg and make enquiries, and write to me at the same time, and I will request Asher write to Hamburg, as well.

Your handwriting is not entirely easy to read and I am therefore not quite sure whether Dr Franz[6] showed you the Leidenfrost experiment[7] here. But I wrote him and asked him to specify the liquids; hopefully I will receive an answer by tomorrow, before this letter is sent and will specify the liquids for you. By the way, Marchand[8] has done quite detailed work on the Leidenfrost experiment using 2 liquids, although the star was not yet known at the time. I was unable to find the work[9] immediately; it may be in volume XIX of his journal, under *Lampensäure.*[10]

You write about my paper: 'Contrary to my expectation the chemical paper[11] pleased me better than the physical one.'[12] If I understand correctly, you did not like the physical one. I had this impression already in Paris, please tell me what you did not like. The physical one seems much better to me, I am aware that the account is somewhat short.

I am now occupied with the liquids and have obtained some quite nice results. Herr Weber[13] has been very useful in this.

What you tell me about Faraday makes me very happy, I came to be very fond of him when we were together in Birmingham, I have always admired him but it is so rare that people who are that eminent in science have such a lovely character; only Berzelius[14] was like that; but not as refreshing as Faraday. Please convey many sincere and friendly thoughts from me and also, my best remembrances to Missis F.[15]

As soon as I see Dubois,[16] I will tell him that a letter from him is awaited in London. He has a severe cold. How did you like his letter on Signor Matteucci?[17] I did, very much.

I ran into Dove today, who is very happy about the honour bestowed upon him through the Copland Medal.[18]

I will give my brother[19] your best wishes, we correspond frequently now, because he is paying for the instruments that I ordered. &.&.&.

I think it is best if Bence Jones's son[20] remains English and my Paul[21] German. But if B.J.[22] wants to send his son here for a while, he could grow up with Paul and go to school, and in return, Paul could go later with him to England. Give many regards to Bence Jones and to everyone who remembers me and write soon again to

Your | sincere | G Magnus.

RI MS JT/1/M/4

---

1. *7[–8] November 1853*: while Magnus dated this letter 7 November, he said that he would wait until the next day to send it.
2. *your letter*: letter missing.
3. *the instruments*: presumably the instruments Tyndall ordered in July (see letters 0787; 0788; 0792; 0793; and 0796). On 11 November, Tyndall noted the receipt of this letter from Magnus and another from Sauerwald, both of which were about his instruments that 'had been sent off a month ago' (Journal, JT/2/13b/635).
4. *Hamburg*: see letter 0769, n. 4.
5. *Nutt*: David Nutt (d.1863), a foreign publisher and bookseller in London, England (*ODNB*).
6. *Dr Franz*: Rudolph Franz (1826–1902), a German physicist known for his collaboration with Gustav Heinrich Wiedemann, with whom he established the Wiedemann–Franz law on the relationship between thermal and electrical conductivity in metal.
7. *the Leidenfrost experiment*: an experiment conducted by the German doctor Johann Gottlob Leidenfrost (1715–94), which demonstrated that when liquid comes into contact with a surface substantially hotter than its boiling point, a layer will form that significantly slows down the evaporation of it. This was called the Leidenfrost Effect.

8. *Marchand*: Richard Felix Marchand (1813–50), a German chemist notable for his work in biochemistry, organic chemistry, and the determination of atomic weights (W. H. Brock, 'Marchand, Richard Felix', *Complete Dictionary of Scientific Biography* [2008], at http://www.encyclopedia.com [accessed 7 December 2015]).
9. *the work*: likely R. F. Marchand, 'Einige Versuche über das Leidenfrost'sche Phänomen', *Journal für Praktische Chemie* 23:1 (1841), pp. 137–43).
10. *Lampensäure*: the German term for a postulated chemical compound found as deposits left behind in lamps. It was eventually determined that it was a variable mixture of acetic acid, acetylaldehyde, and formic acid, after which the term disappeared from use (R. F. Marchand, 'Ueber Bildung der Lampensäure', *Journal für Praktische Chemie* 19:1 [1840], pp. 57–59).
11. *the chemical paper*: likely G. Magnus, 'Ueber die Entstehung von Theer aus ölbidendum Gase', *Poggend. Annal.* 90 (September 1853), pp. 1–12. It was later translated to English as 'On the Origin of Tar in Olefiant Gas', *Phil. Mag.* 6:41 (December 1853), pp. 420–27.
12. *the physical one*: likely G. Magnus, 'Ueber die Verdichtung der Gase an der Oberfläche glatter Körper', *Poggend. Annal.* 89 (August 1853), pp. 604–10. It was later translated to English as 'On the Condensation of Gases by Smooth Surfaces', *Phil. Mag.* 6:40 (November 1853), pp. 334–37.
13. *Herr Weber*: Wilhelm Eduard Weber. For a list of Weber's publications see 'Weber, Wilhelm Eduard', *RS Catalogue*, vol. VI, pp. 293–95.
14. *Berzelius*: Jöns Jacob Berzelius (1779–1848), a prominent Swedish chemist notable for his discovery of a number of new elements, as well as his contributions to atomic theory and chemical nomenclature. Magnus had studied with Berzelius in Stockholm (J. Wisniak, 'Jöns Jacob Berzelius A Guide to the Perplexed Chemist', *The Chemical Educator* 5:6 [2000], pp. 343–50).
15. *Missis F*: Mrs. Faraday.
16. *Dubois*: Emil du Bois-Reymond.
17. *his letter on Signor Matteucci*: see letter 0754, n. 8.
18. *Copland Medal*: misspelling of Copley Medal (see letter 0819, n. 6). Dove received the medal for his work on the distribution of heat over the surface of the Earth.
19. *my brother*: Magnus had five brothers. It is uncertain which brother Magnus was referring to here.
20. *Bence Jones's son*: one of Bence Jones's first two sons, Henry Robert Jones (1844–1912) or Ralph Noel Jones (1845–66). His third son, Archibald Bence Jones (1856–1937) had not yet been born at this time.
21. *Paul*: Magnus's son, Paul Gustav von Magnus (1845–1930), who would grow up to become a banker (M. Reitmayer, *Bankiers im Kaiserreich: Sozialprofil und Habitus der deutschen Hochfinanz* [Göttingen: Vandenhoeck & Ruprecht, 2011], p. 90).
22. *B.J.*: Henry Bence Jones.

From John Peter Gassiot [8–9] November 1853[1] 0821

My dear Dr Tyndall,

Pray do not consider you are under the slightest obligation to me. I have simply done what I considered my duty as a member of the Council[2] in voting the award for what I cannot but consider a most important discovery in Physical Science—One that must have an important bearing in that great question which has yet to be solved—the true cause of the variation of the magnetic needle.[3]

The great difficulty in the awarding of medals arises from the impossibility of fully discriminating between the conflicting claims of the different branches of science. That <u>your</u> discovery[4] is in every respect worthy of the medal[5] is universally acknowledged.

The road to fame is now fairly opened to you. I am confident you do not require such an incentive, but if I should in any way have cheered you on the path you have selected believe me when I say that I am richly rewarded.

Believe me | Yours truly | J.P. Gassiot.

RI MS JT/2/13b/633
LT Typescript Only

---

1. *[8–9] November 1853*: on 9 November, Tyndall noted the receipt of this letter stating, 'This morning the following note was handed to me as I entered the Royal Institution' (Journal, JT/2/13b/633). Since Gassiot wrote from London, it is likely he wrote the letter on 8 November or possibly the morning of 9 November.
2. *the Council*: see letter 0816, n. 7.
3. *a most important . . . the magnetic needle*: Gassiot was referring to the study of diamagnetism, which received significant attention during the mid-nineteenth century (see letter 0710, n. 7).
4. *<u>your</u> discovery*: Tyndall had received the Royal Medal for his 1851 article on diamagnetism, in which he discovered that 'The entire mass of a magnetic substance is most strongly attracted when the attracting force acts parallel to that line which sets axial when the substance is suspended in the magnetic field; and the entire mass of a diamagnetic substance is most strongly repelled when the repulsion acts parallel to the line which sets equatorial in the magnetic field'. Furthermore, Tyndall noted that 'The superior attraction and repulsion of the mass in a particular direction is due to the fact, that in this direction the material particles are ranged more closely together than in other directions; the force exerted being attractive or repulsive according as the particles are magnetic or diamagnetic' (J. Tyndall, 'On Diamagnetism and Magnecrystallic Action', *Phil. Mag.* 2:10 [September 1851], pp. 165–88, on p. 188). See also Volume 3, 'Introduction'; letters 0403 and 0418; and Jackson, 'Tyndall and Diamagnetism'.
5. *the medal*: the Royal Medal (see letter 0815, n. 3).

From Edward Sabine [9] November 1853[1] 0822

My dear Sir,

I have abstained from writing to congratulate you sooner on the award of the Royal Medal,[2] because I have desired that Mr Gassiot, who proposed the award, should be the first person to announce it to you. I will only add the great satisfaction it has given me to witness the general and rapid acknowledgement of your scientific labours, and, in that, the confirmation of the opinion which I expressed to you some time since that the mother country was your proper field.[3]

You will be aware that you will have to attend at the anniversary meeting on the 30th Nov. to receive the Medal from Lord Rosse.

Believe me always | Sincerely yours | Edward Sabine.

RI MS JT/2/13b/633
LT Typescript Only

1. *[9] November 1853*: on 9 November, Tyndall noted the receipt of Sabine's letter after receiving Gassiot's letter (letter 0821) in his journal (JT/2/13b/633). That he received it in the evening suggests that Sabine wrote to Tyndall the same day.
2. *the Royal Medal*: see letter 0815, n. 3.
3. *the opinion which I expressed . . . mother country was your proper field*: see letter 0535.

From Friedrich Ludwig Stegmann 9 November 1853 0823

Marburg, 9 Nov. 1853.

Werthester Freund!

Ich hoffe dass Sie von Ihrer letzten Continental-Reise wohlbehalten wieder heimgekehrt sind. Gar sehr habe ich Verlangen, Ihren lieben Besuch /bald/ einmal erwiedern zu können, und ich hoffe diesen Plan in nicht so langer Zeit auszuführen, da ich in England jetzt so manchen werthen Freund habe. Mittler weile ist von meiner Variationsrechnung, von welcher ich Ihnen bei Ihrem Hiersein Druckproben zeigte, die erste Abtheilung ausgegeben worden und ich habe daher sofort eine sich mir darbietende Gelegenheit benutzt und in einer an unsern lieben Freund Hirst gemachten Zusendung ein für Sie bestimmtes Exemplar beigefügt. Ohne Zweifel wird derselbe Ihnen dasselbe alsbald zugehen lassen und ich bitte Sie herzlichst darin ein Zeichen meiner aufrichtigen Anhänglichkeit gegen Sie erblicken zu wollen. —Außerdem habe ich Hr Hirst ersucht in gefällige Ueberlegung zu ziehen, ob etwa auch Herrn B. Price, welchen Sie vielleicht kennen und welcher gegenwärtig mit der

Herausgabe eines grösseren und allem Anschein nach sehr verdienstlichen Lehrbuchs des Differential-, Integral- und Variations-Calculs beschäftigt ist, ein Exemplar zuzusenden angemessen sei. Ich denke mir dass diesem Gelehrten vielleicht angenehm sein werde, meine Bearbeitung frühzeitig kennen zu lernen, und dass er nur spät auf anderm Wege Kunde davon erhalten dürfte.

Vor acht Tagen hat uns nun unser geschätzter Freund Knoblauch wirklich verlaßen, ein Verlust welcher mich schwer schmerzt. Unterdesßen haben wir wenigstens in so weit einige Genugthuung, dasß unserm Antrage gemäss unser Freund Kohlrausch als außerordentlicher Professor für das Fach der Physik an unserer Universität angestellt worden ist.

Das hier eingeschlossene Briefchen darf ich Sie wohl bitten weiter zu befördern. Herr Dressner, dermalen Professor am R.M.C. in Sandhurst, ist mir der traute Kamerad in meinen Knabenjahren, der treueste Freund im Jünglingsalter gewesen und stets geblieben.

Mit dem Wunsche dass es Ihnen recht wohl gehe und dass Sie an mich manchmal freundlich denken mögen

Ihr ergebenster | Stegmann

Marburg, 9 Nov. 1853.

Dearest friend!

I hope that you have returned home again safe and sound from your last continental trip.[1] I very much desire to be able to return your kind visit *[soon]* sometime, and I hope to carry out this plan in not too long a time, as I now have so many a dear friend in England. In the meantime, the first part of my variational calculus,[2] of which I showed you the proofs while you were here, has been printed, and I have therefore immediately taken an opportunity that presented itself to me and enclosed a copy meant for you in a delivery made to our dear friend Hirst. He will no doubt send it to you immediately, and I ask you whole-heartedly to see in it a sign of my genuine devotion to you. —I have, in addition, requested Herr Hirst to take into willing consideration whether it may by any chance be appropriate to send a copy to Herr B. Price too, whom you perhaps also know and who is currently busy editing a major and, by all appearances, very commendable textbook on differential, integral, and variational calculus.[3] I imagine that this scholar will perhaps be glad to become acquainted with my treatment of the subject early on, and that he might only receive news of it late by another way.

Eight days ago, our treasured friend Knoblauch really left us,[4] a loss which pains me greatly. In the meantime, we have at least some satisfaction, insofar as, in accordance with our request, our friend Kohlrausch has been appointed extraordinary professor[5] for the subject of physics at our university.[6]

May I ask you to forward the little letter enclosed herein.[7] Herr Dressner,[8] at present a professor at the R.M.C at Sandhurst,[9] has been a good chum to me in my boyhood and the most loyal friend in my youth, and has always remained that way.

With the wish that things may go very well for you and that you may think kindly of me sometimes

Your very humble | Stegmann

RI MS JT/1/S/209

---

1. *your last continental trip*: between 6 July and 29 September, Tyndall travelled through Germany, visited Paris, and stopped by Queenwood College before taking up his position at the RI (Journal, JT/2/13b/607–23).
2. *the first part of my variational calculus*: see F. L. Stegmann, *Lehrbuch der Variationsrechnung und ihrer Anwendung bei das Maximum und Minimum*, 2 vols (Kassel: J.G. Luckhardt, 1853), vol. 1.
3. *textbook of differential, integral, and variational calculus*: B. Price, *A Treatise on Infinitesimal Calculus: Containing Differential and Integral Calculus, Calculus of Variations, Applications to Algebra and Geometry, and Analytical Mechanics*, 4 vols (Oxford: Oxford University Press, 1852–60).
4. *Knoblauch really left us*: see letter 0727, n. 13.
5. *extraordinary professor*: a (full) professor rank in Germany, but without a chair and many of the perks that are associated with it. They could, for instance, take fees from students but did not receive a university salary.
6. *our university*: the University of Marburg.
7. *little letter enclosed herein*: not identified, although presumed to be a letter for Dressner (see n. 8).
8. *Herr Dressner*: Carl Dressner, a Professor of German at the Royal Military College, Sandhurst ('Professors and Masters', *The Royal Kalendar, and Court and City Register for England, Scotland, Ireland, and the Colonies for the Year 1853* [London: Suttaby & Co. Stationers, 1853], p. 273).
9. *R.M.C at Sandhurst*: the Royal Military College, a British military academy for training officers located near Sandhurst, Berkshire.

From Henry Thomas De la Beche 10 November 1853 0824

Private | Jermyn Street 10 Nov. 1853.

My dear Sir,

Would you have any objection to have a confidential talk with me about matters which you may remember we spoke about[1] some months since? If you have not shall I call upon you, or would you kindly call upon me, as I am lame just now.

very faithfully | H.T. DelaBeche.

RI MS JT/2/13b/634
LT Typescript Only

1. *matters which you may remember we spoke about*: De la Beche first discussed his wish for Tyndall to lecture at the Government School of Mines in February (see letter 0721). It was one of the options Tyndall was considering before taking up his position at the RI; see letters 0736, n. 5 and 0765. On 10 November, after receiving this letter, Tyndall spoke again with De la Beche about the position at the Government School of Mines, and left 'with the understanding that I was to turn the matter over in my mind' (Journal, JT/2/13b/634–35).

To Thomas Archer Hirst 7 [and 11] November 1853 0825

7th Nov. 1853.

My dear Tom.

I have sent a copy of your lecture[1] to Lyon Playfair. By degrees I shall distribute them all, just as occasion offers, it is better to seize the occasion than to appear to thrust them on people's attention. With regard the the medal[2] I told you all I knew, I have as yet received no official intimation of the thing, nor will the medal be given till the 30th of this month. It is the Royal medal,[3] in reality two medals, one of gold and the other of silver, value 50 pounds. This is all I know about it. The investigation you enquire about is one on the vibrations produced by heated metals, and the man from whom I diverge is Prof. Forbes of Edinburg—He has written a long paper on the subject, and written very confidently, stating certain general laws from which he draws important conclusions—These general laws it has been my fate to annihilate utterly, I have besides expanded the subject far beyond the limits hitherto assigned to it.[4] It will form a beautiful Friday evening lecture,[5] and will also I trust furnish materials for a pretty paper in the Philosophical Transactions.[6] I went to see Dr. Simpson's family[7] yesterday. We had no tea but an hour's pleasant conversation—which Mrs. Simpson kept wholly directed upon herself and your daughter[8] sat on the sofa rather behind and said almost nothing—I like her notwithstanding, just as well as if she had talked more, for the little she did say was quite to the purpose. I said once that you were my son and that the logical conclusion was that I am her grand papa. I think I will call upon them tomorrow again—they leave town on Wednesday. They would have been very glad to see you I have no doubt, we talked a good deal about you and Simpson has a certain theory regarding your passive submission to adverse circumstances which amused me a little. Ginty is wrong in saying that I am growing morose. I myself testify to the contrary and I know better than

Ginty. I take no pains to explain to Ginty the foundations of my actions—he sees their surface only and draws foolish conclusions from them. I ought to have told you that Dove got the Copley medal[9]—I am sincerely glad of it for I know it will greatly gratify a very worthy fellow. I believe the Royal and the Copley are the only two that have been awarded this year.

<Hereafter LT Transcript Only>

11th. Nov[10]

The other leaf has been written sometime. I intended to write more to you to explode that philosophy which you call strange and yet true[11]—but I forbear for the present.

Playfair I learned yesterday is going down to Salisbury[12] on a visit I will write a note to him today recommending him to pay Queenwood a visit.

I send you a note I got from Gassiot[13] a day or two ago and another from Col. Sabine.[14] Send me both again.

Your affectionate | John.

RI MS JT/1/T/577
RI MS JT/1/HTYP/282

---

1. *your lecture*: see letters 0808, n. 2 and 0810, n. 1.
2. *the medal*: the Royal Medal (see letter 0815, n. 3).
3. *the Royal medal*: Hirst had asked Tyndall if he was to receive the Copley medal (see letter 0819, n. 6).
4. *The investigation . . . assigned to it*: see letters 0814, n. 7 and 0819, n. 20.
5. *Friday evening lecture*: 'On the vibration and tones produced by the contact of bodies having different temperatures', *Roy. Inst. Proc.*, vol. 1, 356–59.
6. *pretty paper in the Philosophical Transactions*: J. Tyndall, 'On the Vibrations and Tones produced by the Contact of Bodies having different Temperatures', *Phil. Trans.* 144 (1854), pp. 1–10.
7. *Dr. Simpson's family*: see letter 0819, nn. 2–4.
8. *your daughter*: Anna Martin (see letter 0816, n. 11).
9. *the Copley medal*: see n. 3 and letter 0820, n. 18.
10. *11th. Nov*: This addendum appears to have been sent together with the letter written 7 November.
11. *that philosophy . . . true*: in response to a comment that Tyndall made to Hirst about the scanty letters he had been receiving (see letter 0814), Hirst responded with an explanation in letter 0819.
12. *Salisbury*: a city located in Wiltshire, England about 13 miles from Queenwood.
13. *a note I got from Gassiot*: see letter 0821.
14. *another from Col. Sabine*: see letter 0822.

To Thomas Henry Huxley 11 November 1853 0826

11th. Nov. 1853.

My dear Huxley,

I have just gasped up the ascent of Baker Street[1] having walked all the way from the Institution; and here I find your note,[2] like a glass of Johannisberger,[3] to cheer and to refresh me. I quite agree with your philosophy of the medal.[4] As long as man is man and lives in a world where he is anxious to bear his head erect such things I suppose will have a value for him, for they are the recognition of his manhood in the sphere into which circumstances have cast him. It is exceedingly curious to me what strokes and upliftings you and I have had in common, this time last year you were in my present position,[5] and I learned to day that there is some similarity in the circumstances under which both got the medals. I trust that we may weave a prediction from these omens, and that providence still intends to build us a Toronto[6] at home. That we shall be in closer proximity to each other I sincerely hope and firmly believe.

I was not aware that a speech was expected at the dinner.[7] I hope I shall be half drunk if called upon, for otherwise my courage will fail. I often wish that I could stultify the motor nerves on such occasions—They tickle my muscles into most unpleasant quiverings.

I shall look with joy upon your countenance whenever you enter my door, and were you a smoker I would offer you my best cigar. I have a most confused idea of the geography of the outskirts of London. Were I to be hanged for my ignorance I could affix no precise locality to the term St. John's Wood.[8] I have contented myself ever with the fact that Huxley lived there, and this is the whole significance which the place has ever had for me. By degrees however I trust I shall encompass the question and track you out.

Staite's[9] contrivance is to keep the coal points a certain distance apart and thus give fixity to the electric light. Several contrivances of the kind have been made, the best I think by Duboscq Rue de l'Odéon Paris.[10] I heard Staite lecture once[11] in Manchester,[12] the light was good but my impression at the time was that Mr Staite was a superlative ass. They formed a Committee upon the subject in Manchester but it came to nothing. Joule was on the Committee.[13] I saw him a year or so afterwards and on mentioning the subject I know he gave utterance to something that was quite in harmony with the impression which I have permitted myself the pleasure of expressing above. If you wanted information on the subject it would be best to write to Manchester. I believe the thing exploded there. And now my dear fellow-cub, goodbye

ever Yours most truly | John Tyndall

I reciprocate your congratulations. Next year you will be a power in the state yourself and contribute to deciding as to the medals.

IC HP 8:13
RI MS JT/1/TYP/9/2863
Typescript Only

---

1. *Baker Street*: a street in London, England within the Marylebone district, located to the north of the Thames River. Presumably Tyndall had to walk along Baker Street to get to his house
2. *your note*: letter missing; Tyndall did not note receipt of Huxley's letter in his journal.
3. *a glass of Johannisberger*: a type of white wine produced at Castle Johannisberg, located on the Rhine River (*OED*).
4. *the medal*: the Royal Medal (see letter 0815, n. 3).
5. *you were in my present position*: about the same time in 1852, Huxley learned that he received the Royal Medal for his *Rattlesnake* papers. See in particular Huxley's letter to Miss Heathorn, 13 November 1852, in *Life and Letters of Thomas Henry Huxley*, ed. Leonard Huxley, 2 vol. (New York: D. Appleton, 1991), vol. 1, pp. 109–11.
6. *Toronto*: both Huxley and Tyndall had been recently rejected for positions at the University of Toronto (A. B. Macallum, 'Huxley and Tyndall and the University of Toronto', *The University of Toronto Monthly* 2:3 [December 1901], pp. 68–76). (See also letters 0728, n. 2; 0748, n. 16; and Volume 3.)
7. *the dinner*: the dinner Tyndall was asked to attend on 30 November to receive the Royal Medal (see letter 0822).
8. *St. John's Wood*: a district in the City of Westminster, London, located near Regent's Park. Huxley lived at 4 Upper York Place, St John's Wood.
9. *Staite's*: William Edwards Staite (1809–54), an English inventor notable for his work on the arc lamp.
10. *Duboscq Rue de l'Odéon Paris*: Jules Duboscq (1817–86), a French photographer and maker of scientific instruments whose workshop was located at 35 Rue de L'Odéon in Paris (P. Brenni, '19th-Century French Scientific Instrument Makers XIII: Soleil, Duboscq, and Their Successors', *Bulletin of the Scientific Instrument Society* 51 [1996], pp. 7–16).
11. *I heard Staite lecture once*: see letter 0424, n. 10.
12. *Manchester*: the University of Manchester.
13. *the Committee*: not identified.

To Thomas Archer Hirst 13 November 1853 0827

Sunday 13 Nov. 1853.

My dear Tom

Up to 3 oC. today I lived a life of thought and half strangled my intellect over the Mechanical Theory of Heat.[1] By slow thoughtful plodding—quick jumping and some pausing to take breath I managed to enter the subject here and there and to master a good deal of it. At 3 oC. I went to Anderton's[2] to dinner, had some roast mutton and apple tart—the thought struck me that I would have a glass of brandy and water, I ordered it, it came, and I drank it, wandering meanwhile through some columns of the Weekly Dispatch.[3] But the dose was too much for me, it stultified me, and completely spoiled my evenings work. Though it is now nearly 9 oC. I am unfit for any thing serious, and to kill time I take up my pen to write common-place to you. The moon is up, and I have been just out amid the crowds that cumber the pathways. While looking at the silver disk in the ultramarine of heaven I forgot that I had dined, and was lifted a little way, brandy and all, above the earth. There are the stars too when I move my Venitian blind looking steadfastly and solemnly down upon the mighty anthill with all its petty joys and cares. But the Clod[4] returns and I cannot moralize further. Your daughter[5] was ill with a cold during her stay in London, she went to a theatre and for a sore throat, I called twice upon Simpson being prompted purely by the fact of having heard you speak of him so often and having read those letters from your daughter to you.[6] When I saw her last she was muffled up upon a sofa, but getting better of her cold. They are now I suppose gone—at all events I have heard nothing of them for some days. I have as yet had time only to glance over your strictures upon my lecture,[7] no doubt they will be of service to me, though I dont know whether you kept the fact of the living man uttering the sentence, in view—your correction in some case, would undoubtedly be necessary for the press, but I am not sure whether this is equally the case for a spoken lecture—you ask two questions about radiant heat which I do not quite understand. The conversion of radiant heat into conducted heat presents no difficulty and the heat of conduction I believe to be a motion of the molecules of the body. I read your remarks upon the medal[8] with much pleasure, but when I proposed the Prussic acid[9] I intended that the world should know nothing about it, but should regard my death as orthodox. There is little fear of the medal troubling my mental health. Its chief value to me was that it stirred up my laziness, and compelled me to climb up to a position from which I looked down upon it with the calmness of a philosopher. Strange to say that when a man attains this position he is capable of most enjoying such a thing. I suppose it is just the same with plum pudding. The man who can perfectly control his appetite and can relinquish plum pudding upon occasion, enjoys it more than the pig who gorges himself with it—I think this is a bit of Oliver Cromwell's philosophy[10] and it appears to me to be very sound. It is a strange antithesis—the medal is very pleasant alone and still I dont care a damn about it. dont think that I swear—I merely use a commonplace expression because it seems showy and therefore, suits me. Your remark about the condensing of the labour of years into weeks and thus exploding, rebounds from me like one of your own elastic balls. I feel as strong tonight thank God as 99 per cent of those who know nothing of a taxed brain. There is *[work]* in my muscles still boy—You have asked me once or twice what those experiments are that I have been making[11]—they are on those vibrations and tones produced by the contact of bodies having different temperatures. Forbes of Edinburgh[12] has written a long paper[13] upon the subject and enunciated with a great degree of confidence what he calls the general laws of action. These general laws it has been my lot to annihilate utterly, not leaving them a scrap of logic to stand on. Dont mention the matter, as it will be some time before I can publish my results—The subject will make a capital Friday evening lecture. I saw Mr Best a few days ago and, was much pleased to hear his account of your Andover lecture.[14] You thought it was in some measure a failure, but he told me that he should not desire to hear a better discourse. I think this letter of mine is a practical refutation of that "<u>reflexional</u>" philosophy of yours.[15] Why man, when I write to you I could write by the yard, whereas when I write to others I am often cramped for expression. Your reflexional philosophy is personal and private, and you must beware of offering it to be universal. Perhaps you want to say something good—now I dont, and therefore my pen is free, I am content if there be a grain here and there amid the chaff. The finding of one pearl compensates for rambling through whole bushels of mud, and if there be no pearl, why you can turn the mud to agricultural purposes. I have had a letter from Tommy Grieve[16] which I have copied into my journal side by side with another from Sir Henry delaBeche[17] which reached me at the same time, but giving Tommy's the precedence.[18] If you were near me I would consult you on a point of some importance. I am virtually offered a position under Government[19] which I could hold in connexion with my present one—but I am afraid to accept it. I fear being weaned from my researches. What do you say? Is it not better to work forward, even on short commons, for some years, than to give up, or even seriously interfere with, the time necessary for investigations? supposing Faraday had done so at my age would he be as respectable a man as he is at present. Would Michael Faraday with a heavy Bankingbook be as handsome a figure to the human intellect as Michael Faraday "der größte Physiker"[20]—as some German[21] calls him. I have been tempted once or twice since I came to London to mingle in the ways of the world, but I feel great peace of mind in the fact that I pay no visits and am not interrupted in my work, the bonds between me and Faraday are strengthening daily. We know each other and trust each other. There is certainly a kind of freemasonry between souls, by which the spirit recognizes his fellow. Its a wonderful consummation to me when I think of it how I have gravitated towards Faraday. It is a sweet reward which the good gods had stored up for a hard working son of a mother. All the Fellows[22] are to dine together on the 30th of November—the day the medals will be distributed. I shall be expected to make a speech, I am told. I wish you would sit down and write a short one for me. I am now getting clearer—the clod which weighed upon my brain is dissolved and become volatile, Now if I with this incubus[23] squat upon me could get through these 6 pages without much effort, what could I accomplish with the incubus gone home to the devil, and all my faculties free? There is a rule of three question, and when you have solved it reconcile the result with your "reflexional" philosophy. reflexional, reflexional—let me pause; what is the meaning of this 'tinkling cymbal',[24] it has either no meaning or it is totally inapplicable to the case in point, for I reflect you more truly than any body else, you mean non-reflexional. You are thinking of those sordid dogs whose souls are like the blackened surface of an old kettle, and unto whom when you shoot the rays of thought you get none back, or if you get them at all the rays are 'degraded'. It is pleasant to you, is it? to discharge the liquor of the intellect into such gaping sand-bags! Thou squanderer of heaven's riches, thou giver of medals to those who dont deserve them! It is easier to jump down hill than to jump horizontally and thus you find it easier to write down hill than to write to a man on your own level—from which I infer that the basis of your reflexional philosophy so 'strange' and 'true' is pure laziness. It gives me pleasure to shell those lath bastions behind which you have esconced yourself, and having thus exposed the rottenness of your position I retire to bed with great satisfaction.

John Tyndall

RI MS JT/1/T/578
RI MS JT/1/HTYP/284–284a

---

1. *the Mechanical Theory of Heat*: a theory articulated by the physicist Benjamin Thompson (1753–1814) in 1798, which stated that heat and mechanical energy were equivalent. The theory played an important role in the development of thermodynamics.
2. *Anderton's*: see letter 0814, n. 5.
3. *the Weekly Dispatch*: a radical weekly newspaper published in London between 1801 and 1928.
4. *Clod*: 'a mass formed by the coagulation of anything liquid, esp. blood' (*OED*).
5. *Your daughter*: Anna Martin (see letter 0816, n. 11).
6. *those letters from your daughter to you*: not identified.
7. *your strictures upon my lecture*: with letter 0811, Tyndall sent Hirst his introductory lecture (see letter 0809, n. 6), requesting feedback. Hirst's response is missing and Tyndall did not record receipt of it in his journal.
8. *the medal*: the Royal Medal (see letter 0815, n. 3).
9. *Prussic acid*: see letter 0816, n. 10.
10. *Oliver Cromwell's philosophy*: possibly an allusion to Oliver Cromwell's (1599–1658) puri-

tanism. Cromwell was lord protector of England, Scotland, and Ireland from 1653 until his death, and played an important role in the Puritan Revolution of the mid-seventeenth century (*ODNB*).

11. *those experiments are that I have been making*: see letter 0819, n. 20.
12. *Forbes of Edinburgh*: James David Forbes.
13. *a long paper*: see letter 0814, n. 7.
14. *your Andover lecture*: a lecture Hirst delivered with which he was displeased (see letter 0814, n. 6).
15. *that 'reflexional' philosophy of yours*: this was in reference to letter 0819. Hirst noted in his response to this letter (letter 0829), that Tyndall had misunderstood his philosophy.
16. *a letter from Tommy Grieve*: letter 0818.
17. *another from Sir Henry delaBeche*: letter 0824.
18. *giving Tommy's the precedence*: on 10 November, Tyndall wrote out Grieve's letter above De la Beche's (Journal, JT/2/13b/634).
19. *I am virtually offered a position under Government*: this was a position at the Government School of Mines (see letter 0824, n. 1).
20. *"der größte Physiker"*: German for 'the Greatest Physicist'.
21. *some German*: possibly Heinrich Wallmann, author of 'Wissenschaftliche Mittheilungen. Ueber organische Elektricitätsentwickelung', in Z. Tahrgang (ed.), *Lotos, Zeitschrift für Natur-Wissenschafte* (Prag: Jerabek, Ursulinergasse, Nr., 1852), vol. 2, pp. 4–21.
22. *the Fellows*: the Fellows of the Royal Society.
23. *incubus*: a demon, or something that weighs heavily upon a person (*OED*).
24. *'tinkling cymbal'*: Corinthians 13:1.

To Samuel Hunter Christie 15 November 1853[1] 0828

Royal Institution, Albemarle Street | 15th November 1853.

Dear Sir

From three several communications, for which I am indebted to the kindness of three members[2] of the Council of the Royal Society,[3] I learned that the Council had awarded me the Royal Medal.[4] I did not seek the honour thus conferred upon me, but I regarded it as the spontaneous utterance of the Council's opinion of my labours, and valued it accordingly.

This morning I have learned that some Members of the Council[5] have, on more mature consideration, found reason to doubt the wisdom of the award. They are not convinced of the originality of my labours,[6] nor perhaps of its worthiness, even if original. On the former point the knowledge I possess needs no ratification; but with regard to the latter, I am anxious, without a moment's unnecessary delay, to place the Council in an unpledged position, and to restore to it that perfect freedom of action which it enjoyed before, in a moment of precipitancy apparently, it made the medal mine. I share to the fullest extent the conviction evidently entertained by the Council that such awards ought not to be lightly made, and I should be doubly unworthy of any such distinction were I willing to accept it while coloured by a doubt. While prosecuting those inquiries which have for a moment attracted the notice of the Council, there was no Royal medal before my eyes; it is an object at which I never aimed—a consummation on which I never based a hope. And whatever may have been the pleasure which the award imparted to me, it was, I believe, inferior in kind to that with which I now resign all claim to the medal, and thus restore to the Council the liberty of making a better choice.

I have the honour to be, dear Sir | Your most obedient servant | John Tyndall.

Samuel Hunter Christie Esq. | Secretary of the Royal Society &c. &c. &c.

RI MS JT/1/T/579
RI MS JT/5/12/f23
RI MS JT/2/13b/637

---

1. *15 November 1853*: the original letter is missing. The version relied on here is a copy written in Tyndall's hand (RI MS JT/1/579). There is also a typescript version in his journal (16 November, JT/2/13b/637) and another typescript copy of that version with LT's annotations (RI MS JT/5/12/f23).
2. *three several communications . . . kindness of three members*: see letters 0821 (from John Peter Gassiot) and 0822 (from Edward Sabine). The third member is unidentified.
3. *Council of the Royal Society*: see letter 0816, n. 7.
4. *the Royal Medal*: see letter 0815, n. 3.
5. *some Members of the Council*: on 15 November, Tyndall recorded that Gassiot informed him that some Members of the Council were in objection to Tyndall receiving the Royal Medal. He stated, 'It appears that certain people have set about sifting my claim and have contrived to get up an opposition against me. They say that my investigations were partly conducted along with Knoblauch and partly in the private cabinet of Prof. Magnus in Berlin, and add something regarding Plücker which I don't understand' (Journal, JT/2/13b/636). Tyndall later learned that William Hopkins, a member of the Council, who was not present when the decision regarding the Royal Medal was made, opposed his win. Charles Wheatstone, who was not a member of the Council, was also against Tyndall receiving the Royal Medal (Jackson, 'Tyndall and the Royal Medal', pp. 157–58).
6. *my work*: Tyndall was to receive the Royal Medal for his work on diamagnetism (see letter 0821, n. 4).

From Thomas Archer Hirst 15 November 1853 0829

Queenwood College | Novr 15th /53

My dear John

I have just perused your long letter[1] with very great pleasure. Carlyle mentions somewhere an old King who liked to consult his Barons first when they were sober and next when they were drunk.[2] After profiting by many a letter written by you in a sober state I have enjoyed much this one written confessedly under the influence of a glass of brandy. Only fancy a fellow under the influence of Brandy asking in a solemn manner What is reflection? is it not a mere tinkling cymbal? It reminded me of four lines in Hafis:[3]

> Frage nicht: 'Welch einen Nutz
> Schafft die Trunkenheit?'
> Vom Verstande, wenm du trinkst,
> Bist du rein befreit.[4]

Not that I assert you were drunk, but drunk or sober you have interpreted my reflexional philosophy[5] all wrong. I'll set thee right some day face to face thou old tipler,[6] meanwhile chew it well for I still mean what I said, and did not mean to excuse my laziness. There were other sentences in thy letter for which I could have hugged thee. Especially that about Faraday, I have often thought about the remarkable proof offered in your case to our old precept that like gravitates to like and I have just as often rejoiced about the sweet reward which the Gods are granting to thee thou hardworking son of a mother. I don't think you would have written that sentence without the Brandy although you might have felt it just as much. I thank the Brandy for it. You dont give me particulars sufficient to give my decided opinion about the offered situation under Government.[7] At present I feel inclined to say refuse it and stick to investigation like Faraday. at the same time circumstances might render a divergence from Faradays footsteps advisable; in any case I shall not be afraid of the decision you come to. I will write you no speech for the 30th of November[8] because I dont feel able to tell you what you ought to say. As to you being tempted to mingle in the ways of the world it made me smile. The dinner bell has rung I must run. For the next 5 weeks I see more work before me than I can satisfactorily get through, so spare me your reproofs if I should send short letters.

Yours affectionately | T.A. Hirst

RI MS JT/1/H/184
RI MS JT/1/HTYP/285

1. *your long letter*: letter 0827.
2. *an old King . . . sober and next when they were drunk*: Hirst was likely referring to Carlyle's discussion of William Rufus (1056–1100), King of England from 1087 to 1100, and his Barons who would debate 'everything, as Tacitus describes the Ancient Germans to have done, two times: once sober, and once what he calls "drunk,"—not dead-drunk, but jolly round their big table' (T. Carlyle, *Latter-day Pamphlets* [London: Chapman & Hall, 1850], p. 262).
3. *Hafis*: also spelled Hafiz, the pen name of Shams al-Din Muhammed (1320–89/90), a famous Persian lyric poet (M. Hillman, 'Hafiz: Shams al-Din Muhammed' in F. Magill [ed.], *Dictionary of World Biography: The Middle Ages*, 10 vols [London: Routledge, 1998], pp. 417–20).
4. *Frage nicht . . . Bist du rein befreit*: German. 'Ask not: "What is the benefit of drunkenness?" | From understanding reason, when you drink, | you are purely exempt, released' (for German see 'CXLIV.', in G. F. Daumer [ed.], *Hafis: Eine Sammlung persischer Gedichte* [Hamburg: Hoffman und Campe, 1846], p. 86).
5. *my reflexional philosophy*: see letter 0827, n. 15.
6. *tipler*: misspelling of tippler, 'a habitual drinker of intoxicating liquor (implying more or less excess, but usually short of positive drunkenness)' (*OED*).
7. *the offered situation under Government*: see letters 0824, n. 1 and 0827, n. 19.
8. *speech for the 30th of November*: in letter 0825 Tyndall said that the Royal Medal would be officially presented on 30 November.

To John Peter Gassiot 18 November 1853 0830

18th Nov. 1853.

My dear Sir,

I am anxious to clear up an ambiguity which may possibly attach itself to my letter to Mr Christie[1] and thus prevent the further expenditure of the Council's[2] time upon a subject which I feel has already occupied it too long. The Royal Medal[3] was awarded to me for the discovery of a law[4] which I for my part firmly believe to be all-pervading. But as I am informed that some men of eminent standing in Science, and upon the council,[5] entertain a different opinion, I have to request that you will make no further attempt to confirm the award. For however highly I may value such a mark of distinction, I should, under present circumstances, feel it to be my duty respectfully to decline it.

I remain dear Sir | most truly yours | John Tyndall.

RI MS JT/1/T/579
RI MS JT/2/13b/637–38

1. *my letter to Mr Christie*: see letter 0828.
2. *the Council's*: the Council of the Royal Society (see letter 0816, n. 7).
3. *The Royal Medal*: see letter 0815, n. 3.
4. *the discovery of a law*: see letter 0821, n. 4.
5. *and upon the council*: after sending this letter, Tyndall wrote another version and revised this sentence to read: 'But as I am informed that some eminently qualified judges, men of eminent standing in science, (Fellows of the Royal Society) not upon the Council . . .'. Tyndall refers to this altered version in letter 0831. This correction is also included in his journal entry for 18 November (JT/2/13b/637–38). See also letter 0828, n. 5.

To John Peter Gassiot 20 November 1853 0831

Dear Mr Gassiot,

I think I must ask you to substitute the accompanying note[1] for the former one.[2] It is altered in one particular only, apparently trifling, but involving a matter of conscience to me. I say 'men of eminent standing in science' instead of 'eminently qualified judges' for how high soever their standing may be it is my conscientious conviction that in the matter before us they have not qualified themselves to judge aright and that they are now taking up a position which they will be unable to stand by at a future day. The matter reminds me of Melloni's struggle with the Paris Commissioners[3] and I can well appreciate the smile which the safety of his position enabled him to indulge in when eminent names[4] pronounced against him.

Believe me | very truly yours | John Tyndall.

RI MS JT/2/13b/638
LT Typescript Only

1. *the accompanying note*: this was a different version of letter 0830 that included an important alteration (see letter 0830, n. 5).
2. *the former one*: see letter 0830.
3. *Melloni's struggle with the Paris Commissioners*: reference to Macedonio Melloni (1798–1854), an Italian physicist and political exile in Paris, notable for his work on radiant heat. Melloni struggled to get the Commissioners of the French Academy of Sciences to examine his work for publication, but Faraday recognized its merit and helped Melloni obtain the Royal Society's Rumford Medal in 1834 (J. Tyndall, 'Life and Letters of Faraday. By Dr. Bence Jones', in W. H. Bidwell [ed.], *The Eclectic Magazine, or Foreign Literature, Science, and Art* 12 [1870], pp. 353–57).
4. *eminent names*: see letter 0828, n. 5.

To Tommy W. Grieve 20 November 1853 0832

Queenwood College | November 20th. 1853.

My dear Dr. Tyndall,

I have to stop at home to day because of my cold and there 13 boy becides[1] me and every one has got colds I sapose[2] you will tell me when you will come down here. It is very cold and rainy we dont have lessons before breakfast because of the boys colds. We had an examination in ritting[3] and there was two and in the first I was 22nd and in the other I was 18th so that I have risen three and it was between the third and fourth classes and some had gon down. I dont think that I can write a very long letter for there is not much to say. I am going to read a paper[4] but as I herd[5] that you would come down here I will not read it till you come down and I think you would like to hear it. Buchan sends his kind love to you and so do I

I remain | your affectionate friend | T. W. Grieve | Good Bye.

The holidays are coming Near. Hip Hip Hip Hura.

RI MS JT/1/TYP/1/460
LT Typescript Only

1. *becides*: misspelling of 'besides'.
2. *sapose*: misspelling of 'suppose'.
3. *ritting*: misspelling of 'writing'.
4. *I am going to read a paper*: see letter 0836, n. 4.
5. *herd*: misspelling of 'heard'.

From Thomas Archer Hirst 20 November 1853 0833

Queenwood College | Nov[r] 20th/53

My dear John.

If I cannot always write you long letters I can pass others through your hands and thus let you see for yourself what I am doing when such doings are likely to interest you. Enclosed is Henry Hill's last letter[1] and my answer post the latter to him and the former to me when read. My health has been much better during the week and as I have worked pretty well I am allowing myself a little relaxation to day in the way of letter writing. Edmondson arrived home last night with news from you which was welcome. I have devoted the principal part of the week to my lecture at Abbotts Ann on heat[2] and have already written what will take 2 ½ hours to deliver, as I do not wish to cut it shorter I shall split it into two lectures one this season and the other next. I am inclined to think the two lectures will interest the good folks they are simple and illustrated by continual reference to the daily experiments on Heat which they make for themselves[3] The experiments on the lecture table will be few in number and are all simple & well known ones. I wish I could hit upon a few more neat ones that would not require too much trouble and expense

1st I shall first shew the dilation of solids, liquids, and gases.[4] That of solids by an iron bar fitting exactly into a form when cold—

That of liquids and gases by a simple glass tube and bulb

I want you to send me a couple of suitable tubes

2nd I shall shew the currents in boiling water

3rd The fire syringe[5]

4th Latent heat[6] of water by melting ice and observing the thermometer to be stationary (I do not like this class of experiments, however, for such a lecture it does not appeal to their sense of sight sufficiently) or better by mixing on the one hand a pound of ice at 32° and on the other a pound of water at 32° with a pound of water at 174°

5th Latent heat of steam by shewing the heat disengaged by condensing steam from the vessel A in the cold water of vessel B

6th Dependence of boiling point on pressure By means of a closed flask containing only water and steam at 212° which when immersed in cold water boils and in warm water ceases.

7th The three cases of conductibility in wood illustrated by shewing part of the trunk of a tree so made that bars in the three directions can be lifted from other places and put back again.

8. On Radiant Heat I should here like to shew the experiment of igniting tinder with parabolic reflectors.[7] For this reason I want to know if you can get me a pair of parabolic reflectors If you will suggest any alteration in these experiments I shall be glad at the same time I should like you to procure me the

above apparatus and to have it sent down to me before the end of the month, If you suggest anything further, or can meet with other apparatus that will be serviceable buy the latter and send all together. I have not yet mentioned it to Edmondson but if he objects the apparatus will be my own. You see, if I lecture on physics and make experiments I must have apparatus furnished, for my other work which I consider more suited and important to me, will not allow me the time to construct much apparatus for myself. After all has been provided there will be much which will require my time and constructive powers. At Queenwood all goes on well and smoothly, I never knew 76 better boys. As to the Edmondson family[8] I keep on the same constant terms with them which maybe termed a relation of respectful non-interference. In Edmondson I have seen nothing to blame or praise lately as far as I have been permitted to see he has done his duty and nothing more. As regards M^rs^ and Miss E.[9] who are almost strangers to one another, I rarely see them more than once a week at supper on Sunday Evenings With Mummery I have had some slight difference; I offended him sorely about a week ago, but now he sees that my conduct was merely firm and not offensive. He told me one day with that annoying giggle of his that he was about to purchase a German Doctorship[10] for £25 at first I did not think he was in earnest, but when I saw he was, I did not spare his feelings in expressing my opinion about the matter. He scarcely spoke to me for two or three days after and was evidently in low spirits. I was about to go to him and tell him why I had spoken so harshly when he came to me again, telling me he had almost given up the notion but that he had written to two or three friends[11] about the subject, he still supported the step with his former sophistries but it was evident that his feeling of justice and honour had been touched and had rather shaken his decision. Well, answers came to his letters, and it appears they encouraged him in the matter and hoodwinked him as to its positive dishonesty. Edmondson himself did so and gave him a testimonial to help him to attain it. With more courage than before he came to me again, attacking me stronger than ever with his sophistries which apart from a sense of honour are staggering enough. It was evident that he wanted nothing more now than to win me over to conviction. This he found more difficult than he anticipated and pushed his arguments until they became obtrusive and tedious. We were talking a while together and I had appealed to him and combatted his arguments in vain for a whole hour, at the end of which I cut it short by requesting him to change the subject telling him that he must see from what I had said and repeated that I could never recognize a title so obtained either by him or any other man, and that on his obtaining his Doctorship I should only be acting conscientiously if I wrote to the Times and under my own signature exposed its invalidity. He laughed at my threat and has not spoken about the matter again, but as I hear he is taking steps towards procuring the degree. Of course this affair has lowered him considerably in my estimation though we continue as friendly as before. There is no depth of character in the man at all and not much talent either, he is in general sensible in his conversation but when pushed superficial. He is good natured and honest so far as this is recognized in the world. Beyond this his honesty has little root. It is not the flower of his character as it should be, it is but grafted on the tree of worldly providence and prosperity whose roots sink far deeper into his nature. Once in his day he has worked hard at study but it was less from a humble desire to learn than in order to attain a certain worldly position; this is rendered the more evident because now having reached this position he intends from his own confession to rest and work only so much as is necessary to maintain it. So that, good fellow though he is, sympathy with him does not go far Goodbye, John

As ever | T. A. Hirst

I have spoken to Edmondson he says he has ordered some reflectors so never mind. I will see that the Electro Magnet is sent soon.

RI MS JT/1/H/185
RI MS JT/1/HTYP/287–288a

1. *Henry Hill's last letter*: presumably a letter Hill sent to Hirst regarding his financial difficulties. Hirst eventually helped Hill gain employment with Bevington (see letters 0836; 0848; 0849; 0853; 0855; 0868; and 0898).
2. *my lecture at Abbotts Ann on heat*: see letter 0836, n. 10.
3. *the daily experiments . . . for themselves*: not identified. Hirst was likely suggesting that the experiments he planned for the lecture were related to everyday household tasks involving the use of heat.
4. *dilation of solids, liquids, and gases*: thermal dilation, or expansion, refers to the tendency of matter to change in shape, size, or volume when exposed to heat.
5. *The fire syringe*: 'an instrument by which dry tinder or spunk may be ignited by the heat produced by the sudden compression of a portion of atmospheric air. It consists of a hollow cylinder, closed at one end, and a solid piston fitting it accurately, and having in its under side a cavity to receive the substance to be ignited' (J. Johnston, *A Manual of Chemistry on the Basis of Turner's Elements of Chemistry; Containing, in a Condensed Form, All the Most Important Facts and Principles of the Science*, 6th edn. [Philadelphia: Charles Desilver, 1856], p. 60).
6. *Latent heat*: see letter 0836, n. 5.
7. *parabolic reflectors*: 'a reflector for light, sound, radio waves, etc., made in the form of a paraboloid of circular cross-section so as to reflect parallel rays to a focus, or produce a parallel beam from a source at the focus' (*OED*).
8. *the Edmondson family*: see n. 9. The Edmondsons also had a son, George, who attended Queenwood College.

9. *Mrs and Miss E.*: George Edmondson's wife, Anne Edmondson (see letter 0818, n. 2) and their daughter, Jane Edmondson.
10. *purchase a German Doctorship*: see letter 0817, n. 4.
11. *two or three friends*: not identified.

From George Wynne 20 November 1853 0834

Board of Trade[1] | 20 Nov. 1853

My dear Tyndall,

I am somewhat tardy in offering you my congratulations but they are not the less sincere I assure you on that account, I sympathize most fully in the great gratification you must have felt in receiving so high a distinction[2] from such a quarter and the more gratifying as being unsought for and unexpected and I am delighted at seeing that men are sometimes honored without pushing themselves forward.

My old draftsman[3] as someone is pleased to designate himself will do me at the least the justice to say that I appreciated him long before he came to honour and that I valued him for himself and not for the distinctions he has obtained tho' I may flatter myself a little upon my discernment.

Mrs Wynne[4] in joining me in my congratulations has desired me to say that she will not consider the visit you purpose on the Sunday very wrong and she suggests your taking the train which will just bring you in time to go to church with us and afterwards join our early dinner Any Sunday will find us at home and you a hearty welcome.

Frank[5] has written you a letter of thanks[6] but as the character in which it is written requires a key I thought it best to reserve it until you come.

Ever most truly yours | Geo. Wynne

RI MS JT/1/TYP/5/1851

---

1. *Board of Trade*: see letter 0784, n. 9.
2. *receiving so high a distinction*: the Royal Medal (see letter 0815, n. 3). Wynne was not aware that Tyndall had declined the Medal due to objections from some Fellows of the Royal Society (see letters 0828; 0830; and 0831).
3. *My old draftsman*: Tyndall had worked as a draftsman in Wynne's Division of the Irish Ordnance Survey from 1839 until 1840, when Wynne left the Survey. Tyndall remained with the Irish Survey until 1842, when he was transferred to the English Survey (Volume 1, pp. xl–xlvii).
4. *Mrs Wynne*: Anne Wynne (d. 1864), George Wynne's wife and daughter of Sir Daniel Toler Osborne (1783–1853), an Irish baronet.

5. *Frank*: Francis George Wynne (1846–1929), one of George and Anne Wynne's four children (see letter 0716, n. 2).
6. *a letter of thanks*: letter missing.

From John Peter Gassiot [20–21] November 1853[1] 0835

My dear Tyndall,

I have your note[2] and will attend thereto, I would sooner hold the position you have so promptly taken than be the recipient of 20 medals

Believe me | yours ever truly | John P. Gassiot.

RI MS JT/1/T/579
RI MS JT/2/13b/639

1. *[20–21] November 1853*: on 23 November, Tyndall stated that he received this 'note from Gassiot two days ago', suggesting the letter was written on 20 or 21 November (Journal, JT/2/13b/639).
2. *your note*: this is in reference to letter 0830 and/or letter 0831.

To Thomas Archer Hirst [21 November 1853][1] 0836

My dear Tom

I will post the letter to Henry Hill[2] this day. I would show it to Bevington but he would assuredly be wanting to have a finger in the pie. I was much pleased with the whole packet. You are a good fellow Tom no doubt in the eyes of the world, or would be so if the world knew all, but in my eyes you are one unprofitable servant for you do no more than you ought to do. I think you ought to be very thankful to be in a position to do so much for certainly you are the privileged individual.[3] Tell Tommy Grieve that I will write to him before I go down and that I hope he will put back his paper[4] until I come. Cling to the handles of teapots &c as much as possible to your lectures. To the plumage of birds and so forth—There are combustions to no end & you want to dazzle them—latent heat[5] <u>is</u> a little deep but <u>may</u> be made understandable to them.

I have a good deal to say to you but it is not yet time—This medal affair[6] has taken an extraordinary turn—am to loose it again—the circumstances are not yet ripe for statement but will be at the end of the week

Your affectionate | <u>John</u>

The scoundrels who trade in those Diplomas[7] certainly deserve exposure

& ought to get it—If you could make out the University[8] it would really be worth while to smash it.

You could make some fine experiments with parabolic mirrors[9] but the idea of such things for the Abbotts Ann people[10] is preposterous—I will enquire about the price of them and let you know.

RI MS JT/1/T/580
RI MS JT/1/HTYP/286

1. *[21 November 1857]*: this is a reply to letter 0833, which Tyndall received on 21 November. This letter was likely written on the same day.
2. *the letter to Henry Hill*: a letter Hirst wrote to Hill, who was having financial difficulties (see letter 0833, n. 1).
3. *you are the privileged individual*: Hirst's mother had left him a substantial inheritance upon her death (see letter 0813, n. 3).
4. *his paper*: presumably a paper Grieve was to deliver at a weekly meeting of the Mutual Improvement Society at Queenwood (see letters 0816, n. 4 and 0819 n. 16).
5. *latent heat*: the heat necessary to change a liquid into vapour, or a solid into a liquid or vapour, without the occurrence of a temperature change (*OED*).
6. *This medal affair*: eleven days after Tyndall learned that he had received the Royal Medal (see letter 0815), he found out that some members of the Royal Society objected to his selection (see letter 0828). He declined the award (see letters 0828; 0830; 0831; and 0837).
7. *The scoundrels who trade in those Diplomas*: Mummery had recently 'bought a German Doctoral Diploma in London' (see letter 0817, n. 4).
8. *the University*: not identified.
9. *parabolic mirrors*: a mirror with a cross-section in the curved shape of a parabola.
10. *the Abbotts Ann people*: on 9 December, Hirst gave a lecture on Heat at Abbotts Ann, a village near Andover, England to 'a small, quiet audience' (Hirst, 'Journals', 25 December 1853) (see also letter 0833, n. 2).

To Thomas Archer Hirst [24 November 1853][1] 0837

My dear Tom

The medal affair[2] is now consummated. The enclosed copies of letters[3] will shew you how I have acted. You know that it was a very near tie between another[4] and me. There was some heartburning and disappointment on the other side, and hence grumbling afterwards. They set about scrutinizing the affair; first they found that Thomson was opposed to me, second that my two first papers[5] were joint papers with Knoblauch, third that my third paper[6] had been got up in the private laboratory of Prof. Magnus, fourth that this same paper was published in German[7] before it appeared in English, fifthly that one or two men who have got a name in England[8] thought that I had not quite made out my point. And I could have urged in opposition to all this that they were too late with their objections, that I had been long proposed, say that my proposers had offered to adjourn the discussion of the matter but that they pushed the debate forward and were defeated. But I would not avail myself of the claim which the decision gave me. A medal is a thing so highly esteemed that I suppose they thought it would never enter into my brain to yield it up—But I thought the best way to put down all pettifogging[9] and whispering and thus leave myself a clear and unquestionable path for the future would be to decline the medal altogether—I did so and this day the Council met,[10] and it was resolved that the medal should not be awarded at all this year. I will say no more—the copies of the letters will explain my measures of action.

Affectionately yours | John

Faraday subscribes most cordially to all that I have done—My resolution on the subject just anticipated the advice that he was going to give me. My letter to Christie[11] will stand upon the Council minutes.[12] I was not at Clapham[13] with Faraday & Mrs. Faraday on Tuesday eve[in]g. I left them in the midst of a fog at Charing Cross.[14] but could not get near 'the Cross'.[15] They wanted me to spend the night at Albemarle Street[16] but I anticipated no serious difficulty in getting home. I was glad to run after the rumble of the cab and go home along with them.

Copy of Letter to Mr Christie

[. . .][17]

Copy of note to Mr Gassiot

[. . .][18]

Mr Gassiot's reply

[. . .][19]

RI MS JT/1/T/581
RI MS JT/1/HTYP/287–288a

1. *[24 November 1853]*: most likely written on 24 November as Tyndall stated that he wrote this letter on the day that the Council of the RS met to determine the fate of the Royal Medal for 1853 after he had declined the award.
2. *The medal affair*: see letters 0828; 0830; 0831; 0836, n. 6; and 0876, n. 9.
3. *The enclosed copies of letters*: see nn. 17–19.
4. *another*: August Wilhelm Hofmann. On 4 November, Tyndall noted the following about the Council's decision surrounding the Royal Medal: 'it was found that Cayley had one vote and that Hofmann and I were equal. The choice thus rested between us two. It was again put and the man who first voted for Cayley appears to have voted for me and thus turned the scale in my favour' (Journal, JT/2/13b/631).
5. *my two first papers*: see J. Tyndall and H. Knoblauch, 'On the deportment of crystalline

bodies between the poles of a magnet', *Phil. Mag.* 36:242 (March 1850), pp. 178–83; and J. Tyndall and H. Knoblauch, 'On the magneto-optic properties of crystals, and the relation of magnetism and diamagnetism to molecular arrangement', *Phil. Mag.* 37:247 (July 1850), pp. 1–33.

6. *my third paper*: J. Tyndall, 'On Diamagnetism and Magnecrystallic Action', *Phil. Mag.* 2:10 (September 1851), pp. 165–88.
7. *this same paper was published in German*: J. Tyndall, 'Ueber Diamagnetismus und Magnekrystallinische Wirkung', *Poggend. Annal.* 83 (1851), pp. 384–416.
8. *one or two men who have got a name in England*: likely Charles Wheatstone and William Hopkins (see letter 0828, n. 5).
9. *pettifogging*: petty quibbling (*OED*).
10. *the Council met*: the Council of the Royal Society (see letter 0816, n. 7) met on 24 November to discuss the Royal Medal (see n. 1).
11. *My letter to Christie*: see letter 0828.
12. *the Council minutes*: Tyndall's letter to Christie (letter 0828) was read out at the Council meeting of 24 November (Jackson, 'Tyndall and the Royal Medal', p. 157).
13. *Clapham*: see letter 0795, n. 8.
14. *Charing Cross*: the junction where Strand, Whitehall, and Cockspur Street meet in central London (see also n. 15).
15. *the Cross*: a cross erected by King Edward I (1239–1307) in memory of his wife, Eleanor of Castile (1241–90), for which Charing Cross (see n. 12) was named. However, from 1675, a statue of King Charles I (1600–49) occupied the site of the cross.
16. *Albemarle Street*: Michael Faraday and his wife, Sarah Faraday, were possibly suggesting that Tyndall spend the night at the RI, which was located at 21 Albemarle Street.
17. *[. . .]*: here Tyndall copied out letter 0828.
18. *[. . .]*: here Tyndall copied out letter 0830.
19. *[. . .]*: here Tyndall copied out letter 0835.

To Thomas Archer Hirst [26 November 1853][1] 0838

Saturday

My dear Tom.

I should not like to leave the impression upon your mind that the medal affair[2] was one of mere partisanship and the affair is too long to write about. I will wait till we meet and talk the matter over. You can hardly estimate how little the whole transaction has affected me. I kept my thoughts in clear daylight all through, and simple straightforward action has kept my mind at peace.

Yours | John

RI MS JT/1/T/934
RI MS JT/1/HTYP/291

1. *[26 November 1853]*: the first Saturday following the 'medal affair' (see n. 2) was 26 November.
2. *the medal affair*: see letter 0837, n. 2.

From H. M. Grieve 26 November [1853][1] 0839

Branscholm Park | Hawick | 26th. Nov.

My dear Sir,

I was happy to hear on your own account of the excellent appointment you had got,[2] and I write to congratulate you on such good luck, and I understand from my dear boys[3] that you mostly every summer take a jaunt somewhere. I hope it will be to Scotland next, I shall be extremely happy if you pay me a visit it is only two hours and a half by rail from Edin.[4] to our station at Hawick[5] and I am only three miles to it. We have some attractions in this county to offer to visitors, Melrose[6] and Abbotsford[7] sixteen miles from us, and our hills and glens are a great contrast to the level plains of England, but yours is the rich and productive soil My Pets[8] are no doubt counting every day just now to the holidays, I shall be happy to see them for I am quite alone and have been all summer, I had a variety of not a very pleasing kind, I was confined for twenty four days in a Hotel in Carlisle[9] with Iresiplus[10] in the head and suffered very much from it, I was fortunate in being in such good hands as the Mrs of the Hotel[11] whose attention and care was more than I could have had at home, I am thankful to say I am daily getting better—With my kindest regards and sincere wishes for your prosperity, I am ever

yours very truly | H. M. Grieve | Dr. Tyndall.

RI MS JT/1/TYP/1/468
LT Typescript Only

1. *26 November [1853]*: this letter was clearly written in 1853 because Grieve refers to Tyndall's new position at the RI (see n. 2). Tyndall also noted the reception of this letter on 29 November 1853 (Journal, JT/2/13b/641).
2. *the excellent appointment you had got*: Grieve was referring to Tyndall's position at the RI as Professor of Natural Philosophy (see letter 0762), which he took up in late September.
3. *my dear boys*: one of the boys is likely Tommy Grieve. Not further identified.
4. *Edin.*: Edinburgh.
5. *Hawick*: a town located in the Scottish Borders, near the rivers Teviot and Slitrig.

6. *Melrose*: a town in the Scottish Borders, located south of the River Tweed.
7. *Abbotsford*: Abbotsford House, the former home of the well-known Scottish writer and poet Sir Walter Scott (1771–1832), located near the town of Melrose.
8. *My Pets*: likely a reference to Grieve's children (see n. 3).
9. *a Hotel in Carlisle*: not identified.
10. *Iresiplus*: likely a misspelling of erysipelas, 'a local febrile disease accompanied by diffused inflammation of the skin, producing a deep red colour' (*OED*).
11. *the Mrs of the Hotel*: not identified.

From Thomas Archer Hirst 26 November 1853 0840

Queenwood College | Nov. 26th/53

My dear John

I shall be busy tomorrow and as I wish to say a word I must scribble it tonight. I have been in a state of anxious expectation all the week, and was pleased in no small degree to receive your letter[1] this morning. You have exactly done what I was hoping all the week you would do. That is shewn the limits of your esteem for medals, and your independence of all such things Depend upon it that refusal of yours will be more to your honour than had you gained undisputed possession of the medal.[2] If you die before me (which of course you will) I shall make my own use of that letter of yours to Mr Christie.[3] It was a brave manly letter John and a terrible rebuke to the Council of the Royal Society[4] if they were not too thick skinned to feel it. To see you praised and successful is of course pleasant to me but it is when you meet such cases as these that I am most proud of you. It is true that as you proceed such difficulties will decrease for every one you overcome in the way you have now done, will but lift you out of the sphere of all such. I hate to have to contradict any good news that I may have prematurely circulated, but I had as much pleasure today in telling Edmondson that you were to receive no medal as I had in telling him that it had been awarded to you. I let your letter tell its own tale and even little George[5] was reconciled. Go on and prosper says

T.A. Hirst.

RI MS JT/1/H/186
RI MS JT/1/HTYP/292

---

1. *your letter*: letter 0836.
2. *the medal*: the Royal Medal (see letter 0815, n. 3).
3. *that letter of yours to Mr Christie*: see letter 0828 (see also letter 0837).
4. *the Council of the Royal Society*: see letter 0816, n. 7.

5. *little George*: most likely George Cumming, a student at Queenwood with whom Tyndall had a good relationship.

From Heinrich Gustav Magnus 27 November 1853 0841

Berlin | 27 Nov. | 1853

Werther Freund

Ihr freundliches Briefchen vom 14ten d. ist mir vor einigen Tagen durch Herrn Sauerwald überbracht worden, und hat es mich gefreut aus demselben zu sehn daß Sie mit den Instrumenten zufrieden sind. —Meine Notiz über die Expansion der Gase ist nun auch in den Annales de Chimie, und ich freue mich in so guter Gesellschaft dort zu sein, denn Ihre Abhdl[.] steht dicht dabei, aber M^{r}. Verdet hat die Stelle wegen der verschiedenen Ausdehnung der Gase wohlbedächtig fortgelassen, denn sie könnte für Regnault unangenehm sein; vielleicht hat er es auch gethan weil die Franzosen einen sehr großen Nationalstolz oder Nationaleitelkeit haben, und alle Entdeckungen für ihre Landsleute vindiciren

Die Geschichte der Untersuchungen über die Ausdehnung der Gase ist kurz folgende:

Im Jahre 1802 veröffentlichte Gay Lussac seine Arbeit über die Ausdehnung der Gase, und es ist sehr überraschend zu sehn wie vortrefflich seine Zahlen untereinander stimmen; denn sie liegen alle zwischen 0,374 und 0,375. Es war daher ganz natürlich daß man diese Zahlen als unumstößlich richtig entnahm, und daß kein Mensch daran dachte die Versuche zu wiederholen. Ein Auftrag den Rudberg von der schwedischen Regierung wegen Construction von einem (wenn ich nicht irre) Alcoholometer erhalten, veranlaßte ihn sich mit der Construction der Thermometer zu beschäftigen, und dadurch wurde er auch auf die Untersuchung über die Ausdehnung der atm. Luft im Jahre 1837 geführt; und fand für den Ausdehnungscoefficienten 0,364. Aber Rudberg hat sich nur mit der atmosph. Luft beschäftigt, und am Schlusse seiner Abhandlung (Poggend. Annalen XLI 293.) sagt er „Hieraus kann ich nicht andern schliessen als daß die Ausdehnung der trocknen Luft, und ohne Zweifel auch aller anderen trocknen Gase, zwischen 0° und 100° nur 0,364 bis 0,365 ist." So blieb die Sache bis zum Jahre 1841. In der Zwischenzeit starb Rudberg. Am Schlusse dieses Jahres erschien Regnaults und meine Arbeit fast gleichzeitig, vielleicht Regnaults ein wenig früher. Diese letztere findet sich in den Annales de Chim Bd. IV.5. und daselbst ist pag 52. eine Tabelle, welche die ausführlichen Versuche für verschiedene Gasarten enthält, und die ich Sie anzusehn bitte; unter diesen Gasen hat keins einen höheren Ausdehnungscoefficienten als 0,368; die Schweflichte Säure hat nur 0,3669!! Mir ist es ganz unbegreiflich wie Regnault diese Zahlen hat finden können?!—Unter dessen erschien auch meine Arbeit in Poggend LV.1. also beide Arbeiten in den ersten Heften des Jahrgang 1842, Regnaults in den Annales de Chim, meine Arbeit in Poggend[.] Sie wissen daß ich für den Ausdehnungscoefficienten der Schweflichten Säure 0,385 gefunden habe. —Danach erschien in Bd V der Annales de Chim. eine 2te Arbeit Regnault's „Sur la Dilatation des Gaz" und in der finden sich dann ganz andre Zahlen als in dem vorhergehenden Bd. Bitte vergleichen Sie doch die Tabelle Bd IV.p. 52 mit der Bd V.p. 75. —Pag. 76 erwähnt Regnault daß ich den Ausdehnungscoefficienten des Wasserstoffs kleiner als den der atm. Luft gefunden hätte, aber wo er auf pag 77 von den übrigen Gasarten spricht, nennt er mich nicht; doch ist meine Abhandl[.] in die Annales de Chim aufgenommen Tom IV.330. zwar nur in einem erbärmlichen Auszug, aber doch zwischen beiden Arbeiten Regnault's. —Nun kennen Sie die ganze Geschichte, und sind im Stande zu beurtheilen wie es mir ergangen, was übrigens viel weniger interessant ist als daß Regnault solche Zahlen hat finden können wie die welche er zuerst mitgetheilt hat.

Warum haben Sie die Stelle in Ihrem Briefe ausgestrichen? Weßhalb wollen Sie mich nicht wissen laßen daß Ihnen die Ehre der Royal Medal zu Theil geworden? Sie wissen doch wohl wie sehr ich mich über alles freue was Ihnen Gutes zukommt.

Leben Sie wohl und schreiben Sie bald recht ausführlich | Ihrem | GMagnus.

Die besten Grüße für D^{r}. Faraday und für M$^{sss.}$ Faraday. sowie auch für D^{r}. Bence Jones.

Berlin | 27 Nov. | 1853

Dearest friend

Your kind little letter of the 14th of this month[1] was brought to me some days ago by Herr Sauerwald, and I was happy to see from it that you are satisfied with the instruments. —My note on the expansion of gases[2] is now also in the Annales de Chimie,[3] and I am happy to be in such good company there, for your paper[4] is close to it, but M. Verdet judiciously left out the passage about the regarding the different expansion of gases, for it could be awkward for Regnault; perhaps he also did it because the French have a very great sense of national pride or national vanity, and claim all discoveries for their countrymen

The history of investigations into the expansion of gases is, in brief, as follows:

In 1802, Gay Lussac[5] published his work on the expansion of gases,[6] and it is very surprising to see how splendidly his figures agree; for they lie all between

0.374 and 0.375. It was therefore quite natural that one took these figures to be irrefutably correct, and that no one thought of repeating the experiments. A commission that Rudberg[7] received from the Swedish government about constructing an (if I am not mistaken) alcoholometer[8] prompted him to concern himself with the construction of thermometers, and through this he was led to investigate the expansion of atm. air in 1837; and found the expansion coefficient to be 0.364. But Rudberg was concerned only with the atmosph. air, and at the conclusion of his paper (Poggend. Annalen XLI 293)[9] he says 'From this I cannot conclude other than that the expansion of dry air and, without a doubt, of all other dry gases between 0° and 100° is only 0.364 to 0.365.' This is where the matter remained until 1841. In the meantime Rudberg died. At the end of that year Regnault's[10] and my own work[11] appeared at almost the same time, Regnault's perhaps a little earlier. The latter can be found in the Annales de Chim Vol IV.5. and in there, there is a table on p. 52 which contains the detailed experiments for different types of gases, and which I ask you to look at; among these gases, none has an an expansion coefficient larger than 0.368; sulphuric acid has only 0.3669!! It is quite incomprehensible to me how Regnault was able to come up with these figures?!—In the meantime, my work[12] appeared also in Poggend[.] LV.1., that is, both works in the first issues for 1842, Regault's in the Annales de Chim, my work in Poggend. You will know that I found the expansion coefficient of sulphuric acid to be 0.385. —A 2nd work of Regnault's[13] appeared afterwards in Vol. V of the Annales de Chim., 'Sur la Dilatation des Gaz', and in it quite different figures can be found than in the previous vol. Please do compare the table in Vol. IV, p. 52[14] with that in Vol. V, p. 75.[15] —On p. 76, Regnault mentions that I had found the expansion coefficient of hydrogen to be smaller than that of atm. air, but on p. 77, when he speaks of the remaining types of gases, he does not mention my name; yet my paper is included in the Annales de Chim., Vol. IV, 330, admittedly only in a miserable excerpt,[16] but still between both of Regnault's works.[17] —Now you know the whole story, and are in a position to judge how I fared, which incidentally is much less interesting than the fact that Regnault has been able to find figures like those which he initially communicated.

Why did you cross out the passage in your letter?[18] Why do you not want me to know that you were given the honour of the Royal Medal?[19] You know very well how much pleasure I take about everything good that comes your way.

Live well and write soon very much in full | to your | G. Magnus.

The best of greetings to Dr. Faraday and to Mrs. Faraday, as well as to Dr. Bence Jones also.

RI MS JT/1/M/5

1. *Your kind letter from the 14th of this month*: letter missing.
2. *My note on the expansion of gases*: H. G. Magnus, 'Ueber die Verdichtung der Gase an der Oberfläche glatter Körper', *Bericht über die zur Bekanntmachunggeeigneten Verhandlungen der Königl* (Berlin: Königlichen Akademie der Wissenchaften, 1853), pp. 378–83.
3. *also in the Annales de Chimie*: Magnus's work on the expansion of gases (see n. 2) was translated into French as 'Note sur la condensation des gaz à la surface des corps solides polis', *Annal. Chim. et Phys.* 39 (1853), pp. 344–47. For more on the *Annal. Chim. et Phys.* see letter 0745, n. 13.
4. *your paper*: Magnus was likely referring to J. Tyndall, 'Mémoire sur la transmission de la Chaleur à travers les substances organiques', *Annal. Chim. et Phys.* 39 (1853), pp. 348–54.
5. *Gay Lussac*: Joseph Louis Gay-Lussac (1778–1850), a French chemist and physicist who played an important role in the development of modern physical chemistry, notable for his experiments on gas (J. R. Partington, 'J. L. Gay-Lussac (1778–1850)', *Nature* 165:4201 [1950], pp. 708–9).
6. *his work on the expansion of gases*: J. L. Gay-Lussac, 'Sur la dilation des gaz et des vapeurs', *Annal. Chim. et Phys.* 43 (1802), pp. 137–75.
7. *Rudberg*: Frederik Rudberg (1800–39), a Swedish physicist notable for his work on heat and optics.
8. *alcoholometer*: an instrument used to measure the concentration of alcohol (*OED*).
9. *his paper (Poggend. Annalen XLI 293)*: 'Ueber die Ausdehnung der trocknen Luft zwischen 0° und 100°', *Poggend. Annal.* 41 (1837), pp. 271–92.
10. *Regnault's*: H. V. Regnault, 'Recherches sur la Dilatation des Gaz', *Annal. Chim. et Phys.* 4 (1842), pp. 5–63.
11. *my own work*: G. Magnus, 'Ueber die Ausdehnung der Gase durch die Wärme', *Poggend. Annal.* 55 (1842), pp. 1–27.
12. *my work*: see n. 11.
13. *A 2nd work of Regnault's*: H. V. Regnault, 'Recherches sur la Dilatation des Gaz: Deuxième Mémoire', *Annal. Chim. et Phys.* 5 (1842), pp. 52–83.
14. *Vol. IV, p. 52*: Magnus was referring to Regnault's 'Recherches sur la Dilatation des Gaz' (see n. 10).
15. *Vol. V, p. 75*: Magnus was referring to Regnault's 'Recherches sur la Dilatation des Gaz: Deuxième Mémoire' (see n. 13).
16. *my paper . . . a miserable excerpt*: the excerpt was titled 'Extrait du Mémoire de M. Magnus, sur la dilatation des gaz', *Annal. Chim. et Phys.* 4 (1842), pp. 330–36.
17. *both of Regnault's works*: see nn. 10 and 13.
18. *your letter*: see n. 1.
19. *Royal Medal*: see letter 0815, n. 3.

From E. D. Steuart 28 November 1853 0842

Steuart's Lodge | Nov. 28th 1853

Dear John,

I had been for some time anxiously wishing to hear from you, before your first letter[1] arrived, as I feared your health had not been good previous to your departure for the Continent,[2] and my reason for seeking to know your address was, that I might write to make enquiries about you: I truly hope you have derived benefit from the excursion which must have been an agreeable relaxation to mind and body. I sincerely rejoiced to hear of your having been awarded the gold medal,[3] but your few lines by the late post on Saturday[4] brought disappointment and regret that you had relinquished the honor[5] of retaining it—: No doubt you have good reasons for this, which even if you had told me, I might not have been able to comprehend, but your refusal does not alter the high estimation in which your talents are held, and therefore the gratification of your friends remains the same. I ought sooner to have written to you, but I often find it difficult to collect my ideas, for I have much that is distressing to occupy my mind, and take up a large portion of these short days: Unaccustomed as I have ever been to care or responsibility, I now most sensibly feel <u>all</u> that I am obliged to contend with, but still I <u>try</u> to be contented, and thankful for the blessings still left to me, and though happiness in <u>this</u> world cannot henceforth be mine, I have the comfort of looking forward with hope of never ending glory, where there is no more sin and sorrow: This prospect is <u>all</u> that enables me to bear my <u>great</u> loss[6] and utter desolation, for I have <u>no</u> earthly support, <u>nothing</u> to lean upon or cling to: Dear John I trust it never may be your wretched lot to feel alone in the world, without object or interest, a weary wanderer, longing to be released from care and strife.

My brother[7] is, I believe, at Brighton[8] just now, at least he was about ten days ago. It is likely he may be in London before he returns to Ireland, and I am <u>sure</u> would be very glad to see you: he frequents the Conservative Club,[9] when there, where perhaps you might hear of him: he of course cannot know you are in London, as he has been in Paris for some weeks.

With every kind wish for you welfare, temporal and spiritual, believe me as ever,

my dear John | Your most sincere friend | E. D. Steuart.

Professor Tyndall | Royal Institution | Albemarle | London[10]

RI MS JT/1/TYP/10/3337–3337a
LT Typescript Only

1. *your first letter*: letter missing.
2. *your departure for the Continent*: see letter 0823, n. 1.
3. *the gold medal*: the Royal Medal (see letter 0815, n. 3).
4. *your few lines by the late post on Saturday*: letter missing.
5. *relinquished the honor*: Tyndall declined the Royal Medal (see letter 0837).
6. *my great loss*: the death of Steuart's husband, William Richard Steuart (see letter 0707, n. 2).
7. *My brother*: Steuart had four brothers: John Dawson Duckett (1791–1866), William Duckett (1796–1868), Joseph Fade Duckett (1796–1875), and Thomas Jonas Duckett (1799-?). It is not clear which brother Steuart was referring to. Not further identified.
8. *Brighton*: a popular seaside destination on the south coast of England.
9. *the Conservative Club*: a Conservative London gentlemen's club formed in 1840, which many who could not gain admission to the more popular Carlton Club joined ('St. James's Street, West Side, Existing Buildings', in F. H. W. Sheppard [ed.], *Survey of London* [London: London Country Council, 1960], pp. 472–86, http://www.british-history.ac.uk/survey-london/vols29–30/pt1/pp472–486 [accessed January 1, 2016]).
10. *Albemarle | London*: LT wrote this at the top of the typescript. Presumably it is from the envelope.

From George Wynne 28 November 1853 0843

Board of Trade[1] | 28 Nov. 1853.

My dear Tyndall,

I only received your letter[2] on Saturday evening when I returned from the North—Every one must approve of the independent attitude you have assumed and it will raise you in the esteem of those who do not know you otherwise than by your attainments but it will make no difference to those who have the pleasure of knowing you personally as they could have told beforehand that you would not accept any doubtful mark of distinction or I should rather say a distinction on which a doubt was cast from a quarter commanding respect—I have not acquaintance myself with the subject to form an opinion on the originality of your views which have been questioned, but I am inclined to attach a great deal more importance to the decision of the majority of the Council[3] almost twice expressed than to the querulous objections of some disappointed philosopher who was perhaps anxious to show the society the loss they had in not having him of the Council to keep them straight—You have however all the distinction of the medal[4] with the additional one of shewing that you can afford to do without it.[5] I would go down to the Institution and have a talk with you over the matter but that I have to start to-morrow on a rather prolonged tour of inspection which will occupy me some ten days and I am necessarily very busy. Ginty has sent me a Manchester Guardian[6] with the announcement of his election to the Honorary Secretaryship of the Manchester Athenaeum[7] I send you the paragraph.[8] In making the selection I am sure the society has but studied their own interest. I shall look out for a visit from you the Sunday after next but I cannot offer a bed as my Mother in law[9] and another lady[10] are staying with me and Harry[11] will be at home for the holidays, but you will meet a very jolly party

farewell | Most truly yours | Geo. Wynne.

RI MS JT/1/TYP/5/1852
LT Typescript Only

1. *Board of Trade*: see letter 0784, n. 9.
2. *your letter*: letter missing.
3. *the Council*: the Council of the Royal Society (see letter 0816, n. 7).
4. *the medal*: the Royal Medal (see letter 0815, n. 3).
5. *afford to do without it*: Tyndall declined the Royal Medal (see letter 0837).
6. *Manchester Guardian*: a British newspaper founded by John Edward Taylor (1791–1844) in 1821, published weekly, and then daily from 1855 (W. H. Mills, *The Manchester Guardian*: *A Century of History* [New York: Henry Holt and Company, 1922]).
7. *Manchester Athenaeum*: see letter 0819, n. 12.
8. *the paragraph*: not identified.
9. *my Mother in law*: Harriette Trench (1785–1855), mother of Anne Wynne (see letter 0834, n. 4).
10. *another lady*: not identified.
11. *Harry*: George Wynne's son Henry (see letter 0716, n. 2).

From Henry Thomas De la Beche 1 December 1853 0844

Jermyn Street | 1st Dec. 1853.

Confidential

My dear Sir,

As the subject[1] on which we conversed the other day has become a little pressing would you kindly tell me what you think of it.

very faithfully | H.T. DelaBeche

RI MS JT/2/13b/641
LT Typescript Only

1. *the subject*: De la Beche wished for Tyndall to lecture at the Government School of Mines (see letter 0824, n. 1).

To Thomas Archer Hirst [1 December 1853][1] 0845

My dear Tom.

I am dreadfully crossgrained[2] tonight I have been working all day and could get no result. This morning I set out beautifully and expected a clear way before me. I got into entanglements—petty contemptible entanglements—afterwards and I failed to get out of them the entire day. I do not feel like a fellow that had worked at all. I feel jaded and tired, but not healthily tired, and in this state I try and converge my thoughts upon you, hoping thereby to gather together and render my mind transparent. At present it is */between/* shattered crystals—full of veins and interior reflexions, so that no light can pass through it. I was at the anniversary meeting of the Royal Society[3] yesterday though not at the dinner afterwards—after the presentation of the medals one of the fellows—Huxley—got up and asked why he did not see his friend among those to whom medals were handed. 'I can answer you in one minute said Lord Rosse. I hold here in my hand two letters[4] from Mr Tyndall declining the medal—so you see there was no course left to us but let the matter rest.' 'Might I request those letters to be read your Lordship' demanded the speaker—Christie read them both—there was great silence, and a general rumble of applause at the end. Some think I ought not to have written the second letter to Gassiot.[5] Others think it was quite right that I should do so—for my part I enjoy the calm consciousness of having done what was most consistent with my own constitution.

I now want to ask you a question and I do it with some reluctance. It is regarding money. It has sometimes appeared to me that you have no more money than you can spare.[6] I remember before I went to Germany you told me that you had plenty,[7] but you probably made a miscalculation and I am dreadfully afraid that I */bore/* you—I want ten pounds now. and I may want more before I receive my salary for I dont know when it is paid. Can you let me have them without inconvenience to you? I will make no promises, but I have a strong hope before this day 6 months to have your account completely sponged away. When I look at myself writing thus it is half an inducement to me to act upon a letter which I received today[8] from Sir Henry de la Beche*—it would at least put me above the necessity of borrowing—If you have not the money plentiful let me know—I can get what I want elsewhere.

Your affectionate | John |

* Dont mention this—it is 'confidential.'

RI MS JT/1/T/582
RI MS JT/1/HTYP/293

1. *[1 December 1853]*: Hirst wrote 'Dec[r] 3rd' on this letter reflecting the day he received it. Not only did Tyndall mention that he wrote to Hirst in his journal entry of 1 December, but the meeting of the Royal Society (see n. 3) that Tyndall noted as occurring 'yesterday' took place on 30 November (Journal, JT/2/13b/641).
2. *crossgrained*: 'of opposed nature or temper; difficult to deal with, intractable' (*OED*).
3. *the anniversary meeting of the Royal Society*: the Royal Medals were presented at this meeting, which took place on 30 November (Journal, JT/2/13b/641). (See also letter 0822.)
4. *two letters*: for the first, see letter 0828; for the second, see letters 0830 and 0831.
5. *the second letter to Gassiot*: letter 0830 (see also letter 0831). Tyndall recorded in his journal that 'Dr. Booth told me afterwards that he regretted I had written the second letter, because he thought I ought to accept the medal if awarded me a second time (30 November 1853, JT/2/13b/641).
6. *you have no more money than you can spare*: Tyndall may be referring to letter 0791, in which Hirst told Tyndall he only had £20 for Tyndall's use; in letter 0793, Tyndall responded by stating, 'I thought you were rich'. For more on Tyndall and Hirst's financial relationship, see letter 0813, n. 3.
7. *before I went to Germany . . . you had plenty*: Hirst may have told Tyndall this in person or in a missing letter. Alternatively, Tyndall may have been referring to an earlier trip to Germany. Leaving for Germany in October 1850, Hirst wrote to Tyndall in a letter dated 1 September stating, 'I can lend you £40 as easily as I could lend you a penny' (see letter 0433).
8. *a letter which I received today*: letter 0844.

To Henry Thomas De la Beche [2 December 1853][1] 0846

Dear Sir Henry,[2]

Our last interview[3] gave me an approximate notion of your wishes, and allowed me an opportunity of stating to you my general feelings in matters of science. But it did not, as you are aware, furnish data sufficiently distinct to enable me to make a distinct statement. If you would be kind enough to let me know the nature and exact amount of the duties which would be required of me I shall be able to inform you without delay whether it is in my power to engage in such duties. The question is altogether one of time, and on this point I shall be able to come to a prompt decision when once in possession of the elements necessary to its formation.

I have the honour to be | your very obedient Servant | John Tyndall

RI MS JT/2/13b/642
LT Typescript Only

1. *[2 December 1853]*: Tyndall noted writing and sending this letter in his journal on 2 December (JT/2/13b/642).
2. *Dear Sir Henry*: an earlier draft of this letter is in Tyndall's journal (1 December 1853, JT/2/13b/641). In the draft, Tyndall asked for a 'plain statement' of both the duties required and the income, while in this version Tyndall did not directly speak of income, and emphasized that the 'question is altogether one of time'.
3. *Our last interview*: De la Beche and Tyndall last met on 10 November to discuss the possibility of Tyndall lecturing at the Government School of Mines (see letter 0824, n. 1).

To Henry Thomas De la Beche [3 December 1853][1] 0847

Dear Sir Henry,

You requested me to call upon you at 12 o'clock on Monday. I shall undoubtedly do so, though the more I think on the matter pending between us, the less do I see that you have given me any grounds on which to base a decision. I really feel ashamed of taking up so much of your time to no apparent purpose, and I earnestly hope that you will be in a position on Monday to make a distinct proposal to me,[2] and thus place it in my power to give you a distinct answer.

I have the honour to be | your very obedient Servant | John Tyndall

RI MS JT/2/13b/642
LT Typescript Only

1. *[3 December 1853]*: Tyndall copied this letter into his journal entry for 3 December 1853 (JT/2/13b/642).
2. *a distinct proposal to me*: De la Beche wished for Tyndall to lecture at the Government School of Mines. However, De la Beche had not been clear about the number of lectures Tyndall was to deliver or the amount of remuneration he would receive (see letters 0824, n. 1 and 0846).

From Thomas Archer Hirst 4 December 1853 0848

Queenwood College | Decr 4th 53

My dear John.

A week ago I told Wright that he need not trouble himself about any money for me this Christmas for I should not want it. Yesterday I told him to arrange to let me have £20 which he promised to do. In my purse I have this moment £5 which is of no mortal use to me if you would like it before Wright will have the £20 ready tell me. On the 1st of January I receive my Quarters Interest[1] of £30 which after paying my Christmas debts will leave me amply sufficient until April next when another £30 will come for which I foresee no use now John such are the plain figures which will be a better argument with you than anything else I can say for you know as well as I do that it would pain me very much if you did not use it as freely as if it were your own. But you will. The reluctance of which you speak sprung merely from the thought that you were pinching me, and I am sure that inducement to accept Sir H. dela B–'s proposal[2] vanished before you had finished writing it

I am glad Huxley mentioned your affair at the Royal Society[3] and that your two letters[4] were read before all the members. The way you have treated the whole case both as regards yourself and everything else leaves not a syllable to be said except in pure praise. Ginty has heard of it (through Capt. Wynne) in a letter to me[5] the other day he tells me all the circumstances as if I had not heard a word about them! And truly they were as interesting to me as if I had been ignorant for they shewed me how such fellows as Ginty view the matter and that is worth something he says 'Tyndall came to their (the R. S) rescue by at once releasing them from their oversight for honor—this will reflect more to Tyndall's honor that 10 such medals'. I am expecting you down here on Saturday the 17th (a week next Saturday) to be ready for our stir[6] on the Monday and to attend the last Saturday night's meeting of the Improvement Society.[7] Tommy Grieve is reserving his paper[8] for you I believe. Enclosed is another letter from and answer to Henry Hill.[9] Post mine to him and return me his for as I think Bevington can help us now I will write to him shortly and tell him what I know of Henry shewing his letters.[10] Perhaps Bevington can find employment for him at the Neckinger Mills[11] who knows? From his knowledge of Leather & Leather cutting one would think he might be of service in a Tanning manufactory at any rate I will see.

Debus for the last day or two has withdrawn into himself in other words he has taken umbrage at something or other. He has been indulging in that unfortunate twaddle of his too much. Haas I hear cut him rather short one night at supper and on relating the matter to me at the Lodge[12] I in perfect good humour told him that it all originated with himself (I have often told him so before) This coupled with a peculiar physical indisposition I'm afraid has thrown him into one of his sullen fits. As soon as I saw it I told him quietly that as no body had the slightest intention of hurting his feelings the offence was entirely on his own side, after which I have taken no notice whatever of his sulks but forced him to talk to me civilly. He is thawing evidently and in a day or two if well let alone will come around.

Yours affectionately | TA Hirst

RI MS JT/1/H/187

1. *Quarters Interest*: interest paid out four times yearly by the bank based on the amount invested and held in one's account.
2. *Sir H. dela B–'s proposal*: Henry De la Beche was negotiating with Tyndall concerning a position at the Government School of Mines (see letters 0845, n. 8 and 0844).
3. *your affair at the Royal Society*: this is in reference to Tyndall winning and then declining the Royal Medal (extensively discussed in letter 0837). Tyndall also wrote about the matter in his journal entries of 3 November through 30 November 1853 (Journal, JT/2/13b/631–41).
4. *your two letters*: see letter 0845, n. 4.
5. *in a letter to me*: not identified.
6. *our stir*: the annual Christmas concert at Queenwood College. In his journal entry of 19 December, Tyndall noted his attendance at the concert (Journal, JT/2/13b/647). Furthermore, Hirst noted in his journal that the 'final party on the 19th' 'was the best concert ever given at Queenwood' (Hirst, 'Journals', 25 December 1853).
7. *the Improvement Society*: see letter 0819, n. 16.
8. *his paper*: see letter 0836, n. 4.
9. *another letter from and answer to Henry Hill*: Hill had been corresponding with Hirst about his financial difficulties (see letter 0833, n. 1).
10. *his letters*: the letters Hill sent to Hirst (see n. 9).
11. *the Neckinger Mills*: a leather factory located in Bermondsey, South London and owned by Bevington; the site was formerly a paper mill ('A Day at a Leather-Factory', *The Penny Magazine of the Society for the Diffusion of Useful Knowledge* [London: Charles Knight & Co, 1842], vol. 11, pp. 209–16).
12. *the Lodge*: a lodge that stood at the entrance to the grounds of Queenwood College. Not further identified.

To Thomas Archer Hirst [5 December 1853][1] 0849

Monday

Dear Tom

I am in the midst of a series of experiments[2] & therefore can only say a word or two. Send me the £5—I had half resolved this mor[nin]g not to go to Queenwood, it will be 25/ out of my pocket. but I hardly think I can let the consideration stop me—for I should like to be present at your Society[3] & at your Concert[4] Write to Bevington by all means give him a full account of the kind of a chap your friend Henry[5] is—in all probability he will be able to help him. <u>Dont hear of a money offer</u>. I shall be able to render help in these cases myself at a future day and until then you must bear the brunt alone—I mentioned Hill I think once to Bevington. send him over two of Henry's letters.[6] Your letter to him[7] is extremely good and wise—he is suspicious as most fellows like him and this he must *[1 word illeg]* of.

Debus has a spice of the same—give him time he will come *[from]*—his ills are more than half physical. I think I shall get a reputation here for straightforwardness for I have commenced with the resolution to venture plain, manly commonsense and Truthfulness in approach to the diplomacy of the world. If a man fails by following this principle he has something to encourage him in defeat.

ever yours affectionately | John

RI MS JT/1/T/583

1. *[5 December 1853]*: Hirst wrote 'Dec 6—1853' on the top of this letter, which was presumably the day he received it. 'Monday' was 5 December.
2. *series of experiments*: not identified.
3. *your Society*: Queenwood's Mutual Improvement Society (see letter 0819, n. 16).
4. *your Concert*: see letter 0848, n. 6.
5. *Henry*: Henry Hill.
6. *two of Henry's letters*: see letter 0848, n. 9.
7. *Your letter to him*: a letter Hirst wrote to Henry Hill, which Hirst included with his last letter to Tyndall (see letter 0848).

To Thomas Archer Hirst 6[–7] December [1853][1] 0850

My dear Tom.

I enclose you a letter from our good friend Stegmann.[2] I hope you will feel inclined to write to him and say whether you think it well to send a copy of his work[3] to the person mentioned[4]—He is a man for whom I entertain a high esteem and it was with true pleasure that I sat down this evening to reply to his letter. The intelligence of Kohlrausch's success[5] was very pleasant to me. He is a lovable fellow and extremely ingenious. I have been obliged to stop at home today by a cold. I moved on Saturday from the warm laboratory into a cold moist room for the sake of securing a constant temperature but the cold and moisture made an experiment on me and sent me home at night with a shiver and headache. Sunday I remained within doors, but yesterday I planted myself in the same room and though constantly admonished by a violent headache remained there all day. I have paid for it, for today I have been able to do nothing. I am better tonight and tomorrow I will try it again. It is a flagged room with moist walls, but I hope to be able to make myself used to it.

Affectionately your | John | 6th Dec.

The following is my theory of Debus.[6]

He thinks it incumbent on him to say something in company and this requires a certain tact acquired by practice—he has never been used to company and he knows nothing of conversational tactics. I remember very well 8 or 10 years ago that I was capable of saying very silly things merely for the sake of saying something. I cannot imagine that Debus thinks he is speaking wisely when he indulges in what you call twaddle. He merely imagines that he is dealing in the stuff common to conversation and as this is stuff of whose chemistry he is totally ignorant he makes mistakes. If he were an observant man he would soon correct this, but he is not observant in such matters, and he cannot see the intrinsic difference between his twaddle and what he believes to be other people's twaddle. He has given me such proofs of wisdom and penetration as forbid me to believe that he attaches any value to what he says on such occasions. He has as keen a glance into some of the deeper recesses of human nature as most men that I have met. when in good health and flushed with the spirit from on high his words are worth listening to. I have heard him say things fit to be uttered by the greatest men—but he is unequal; the victim of a deranged body and a mind so sensitive that it sometimes borders on the suspicious. His gold however is at heart—his drop is superficial.

J.T.

<Hereafter LT Transcript Only>

Dear Tom

Dont send me the money.[7] I have just got a cheque for £100.[8]

Your affectionate | John

RI MS JT/1/T/584
RI MS JT/1/HTYP/296

1. *6[–7] December [1853]*: while the first section of this letter was written on 6 December, it is possible that the section on Debus and the final part beginning 'Dear Tom' were written on 7 December. Hirst wrote 'Dec. 7' at the top of the letter, which could indicate the day he received it.
2. *a letter from our good friend Stegmann*: most likely letter 0823.
3. *a copy of his work*: see letter 0823, n. 2. Hirst recorded receiving Stegmann's book in his journal (Hirst, 'Journals', 25 December 1853), and the work is further discussed in Hirst's response to Tyndall (see letter 0851).
4. *the person mentioned*: Bartholomew Price (see letter 0823, n. 3).
5. *Kohlrausch's success*: a reference to Kohlrausch becoming Extraordinary Professor of Physics and Mathematics at the University of Marburg (see letter 0823, n. 5).
6. *my theory of Debus*: this is in response to Hirst's complaints about Debus's 'sullen fits' (see letter 0848).

7. *Dont send me the money*: in letter 0848, Hirst noted that he had £5, which Tyndall was free to borrow. Tyndall had asked Hirst to send this money in letter 0849.
8. *cheque for £100*: this was Tyndall's half year salary from the RI which was authorized on 5 December (Minutes of the Managers' Meeting, RI MS AD/02/B/02/A/11/38).

From Thomas Archer Hirst [8 December 1853][1] 0851

Queenwood College | Thursday. Noon

My dear John

I received your letter[2] with Stegmann's[3] this morning Yesterday I received one from Stegmann[4] himself and another from the Dead Letter office[5] telling me there was a german parcel ready for me as soon as I should think fit to send 8/4 This of course I have done and expect the parcel tomorrow. Rest assured I shall attend to what good old Stegmann says, nay more I will ask advice about translating his book[6] which I have no doubt merits it, of this we will speak soon. Before writing to Price I should like to know as much about him as possible whether his work[7] is to be a compilation merely or contain new investigations of his own Get to know as much about him as possible and ask Mr Gyde[8] if you please to send me the 1st vol of his work[9] which was published a short time ago Its title is 'A treatise on infinitessimal Calculus containing differential and integral calculus, Calculus of Variations and Applications to Geometry'. By B. Price. Published by J. H. Parker[10] in Oxford. As yet I know nothing of Mr Price.

Enclosed I send you a letter[11] received by the same post as brought Stegmanns and which has pleased me still more perhaps. It is the first letter I have ever received from my old Tutor[12] who treated me so nobly when a lad at school and when I ran away there to Germany. I have no doubt that even from this short letter you will get a glimpse into one of the best hearted noble fellows I know. As an educator I never met his equal and daily I recall his example for my own use. He is now about 45 years old I should say, and was a wrangler[13] at Cambridge as far as I remember. With all my heart I wish you knew him. It will be a week or two yet before I shall be able to answer Stegmann What you say of Debus[14] contains much truth, as I predicted we were never better friends than we are at present again.

Yours affectionately | T. A. Hirst

P.S. Tomorrow I lecture at Abbotts Ann on Heat[15] Yesterday Debus lectured at Andover on Electro-Chemistry[16] I was with him. Edmondson is in London again on very important business, he told me, for I objected that he was leaving home too much at present. What his important business is I don't know. If I mentioned a word about your cold and the way you are treating it you would not be well pleased so I wont. | TAH

RI MS JT/1/H/485
RI MS JT/1/HTYP/297

1. *[8 December 1853]*: dated by LT. Her dating is correct as 8 December was a Thursday and Hirst lectured at Abbotts Ann on 9 December (see n. 15).
2. *your letter*: letter 0850.
3. *Stegmann's*: letter 0823.
4. *I received one from Stegmann*: not identified.
5. *another from the Dead Letter office*: letter not identified. A Dead Letter Office is a department within the post-office where unclaimed or undeliverable letters are examined, being returned to the writer or destroyed after a certain time (*OED*).
6. *his book*: see letters 0823, n. 2 and 0850, n. 3.
7. *his work*: see letter 0823, n. 3.
8. *Mr Gyde*: see letter 0761, n. 4.
9. *the 1st vol of his work*: see n. 8.
10. *J. H. Parker*: John Henry Parker (1806–84), a writer on architecture who became a bookseller at Oxford in 1821 ('Parker, John Henry', *The Encyclopedia Britannica: A Dictionary of Arts, Sciences, Literature and General Information*, ed. H. Chisholm, 11th edn., 29 vols [Cambridge: Cambridge University Press, 1911], vol. 20, pp. 827–28).
11. *a letter*: see n. 13; not further identified.
12. *my old Tutor*: in his journal, Hirst noted the letter was from his Wakefield schoolmaster, Reverend Richard Garvey (1785–1861) (Hirst, 'Journals', 25 December 1853).
13. *a wrangler*: 'The name for each of the candidates who have been placed in the first class in the mathematical tripos at Cambridge University' (*OED*).
14. *What you say of Debus*: see letter 0850, n. 6.
15. *lecture at Abbotts Ann on Heat*: see letter 0836, n. 10.
16. *lectured at Andover on Electro-Chemistry*: not identified.

From Heinrich Gustav Magnus 10 December 1853 0852

Berlin | d. 10 Decemb | 1853.

Werther Freund

Schon mache ich mir Vorwürfe daß ich Ihren freundlichen Brief vom 1ten. noch nicht beantwortet habe, denn ich fühle wohl, daß man nach einem solchen Schritt, wie der den Sie gethan auch gern wissen möchte was Andere dazu sagen. Doch wissen Sie schon was ich darüber denke: Ob Sie zu schnell gehandelt; kann ich nicht beurtheilen, da ich die Verhältniße nicht hinreichend kenne, aber daß Sie keine Auszeichnung wollen an der auch nur der geringste Makel haftet, das kann ich wohl nicht anders als gut heißen. Sie werden sich hoffentlich über den Verlust zu trösten wissen. Meine Ansicht ist die: Wenn ich etwas geleistet habe und es wird anerkannt, so ist das um so besser, wird es aber nicht anerkannt, so ist die Leistung darum nicht geringer. Sehn Sie um sich wie es Andern geht. Denken Sie nur wie es mir gegangen ist, mit der Ausdehnung der Gase, mit der Spannkraft der Wasserdämpfe, mit der Respiration. Ich liebe es nicht von meinen Dingen zu reden, aber ich hoffe Sie werden mich nicht mißverstehen, wenn ich sage daß nach meiner Ansicht nicht leicht eine so einflußreiche Aufklärung gegeben worden als die Theorie welche ich von der Respiration geliefert. Niemand hat früher daran gedacht daß im Arteriellen Blute Kohlensäure und im venösen Sauerstoff sei; Niemand wußte wie diese Gase in die entferntesten Theile des Körpers gelangen; Niemand glaubte daß der Oxidationsproceß überall im Körper statt findet. —Aber mein Verdienst ist nirgends zu Tage gekommen, man hat die Theorie angreifen wollen. Gay-Lussac hat sich zuerst blamirt und ich habe ihn zurecht gewiesen, aber gegen Liebigs unzählige Theorien von der Respiration bin ich nicht aufgetreten eben so wenig wie gegen Mulder's Unsinn den er in seiner physiologischen Chemie vorbringt. —Aber ich bilde mir ein daß deßhalb mein Verdienst um die Aufklärung dieses Processes nicht geringer ist; wenn auch keine wissenschaftliche oder medicinische Gesellschaft es anerkannt hat, mit Ausnahme der Königsberger medicinischen Facultät die mich zum Doctor der Medicin gemacht hat. —Doch ich schwatze von mir, was ich gar nicht thun wollte, ich wollte Ihnen nur ein Vorbild für Ihren Trost liefern.

Da Sie mir aber ein Mal so viel über den Council der R.S. erzählt haben, so sagen Sie mir wohl noch etwas mehr. Welches sind die zwei Partheien? wer gehört zu der einen? wer zur andern? Gehört Graham zu den Ihrigen? Ist derjenige auf den Sie in Ihrem Briefe an Mr Gassiot anspielen Wheatstone?

Du Bois ist unwohl, ich habe ihn deßhalb in der letzten Sitzung der Academie nicht gesehn, und kann Ihnen seine Ansicht noch nicht mittheilen.

Statt der Royal Medal wünsche ich Ihnen eine recht schöne Entdeckung, die ist mehr werth.

Meine Empfehlung an D^r. Faraday und M^sss Faraday; an D^r. Bence Jones, und an M^r. Gassiot den Sie ja öfter zu sehn scheinen so wie an seine liebenswürdige Tochter. Grüssen Sie Dr. Francis und schreiben Sie mir recht bald

von Herzen | Ihr | GMagnus.

Viele Gruße von meiner Frau und von den Kindern, die sich Ihrer häufig erinnern.

Berlin | d. 10 Decemb | 1853.

Dear friend

I am already reproaching myself for not yet having answered your kind letter of the 1st,[1] for I probably suspect that after taking such a step as the one you have,[2] one would also like to know what other people are saying about it. But you already know what I think about it; Whether you acted too quickly, I cannot judge, as I do not know the circumstances sufficiently well, but the fact that you do not want any award to which even just the slightest taint clings is something I can probably do no other but approve of. You will hopefully know how to console yourself about the loss. My view is this: When I have accomplished something and it is acknowledged, then so much the better, but if it is not acknowledged, then the achievement is therefore none the lesser. Look around you at how others are doing. Just think of how it was for me with the expansion of gases, with the tension of steam, with respiration. I do not like to talk of my affairs, but I hope that you will not misunderstand me when I say that, in my view, such an influential explanation has never been as easily given as the theory which I provided on respiration.[3] No one thought earlier that there may be carbonic acid in arterial blood and oxygen in venous; no one knew how these gases reach the most distant parts of the body; no one believed that the oxidation process takes place everywhere in the body. —But my credit for this has never accrued anywhere, and people wanted to attack the theory. First Gay-Lussac[4] made a fool of himself[5] and I put him in his place,[6] but I did not speak out against Liebig's countless theories of respiration[7] as little as I did against Mulder's nonsense[8] that he puts forward in his physiological chemistry. —But I fancy that for this reason my credit for explaining this process is none the lesser; even if no scientific or medical society has acknowledged it, with the exception of the medical faculty in Königsberg,[9] which made me a Doctor of Medicine. —But I am babbling on about myself, which I did not want to do at all, I only wanted to provide you with an example of how to console yourself.

As, however, you told me once so much about the Council of the R.S.,[10] please tell me something more. Which are the two parties? Who belongs to the one? Who to the other? Does Graham[11] belong to your men? Is the man whom you allude to in your letter to Mr Gassiot[12] Wheatstone?

Du Bois is not well, for this reason I did not see him at the last session of the Academy,[13] and cannot give you his view yet.

Instead of the Royal Medal,[14] I wish you a really nice discovery; that is worth more.

My compliments to D^r^. Faraday and Mrs Faraday; to D^r^. Bence Jones, and to Mr Gassiot, whom you seem to see once in a while, as well as to his lovely daughter.[15] Give my greetings to D^r^. Francis, and write to me very soon

sincerely | Your | GMagnus.

Many greetings from my wife[16] and from the children,[17] who often think of you.

RI MS JT/1/M/6

1. *your kind letter of the 1st*: letter missing.
2. *such a step as the one you have*: Tyndall had recently declined the Royal Medal (see letter 0837).
3. *the theory which I provided on respiration*: see G. Magnus, 'Sur les Gaz que contient le Sang: Oxigène, Azote et Acide Carbonique', *Annal. de Chim. et de Phys.* 65 (1837), pp. 169–92.
4. *Gay-Lussac*: see letter 0841, n. 5.
5. *made a fool of himself*: Magnus was referring to Gay-Lussac's review of his work on respiration (see n. 3): J. L. Gay-Lussac, 'Critical Observations on the Phenomena of Respiration', in C. Watt and J. Watt (eds.), *The Chemist* (London: Alexander Watt, 1844), pp. 243–48.
6. *I corrected him*: Magnus responded to Gay-Lussac's review (see n. 5) in a talk at the Royal Academy of Berlin on 17 June 1844 ('On Respiration', in C. Watt and J. Watt [eds.], *The Chemist* [London: George Peirce, 1845], p. 62).
7. *Liebig's countless theories of respiration*: Justus Liebig (1803–73), a German chemist notable for his work on organic chemistry. Examples of his work on respiration include 'The Theory of Respiration', *Provincial Medical Journal and Retrospect of the Medical Sciences* 4:24 (September 1842), pp. 477–80 and *Animal Chemistry, or Organic Chemistry in its Applications to Physiology and Pathology* (London: Taylor and Walton, 1842). For more detail on Liebig's work on respiration, see W. H. Brock, *Justus vov Liebig: The Chemical Gatekeeper* (Cambridge: Cambridge University Press, 1997), p. 183–214.
8. *Mulder's nonsense*: Gerrit Jan Mulder (1802–80) was a Dutch chemist who had many disagreements with Liebig (see n. 7), and taught at the University of Utrecht (H. A. M. Snelders, *The Letters from Gerrit Jan Mulder to Justus Liebig (1838–1846)* [Amsterdam: Rodopi, 1986], pp. 9–11).
9. *Königsberg*: The University of Königsberg, founded in 1544 and located in Prussia. The university had an excellent School of Mathematics.
10. *the council of the R.S.*: see letter 0816, n. 7.
11. *Graham*: likely a reference to Thomas Graham (1805–69), a Scottish chemist notable for his work on the diffusion of gases; Graham became a fellow of the Royal Society in 1836 (*ODNB*).
12. *your letter to Mr Gassiot*: see letter 0830. Tyndall, however, had asked Gassiot to revise this letter (see letter 0831), which was read at the 30 November meeting of the Royal Society.
13. *the Academy*: likely a reference to the Königlich-Preußische Akademie der Wissenschaften (Royal Prussian Academy of Sciences), established in 1700 in Berlin. It was also referred to as the Academy of Sciences of Berlin.
14. *Royal Medal*: see letter 0815, n. 3.
15. *his lovely daughter*: not identified.
16. *my wife*: Bertha Magnus (1820–1910).
17. *the children*: Christine von Magnus (1842–1936), Paul Gustav von Magnus (1845–1930), and Anna Magnus (1841–75).

To Thomas Archer Hirst [11 December 1853][1] 0853

Sunday

My dear Tom

Your bill[2] shall be duly paid.

I have requested Gyde[3] to purchase the book[4] for you and send it down to Queenwood. I shall probably see Francis in the course of the week and learn something more of Mr Price.

I wish you would tie the Scientific Memoirs[5] together along with Stegmanns book.[6] as if you do not want them I shall bring them up with me.

I have read your letter to Bevington[7] and have been impudent enough to send it back to you.

1. I would suggest the total omission of the first paragraph it has not its melody of a prelude in music and it adds nothing whatever to the information regarding Henry Hill. It is mechanical and therefore I dislike it.
2. It would be almost better to omit commas altogether than to use them in such profusion as you do. It stiffens your writing by giving it a studied air.
3. I would omit the word 'most' before 'distinguished' in the second page. In considering what you have said of your other pupils, he might have excelled them and be nothing particular after all. Next sentence 'He was then a currier and leather cutter' is awkwardly brought in—you say so afterward and it ought to be omitted here.
4. Your description of Henry's way of shewing his friendship by hiding round corners pleases me exceedingly. It is the simple statement of a pleasant fact, and is therefore interesting.
5. 'I found him as honest as the day' I dont know that the day is particularly honest although thoughtless people have been pleased to invent the above simile—such similes however tolerated in conversation ought to be banished from writing. The comma after 'same' spoils the sense.
6. I object most gravely against the expression 'although a beggar'. If I have rightly understood your description of Henry Hill the statement is absolutely untrue. You mean to say he was poor. Why not say so? There is an infinite difference between both—and the expression is quite inconsistent with what follows. Have you not dwelt upon Hill's obstinacy a little too much. Did I know nothing about him I should rather fear him from what you have said upon this point.
7. I think it is perfectly within your power so to massage your description that the thought of employing him at Neckinger[8] should arise spontaneously in Bevington's own mind—and the proposal will come far better from him than from you. If he can do it he will and I would leave him the

credit of making the proposition himself. I have little doubt that either he or his friend Bennett[9] will be able to do something for Henry.

On the whole I think this is an important letter and therefore is worth being well done. I frankly confess my belief that if you took a little more time you could state your case in a more pleasing manner than you have here succeeded in doing. I dont like to accept haste as an excuse in a case like the present. If I can make a guess I would venture to say that the letter is not quite such as you like yourself. If you modify the parts I have referred to then I think it will answer very well, although it will still remain in parts a head production, and not connected with that fountain from which truly good writing must ever spring.

Your affectionate | John

RI MS JT/1/T/942

1. *[11 December 1853]*: Hirst wrote 'Dec. 1853' at the top of the letter and 11 December is the most likely Sunday that it was sent given the content of the letter.
2. *Your bill*: not identified.
3. *Gyde*: see letter 0761, n. 4.
4. *the book*: in letter 0851 Hirst had asked Tyndall to send him Bartholomew Price's work on infinitesimal calculus (see letter 0823, n. 3).
5. *the Scientific Memoirs*: see letter 0759, n. 13.
6. *Stegmanns book*: see letters 0823, n. 2 and 0851, n. 3.
7. *your letter to Bevington*: on 4 December (see letter 0848, n. 10), Hirst noted to Tyndall that he would soon write to Bevington regarding the potential employment of Henry Hill at Neckinger Mills. It is not clear when Hirst sent the letter to Tyndall, though it must have been between 6 and 11 December.
8. *Neckinger*: Neckinger Mills (see letter 0848, n. 11).
9. *Bennett*: not identified.

From Heinrich Debus 13 December 1853 0854

Queenwood, 13 Dec. 1853.

Mein lieber John!

Gern hätte ich Dir gestern geschrieben; unser Examen und einige andere Arbeiten haben mich aber so beschäftigt dass ich jetzt erst an die vollständige Beantwortung Deiner Briefe gehen kann.

Palladium.

Das Palladium bildet zwei Reihen von Salzen. Die eine Reihe entspricht dem Oxydul, die andere dem Oxyde. Nur die dem Oxydul entsprechenden

Salze sind bekannt. Dieselben haben eine gelbe oder bräunlichgelbe Farbe; ihre Formel ist z. B. Pd_2Cl für Haloidsalz
$Pd_2\,O\,SO_3$ für Sauerstoffsalz

Das Platin bildet ebenfalls zwei Arten von Salzen

Platinchlor	Pt Cl	\
		grünlich gelb oder roth.
Schwefels: Platin-oxydul	$PtO\,SO_3$	/
Platinchlorid	$Pt\,Cl_2$	\
		gelb oder orange.
Schwefels: Platin-oxyd.	$PtO_2, 2SO_3$	/

Iridium.

Die Iridiumoxydulsalze sind dunkelgrün oder grünlichbraun.

Ir O = Oxydul.

Die Iridiumsesquioxydulsalze sind sehr dunkelbraun.

$Ir_2\,O_3$ = Sesquioxydul.

Die Iridiumoxydsalze haben eine schwarzrothe farbe. Ir O_2 = Iridiumoxyd. Die Haloidsalze haben dieselben Farben.

Die Salze des Osmiums sind sehr wenig bekannt. Die Sauerstoff- sowohl als die Haloidsalze haben eine grüne, gelbe braune oder rothe Farbe. Osmiumchlor ist dunkelgrün. Das Osmiumchlorid Os Cl_2 ist roth. Das Schwefelsaure Osmiumoxydul ist braun Os O SO_3.

Gold.

Vom Golde sind nur die Haloidsalze gut untersucht. Alle Salze sind gelb oder orangegelb.

Titan.

Die Haloidverbindungen des Titans sind farbenlose Flüssigkeiten. Die Titansäure Ti O_2 ist weiss. Sauerstoffsalze sind nicht bekannt.

Wolfram (Tungsten)

Das Chlor (Wo Cl_2) & das Chlorid (W Cl_3) ist fast farblos & ebenso ihre Salze mit den stärkern Basen.

Salze von Molybdän.

Das Molybdän bildet drei Reihen von Salzen

1. Molybdänoxydulsalze. Sie sind schwarz oder purpurfarben.
2. Molybdänoxydsalze roth oder schwarz.
3. Die Molybdänsäure welche mehr Sauerstoff als das Oxyd enthält bildet gelbe oder farbenlose Salze.

Vanadin.

Die Auflösung der Salze des Vanadinoxyds in Wasser ist schön blau. In festes Form ist ihre Farbe blau oder grünlichblau.

Die Salze der Vanadinsäure sind roth oder gelb.
VO_2 = Vanadinoxyd VCl_2 Vanadinchlorid
VO_3 = Vanadinsäure VCl_3 Vanadinsuperchlorid.

Chrom.

Die Salze des Chromoxyduls sind grünlichweiss
Die Salze des Chromoxyds sind grün oder violett
Die Salze der Chromsäure sind roth oder gelb.
CrO Chromoxydul
Cr_2O_3 Chromoxyd
CrO_3 Chromsäure
CrCl Chromchlorür
Cr_2Cl_3 Chromchlorid
$CrCl_3$ Chromsuperchlorid

Uran.

Das Uranoxydul (Ur O) bildet grüne
Das Uranoxyd (Ur_2O_3) schön gelbe Salze
Die Haloidsalze haben dieselbe Farbe.

Tellur.

Chlortellur ist farblos; Tellurbromid gelb; Tellurjodid, schwarz.

Tellursäure ist weiss ebenso ihre Verbindungen mit farblosen Metall- & Metalloxyden.

Ist dies gut so? Entspricht es Deinen Zwecken? Oder willst Du es ausführlicher haben? Sehr gern werde ich es so schreiben wie Du wünschest.

Deinen letzten Brief will ich mündlich beantworten. Wir sind hier vollauf mit dem Examen beschäftigt. Eben war Mr Best hier um uns zu Vorlesungen zu engagiren.

Lebe wohl mein lieber John und denke zuweilen an
Deinen | H. Debus.

Queenwood, 13 Dec. 1853.

My dear John!

I would gladly have written to you yesterday; however, our exams and some other jobs have so occupied me that I can only now get around to answering your letters in full.[1]

Palladium.

Palladium forms two series of salts. One series corresponds to the lower oxide, the other to the higher oxide. Only the salts corresponding to the lower oxide are known. They have a yellow or brownish-yellow colour; their formula is e.g. Pd_2 Cl for the haloid salt

$Pd_2 O SO_3$ for the oxysalt

Platinum also forms two types of salt

Platinum chloride	Pt Cl	\
		greeny-yellow or red.
Sulphur: Platinous oxide	$PtO SO_3$	/
Platinic chloride	$Pt Cl_2$	\
		yellow or orange.
Sulphur: Platinic oxide.	$PtO_2, 2SO_3$	/

Iridium.

The iridious oxide salts are dark green or greenish-brown.

Ir O = lower oxide

The iridium sesquioxide salts are very dark brown.

$Ir_2 O_3$ = Sesquioxide.

The iridium oxide salts have a black-red colour. Ir O_2 = Iridic oxide. The haloid salts have the same colours.

The salts of osmium are very little known. The oxy- as well as the haloid salts have a green, yellow brown or red colour. Osmium chloride is dark green. Osmic chloride Os Cl_2 is red. The sulphuric acid-osmius oxide is brown, Os O SO_3.

Gold.

Only the haloid salts of gold are well investigated. All the salts are yellow or orange-yellow.

Titanium.

The haloid compounds of titanium are colourless liquids. The titanic acid Ti O_2 is white. Oxysalts are not known.

Wolfram (Tungsten)

The chlorides (W Cl_2) & (W Cl_3) are almost colourless & likewise as are their salts with the stronger bases.

Salts of molybdenum.

Molybdenum forms three series of salts.

1. Molybdous oxide salts. They are a black or purple coloured.
2. Molybdic oxide salts are red or black.
3. Molybdic acid, which contains more oxygen than the higher oxide, forms yellow or colourless salts.

Vanadium.

The solution of the salts of vanadium oxide in water is a beautiful blue. In solid state their colour is blue or greenish-blue.

The salts of vanadic acid are red or yellow.

VO_2 = Vanadic oxide VCl_2 Vanadium chloride

VO_3 = Vanadic acid VCl_3 Vanadium superchloride.

Chromium.

The salts of chromic oxide are greenish-white

The salts of chromium oxides are green or violet

The salts of chromic acid are red or yellow.

CrO Chromous protoxide

Cr_2O_3 Chromic oxide

CrO_3 Chromic acid

CrCl Chromium protochloride

Cr_2Cl_3 Chromic chloride

$CrCl_3$ Chromium superchloride

Uranium.

Uranous oxide (Ur O) forms green salts, uranic oxide (Ur_2O_3) beautiful yellow ones.

The haloid salts have the same colour.

Tellurium.

Chloride of tellurium is colourless; Tellurium bromide yellow; tellurium iodide, black.

Telluric acid is white, likewise its compounds with colourless metals and metal oxides.

Is that all right like that? Does it meet your purposes? Or do you want to have it in more detail? I shall be very happy to write it as you wish.

I will answer your last letter[2] in person. We are completely occupied here with the exams. Mr Best was just here to engage us for lectures.

Goodbye my dear John and think occasionally of

Your | H. Debus

RI MS JT/1/TYP/7/2407–2408

1. *your letters in full*: letters missing.
2. *your last letter*: likely one of the letters mentioned above (see n. 1).

To Thomas Archer Hirst [14 and 15 December 1853][1] 0855

Wednesday night.

My Dear Tom.

Here I am a regular invalid, sitting by my lonely fire, clasped by my four walls and looked mutely down upon by the pictures which hang against them. I spent Sunday with Capt. Wynne at Pinner[2] and think I got cold there. not through want of heat and comfort but through the fact of my breeches being lighter and the day being colder than ordinary. I am sometimes impatient to find this body of mine so easily put out of order, but this feeling is checked by the whispering of a higher wisdom which tells me that my duty is to make the most of my body such as it is, and not to tamper with so valuable an instrument. Today I have given up science, and tonight my wandering thoughts, like the dove of Noah seeking for a rest,[3] alighted upon 'Representative Men.'[4] I have read the lecture on Plato,[5] that one which at the commencement fastened upon me with paramount attraction—and still retains its hold of me. It was as fresh to me as when I first read it and in some respects clearer. I like Plato above all others: not that I will pretend to say that he was the greatest of men, but he is that man which comes nearest to the expression of my private constitution. I like his daring and I like his prudence, I like his abandonment and I like his control. That splendid figure of Emerson's flashes over the man with penetrative radiance and reveals him fully to the mind that is able to realize its force and applicability 'His strength is like the momentum of a falling planet and his discretion the return of its true and perfect curve.'[6] I heard two sermons last Sunday from a very devoted man:[7] I predicted the kind of sermons they would be by observing the sparkle of the man's eye as he read the creed during a portion of the service. He delivered himself up wholly to his theme, his individuality seemed completely annulled and he became a mere organ-pipe through which the spiritual afflatus[8] rendered itself audible. Such men have immense power and fascination; Still I never resign myself wholly into their hands. I go with them but I am not carried and I feel a certain quiet, perhaps egotistic, satisfaction in the thought that I possess a certain mastery over my enthusiasm, even at its highest points. This is the portion of Plato's character in which I behold the magnified outline of my own; I look upon him wonderingly and say like the woman of Samaria[9]—'Here is a man who can tell me all things that ever I did.'[10] I respect action but the action of the world is low and vulgar; it is shaped by external motives. But Plato's action is a tree built up by the inner and unknown forces of nature. I care but little for the man who cannot at intervals feel a great sobriety of soul which places him above the ordinary intoxications of society: this is only to be attained by reversion to that source from which Plato derived his strength; and Plato is to me admirable because he demonstrates the compatibility of this strength with the requirements of man as a social being.

Thursday:[11]—I have been within all day. My candles are now lighted and I have shut out the gleam of the white snow which covers the street. I dont know whether you have written to me today for I have not been at the institution. You may have chastised me for that impudent letter of mine regarding Henry Hill[12] but if so

'Before the chastener humbly will I bow.'[13] I trust still to be able to get down to you on Saturday. Francis was talking about going down but he has now decided the other way. It is now near post time & I will end

ever affectionately | John.

RI MS JT/1/T/943
RI MS JT/1/HTYP/300

---

1. *[14 and 15 December 1853]*: Hirst wrote 'Dec. 14th' at the top of this letter, though it was written over two days, 'Wednesday' 14 December and 'Thursday' 15 December (see n. 11).
2. *Pinner*: a village in the county of Middlesex, located in southeast England.
3. *the dove of Noah seeking for a rest*: a reference to Genesis 8:9. During the Great Flood, Noah sent out a dove to find land. The first time the dove was sent out, it found no land and thus no rest. The second time, the dove returned with an olive branch.
4. *Representative Men*: R. W. Emerson, *Representative Men: Seven Lectures* (London: George Routledge & Co., 1850) (see also letter 0400).
5. *the lecture on Plato*: the second essay in *Representative Men* (see n. 4), titled 'Plato; Or, the Philosopher'. Plato (428/427–348/347 BCE) was an ancient Greek philosopher who wrote on subjects such as government, art, and metaphysics.
6. *His strength is like the . . . true and perfect curve*: Emerson, *Representative Men*, p. 34.
7. *a very devoted man*: not identified.
8. *afflatus*: 'The communication of supernatural or spiritual knowledge' (*OED*).
9. *the woman of Samaria*: a figure in the Gospel of John encountered by Jesus while travelling through Samaria. She believed Jesus was the Messiah and encouraged her town to believe the same.
10. *Here is a man who can tell me all things that ever I did*: John 4:29. The woman of Samaria spoke this line in reference to Jesus, who knew that in the past she had five husbands.
11. *Thursday*: 15 December.
12. *that impudent letter of mine regarding Henry Hill*: see letter 0853.
13. *Before the chastener humbly will I bow*: a misquotation of the line 'Before the chastener humbly let me bow' (L. Byron, *Childe Harold's Pilgrimage: A Romaunt* [London: John Murray, 1812], p. 109, II.lxxxviii.832). See also letter 0795, n. 2.

From Thomas Henry Huxley 16 December 1853 0856

4 Upper York Place, | St. John's Wood. | Dec. 16th. 1853.

My dear Tyndall,

Chatting with Forbes[1] to day he informed me that there was some proposition[2] or other between you and Sir H. delaBeche—Sir Henry is as we know the best of men but instead of going so — towards his point he generally prefers  so, and Forbes has found it quite impossible to make out exactly from his account what is the exact nature of your wishes. Now I fancy that he and his colleagues are very desirous that there should be no unnecessary difficulties in your way and he asked me to write and beg that you would write to him privately—stating exactly the nature of your proposition or of your wishes—so that he may understand what to be after.

I need not tell you that Forbes is a great well-wisher of yours and a man thoroughly to be trusted.

He is obliged to go into the country to morrow but if you will write to him at the Museum of Econ. Geol: in Jermyn Street[3] and put 'to be forwarded' on the letter he will get it at once.

I hear from Francis that you have no end of work on hand and mean to astonish our nerves one of these fine days. Don't go and wear yourself out as there is a great deal to be done.

Believe me | Ever yours faithfully | T.H. Huxley.

N.B. Of course the above is <u>private</u>

IC HP 8:14
Typescript Only

---

1. *Forbes*: Edward Forbes.
2. *some proposition*: De la Beche wished for Tyndall to be a lecturer at the Government School of Mines, but Tyndall was not satisfied with the proposal (see letters 0824; 0844; 0846; and 0847).
3. *Museum of Econ. Geol: in Jermyn Street*: see letter 0721, n. 9.

To William Pare[1] [20 December 1853][2] 0857

My dear Sir,

I have just returned from a short visit to Queenwood;[3] and before the engagements which are thickening upon me at the commencement of our session here divert my thoughts into other channels I will take the liberty of troubling you with a word upon a subject which has long been one of interest to me. I went down to Queenwood on Saturday and was gladdened by the sight of a houseful of as fine boys as any man could wish to have committed to his guidance. I am sure it would have gratified you to see them, although the strong personal affection, resulting from acquaintance which I entertain towards many of them would, of course, in your case be wanting Perhaps also the reflection may have mingled with my feelings that I had myself lent a helping hand towards rescuing Queenwood from the state of derangement in which it was plunged three years ago[4] and placing it in its present prosperous condition. Before I left Queenwood, however, it was my fate to learn that the beauty and the happiness and the prosperity which I saw before me were, on commercial grounds to be sacrificed and superseded by a state of things which, in prospect at least, stands out in dismal contrast to the present. I felt a certain sickness at heart as the matter was fully made known to me.[5] If you have acted with your eyes perfectly open, if you are accurately informed of the prospects of Queenwood and still consider the cash stipulated for an equivalent for the interests you destroy, then I have nothing to say. Mr Edmondson indeed did tell me that you have been accurately informed, that the figures were already laid before you. But you may be justified in doubting figures, you have already seen Queenwood advance and again decline. The figures in 1848–9 may nearly cope with the present ones but there is this capital difference which the statistics cannot reveal: at the time alluded to Queenwood, though fair outside, had a worm at its core; while at present it is sound within and without, and the men engaged in it are able to preserve it so. For my own part I must believe that higher considerations than mere pecuniary ones ought to have entered into stipulations of such moment; but even on money grounds I take the liberty of doubting whether if Queenwood had lived a year or two longer you would not hesitate before taking your present step. But I will end. If you demand my object in writing this letter I can give you none. 'A sigh is sometimes a relief to the human breast[6] and you may interpret these pages as a sigh from one who seen in your present act the hope of an honourable future for Queenwood frustrated.

I am | very sincerely yours | J. Tyndall.

RI MS JT/2/13b/647–48
LT Typescript Only

---

1. *William Pare*: William Pare (1805–73), a co-operative movement activist who played a role in the lease of Queenwood to George Edmondson (Brock, 'Queenwood', p. 16).
2. *[20 December 1853]*: Tyndall wrote a copy of this letter in his journal entry for 20 December (JT/2/13b/647–48).
3. *a short visit to Queenwood*: Tyndall left for Queenwood on 17 December to attend Queenwood's annual Christmas concert (see letters 0848, n. 6 and 0849, n. 4). Tyndall returned to London on 20 December (Journal, JT/2/13b/646–47).
4. *the state of derangement in which it was plunged three years ago*: when Tyndall began his second round of teaching at Queenwood in 1851 (his first round was in 1847–48), the school only had 33 students. When Tyndall left the school to take up his position at the RI in July 1853, it had grown to 76 students (Brock, 'Queenwood', pp. 13–14).
5. *the matter was fully made known to me*: on 20 December, Tyndall noted that Edmondson told him 'Queenwood was to be sold for £27000 to a community of American shakers' (Journal, JT/2/13b/647).
6. *A sigh is sometimes a relief to the human breast*: a common idiom used at the time.

To Thomas Henry Huxley [22 December 1853][1] 0858

Thursday night

My dear Huxley

The stupid postman left your letter[2] next door, and this evening by the purest accident it has come into my hands—otherwise of course you should have had a reply to it before this. I understand the curve of double flexure perfectly. It would have been a far simpler matter had he condescended to walk in a straight line.[3] I have done all I could to get a distinct proposal but thus far I have failed. My wishes were simply to have a distinct proposal and my answer would have been prompt. Could I obtain this my answer would be prompt. I will write to Forbes[4] not because I am anxious about the matter but that I esteem the man and should like to do as he requests. You know it is not my place to make a proposition. I have distinctly stated over and over again that I was ready to undertake the lectures provided they did not require too many—that I never could learn how many would be required. The matter is so exceedingly simple that I cannot imagine how so many interviews have taken place about it without any distinct result. Half an hour's frank conversation would settle the whole matter.[5]

Believe me dear *[1 word illeg.]* | always yours | J Tyndall

IC HP 8:15
RI MS JT/1/T/1026

---

1. *[22 December 1853]*: Tyndall was responding to a letter (0856) that he did not receive until 22 December as noted in his journal (JT/2/13b/648), as well as in letter 0859, n. 2. He also noted here (see n. 2) that he would have replied sooner, had the post not been misdelivered. Morever, 22 December was a Thursday.
2. *your letter*: letter 0856.
3. *had he condescended to walk in a straight line*: reference to De la Beche, who wished for Tyndall to lecture at the Government School of Mines, but despite numerous meetings had not given Tyndall a clear proposal for the duties required for the position (see letter 0856, n. 2).
4. *write to Forbes*: Edward Forbes (see letter 0859).
5. *would settle the whole matter*: the final line of the typescript read, 'But I will write to Forbes, perhaps this may bring matters to a crisis' (see n. 4). This statement was not part of the handwritten letter.

To Edward Forbes [23 December 1853][1] 0859

A mistake of the postman's kept Huxley's note[2] out of my hands until last night, otherwise your request would have been attended to before now. Looking at the matter in hand[3] I can hardly imagine how so many fruitless interviews could have taken place about it: half an hour's frank conversation might have settled the whole. Huxley says that you wish to know exactly the nature of my proposition or of my wishes: Now I have made no proposition, for I conceive that my part in the transaction was to listen to one. The fact of my going to Jermyn Street[4] so often at Sir Henry's[5] request proves that I was prepared to entertain a proposition, and the only question is whether the requirements of the Museum[6] could be made to harmonize with what I must for the present consider the prime question of my life—my researches. I have repeatedly stated my willingness to lecture at Jermyn Street and have used my best endeavours to ascertain the extent to which my services would be required. By word and note I have sought to obtain distinct information upon this point, but have hitherto been unsuccessful. I declare to you the perfect truth when I say that I have no wish in the matter: or if the wish exist it is like an equilibrated force, balanced by considerations equal and opposite. I am still ready to give a frank reply to any proposal that may be made to me. I am ready to lecture, but not to sell myself to lecturing. In Sir Henry's first intimation to Faraday he mentioned thirty as the number of lectures to be given before Christmas. Is he prepared to make this number the basis of negotiations. What I want is a distinct proposal. I have written thus much simply because you wished me to write to you. I know your intentions are those of kindness to me, and however this matter ends I trust you will see reason to think kindly of me; for if my instincts of human character do not mislead me your friendship is a thing which I should like to deserve

Believe me | Yours very truly | John Tyndall.

RI MS JT/2/13b/648
LT Typescript Only

---

1. *[23 December 1853]*: Tyndall transcribed this letter on 23 December (Journal, JT/2/13b/648).
2. *Huxley's note*: letter 0856.
3. *matter in hand*: the possibility of taking up a position at the Government School of Mines (see letters 0856, n. 2 and 0858).
4. *Jermyn Street*: a street in London, England on which the Museum of Practical Geology and the Government School of Mines were co-located (see letter 0721, n. 9).

5. *Sir Henry's*: Henry De la Beche.
6. *the Museum*: the Museum of Practical Geology (see n. 4).

To Thomas Archer Hirst 23 December 1853 0860

23rd Dec. 1853

My dear Tom.

I dont want you to love Henry Hill—I dont want you to love any body but me and your wife when you get one. I never meant that you were to expend your love in this general way. God forbid that I should be so prodigal of so precious an article. I did not use the term 'heart' as the organ of the affections in my note.[1] I believe I did not use it well—I spoke of the note[2] being a head production, but I did not mean to put the head forward as antithesis to the heart. I can scarcely tell you what I meant but if you have made the experience you will understand me. I find myself sometimes by a kind of unconscious brain process writing without having realised with perfect clearness what I intend to write about In such cases I am very apt to repeat expressions which I have already heard, without being conscious of their exact weight and meaning. Again I find by looking at any subject that I establish an affinity between myself and it—I see it all round and every word I write about it seems to hit what I want it to hit—while at other times I beat about the bush and make oblique strokes—writing rather of memory than by present insight. It was in some such sense as the latter that I used the term head production—and I may at the time have been wholly unfit to judge of the matter, as I remember well, squadrons of other thoughts were lying around me. I have written a note[3] along with yours[4] to Bevington—There will be a Christmas box to him although they will not reach him until Christmas is gone. My cold is almost gone and I am getting very strong and hardy. I have just finished a digest of Riess's two volumes,[5] which has occupied me these two days. The matter is now clear off my hands and the next step is to smelt up the stuff and cast it into my own moulds. Debus I have not yet seen—I told him that a bedroom was ready for him if he liked to avail himself of it, but he is foolish I suppose and stays away. All you can say in its defense cant beautify that ugly figure 'as honest as the day'[6]—It tempted me out without an umbrella some time ago and wet me before I came back. Day is a fop—a blusterer—often a conspicuous liar—a mere play-actor without earnestness or soul—night has both—Sweet stars! Look smilingly down on me your defender

Your affectionate | John

RI MS JT/1/T/585

1. *my note*: letter 0853.
2. *the note*: a letter Hirst wrote to Bevington, which he showed Tyndall. The letter was regarding Henry Hill's financial problems (see letter 0853).
3. *a note*: letter missing.
4. *yours*: refers to the letter Hirst wrote to Bevington (see n. 2).
5. *a digest of Riess's two volumes*: see letter 0727, n. 14.
6. *as honest as the day*: Tyndall was referring to his critique of Hirst's use of the phrase in the letter to Bevington sent to Tyndall for feedback (see n. 2).

From Edward Forbes 27 December 1853 0861

2 Marine Parade | Hythe | Kent.[1] | Dec. 27. / 53.

Dear Dr Tyndall.

I am much obliged to you for writing.[2] I could not make out from Sir Henry[3] what had passed between you & him.[4] I know him so well that I deemed matters to be exactly as you have stated. He is the very type of an indefinite dodger & requires to be dealt with as such. To get any decided arrangement out of him without first frightening his wits into concentration is impossible. You or you in conjunction with us, already at Jermyn Street,[5] must come to an understanding & make Sir Henry assent. Of himself he never will originate a distinct proposal, unless with a view of distinctly shirking it as soon as he can.

Now as far as I can gather, the duties which would really be required would be about the following—a course of <u>about</u> thirty lectures—i.e. one or two more or one or two less as might suit yourself—to be given either before Christmas or during the months of April, May & June as might be arranged in council;[6] I believe the former time would be preferable. Besides the lectures occasional examinations would be held, much according to your discretion & in agreement with the general guidelines of the place, & in undertaking the lectures you would of course undertake to assist any arrangements necessary for their success. Besides the school lectures you would be expected to take your turn in the working man's evening course[7] a patriotic act the precise arrangement of which depends of the particular year & on ourselves in council assemblies. You would attend as member of the council whenever you possibly could. I know of no other duties than these that would be asked for. What think you?

ever most sincerely | Edward Forbes

I do not write officially but as one of a body who would like much to have you with us & lecture here & quote the general sentiments being very *[collegial]*.

RI MS JT/1/F/26

1. *Hythe | Kent*: Hythe was a small market town located on the southeast coast of England, in the county of Kent.
2. *obliged to you for writing*: see letter 0859.
3. *Sir Henry*: Henry De la Beche.
4. *between you and him*: Tyndall had been (unsuccessfully) negotiating with De la Beche about taking up a position at the Government School of Mines (see letters 0824; 0844; 0846; 0847; and 0856, n. 2). See also letters 0858 and 0859, which document his frustrations with the process.
5. *Jermyn Street*: see letter 0859, n. 4.
6. *council*: Council of the School of Mines.
7. *the working man's evening course*: beginning in the early 1850s, the Museum of Practical Geology began offering 'working men's lectures'; the admission charge was six pence to attend a series of six lectures (H. De la Beche, 'Museum of Practical Geology', *Accounts and Papers*: *Session 4 November 1852–20 August 1853* [London: House of Commons, 1853], vol. 58, pp. 21–23).

From William Hopkins 29 December 1853 0862

Cambridge | Dec. 29 53.

My dear Sir,

I should much earlier have thanked you for the memoirs[1] you were kind enough to send me some time ago, had I not expected to be able to read them with the attention they deserve much sooner than engagements and, latterly, indisposition have permitted. I have now had the satisfaction of reading them with considerable care, and I may add with great pleasure. Your refutation of Plücker's views[2] as opposed to your own appears to me complete, and your proofs that Magnetic and Dia-magnetic action follow the same law,[3] and the getting rid of Faraday's optic axis force[4] I think excellent generalizations, but I am not satisfied with the proofs you offer of your conclusion that the greater intensity of magnetic action in a particular direction is due only to the closer proximity of the constituent molecules of the body in that direction. There is strong presumptive reason to believe it probable that slate rocks presenting on every distinct cleavage may have been subjected to enormous pressure in direction approximately perpendicular to the planes of cleavage, which would bring the constituent particles more closely in contact in that direction. But I fancy you have somewhere shown that in a piece of Coal-shale[5] the action is least in that direction. In such case the influence of subsequent crystallization must overbalance that of pressure, but how this has been is the mystery.

Again your reasoning p. 25 in the memoir on the Magneto-optic properties of Crystals,[6] is not, I think, conclusive. Admitting the force of it, even, as you have used it, it only assigns a reason why the shorter axis of the rhombic slice should be the line of electric polarity rather than the larger one. The distance between the constituent rhombs will be smaller in directions parallel to the sides of the rhomb than along its shorter diagonal. Why should not this be the line of electric polarity? Our ignorance too of the molecular constitution of bodies throws doubt on the conclusiveness of any reasoning of this kind. But I should like to have some further correspondence with you on this part of your theory when I can devote more time to its consideration. I may possibly have suggestions to make to you respecting it. Any thing more that can be done in this part of your subject is well worth your attention and thought. That you have already established a probability, perhaps a strong one, in favour of your views, cannot I think be doubted.

I some time ago wrote to you[7] respecting some experiments on the effect of great pressure on the conductive power of substances, in which you were kind enough to say you would lend me some assistance. I am still continuing my experiments and am now about to set out for Manchester for that purpose. I shall also consider how far they may be made subservient to the establishment of your views on magnetism and dia-magnetism.

Believe me | Yours very truly | W. Hopkins

RI MS JT/1/TYP/2/604–605
LT Typescript Only

---

1. *the memoirs*: Tyndall sent Hopkins his published papers on 26 November (Journal, JT/2/13b/633) because he discovered that Hopkins was one of the members of the Royal Society who opposed him being awarded the Royal Medal (see letter 0827, n. 5), and this despite the fact that Hopkins had apparently 'never read a single one of my papers' (Journal, 25 November, JT/2/13b/640). The papers Tyndall sent would have likely included J. Tyndall and H. Knoblauch, 'On the magneto-optic properties of crystals, and the relation of magnetism and diamagnetism to molecular arrangement', *Phil. Mag.* 37:247 (July 1850), pp. 1–33; Tyndall, 'On Diamagnetism and Magnecrystallic Action', pp. 165–88; and J. Tyndall, 'On the polarity of bismuth, including an examination of the magnetic field', *Phil. Mag.* 2:12 (November 1851), pp. 333–44.
2. *Your refutation of Plücker's views*: Tyndall refuted Plücker's idea that it was solely the optic axis that determined the orientation of crystal in a magnetic field. Furthermore, he and Knoblauch 'showed that Plücker's new law of the behavior of optically positive and negative crystals was also invalid', and 'demolished Plücker's argument that the magnetic attraction decreases in a "quicker ratio" than the repulsion of the optic axis' (R. Jackson, 'Tyndall and the Royal Medal', p. 154).

3. *your proofs that Magnetic and Dia-magnetic action follow the same law*: in his second Memoir, Tyndall demonstrated that 'The repulsion of a diamagnetic substance placed at a fixed distance from the pole of a magnet is governed by the same law as the attraction of a magnetic substance' (Tyndall, 'On Diamagnetism and Magnecrystallic Action', p. 188).
4. *Faraday's <u>optic axis force</u>*: This concept is based on Plücker's studies on the optic axis, and is discussed by Faraday in 'On the crystalline polarity of bismuth and other bodies, and on its relation to the magnetic form of force', *Phil. Trans.* 139 (January 1849), pp. 1–41, on p. 32.
5. *Coal-shale*: a type of rock formed next to coal deposits, sometimes containing fossils (*OED*).
6. *p. 25 in the memoir on the Magneto-optic properties of Crystals*: see n. 3.
7. *I some time ago wrote to you*: letter 0753.

1854

To Edward Forbes 1 January 1854 0863

Dear Prof. Forbes,

Your letter[1] has set me pondering, and as is usual with me under such circumstances, a stable resolution has been preceded by sundry oscillatory movements which must first swing themselves out before any thing definite can be said. This is the cause of my not replying sooner to the query at the end of your letter.[2]

The more I thought upon your statement the more clearly did one great obstacle arise before me. I was not frightened by the number of lectures, for this is so near what I should be willing to undertake that I hardly think a serious difficulty could arise here. But it is the other points you mention which, though apparently of minor importance, reveal to me the difficulty which lie in the way of an arrangement. Were I free I do not know a position which I should prefer to one at Jermyn Street;[3] nor, if I except Faraday, do I know a man in England at whose side I should work with greater goodwill than at your own. But I have plighted my faith to the Royal Institution, and I am sure it would not exalt me in your estimation if I showed a willingness to enter into other engagements, the direct tendency of which must be to weaken my fealty. I know my liability to be grappled by subjects which interest me, and I could not meddle in the discussion of questions bearing upon the general interests of the School of Science[4] without giving myself up wholly to them. My being a member of your Council[5] would have this result, and hence it is that I could not conscientiously accept this otherwise enviable position.

I know that you will not misunderstand me here, and I trust your colleagues will accept the assurance that were I not held back by other engagements I should feel a pride in working along with them with all my strength towards the noble future which I trust awaits the School of Science. But my duty is distinct, beyond lecturing I cannot go. My wedded spouse is at Albermarle Street and if I go elsewhere it must be as a visitor.

Believe me | ever yours sincerely | John Tyndall

RI MS JT/2/13b/652
LT Typescript Only

1. *Your letter*: see letter 0861.
2. *the query at the end of your letter*: after outlining the requirements of the lectureship at the Government School of Mines Forbes asked, 'What think you?' (see letter 0861).
3. *one at Jermyn Street*: Tyndall was offered a position as lecturer at the Government School of Mines (see letters 0824, n.1 and 0827, n. 19).
4. *the School of Science*: the Government School of Mines (see n. 2).
5. *your Council*: see letter 0861, n. 6.

From Heinrich Daniel Ruhmkorff 2 January 1854 0864

Paris den 2ten Januar 1854

Geehrtester Herr Tyndall

Nach Ihren Wunsch gemäs habe ich Herrn Wertheim gebeten mir den Apparath um den Schal in einen Eisernen Draht wenn der elektrische Strom unterbrochen wird zu weisen, welches er mit dem grösten guten Willen gethan hat, und Ihnen auch vielmahls grussen läst. Der Apparath bestet aus einen 2 Metre langen Kästen auf welchen der Draht auf jeder Seite mit Zey Klammern fest gespant ist eine Rolle mit Draht von ½ Meter lang kann man auf dem Drathe schieben, der Schal den mann war nimmt ist nicht sehr stark und ich glaube das es nicht genug Lärm macht um von weiten gut gehört werden kann, ich bitte mir zu entschuldigen nicht schon früher Ihnen benachrichtigt zu haben. Ich empfehle mich Ihnen ergeben und verbleibe mit aller Hochachtung

Ihr | ganz ergebenster Heinr. | Ruhmkorff

Paris 2nd Jan 1854

Most honourable Mr. Tyndall

According to your wishes, I have asked Mr. Wertheim to wrap the apparatus for sound[1] in an Iron Wire, when the electrical current is interrupted, which he did with much good will, and also sends you many greetings. The apparatus consists of a box about 2 meters long. On each side a wire is stretched and fastened with [two] clamps. A roll with wire ½ meter long can be pushed along the wires, the perceptible sound is not very strong and I do not believe that it makes enough noise to be heard well at a distance. Please excuse me for not having notified you sooner

Your | humble servant Heinr. | Ruhmkorff

RI MS JT/1/R/40

1. *the apparatus for sound*: possibly Ruhmkorff's induction coil, an apparatus that consists of two coils of wire, and which produces electrical currents through induction (*OED*).

From Margaret Allen 4 January 1854 0865

Stanley Terrace January 4th/54.

My dear Tyndall,

I am exceedingly obliged for your kind note[1] and wishes.

I assure you, you were not forgotten on the opening of the new year, my warmest wishes were expressed for your happiness & success.

It was only a day or two ago or at least before I received your note that I had been working over some of your epistles and wondering what was the reason I did not hear from you as usual. I presume the wish to hear has been mutual, and had it not been for circumstances I should have *[restated]* a line trusting to Providence for its safe arrival.

The world I think is using me rather severely—You will be sorry to learn that my beloved Sophy[2] is in a similar condition to that in which you saw my beloved Charley.[3] Consumption that bane of all that is lovely is nipping my last cherished flowers, I have no hope given me of her restoration; thus I shall be left desolate, joys depart one by one but those of sorrow, perhaps because they pierce so deeply, last. Anything like happiness appears for ever to have taken flight—but it will not do to repine and therefore I must not indulge in a strain which is becoming more & more familiar to me. I feel I have a very great deal of matter that I should like you to be acquainted with—I had almost hoped that the revolving year would have brought my friend with it, the time appears long since you last visited us; but it is a mercy you are still alive, and not forever unnerved as your last note but one made me rather apprehensive would ever be the case. Dear John I have almost taken a final determination that I shall never join Robert[4] again; you know well that matters may often reach that climax when forbearance no longer holds good as a motto. I feel it is a difficult subject to deal with, in paper but I believe if you knew what I could tell you, you would say 'you do right my girl.' I will answer any questions you propose to not be afraid of grieving me by any of them for they will not—I feel I have scarcely a particle of even Esteem left for him, it has taken time to work this change, as you well know some of my thoughts in an opposite direction. 'Tis a bitter feeling, but it must & shall be endured—I do not hate but I love not, I esteem not. I am only complying with the advice of his own Father, who is ashamed of his Son, and regrets almost that he was born. This will give you some little insight into my present position, far from an agreeable one and you will also judge what has been the conflict of mind which I have been called to endure. Often in the deepest sorrow I have thought of you, of your kindness your brotherly kindness to me, I know I have reason to be grateful to you, you have my gratitude, and I trust the time will come when my mind will be relieved of the burden which ought long since to have been discharged. I trust you will soon be quite at home in your new sphere of duty[5]—I am obliged for the syllabus of the lectures.[6] I thank you also for a number of the 'Critic'[7] and I also thank you, and the young gentleman[8] who directs the Queenwood Observer[9] to me. I have been highly delighted with those translations by Hannchen.[10]

Believe me | Yours sincerely | Margaret A.

RI MS JT/1/A/73

1. *your kind note*: letter missing.
2. *Sophy*: the daughter of Robert Copeland Allen and Margaret Allen; Sophy (sometimes spelled 'Sophie') survived this bout of illness and was mentioned again in letters 0872; 0975; and 0976.
3. *Charley*: Charles Edward Allen, who died of consumption on 11 January 1853 (see letters 0698, n. 5; 0699; and 0700).
4. *Robert*: Robert Copeland Allen (see letter 0779, n. 2).
5. *your new sphere of duty*: reference to Tyndall's position as Professor of Natural Philosophy at the RI, which he took up at the end of September 1853.
6. *the syllabus of the lectures*: the list of the twelve lectures on Heat Tyndall would give at the RI, as well as the six lectures on Electricity and Magnetism he would give at the London Institution in 1854 (A. S. Eve and C. H. Creasey, *Life and Work of John Tyndall* [London: Macmillan & Co. Ltd], p. 44). Also see letters 0726; 0783; and 0789, n. 2.
7. *the 'Critic'*: *The Critic of Literature, Science, and the Drama*, a London-based literary magazine which ran from 1843 to 1863, and was edited by the journalist and translator, James Lowe (d.1865) (*ODNB*).
8. *the young gentleman*: likely Thomas Archer Hirst.
9. *the Queenwood Observer*: see letter 0816, n. 2.
10. *Hannchen*: see, for instance, Hannchen, 'A Midsummer Night's Dream', *The Queenwood Observer* 9 (15 December 1853), pp. 70–71.

To Thomas Archer Hirst 6 January 1854 0866

6th. Jan. 1854.

My dear Tom

I ought to be ashamed of myself. Not until to day have I been attentive enough to send off the book[1]—I have been so full of other matters that this duty was neglected. I rely upon you coming to me in a few days, remember my first lecture is on the 17th.[2] And I want you to look over the corrected proof before hand. Debus has spent a week with me[3]—I lectured him awfully[4] and he is gone back a more sensible man.

Yours affectionately, | John.

RI MS JT/1/HTYP/303b
LT Typescript Only

1. *the book*: Price's *A Treatise on Infinitesimal Calculus* (see letter 0823, n. 3). Hirst requested Tyndall to send him the first volume of this book in letter 0851, n. 9.
2. *my first lecture is on the 17th*: see letters 0867, n. 1 and 0809, n. 6.
3. *Debus . . . with me*: Debus visited Tyndall sometime at the end of December 1853. On 24 December 1853, Tyndall noted he 'expected Debus but he in all probability is silly and stays away' (Journal, JT/2/13b/648). He also noted that 'Debus returned to Queenwood' on 31 December 1853 (Journal, JT/2/13b/652).
4. *lectured him awfully*: possibly in reference to Debus's 'sullen fits' (see letters 0850, n. 6 and 0848).

To Thomas Archer Hirst [8 January 1854][1] 0867

Sunday morning

My dear Tom

My head is weary and I will rest it by writing to you. I was in fine working trim a few days ago but I pushed things a little too far and am now compelled to draw up. My general health is good and I know when my brain is rested that I shall be a strong man once more. But what displeases me is that it requires too much rest. If you come over a few days before my first lecture[2] commences your presence will be medicinal to me for I find mere diversion of thought of inestimable value to me. While Debus was here[3] I worked but little and during the time my strength was continually increasing, until at his departure I was almost as strong as ever I was in my life—Now I know the success of my lectures[4] will depend chiefly on the state of my health and I am therefore determined not to think too much about them, and your presence here lessen the great temptation I should experience if I were alone I want to ask you a business question just now. Don't you think you could undertake a few translations for the Phil. Mag? You will find them of advantage to you in point of information and the funds they bring in will keep you in cigars and pocketmoney. You will be paid as I used to be—I feel an interest in this question now, as my name appears on the wrapper as one of the conductors of the Magazine.[5] This is no gain to me. Francis requested me to permit it to be placed there, but as it is there I should not like it to be a mere sham. If I can improve the Magazine I will and I know you will help me to do so.

Affectionately yours | John

If you come soon I will set you to work translating—you must do something for the next number.

RI MS JT/1/HTYP/303a
LT Typescript Only

---

1. *[8 January 1854]*: give Tyndall's reference to his upcoming lecture on Heat (see n. 2), this letter was presumably written on 8 January, which was a Sunday.
2. *my first lecture*: Tyndall's introductory lecture on Heat, which took place on 17 January 1854 (see letter 0809, n. 6).
3. *While Debus was here*: see letter 0866, n. 3.
4. *my lectures*: Tyndall refers to his concerns that his condition may interfere with his ability to deliver his series of lectures at the RI (see letters 0726; 0783; 0789, n. 2; and 0865, n. 6). Tyndall had not been feeling well for much of December 1853 (Journal, JT/2/13b/642–52).
5. *one of the conductors of the Magazine*: Tyndall was likely recently asked to be a conductor or editor of the *Phil. Mag.* He was co-editor of the *Phil. Mag.* from 1854 until 1863.

From Thomas Archer Hirst 8 January 1854 0868

Bristol | Jan^{ry} 8th / 54

My dear John

Your letter[1] and Price's book[2] have arrived safely. Next Sunday I hope to spend with you on Thursday and Friday I must be here in Bristol but on Saturday if possible I shall leave. I have forgotten where your lodgings are though and should I arrive at a time when you are not at the Institution I should not know where to drive to. Send me your address. On the whole I am satisfied with my Bristol visit I have had many quiet days to myself and have got through a good amount of work one chief item of which is I have worked carefully through Stegmann's book.[3] It has more than fulfilled my expectations—I know no Mathematical Book that surpasses it in clear lucid explanation. The several problems are admirably enunciated and as admirably solved so that one can find in the end, little superfluous, less incomplete, and nothing at all doubtful or incorrect. What more can we desire in any such treatise? I am not sufficiently acquainted with the History of the subject to judge how much of original matter it contains, nor has he thought proper to state the same. Certain I am, however, that few, if indeed any, are better informed on the subject, and of those few none could have written a more useful treatise I am still of opinion that it is worth translating before deciding on doing so, however, it will be better to wait for the second volume, which is to contain the most intricate parts of the Calculus of Variations[4] Price has written to thank me (or Stegmann) and expresses himself well pleased with the part of the book he has read. From his letter[5] I should guess that his own treatise on the matter will be far less complete He says that his own treatise[6] is already written and 'very much akin' to one part of Stegmann's and that it is too late to alter anything even if he felt inclined.

Henry Hill and Bevington have come to an agreement[7] the former will be in London towards the end of the week, he writes to me today well pleased and very thankful. I am just as much pleased and my only hope is that both parties may be satisfied in the end.

But this and other matters I will reserve for our talk when I arrive. What do you think? My Brother[8] possesses a copy of Carlyle's Past and Present![9] Though I do not believe he has yet read it. I have however and with no small satisfaction. Do you remember my first attempt to read it, and what a failure it was?[10] Although every sentence is now as clear to me as my own speech I am not surprised that it was at that time incomprehensible. I'll tell you why when I come. For the present Good Bye

T. A. Hirst

RI MS JT/1/H/189
RI MS JT/1/HTYP/304

1. *Your letter*: letter 0866.
2. *Price's book*: see letter 0866, n. 1.
3. *Stegmann's book*: see letter 0823, n. 2.
4. *the Calculus of Variations*: Price's book (see n. 2).
5. *his letter*: not identified.
6. *his own treatise*: likely the second volume of Price's *Treatise on Infinitesimal Calculus*, which was published in 1854 (see n. 1).

7. *Henry Hill and Bevington have come to an agreement*: see letter 0833, n. 1.
8. *My Brother*: Hirst had three older brothers. Here he was referring to John Henry Hirst (1826–82), who worked as an architect in Bristol ('Obituary', *The Furniture Gazette* [15 July 1882], p. 45). See also letter 0869, n. 7.
9. *Carlyle's Past and Present*: T. Carlyle, *Past and Present* (London: Chapman and Hall, 1843).
10. *what a failure it was*: see letter 0369, n. 5

To Thomas Archer Hirst [10] January 1854[1] 0869

Royal Institution | Jan. 1854.

My dear Tom

My address is No. 9 Chadwell Street, Myddleton Square Islington. It is not far from the Angel Inn[2] which in London is as well known as Charing Cross.[3] Your letter[4] was most interesting one to me. I am glad to hear such an account of Stegmann's Book,[5] glad to learn that you have had so far such a profitable and quiet vacation. Gladder still to learn that Henry Hill and good Bevington have come to terms[6]—and perhaps most pleased of all to find that John[7] has possessed himself of Past and Present.[8] I shall not be surprised however if he pronounces it a pack of damned transcendental nonsense. Amid all the drapery and painting however there is a wonderful solidity in that book. To me it will continue to be a subject of reverence and if I had it beside me I think I could read it all over again. I bought a book[9] at Bishopstoke station[10] as I came from Queenwood written by some shallow methodist[11] against Carlyle. No doubt the writer is a burning and a shining light among his party—to me he was a hurdy gurdy[12] grinding out tunes to which his education has set him but knowing just as little about them as the above instrument

Affectionately | John.

You must prepare yourself to translate some during your stay in London.

RI MS JT/1/HTYP/305
LT Typescript Only

1. *[10] Jan. 1854*: On 10 January, Tyndall recorded the reception of letter 0868, which suggests that he also wrote his response on this day (Journal, JT/2/13b/654).
2. *the Angel Inn*: a tavern and hotel established in the early seventeenth century, and located on Islington High Street in London ('The Angel and Islington High Street', in Philip Temple [ed.], *Survey of London: Volume 47, Northern Clerkenwell and Pentonville*, 47 vols [London: London County Council, 2008], vol. 47, p. 439–55, http://www.british-history.ac.uk/survey-london/vol47/pp439-455 [accessed 9 February 2016]).

3. *Charing Cross*: see letter 0837, nn. 14 and 15.
4. *Your letter*: see letter 0868.
5. *Stegmann's Book*: see letter 0823, n. 2.
6. *Henry Hill and good Bevington have come to terms*: see letter 0868, n. 7.
7. *John*: John Henry Hirst (see letter 0868, n. 8).
8. *Past and Present*: see letter 0868, n. 9.
9. *a book*: most likely T. McNicoll, *Critical Essay on the Writings of Thomas Carlyle* (Liverpool: Edward Howell, 1853), which critiqued Carlyle's deism. The book was originally published in installments as T. McNicoll, 'The Writings of Thomas Carlyle,' *Wesleyan Methodist Magazine*, 75 (1852), pp. 489–93, 696–702, 802–5, 999–1008.
10. *Bishopstoke station*: a train station built in 1838 located near Bishopstoke in the county of Hampshire, England. It was about 30 miles east of Queenwood College.
11. *some shallow methodist*: Thomas McNicoll (1822–63), a Methodist writer and literary critic.
12. *hurdy gurdy*: a rustic musical instrument resembling the lute, which produces a droning sound through the simultaneous revolution of a wheel that moves the strings and the depression of keys to halt certain strings' movement (*OED*).

From Heinrich Gustav Magnus 14 January 1854 0870

Berlin | 14 Januar | 1854

Werther Freund,

Ihre beiden Briefe vom 1[t]. und 7[t.] d. M. liegen vor mir zur Beantwortung. Ich beginne mit dem letzten, dem geschäftlichen.—Gern würde ich die Ausführung der Instrumente überwachen, aber Sie sind wohl so gut, wenn es zu einer Bestellung kommen sollte mir den Namen des Colonel so zu schreiben daß ich erfahre wie er heißt, denn es ist mir nicht möglich die Buchstaben zu entziffern. Das Preisverzeichniss füge ich hier bei, in etwa 2 Monaten können sämmtliche Gegenstände vollendet sein. Vielleicht genügte es nur den ersten Theil meiner Abhandl. über die Abweichung der kugelförmigen Geschoße zu erläutern weil dieser leichter zu verstehn ist, dafür würde es nur der Aparate N°. 1 No: 2 & 3 bedürfen. Für die Erläuterung des zweiten Theils, der Abweichung der Cylindrokonischen Geschoße halte ich den Aparat mit den zwei Scheiben der in meiner Abhandlung abgebildet ist, für sehr zweckmäsßig. Ich habe denselben jetzt verbessert und im Febr. Heft von Pogg. wird eine Abbildung und Beschreibung erscheinen. Man kann sehr viele verschiedene Erscheinungen mit demselben erläutern und ich beabsichtige ihn ein Polytrop oder Autotrop zu nennen. Den Preis des Polytrop [oder] Autotrop habe ich auch auf dem beiliegenden Blatt eingegeben.

Sie fragen mich ob ich der Ansicht bin daß sich der Gegenstand für eine Freitags Vorlesung eignet. Mir scheint es kommt darauf an was man hineinzieht. Sie kennen das bekannte sehr alte Phaenomen das in neuerer Zeit in Manchester in der Fabrik von Sharp brothers wieder beobachtet wurde, und von dem die Franzosen behaupten daß Clement Desormes oder Thenard es beobachtet habe, daß nämlich eine Platte, die der Mündung eines Rohrs, an dessen Ende sich ebenfalls eine Platte befindet, gegenüber gehalten wird, sich dieser Mündung nähert wenn ein Luftstrom durch die Röhre geht. Dies und die damit zusammenhängenden Erscheinungen, daß eine Flamme sich gegen eine Wand bewegt wenn man einen Luftstrom gegen diese Wand richtet, sind für ein größeres Publicum sehr interessant und überraschend. Sie sind es auch die meine erste Arbeit über die Bewegung der Flüßigkeiten hervorgerufen haben. Sie sind so viel ich weiß noch nicht anders als in jener Abhandlung, und namentlich in der Anmerkung zu der Abhandl<.> über die Abweichung der Geschoße erklärt. Diese Erscheinungen können wie ich mir denken sehr gut eine allgemein interessante Vorlesung liefern, zumal sich sehr hübsche Experimente dabei anbringen lassen. Bei Gelegenheit dieser Erscheinungen könnte man dann von der Abweichung der kugelförmigen Geschoße handeln, da diese nur hierauf beruht, und vielleicht die erste Anwendung dieser merkwürdigen Vorgänge zur Erklärung einer praktischen Beobachtung ist. Einer Beobachtung über die dicke Bücher und Abhandlungen geschrieben sind, weil die Verfasser immer von falschen Principien ausgingen. Der zweite Theil meiner Abhandl., über die Cylindroconischen Geschoße ist etwas schwerer für ein großes Publicum, vielleicht ginge er mit den Polytrop [oder] Autotrop zusammen.—Meine Versuche über die Flüßigkeiten werden wohl nicht vor dem Herbst *[erscheinen]*, da ich schon seit einiger Zeit mit einem ganz andern Gegenstande beschäftigt bin, der erst fertig werden muß.

Gern beantwortete ich noch Ihren Brief vom 1ten. aber mein Papier und meine Zeit gehn zu Ende. Nächstens schreibe ich Ihnen über die Respiration. Wenn Sie mir gleich antworten und mir sagen ob die Instrumente bestellt werden oder nicht, was ich, wegen anderer Arbeit des Mechanicus wohl zu wissen wünschte, so schreibe ich Ihnen dann sogleich wieder. Lesen Sie doch Clausius Arbeit über Thermoströme, ich möchte wissen was Sie dazu denken. Meine Ansicht sollen Sie nächstens erfahren.

Ich wünschte daß C. anderes machte Ihr GMagnus.

Magnus 14 Jan. 1854 (appendix)
Preisverzeichniß der
Aparate für die Versuche
über die Abweichung der Geschoße.

1.	Centrifugalgebläse	25 [Rthl.]—[Sgr.]
2.	Zwei Windfahnen nebst Stativen	3. 20.
3.	Cylinder auf eine Rotations maschine zu setzen, mit einer Vorrichtung um ihn excentrisch rotiren zu lassen.	2. 20.
4.	Aparat mit cylindroconischen Körper in 3 Ringen	22.
5.	Eine Platte mit 484 Löchern vor die Windlade eines Gebläses zu schrauben	2.
		55. 10.

Die Windlade besteht aus einem luftdichten Kasten aus Holz, und kann am Orte angefertigt werden.—Da alle physikalische Sammlungen mit Rotationsaparaten versehen sind, so wird solcher als vorhanden angenommen, es ist dann nur nothig den Cylinder dort auf denselben aufzupassen. Doch kann auch noch ein besonderer Rotationsaparat geliefert werden.

Ein Polytrop oder Autotrop. 24 [Rthl.]

Berlin | 14 January | 1854

Dear friend

Your two letters[1] of the 1st and the 7th of this month lie before me to be answered. I shall begin with the last one, the business one.—I would be happy to supervise the completion of the instruments, but if it should come to an order, please be so good as to write the name of the Colonel[2] for me in such a way that I will see what his name is, for it is not possible for me to decipher the letters. I am enclosing the pricelist herewith; all the items can be completed in about 2 months. Perhaps it would suffice to explain only the first part of my paper on the deviation of spherical projectiles,[3] because this is easier to understand, and for this it would only require the apparatuses No. 1, No. 2 & 3.[4] For explaining the second part,[5] the deviation of cylindro-conical projectiles, I think that the apparatus with the two disks that is illustrated in my paper is very suitable. I have now improved it and an illustration and description will appear in the Febr. issue of Pogg.[6] One can explain very many different phenomena with it and I intend to call it a Polytrope or Autotrope.[7] I have included the price for the Polytrope [or] Autotrope[8] on the accompanying page also.

You ask me whether I am of the view that the topic is suitable for a Friday lecture.[9] It seems to me that it depends on what one draws into it. You will know the well-known, very old phenomenon that was recently observed again in the Sharp brothers' factory[10] in Manchester, and about which the French claim that Clement Desormes[11] or Thenard[12] has observed it, namely that a plate that is being held opposite the opening of a pipe, at the end of which there is also a plate, nears this opening when a current of air passes through the tube. This and the related phenomena that a flame moves towards a wall if one directs a current of air towards this wall are very interesting and surprising for a larger audience. They are also what gave rise to my first work on the motion of liquids.[13] They have, as far as I know, still not been explained, other than in that paper, and in particular in the footnote to the paper on the deviation of projectiles. These phenomena can, I think, provide a generally interesting lecture very well, especially as some very nice experiments can be done with them. When these phenomena occur, one could then deal with the deviation of spherical projectiles, as this is caused only by them, and is perhaps the first use of these strange processes for explaining a practical observation. An observation about which thick books and papers have been written, because the authors have always started out from false precepts. The second part of my paper, on the cylindro-conical projectiles,[14] is somewhat more difficult for a large audience, perhaps it would go together with the Polytrope [or] Autotrope.[15]—My experiments on fluids[16] will probably not *[appear]* before autumn, as I have already been busy for some time now with a completely different subject that must finished first.

I would still gladly answer your letter of the 1st, but my paper and my time are running out. I shall write to you shortly about respiration. If you reply to me right away and tell me whether the instruments are being ordered or not, which I well wish to know because of other work the instrument-maker is doing, I shall write to you again at once. Do read Clausius' work on thermo-currents,[17] I would like to know what you think of it. You shall learn of my view shortly.

I was wishing that C.[18] was doing other things, your G. Magnus

Magnus 14 Jan. 1854 (appendix)[19]
Pricelist of the
Apparatus for the Experiments
on the Deviation of Projectiles

1.	Centrifugal impeller	25 [Rthl.]—[Sgr.]
2.	Two wind vanes with supports	3. 20
3.	Cylinder, to be put on a rotation machine, with a device to let it rotate eccentrically.	2. 20

4. Apparatus with a cylindro-conical
   body in three rings 22.
5. One plate with 484 holes,
   to be screwed in front of
   the wind chest of an impeller 2.

———————

55. 10

The wind chest consists of an airtight box made of wood, and can be made on the spot.—As all physical laboratories are equipped with rotation apparatus, it is assumed that one is already present, it is then necessary only to fit the cylinder on to it there. However, a special rotation apparatus can still be delivered as well.

A polytrop or autotrop.[20] 24. [Rthl.]

RI MS JT/1/M/7

1. *Your two letters*: letters missing.
2. *the name of the Colonel*: presumable Colonel Philip James Yorke.
3. *my paper on the deviation of spherical projectiles*: see G. Magnus, 'Über die Abweichung der Geschosse', *Physikalische Abhandlungen der Königlichen Akademie der Wissenschaften zu Berlin* (Berlin: Königlichen Akademie der Wissenschaften, 1852), pp. 1–24.
4. *apparatus No. 1, No. 2 & 3*: likely a reference to the first three instruments of Magnus's appended list of instruments in this letter: the 'Centrifugal impeller', 'Two wind vanes with supports', and the 'Cylinder'.
5. *the second part*: the second part of Magnus's article on the deviation of spherical projectiles (see n. 1), entitled 'Über eine auffallende Erscheinung bei rotirenden Körpern', on p. 11.
6. *the Febr. issue of Pogg.*: The article was published as G. Magnus, 'Verbesserte Construction eines Apparats zur Erläuterung verschiedener Erscheinungen bei rotirenden Körpern', *Poggend. Annal.* 91 (1854), pp. 295–99.
7. *Polytrope or Autotrope*: in the handwritten letter, Magnus, who was unsure what to call the instrument that he described, inserted 'oder Autotrop' at this point in the letter.
8. *Autotrope*: 'Autotrop' has again been inserted in the handwritten letter (see n. 7).
9. *Friday lecture*: see letter 0788, n. 2.
10. *the Sharp brothers' factory*: a factory located in Manchester, England that manufactured textile machinery and tools. It was founded in 1821 by Thomas Sharp (1779–1841) (L. Jewitt, *The History of Ceramic Art in Great Britain From Pre-Historic Times Down Through Each Successive Period to the Present Day* [New York: Scribner, Welford, and Armstrong, 1878], p. 156).
11. *Clement Desormes*: Nicolas Clément (1779–1841), also erroneously known as Nicolas Clémont-Desormes, a French chemist and physicist notable for defining the calorie as a unit of heat (P. Lervig. 'Sadi Carnot and the Steam Engine: Nicolas Clément's Lectures on Industrial Chemistry 1823–28', *British Journal for the History of Science* 18:2 [1985], pp. 147–96).

12. *Thenard*: Louis Jacques Thénard (1777–1857), a French chemist notable for his work on esters and for discovering hydrogen peroxide (M. P. Crosland, 'Thenard, Louis Jacques', in *Complete Dictionary of Scientific Biography* [Detroit: Charles Scribner's Sons, 2008], pp. 309–14).
13. *my first work on the motion of liquids*: possibly G. Magnus, 'Ueber die Verdunstung von Flüssigkeiten aus Haarröhrchen', *Poggend. Annal.* 26 (1832), pp. 463–64.
14. *the second . . . cylindro-conical projectiles*: see n. 5.
15. *Polytrope [or] Autotrope*: in the handwritten letter, Magnus has inserted 'Polytrop' above 'Autotrop' to signify that he is not sure what the instrument should be called (see n. 7).
16. *My experiments on liquids*: possibly G. Magnus, 'Ueber die Spannkraft der Dämpf von Mischungen zweier Flüssigkeiten', *Poggend. Annal.* 93 (November 1854), pp. 579–81.
17. *Clausius' work on thermo-currents*: likely R. Clausius, 'Ueber die Anwendung der mechanischen Wärmetheorie auf die thermo-elektrischen Erscheinungen', *Poggend. Annal.* 90 (December 1853), pp. 568–79.
18. *C.*: Rudolf Clausius.
19. *(appendix)*: this appendix appears to be written in another, unidentified hand.
20. *A polytrope or autotrope*: see n. 7.

To Christian Ludwig Gerling 15 January 1854 0871

My dear Professor.

Will you allow me to introduce to you the Messrs. Bennett[1] two young Englishmen who are anxious to learn something about German Universities. Mr Edward Bennett[2] has pursued Chemistry for some time his brother[3] devotes himself to literature. If you would kindly assist them with your advice should they require it I should feel much obliged—with kind wishes to M^{rs}. Gerling[4] the young ladies[5] and yourself

Believe me | most truly yours | John Tyndall

15$^{\underline{th}}$ Jan. 1854.

UB Mbg 319:737

---

1. *the Messrs. Bennett*: see n. 2 and n. 3.
2. *Mr Edward Bennett*: possibly Edward T. Bennett, who published a small number of articles in the fields of chemistry and phytology during the early 1850s (see 'Bennett, Edward T', *RS Catalogue*, vol. 1, p. 273). Not further identified.
3. *his brother*: not identified.
4. *Mrs. Gerling*: Christiane Wilhelmine Suabedissen, who married Christian Ludwig Gerling in 1814.
5. *the young ladies*: presumably the daughters of Christian Ludwig Gerling. Not further identified.

From Margaret Allen 21 January 1854 0872

Stanley Terrace January 21st

My dear John,

Perhaps you will just think that is very familiar, certainly you will think rightly, but at the same time I trust you will also think with me that some degree of the formality which has existed in our addressing each other may very sensibly be dispensed with. From your own permission I am allowed to call you Brother, from that permission I have thus addressed you and I trust for the future, I shall receive a similar mark of your respect by your dispensing with the title you generally commence your notes with.

I am exceedingly anxious to learn concerning the success of your first lecture.[1] I remembered the day, marked the hour, and in fancy beheld you expounding some of the great laws of Nature, to I hope an interested and delighted audience. Yours must certainly be an enviable position, just imagine how you have the great, the learned, the noble of the land, at the *[hest]* of one man, listening to and discussing the topics of your lectures. I cannot but think this must be some repayment however slight for the many days and nights of study you have spent. I should not mind if I heard the same declaration concerning you which the Pythian oracle uttered concerning Lycurgus 'Beloved of the gods, and rather a god than a man.'[2] You are elevated to a high position, and I rejoice in it; but sometimes in pondering the matter over I feel just as infinitely small as the little mite which has to be *[viewed]* with the microscope, in comparison to the great cheese of which it once made part. Do not smile. I often pour a profusion of regrets on the senseless air and after all I might just have contented myself and sat quietly down with my book, but I have read somewhere, that only mankind together makes the true man, and that that individual is the most happy who can with courage view himself in the great Whole. Thus you see your little *[mouse]* finds some consolation. It would appear from your letter that I have a friend[3] I know nothing of I perhaps should not say 'nothing' for I was aware he had taken your place at Queenwood and where he had studied and moreover that he was a great friend of yours, but did not know that he was acquainted with me, I assure you I sincerely thank him & hope he will still fulfil his self-imposed duty—I wish you would come to Preston[4] this summer & bring him with you, we shall be able to accomodate you, as we are leaving Stanley Terrace[5] in May for a very large house on the Moor close beside the Broughton Toll bar,[6] you will remember the situation on the road to Goosnargh;[7] now when you do come to Preston remember you are not to make Mr. Hebdon's[8] your home.

I feel your kindness to me is very great but with respect to the last subject of your letter[9]—I am at a loss what to write—when I consider what is past and how ungrateful some people have been I shudder and this was the matter I reffered to in my last, but you will not be angry with me; though I have at present no hope of what shall I say—repaying you I trust someday will find matters adjusted. You are well aware of the position in which I am placed,[10] I shall therefore leave the matter entirely with you—Did I know your thoughts concerning myself as well as I know my own concerning you—I should say do this, or that—but not knowing I leave the matter as I said before to yourself. I have long had a project in my head which I think would bring me a considerable amount of happiness were it accomplished; and would make me independent of my Father. I have certainly no need, but you know me well enough to believe that my present position is most irksome in many respects—however happy I may be there is the continual feeling of obligation—& then again if I could be actively employed many disagreeable subjects of thought would be banished. When you have a little leisure just think the subject over & give me your thoughts. Send me word if those shirt collars are all worn out, you will say you can buy them in London cheap—yes, I know but they are not well made, say if you like them rounded or with points. You will be gratified to hear that my little Sophy is considerably better, and we have some little hope of her getting strong again.[11] This epistle being written without premeditation you must excuse all defects, from

Yours Sincerely, Gretchen.[12]

RI MS JT/1/A/74

1. *your first lecture*: see letter 0867, n. 1.
2. *the Pythian oracle . . . 'Beloved of the gods, and rather a god than a man'*: in Plutarch's biography of Lycurgus, the famous law-giver of Sparta, a priestess at the Oracle of Delphi speaks these lines to Lycurgus (J. Langhorne and W. Langhorne [eds.], *Plutarch's Lives, Translated from the Original Greek; with Notes, Critical and Historical, and a Life of Plutarch*, 6 vols [London: Davison and Whitefriars, 1809], vol. 1, p. 129).
3. *a friend*: Thomas Archer Hirst.
4. *Preston*: see letter 0698, n. 4.
5. *Stanley Terrace*: see letter 0699, n. 1.
6. *Broughton Toll bar*: a location at a road or bridge where a toll is collected in Broughton, a village in the borough of Preston, England.
7. *Goosnargh*: a village in the borough of Preston, England.
8. *Mr. Hebdon's*: possibly John Henry Hebdon (b. 1822), a friend of Tyndall who lived in Preston and worked as a banker's clerk.
9. *your letter*: letter missing.

10. *the position in which I am placed*: this is likely in reference to her husband, who was an irresponsible husband and father (see letters 0779, n. 2 and 0865, n. 4).
11. *Sophy . . . getting strong again*: Allen had written to Tyndall regarding her concerns about Sophy's recent bout of consumption (see letter 0865, n. 2).
12. *Gretchen*: a nickname for Margarethe, the German form of the name Margaret. By taking up the nickname Gretchen, Allen seemed to be aligning herself with the character Margaret—nicknamed Gretchen—in the legend of Faust. In the tale, Gretchen is the woman Faust seduces, and whose life he ruins. In the end, though, Gretchen is saved and goes to heaven. Allen's relationship with her husband (see n. 10), and the recent deaths of two of her children (see letter 0701, nn. 2 and 3), likely caused her to identify with the misfortunes of the character Gretchen. It was clear that Allen was familiar with the story of Faust (see letters 0699, n. 5 and 0918, n. 7).

From James Craven 27 January 1854 0873

Halifax | 27th. January 1854.

Dear Tyndall,

How many years is it since I wrote you, and how many centuries does your last letter[1] to me bear date—I could were I not lazy give myself an answer to this question but cui bono?[2] All I should get would be that the time had been very long indeed. I remember something about a promise connecting Bolton Woods[3] and explorations, when is this to be realized? I hope this summer when I should be once more glad to peregrinate[4] under your wing, and for the nonce[5]—should you desire—forswear whisky toddy—cigars—women and the whole catalogue to give place to beautiful and fantastic ideas and speculations as to Death Immortality—the Inhabitants of the Moon &c. &c.—Speculate! think over these things! and dream over them, as I sometimes do—then of human nature with its manifold gradations—society high and low. I often have at times a hankering to get out of leading strings and set off on some erratic course—wending my way if needs be as a shoeblack[6] but seeing all, doing all, and enjoying all. At other times how I despise and ridicule so extravagant an idea and wonder how it could have entered into my head—Well, well, it seems to me to be innate in me and I dream and live in dreamy contradictions. These strange fantastic dreams—ideas or aught else you like—when one comes to settle and think soberly—earnestly yet materially on the subject, how little can one carve out distinctly as to what one can say one believes. You know my old subject. I am as I always was, unsatisfied with any of the dogmas (so far as I am aware of them) of the day and know not what religious tenets so far as doctrine is concerned I hold!! Strange that this leading point which people tell us we were formed for—viz. to worship

God—should be so shadowy and indistinct as to the mode, and that so many theories all professing infallibility should exist. Kingsley[7] in his Yeast[8] broaches the subject in many lights—he sees the disease but I can find no remedy suggested in any portion of his work to satisfy. He leaves it a problem for others to solve—Strange is it not, that after all the stir, fuss and bother that has been made about the glories revealed in Scripture—its mysteries and events—one should be ignorant and doubtful. Surely to me it has been of no use, but had the ------[9] I will drop the subject.—You're laughing and chuckling in your old way & think now I have got on my hobby[10] I shall proceed for ever—with it. Well—well, I must apologize but when I write to you I always like to open my mind and thoughts to you and I am thus lead to think over these strange matters more deeply, as I have no one else here to unburden myself of them—once I did, but I shall not repeat it.

I have just finished reading D'Israeli's work 'Venetia'[11]—it is an old work of his but I had not read it. He has painted his heroes as wild flighty poetic geniuses who speculate largely on subjects of curiosity—as death—Pontius Pilate[12] &c. &c. rum subjects indeed for the author to think about certainly. There are some good noble sentiments expressed and the conversation introduced between the characters (without entering into *[particulars]*[13]) is terse and logical. There is a deal of poetic rigmarole[14] and what you would call '<u>bombs</u>' interspersed especially among scenery and lackadaisical passages, which the critic on this occasion slipped. Poor Dizzy[15] I am afraid by not acting up to the sentiments so expressed in his book[16] has done for himself, however he will turn radical again yet, but must be content meanwhile to chew his cud for some time to come. I am sorry to say we are not so busy as we have been, however we can improve ourselves in the interim. J. Perkinton[17] is as usual—he has gone to a dance where I hope pray and trust he may pick up a wife as I am sure wedding would do him good. Think of Ellershaw and the twins,[18] and his better half with another compliment in store! How by the way are you getting on herein—surely you are not arrived at so ascetic a state in your philosophical labours as to entirely pass by the gentler sex—I thought some enchantment existed for you at Owen Darwen[19] or in that neighbourhood. Now I know your address—got from Tom Hirst[20]—I shall if at any time in London give you a call, though at the risk of having the scowl and jeers of the other Professors who are not and cannot be so amiable and accessible as you. When are you about to visit this infernal part of the country and give me a look of your lean pale faced jaws with the prominent nasal organ very largely developed. Have you got bald yet—have you commenced wearing spectacles—and show other signs of approaching old age. At all events I hope the inner and outer man are well and that nothing remains of the illness commenced in Northgate Halifax[21] and with brief interruptions continued until the Autumn of last year when the air of Bonn was sought to alleviate it, and I hope most sincerely it has. A great friend told Perkinton the other day when having both imbibed freely that he should live to see him hanged yet—So I may say I sincerely believe I shall see your name handed down to posterity as the greatest and worthy to be ranked above all the Newtons Franklins[22] Playfairs and all the other literary *[1 word illeg]* in existence. Perpetual motion—aqua vitae[23]—(not aqua forte)[24] or some devilment I expect you will discover some thing of this kind one of these days.

I have no news that will interest you. Carter is well and I believe busy I never see him—the lovely S.[25] whom Perkinton as usual gives the euphonious title of Bitch is still a virgin to the best of my knowledge. She talks when I see her learnedly about you and dimples in her cheeks make many appearances while her eyes sparkle away famously. She was I think in love with you and if it is any interest to you she is yet disengaged—N.B. I don't particularly wish you to have her. How is Tidmarsh?[26] When did you hear from him? Has he got a rich harvest of the Australian gold? How is his wife[27] for I understand he had one. Has he a family. Tom Hirst writes me he has had the pleasure of having seen and staid with you. What a noble fellow he is! My brother[28] whom you may remember has left London and has commenced on his own fundament at Sowerby B^rg[29]—as a Solicitor—he is well and in tolerable spirits. I hope and trust we are going to have a jolly good war[30] and that those crafty devils the Russians will get a h---l[31] of a licking such as they never had before. the Czar[32] should be hanged cut in eights and distributed about by way of example—and Richard Cobden[33] should be his successor—only the kingdom or Empire should be materially lessened in extent—the allied powers[34] should take all they can get from the devils by way of compensation and payment for expenses. Such is my idea of Russia and Turkey of which all the world talks. You must write me please when you have some loose spare time

I am | yours as usual | Jas. Craven.

RI MS JT/1/TYP/11/3552–3554
LT Typescript Only

1. *your last letter*: letter missing.
2. *cui bono*: a Latin phrase meaning, 'To whom for a benefit?' (*OED*).
3. *Bolton Woods*: a community located in Bradford, England.
4. *peregrinate*: to travel or journey (*OED*).
5. *for the nonce*: 'for the particular purpose' (*OED*).
6. *shoeblack*: 'A person who cleans boots and shoes for a living' (*OED*).
7. *Kingsley*: Charles Kingsley (1819–75) was a novelist and clergyman of the Church of England who advocated social reform (*ODNB*).

8. *Yeast*: C. Kingsley, *Yeast: A Problem. By the Author of Alton Locke* (London: John W. Parker, 1851), a 'Condition of England' novel focusing on the Chartist movement in rural England.
9. ------: when this occurs in an LT transcription, it often means she has excised something, a word or three words or a phrase.
10. *on my hobby*: likely a reference to 'hobbyhorse', an allusion to L. Sterne, *The Life and Opinions of Tristram Shandy, Gentleman*, 9 vols (York: Ann Ward, 1759–67). In Sterne's novel, the main male characters all have infatuations or obsessions that they 'ride' called 'hobbyhorses', which contribute to the digressive form of the novel.
11. *D'Israeli's work 'Venetia'*: B. Disraeli, *Venetia*, 3 vols (London: Henry Colburn, 1837), a novel that explores social class with characters based on the Romantic poets Lord Byron (1788–1824) and Percy Bysshe Shelley (1792–1822) (see also n. 13).
12. *Pontius Pilate*: the fifth prefect of Judea, and the governor under which Jesus was crucified (H. Bond, *Pontius Pilate in History and Interpretation* [Cambridge: Cambridge University Press, 1998], p. xi).
13. [*particulars*]: in the typescript, LT has written 'partars (sic) [?particulars?]'.
14. *rigmarole*: 'An unduly protracted, involved, or diffuse piece of speech or writing' (*OED*).
15. *Dizzy*: nickname for Benjamin Disraeli (1804–81), a politician and novelist who was Prime Minister of the United Kingdom in 1868 and from 1874 to 1880 (*ODNB*).
16. *his book*: see n. 11.
17. *J. Perkinton*: likely a typing mistake by LT. The original letter most likely read T. Perkinton.
18. *Ellershaw and the twins*: Mrs. Ellershaw was one of Tyndall's colleagues on Richard Carter's surveying team. In his 26 February 1846 journal entry Tyndall wrote that Queen Victoria, from her coach 'fixed her clear blue eyes right upon Mr Carter, Mr Ellershaw and myself, and acknowledged our bows with a gracious nod' (JT/2/13a/108). Tyndall worked for Carter in the mid-1840s (see Volume 2). Not further identified.
19. *Owen Darwen*: misspelling of Over Darwen, a town in Lancashire, England. Tyndall's close friend, Josiah Singleton, ran a school at Spring Brooke, Over Darwen. Tyndall helped teach there in the summer of 1850 (see letter 0424, n. 5).
20. *address . . . from Tom Hirst*: Hirst asked Tyndall for his address in letter 0868 and Tyndall provided it in letter 0869.
21. *the illness commenced in Northgate Halifax*: possibly a reference to the illness Tyndall developed soon after his father's death in April 1847, while he was in Halifax, UK. Tyndall noted the commencement of this illness on 29 April 1847 (Journal, JT/2/13a/193).
22. *Franklins*: reference to Benjamin Franklin (1706–90), an American writer, natural philosopher, and politician notable for his public-spiritedness and for inventing the lightning rod (*ODNB*).
23. *aqua vitae*: Latin, 'water of life', an alchemic term for alcohol (*OED*).
24. *aqua forte*: Latin, likely a misspelling of *aqua fortis*, an alchemic term for nitric acid (*OED*)
25. *the lovely S.*: not identified.

26. *Tidmarsh*: John Tidmarsh (1824–1906), a surveyor who worked alongside Tyndall in Halifax during the mid-1840s.
27. *his wife*: not identified.
28. *my brother*: John Craven, likely James Craven's younger brother (see letter 0389, n. 5).
29. *Sowerby Brg*: Sowerby Bridge, a village in the county of West Yorkshire.
30. *a jolly good war*: the Crimean War (1853–56), fought between the Russian Empire and the allied forces of Great Britain, France, Sardinia, and Turkey. The two sides fought for control over the land under Ottoman rule (*OED*).
31. *h---l*: see n. 9.
32. *the Czar*: Nicolas I (1796–1855), Czar of Russia from 1825 to 1855.
33. *Richard Cobden*: Richard Cobden (1804–65), a politician and manufacturer. Cobden was against the Crimean War, and believed that Britain's support of Turkey was misguided (*OED*).
34. *the allied powers*: see n. 30.

From Thomas Archer Hirst 29 January 1854 0874

Queenwood College. | Jan[ry] 29th / 54

My dear John

You hardly deserve a letter you rascal for turning me out so shabbily. You pretend to be a disciple of S[t] Paul[1] in the matter of hospitality and yet you could let me sleep on while you were getting breakfast and then cut away without seeing me or setting me on my journey respectably. Fie John!! You say I am to fond of Bed, and I am not inclined to dispute it but you ought not to judge of me in that respect when I am visiting. In ordinary cases your stern example is of great use to me but there are occasions when you ought to unbend a little. At first I was a little vexed, at myself most and then at you but I soon forgot it and came away reconciled but with the determination to scold you in my first letter and here you have it. Most sincerely do I wish that you were as much master of sleeping as I am of rising. It would clear away many a sad misgiving I sometimes have for you. I don't think you would say I was too fond of my bed if you saw me in my settled habits going to bed at 1 AM and rising at 7 ½ that I have done and must begin again to do. All the teachers are already here and about 25 boys. Mummery has actually got his diploma[2] Edmondson Haas and Debus have seen it. I am sorry for him and for us for it will be a cause of much unpleasantness I am afraid. There is less chance of our pulling well together, for of course I for one cannot recognize his title. Of one thing I am decided He must leave Queenwood and shall if I have any influence of course he will not rest satisfied with his unacknowledged title he will make it known and—he will injure Queenwood, it will soon spread that he is a sham doctor, people will naturally suspect we are all sham doctors and that Queenwood altogether is a sham. No, no, he must go.

Enclosed is a letter from Atkinson[3] tell me your opinion about the contents. I must say I don't feel inclined to take any steps in the matter. I think it premature and that it would end in a failure if I were to try for the situation.[4]

On Feb. 20th I give my lecture at Salisbury.[5] Now I have a request to make. Get Newman[6] to make a spiral like yours[7] and to send it not later than the fourteenth. From your syllabus[8] I judge that you will want the Electro Magnet on Feb. 2nd and on March 2nd Get Anderton[9] to send it to Dunbridge[10] before the 15th and I will take care that you have it again before March 2nd In a few days I shall send you Weber's paper[11]

Yours affectionately | T. A. Hirst

On Enquiry I find all are unanimous about Mummery's Doctorship so that it is probable he will keep it quiet at Queenwood and make use of it only when he leaves, which perhaps he will do on his own accord.

Send Atkinson's letter back. Don't forget my advice about your health there's a good fellow.

RI MS JT/1/H/190
RI MS JT/1/HTYP/306

---

1. *St Paul*: Paul the apostle (*c.* 4 BCE–68CE), who spread the gospel of Christ during the first century.
2. *Mummery . . . diploma*: Mummery had recently purchased a German doctoral diploma, which Hirst, Debus, and Tyndall were all highly displeased about (see letter 0817, n. 4).
3. *a letter from Atkinson*: Tyndall recorded part of this letter in his journal on 30 January: 'Professor Williamson of London has been here looking out for a German assistant. I was introduced to him and had some talk about the Royal Society's medal. He said it was originally intended to be given to Hofmann, but a party set up Frankland in opposition to him and not being able to decide to which it should be given it was awarded to Tyndall that Bence Jones had said the following about it: The Lion and the Unicorn | Were fighting for the crown | When up came the little dog | And knocked them both down' (Journal, JT/2/13b/655–56). This is a play on the traditional British nursery rhyme 'The Lion and the Unicorn', which features a battle for a crown by the two beasts (I. Opie and P. Opie, *The Oxford Dictionary of Nursery Rhymes*, 2nd edn. [Oxford: Oxford University Press, 1997], p. 442–43). The rhyme was being used metaphorically to symbolise the two groups fighting over who should win the Royal Medal.
4. *try for the situation*: while Tyndall recorded the 'Lion and Unicorn' reference from Atkinson's letter (see n. 3), it seems that here Hirst was referring to a different part of the letter that may have referred to a possible job prospect in Manchester (see letter 0876, n. 5).

5. *my lecture at Salisbury*: Hirst did not note the topic of this lecture in his journal, but stated after the event that he thought he 'never delivered a better lecture', there being 'about 300 present' (Hirst, 'Journals', 21 February 1854).
6. *Newman*: see letter 0717, n. 3.
7. *a spiral like yours*: instrument, not identified (see letter 0876, n. 7).
8. *your syllabus*: see letter 0865, n. 6.
9. *Anderton*: misspelling of Anderson, Faraday's assistant. Hirst also made this spelling error in a journal entry discussing Tyndall's introductory lecture on Heat (see letter 0809, n. 6) (Hirst, 'Journals', 21 January 1854).
10. *Dunbridge*: see letter 0784, n. 25.
11. *Weber's paper*: not identified. Weber published numerous papers (see letter 0820, n. 13).

From Thomas Archer Hirst [29–30 January 1854][1] 0875

I have been greatly astonished, indeed indignant about what Williamson[2] told Atkinson concerning the medal.[3] I cannot reconcile it with what I know of Bence Jones, and am inclined to think it untrue. If so the untruth must be placed at Williamson's door. Did I not know that you are perfectly above being annoyed about the matter I should not have sent the letter,[4] as it is I think you ought to see it. Time will shew who is the lion and who is the little dog,[5] will it not?

RI MS JT/2/13b/655
LT Typescript Only

---

1. *[29–30 January 1854]*: Tyndall noted the reception of this letter on 30 January, suggesting Hirst wrote it on 29 or 30 January (Journal, JT/2/13b/655). It is unclear whether this letter was sent with letter 0874 or soon after.
2. *Williamson*: Alexander William Williamson (1824–1904), an English chemist notable for his work on etherification (*ODNB*). Williamson had told Atkinson that Bence Jones made a short rhyme about Tyndall and the Royal Medal (see n. 3), which Atkinson relayed to Hirst in a letter. Hirst sent Atkinson's letter to Tyndall with letter 0874, n. 3.
3. *the medal*: the Royal Medal, which Tyndall declined (see letter 0837).
4. *the letter*: see letter 0874, n. 3.
5. *who is the lion . . . dog*: see letter 0874, n. 3.

To Thomas Archer Hirst [30 January 1854][1] 0876

My dear Tom.

I received your letter[2] this morning—Faraday was in the laboratory at the time, and I dare say if he looked at me he could see my eye light up with pleasure. I read your 'scolding'.[3] I thought on Saturday that I had the right to be the scolder, that you ought to have been up to see me before I left; but looking backward from this distance I believe you have all the right on your side. Dear Tom if it be only a holiday lounge I should not object to your lying seven days at a time provided you could find enjoyment in it. You know that I have ever looked to tendencies instead of to isolated acts, and as for want of hospitality, heaven knows that I shall be always ready to stuff you with the best I can lay my hands on whenever you come to see me.—As I came home on Saturday and looked at my tenantless room I felt like a man who had become a widower—there was a vacant seat by my fireside and I felt a slight vacancy about my heart also. So you must not think bitterly of your poor brother. He may be stern and perhaps churlish sometimes, but give him time and the tide of humanity will return and you will find that there are milksprings of kindly feeling bubbling through his rugged constitution. This is an exceedingly bad figure, if you used such a one when writing to me I would doubtless be impudent enough[4] to criticize you for it. The defect of the composition is the mixing up of the 'tide' with the 'milk springs'—either of itself is good but both conjointly beget a monstrous idea. With regard to the Manchester affair[5] I can only subscribe to your own notion of the matter. I would make no more in this affair—I say so because I do not see what possible good could result from it. Of course Mr Edmonson's duty and interest is not to permit the fraud you mention a moment's existence at Queenwood. If such things are tolerated there then I prophesy a practical doom to Queenwood.

From being an object of the greatest interest to me I confess it would because the object of my contempt and you may make this my feeling known to Mr Edmondson if you think fit. The thing is a public cheat and Mr M.[6] has shewn any thing but prudence in putting his character in such peril. But the law I suppose is that there is no real prudence when there is no high moral principle—a law which often defeats the calculations of the worldly wise.

I have ordered a spiral for you[7] this day and you shall have it before the 14th—am I not a charming boy?

I copied a scrap of the letter you sent me[8] and sent it up to Faraday—We regarded the matter as simply absurd and indeed it is so—why I was proposed in June as a candidate for the Royal Medal[9]—Tyndall, Frankland, Lindley,[10] Ehrenberg,[11] Sylvester, Bunsen were all proposed the same day. I suppose those who proposed Hofmann 'intended' that he should have the medal, but the majority of the council[12] it appears intended otherwise—Bence Jones declares to me that by the Lion and the Unicorn[13] he meant Hofmann and myself and that the thing would be simply nonsensical were he to mean any thing else, for nobody else had the smallest chance of the medal, and as we were equal somebody happened to suggest that the best plan would be to give the medal to a third party. Do not think that I wish to decry Frankland's best merit when I say that he had not a single vote: How then can it be said that it was owing to the inability of the Council to decide between Frankland & Hofmann that I got the medal—The whole story bears its own refutation on the face of it. I wonder at any man possessing a particle of reflection giving currency to such unadulterated nonsense. I have written this much more to please you than to please myself, I hope I shall never hear another word about the Royal Medal.

My Friday evening lecture[14] appears to have made a good impression—far better indeed than I could have anticipated—I will take your advice as far as I can with regard to my health. I have had some headache today, but employing the evening as I do in writing to you must have a tendency to recover it.

Say to Heinrich that I will examine his crystal tomorrow after my lecture[15] is finished and write to him tomorrow night.—Accept these kisses on one your forehead and one on each cheek from

Your affectionate | John

RI MS JT/1/T/586
RI MS JT/1/HTYP/307

---

1. *[30 January 1854]*: Hirst wrote 'Feb. 1st' at the top of this letter, but that was presumably the day he received it. Tyndall wrote here that he received 'Hirst's letter this morning' (see n. 2), a letter he recorded that he received in his journal entry of 30 January (JT/2/13b/655).
2. *your letter*: letter 0874.
3. *your 'scolding'*: Hirst was upset at Tyndall for not waking him up or sending him off after his visit to London (see letter 0874).
4. *only a holiday lounge . . . be impudent enough*: the transcription for this section of Tyndall's letter is based on LT's typescript, as the handwritten copy is missing.
5. *the Manchester affair*: see letter 0874, n. 4. Not otherwise identified.
6. *Mr M.*: John Mummery. He had purchases a German doctorate (see letter 0817, n. 4).
7. *I have ordered a spiral for you*: Hirst had requested Tyndall to order a spiral for him on 29 January (see letter 0874, n. 7).
8. *the letter you sent me*: see letter 0874, n. 3.
9. *the Royal Medal*: Tyndall was notified that he won the Royal Medal at the beginning of November 1853, but because of controversies surrounding his selection, he declined the award (see letter 0837, esp. n. 2).
10. *Lindley*: John Lindley (1799–1865), an English botanist and horticulturalist notable for

his work in botanical nomenclature. Lindley provided scientific counsel to various British organizations and government bodies, and won the Royal Medal in 1857 (*ODNB*).

11. *Ehrenberg*: Christian Gottfried Ehrenberg (1795–1876), a German naturalist notable for his work on microorganisms and fossilization (W. Sarjeant, 'Hundredth Year Memoriam: Christian Gottfried Ehrenberg 1795–1876', *Palynology* 2 [1978], pp. 209–11).
12. *the council*: see letter 0816, n. 7.
13. *the Lion and the Unicorn*: see letter 0874, n. 3 and 0875, n. 5.
14. *My Friday evening lecture*: 'On the vibration and tones produced by the contact of bodies having different temperatures', *Roy. Inst. Proc.*, vol. 1, 356–59. Tyndall delivered this lecture at the RI on 27 January.
15. *my lecture*: the first of six lectures Tyndall delivered at the London Institution in 1854. Tyndall noted that he delivered this lecture on 2 February on the topic of magnetism (Journal, JT/2/13b/656).

From Heinrich Gustav Magnus 3 February 1854 0877

Berlin 3. Febr. 1854.

Verehrter Freund,

Der Ueberbringer dieser Zeilen Herr Hansen, Sohn des berühmten Gothaer Astronomen, ist Ihnen wie ich weiß von hier schon bekannt, können Sie für denselben etwas thun um ihm seinen Aufenthalt in London nützlich und angenehm zu machen, so würden Sie mich dadurch verbinden. Sie werden in Herrn H. nicht nur einen sehr liebenswürdigen, sondern auch einen sehr unterrichteten und höchst talentvollen jungen Mann kennen lernen.

Noch erwarte ich eine Antwort auf meinen letzten Brief, sobald ich solche von Ihnen erhalten, werde ich Ihnen sogleich auch noch in Betreff der Fragen antworten die Sie an mich gerichtet haben.

Von Herzen | Ihr | GMagnus.

Berlin 3. Feb. 1854.

Honoured friend,

The bearer of these lines, Herr Hansen, son of the famous Gotha astronomer,[1] is, as I know, already known to you from here. If you can do something for him to make his stay in London useful and pleasant for him, then you would oblige me by doing so. In Herr H. you will make the acquaintance of not only a very amiable, but also a very informed and highly talented young man.

I am still awaiting a reply to my last letter.[2] As soon as I have received it from you, I shall reply to you at once as well with respect to the questions that you have sent me.

Cordially | Yours | G. Magnus

RI MS JT/1/M/8

1. *Herr Hansen . . . Gotha astronomer*: the son of Peter Andreas Hansen (1795–1874), a German astronomer notable for his work on revisions to lunar theory. Hansen had been the Director of the Seeberg Observatory, located near Gotha, Germany since 1825 (W. T. L. 'Peter Andreas Hansen', *Monthly Notices of the Royal Astronomical Society* 35 [February 1875], pp. 168–70).
2. *my last letter*: letter 0870.

To Thomas Archer Hirst [4 February 1854][1] 0878

I was in the library with many more before the lecture.[2] Five minutes before the commencement, I was talking to somebody, when I heard Faraday's voice saying, I am looking for Tyndall,' 'Here I am!' I replied. 'I am going in,' he said, 'will you come with me?' He took me by the hand and we walked together towards the theatre. He went in and I followed. The house was full. He stopped short in the midst of the space before the lecture table, and commenced talking to me about the apparatus upon it.[3] 'Do you know Herschel? he asked. On learning that I had not been introduced to him, he singled him out on one of the benches, and introduced me to him. This occurred before the eyes of the assembled congregation, dukes included. I saw in a moment what his object was: it was, in his own quiet way, a public recognition of your palefaced brother—a kind of baptism of a scientific infant. Small as the incident was, I don't know that England could do anything more gratifying to me.

RI MS JT/5/10/664
LT Typescript Only

1. *[4 February 1854]*: this is a fragment of the original letter that was transcribed and included in LT's biographical notes on Tyndall. The full, handwritten copy is missing. LT noted that it was written on 'Feb. 4./54', and the content follows entirely from Tyndall's journal entry of 3 February, which was written at the end of the day 'past 12 o'clock' (JT/2/13b/656).
2. *the lecture*: William Robert Grove's Friday evening lecture at the RI entitled, 'On the Transmission of Electricity by Flame and Gases' took place on 3 February 1854 (*Roy. Inst. Proc.*, vol. 1, pp. 359–62).
3. *about the apparatus upon it*: the RI's large galvanometer discussed in Grove 'On the Transmission of Electricity by Flame and Gases', *Roy. Inst. Proc.*, vol. 1, p. 362. Tyndall also explicitly referred to the galvanometer in his journal entry for 3 February (JT/2/13b/656).

From Thomas Archer Hirst 5 February 1854 0879

Queenwood College | Feb. 5th 1854

My dear John,

The Electro-magnet arrived yesterday evening for which I am obliged. The explanation of the 'little Dog' business[1] although not quite satisfactory, makes matters a little better. Either Bence Jones or Williamson,[2] however, have been guilty of considerable narrow-mindedness and as you say the best way is to leave the matter entirely to them. From your letter to Debus[3] I see you have given your first lecture at the London Institution[4] and that at the time of writing it you were not in very cheerful spirits. Such fluctuations are to be expected in you for the next six months. Being kept away from your legitimate work and taken continually off your feet in this whirlpool of popular lecturing would make anyone's spirits fluctuate. But you cannot help it at present and you must have recourse to efficient antidotes. I hope an occasional letter from me will have such an effect, as I am sure your occasionally sitting down to write to me must have.

Queenwood is filling splendidly It is now certain that we shall have 85 boys almost 80 have arrived. We commence full work to-morrow, i.e. with properly arranged classes. I have already given two lectures as well as two lessons per day. I fully agree with what you say about Mummery[5] and if necessary I shall make use of your letter.[6] I have my eye upon him and I am afraid I shall astonish him before the close of the half year. I am exceedingly sorry to be on such terms with any man, to feel no personal enmity and yet to be watching all his motions, to be plotting against him and to be ready at the moment he commits himself to pounce upon him; this is a miserable state to be in, but my interest in Queenwood demands it. It would be far more comfortable for me to shut my eyes to it all in indifference and contempt, I would gladly do so if we had not a Principal at Queenwood[7] who was not only blind to his own interests but to his own duties. Everybody at Queenwood is indifferent about the matter, as long as he does not require them to recognise it at Queenwood they think it of no importance. I have spoken to Edmondson in very plain terms about it and elicited from him his entire approval (which as you know is not worth an entire straw). He was afraid it would end in a rupture: to which I answered that he must be afraid of no no such thing for it must and will end in one. A day or two after came a letter addressed Dr Mummery many more have come since. 'There's the first symptom' said I. 'Boys will see these letters They will inquire about the matter, address him thus in short 'Murder' will out'[8] and you must make up your mind to act very promptly'. Armer Teufel,[9] he crouched immediately and stammered a cowardly 'I'm afraid we must let it take its course'. I felt inclined to reply 'I'm damned if it shall' but I modified it to 'I'm just as certain that we must not let it take its course'. I see my course perfectly well. I have made up my mind what to do if he goes on vacillating (I see well enough that if he was let alone he would not only wink at it all but in a months time from now he would himself give Mummery his title) As a last resource I shall put him in a terrible fix. I have no fear about the upshot of it but when I use the threat I shall do so perfectly prepared to abide by the decision whatever it may be. It will be simply this. 'Mr Edmondson either you give Mr Mummery notice to leave or I shall give you notice to look out for another mathematical master.' The other evening we had a consultation about new classes and promotions; it was very instructive as a total refutation of Edmondson & Co's notions about statistics and consistency to rule. The question was whether a boy or two in the 2nd class whose abilities I knew should go up into the first. They were not first in the list of examinations and Edmondson considered it would be a great injustice not to abide by these numbers! He deemed numbers an infallible test without ever dreaming that the hand that wrote them was perhaps guided by very fallible opinions. Having studied the subject a little before-hand I analyzed the whole system and proved that these same numbers were deceptive in the highest degree, that they were even inconsistent towards their own numerical standards of ability, not to speak of the imperfections of the latter in themselves. Yesterday evening I invited all the masters to a teetotal[10] meeting (!) at the Lodge[11] and tried to infuse as much good will and energy into them as circumstances will admit.

Yours affectionately, | T. A. Hirst

P.S. Send me Atkinson's letter.[12]

RI MS JT/1/H/191
RI MS JT/1/HTYP/309–310

1. *the 'little Dog' business*: see letters 0874, n. 3; 0875 n. 5; and 0876, n. 13.
2. *Williamson*: see letter 0875, n. 2.
3. *your letter to Debus*: letter missing.
4. *your first lecture at the London Institution*: see letter 0876, n. 14.
5. *what you say about Mummery*: see letter 0876, n. 6. Mummery had recently purchased a spurious German doctoral diploma; Tyndall and Hirst were concerned about its potential impact on Queenwood's reputation (see letter 0874, n. 2). For a full list of references see letter 0817, n. 4.
6. *your letter*: letter 0876.
7. *Principal at Queenwood*: George Edmondson.
8. *'Murder' will out*: 'murder cannot be hid; something surprising is revealed or explained' (*OED*).
9. *Armer Teufel*: German, 'Poor Devil'.

10. *teetotal*: absolute abstinence from the consumption of alcoholic beverages (*OED*).
11. *the Lodge*: see letter 0848, n. 12.
12. *Atkinson's letter*: see letter 0874, n. 3. This postscript appears in the very top left corner of the handwritten letter, above the date and salutation.

From Johann Konrad Bromeis[1] 7 February 1854 0880

Aachen den 7 Februar 1854.

Mein lieber Freund!

Seit Jahren habe ich nur auf indirectem Wege Nachrichten von Ihnen bekommen und auch nur auf dieselbe Art meine Grüße an Sie gelangen lassen. Heute bietet sich mir die Gelegenheit einer directen Unterhaltung mit Ihnen durch die Vermittelung einer Freundin, sodaß ich mich gern beeile, aus meiner andauernden Schweigsamkeit hervorzutreten.

Seit unserer Trennung von dem lieben Marburg sind mancherlei Veränderungen in unserem beiderseitigen Lebensläufen eingetreten und es ist mir eine eben so große Freude wie Genugthuung, sagen und einsehen zu müssen, daß der Erfolg unserer Bestrebungen ein recht befriedigender genannt werden kann. Bei meiner letzten Anwesenheit in Marburg im September des verflossenen Jahrs hörte ich, daß Sie einen neuen und sehr ehrenvollen Wirkungskreis in London erhalten, der Ihnen alle diejenigen Mittel in reichlichem Maaße zu Gebote stellt, ohne welche Ihre rastlose wissenschaftliche Thätigkeit gewiß nur eine sehr gehemmte sein konnte. Ich gratulire also, wenn auch etwas sehr spät, zu dieser Ihrer glänzenden Errungenschaft.

Debus läßt mich mit Briefen, scheint es, ganz im Stich, wahrscheinlich aus Mangel an Zeit, welche ihm seine philosophischen breiten Studien sehr kostbar machen werden. Ich habe Furcht, daß er ein sehr närrischer Mensch geworden ist, denn seine letzten Briefe an mich berechtigen mich beinahe zu dieser Meinung. Uebrigens beneide ich ihn wie Sie um die Fülle von Zeit, welche Ihr der Erstrebung wissenschaftlicher Resultate widmen könnt. Obwohl mit meiner Stellung in der That recht zufrieden, bin ich doch so sehr in Anspruch genommen, daß meine privatlichen Arbeiten, deren ich mehrere unter Händen habe, im Augenblick mehr stille stehen als fortschreiten. Es hat das seinen Grund in den Ihnen gewiß nicht unbekannten Schwierigkeiten, mit welchen man in Deutschland bei der Gründung und Einrichtung wissenschaftlicher Anstalten zu kämpfen hat. Ich habe mir ein ganz neues chemisches Laborat. seit Herbst vorigen Jahrs eingerichtet und bin mit dieser Schöpfung sehr zufrieden. Nun wachsen mir aber schon die Arbeiten sowohl seitens der Gewerbeschule, an welcher ich außer der Chemie auch noch Physik Mineralogie und technische Chemie zu vertreten habe, als auch seitens der hiesigen Fabrikanten so über den Kopf, daß ich fast täglich davon gänzlich in Anspruch genommen bin. Hier kann ich erst von der Zeit eine bessere Regelung meiner Lage erwarten.

Pfingsten gedenke ich, wenn die europäischen Verhältnisse nicht aus dem Gleichgewicht kommen, zu heirathen.

Es wird unnöthig sein, lieber John, Ihren Schutz für diejenige Dame zu erbitten, welche London zu ihrem nächsten Aufenthaltsort zu machen im Begriffe steht, und welche, wie ich weiß, Ihre Achtung und Ergebenheit nicht minder besitzt, als die meine und die meiner Braut. Fraeulein Hermine Ballin wird sicher stehen unter John Tyndalls protection!

Sollte Ihnen einmal ein gutes Buch (Physik oder Chemie oder Technologie etc.) begegnen, das der Uebersetzung würdig wäre, so theilen Sie es mir doch mit. Im Uebrigen werden Sie mir eine große Freude machen, wenn Sie mich recht bald einmal mit einigen Reihen bedenken wollten.

Für heute leben Sie wohl, lieber John und nehmen Sie zum Schluß den freundlichsten Gruß

Ihres | JK. Bromeis.

Schicken Sie mir ein Paar junge Engländer mit gefüllter Börse als Practikanten in mein Laboratorium.

Empfehlen Sie mich Hofmann und Stenhouse. Letzterer besuchte mich hier.

Aachen, 7th February 1854.

My dear friend!

For years I have received news from you only indirectly and that has also been the only way I let my regards reach you. Today I have the opportunity for a direct communication with you through the agency of a friend and so I gladly hasten to break my continued silence.

Since our departure from dear Marburg, many changes have occurred in our paths of life and it is both a great joy and gratification for me to say and to realise that the success of our endeavours can be called a quite satisfactory one. During my last visit to Marburg in September of last year I heard that you attained a new and very honourable sphere of activity in London,[2] which offers you in great measure those means without which your indefatigable scientific activity would certainly be hampered considerably. So I congratulate you, albeit rather late, for this brilliant achievement.

Debus, it seems, is letting me down with his correspondence, probably because of a lack of time, which his broad philosophical studies make very valuable to him. I fear he has become a very foolish man, as his recent letters to me almost entitle me to say. Incidentally, I envy him and you for the abundance of time you can devote to the pursuit of scientific results. Even though I am indeed reasonably happy with my post, I am very busy, so that my private pursuits, of which I have several, are currently standing still rather than moving forward. The reason for this is the difficulties, certainly not unknown to you, which one has to deal with when establishing and equipping scientific institutions in Germany. Last year I set up a completely new chemical laboratory and I am very happy with this creation.[3] But now the various tasks not only at the trade school, where in addition to chemistry I also have to teach physics, mineralogy and technical chemistry, but also those given to me by local factory owners are all becoming too much for me, so much so that I am completely absorbed by work almost every day. It is only from the passage of time that I can expect an improvement of my situation.

If the European situation[4] does not get out of hand, I intend to get married on Whitsunday.[5]

It will be unnecessary, dear John, to ask for your protection for that lady who is about to make London her next place of residence and who owns your esteem and loyalty no less than mine and that of my bride. Miss Hermine Ballin[6] will be safe under John Tyndall's protection!

Should you ever come across a good book (physics or chemistry or technology etc.), which is worthy to be translated, would you please let me know. Incidentally, you would give me great pleasure if you honoured me with some lines in the near future.

For the time being I bid you well, dear John and do accept my kindest regards

Your | JK.[7] Bromeis.

Send me a couple of young Englishmen with a full purse as trainees in my laboratory.[8]

Give my regards to Hofmann and Stenhouse.[9] The latter paid me a visit here.[10]

RI MS JT/1/B/133

1. *Bromeis*: Johann Conrad Bromeis (1820–62), a German chemist who studied under Bunsen at Marburg (see letter 0728, n. 14). Bromeis was made 'privatdozent' in 1853, and later named Professor of Chemical Technology at Marburg. In his journal entry for 24 February, Tyndall noted receipt of this letter that was enclosed in a note from Hermine Ballin (note missing), who had recently arrived in London (JT/2/13b/660) (see also letter 1025).
2. *you attained . . . in London*: Tyndall was appointed Professor of Natural Philosophy at the RI on 4 July 1853. See letter 0762 for the offer of appointment.
3. *Last year . . . this creation*: not identified.
4. *European situation*: perhaps a reference to the Crimean War (see letter 0873, n. 30).
5. *Whitsunday*: see letter 0759, n. 2.

6. *Hermine Ballin*: a female friend of Tyndall's from Marburg, who he seems to have helped acquire a position as a governess at a school for girls in Stamford Hill run by Jemima Hornblower (d. 1861) (see letter 0885). Tyndall first noted receiving a letter from Ballin on 23 November 1853 (Journal, JT/2/13b/639), and shared an unidentified letter from Ballin with Faraday and his wife on 21 December 1853 (Journal, JT/2/13b/648). It is likely that Tyndall's connecting of Ballin with the Faradays resulted in her appointment at Stamford Hill, as three of their nieces attended school there (G. Cantor, *Michael Faraday: Sandemanian and Scientist* [London: The MacMillan Press, 1991], p. 48).
7. *JK.*: abbreviation for Johann Konrad.
8. *Send me . . . my laboratory*: this note appears in the left margin of the first page of the letter.
9. *Stenhouse*: John Stenhouse (1809–80), a Scottish chemist and inventor who had studied under Playfair. In 1854, Stenhouse was a lecturer on chemistry at St. Bartholomew's Hospital in London (*ODNB*).
10. *Give me . . . visit here*: this note appears in the left margin of the third page of the letter.

To Thomas Archer Hirst 8 February 1854 0881

Royal Institution | 8th. Feb. 1854.

My dear Tom

I have just arranged my notes for the lecture at the London Institution tomorrow[1] and as I have an hour's breathing time I will expend a portion of it in your service. Were it your object to make me knit my brow and feel stern at heart your letter[2] was effectual. In fact the interest I took in your account of how the fraud[3] is prospering excited my interest so much as almost to inflict upon me the very thing which your letter was kindly intended to dissipate, namely—a good headache. I pictured that faltering little man[4] when his duty was pointed out to him crouching as you say and whimpering that things must take their course. Certainly had things taken their course with him, had he not been already two or three times rescued from the destruction to the verge of which his own folly had brought him he would not now enjoy the privilege of getting wrong again. Of course the string is a mouldered one and it is waste of time to harp upon it the man has neither insight nor principle and would be a scoundrel if tempted and not checked by motives external to himself. I wrote to him[5] about Mummery and imagine that I have stirred his choler or his fear a little. I fully agree with you that the semblance of such fraud must not be tolerated at Queenwood and it is pure shortsightedness on the part of your colleagues which prevents them from seeing the thing in the same light as you do. The stolid stultified little materialist, it has never occurred to him what an effect a straightforward decisive honorable mode of action at such a juncture must have on the whole dynamics of his establishment. He poor blind worm only looks at the fact of losing a teacher and is unable to see the forces which his own vacillation tends to damp and to destroy. I have taken him in my claws and fluttered him as an eagle would a wren. It is only by such a process that you can make his duty clear to him. When he sees that you are in earnest he begins to fancy that you must be right. Goodbye

Yours affectionately | John

RI MS JT/1/HTYP/311
LT Typescript Only

---

1. *the lecture at the London Institution tomorrow*: presented on 9 February, this was the second of six lectures Tyndall gave at the LI in 1854 on Electricity and Magnetism (see letter 0865, n. 6).
2. *your letter*: letter 0879.
3. *the fraud*: reference to John Mummery's purchase of a German doctoral diploma (see letter 0879, n. 5).
4. *that faltering little man*: George Edmondson.
5. *I wrote to him*: letter missing.

To Thomas Archer Hirst [12 February 1854][1] 0882

Sunday night

My dear Tom.

I received a letter[2] from you a week ago. It came to me like the opening of a radiant dawn full of promise, the foretaste of a splendid day. "Good fellow" I exclaimed as I read of your purpose of writing to me—"his letters will be a medicine to my soul." I give you credit for the best intentions when you wrote that letter—no doubt in one of those sunny moments when the heart is uppermost you conceived the amiable thought of lightening the weariness of your friend, by pouring the balm of friendly counsel and communion into his hot brain; but the week is gone and there has been no realization of the promise. But I do not blame you no more than I blame the sun of April for being occasionally hidden behind a cloddish cloud. I must content myself with those dull atmospheric darkenings which belie the promise of the morning, and wait patiently for the time when the celestial blue will shew itself once more and the sun can radiate unhindered. Nor do I take credit to myself for writing to you tonight—indeed I rather regret that I should find it needful, that this body of mine renders me afraid of taxing it and drives me to write letters merely with a view of killing time. I might write to somebody else, but there is nobody to whom I can write with so much freedom. I swear there is no body who will understand me so well when I wander hither and thither over the regions of thought as yourself, and therefore through pure love of liberty I choose you as a but against which to discharge my epistles. I was almost tempted to go out to Bevington this afternoon; I was down at Anderton's[3] and the thought occurred to me as I returned. When I reached home and found myself warm and well I regretted that I had not gone, but the regret soon vanished as my regrets usually do, being swallowed up speedily by something else. I think you are mistaken in deeming that my life fluctuates so much—in fact considering that I am an Irishman I consider my life a remarkably level one, and I feel a certain contempt for these fluctuations to which you refer. What I want is steadily to get my work done with the straightforward persistence of a natural force, and not to heap up the little strength I possess into eminences which are checked by hollows of corresponding depth. Well what am I going to say next? Myself appears to be my principal theme in these latter days—but I think I am not altogether to blame for this, for the kind solicitude of my friends has in some measure forced myself on myself as an object of contemplation, and hence I am likely to become an egotist if I do not take care. Francis read a letter of yours[4] for me yesterday—Francis can read a letter very nicely, and I suppose this was the cause of the conclusion that I came to regarding it, that it was well and wisely written. He intends coming up to me one night a week I suppose with the laudable intention of driving away the scientific blue devils[5] which he imagines haunt me. I shall be glad to see him at all events. I get on with Faraday just as usual, last Friday evening at Owen's lecture[6] I found myself planted between him and the Duke of Argyl.[7] The relation between us is exceedingly frank and pleasant. I feel it a privilege to be able to attest my loyalty to such a man in his declining years, and I trust I shall never do any thing which he might wish I had not done; for I think the noblest reward I can offer to such a man is the free reverence of a free soul; and it is a benefit to myself to find an objective outflow for this reverence. On Sunday evening he told me that some scientific bookseller[8] had asked him for an introduction to me with a view of engaging me to write for him, but in the bustle after the lecture I made my way towards Bevington and forgot the bookseller. The managers[9] have a scheme for a course of 4 lectures[10] after Easter to be given by several different men: The subject will be education; among the number spoken of are Lyell,[11] Playfair Dawes[12] and your friend John.[13] An older man would speak with more authority but they seem to desire that I should be one of the lot. I shall endeavour to make the little experience I had at Queenwood useful and if proper occasion offered would mention the place; but supposing the Doctor sham[14] to prosper there do you think that I could mention it?—I should on the contrary rather feel inclined to disclaim all sympathy with a place where such things were tolerated. I have little fear however that you will permit this weed to flourish there. Trample it down by all means my boy and make the light heads and hearts around you feel that there is a focus of stern sincerity among them. I may possibly find a page or two of your journal[15] of service to me also—that part where you describe taking that little boy with the eloquent eyes into the fields and teaching him Euclid[16] with bits of stubble. I should like to introduce it, if the general tenor of the lecture would permit of it, for I believe the thing is a record of true teaching. Last night I had a dull headache which I believe was brought on by the cold north wind at present prevalent. I sent out for a cigar and at the risque of fumigating Faraday if he should call to see me I smoked the half of it—strange to say it cured the headache—thus the thing which under ordinary circumstance, would give me headache in this instance cured it. Will D^{r}. Fox[17] accept this as a proof of the sound uses of homeopathic principles? My lectures thus far have gone on pretty prosperously—last Thursday at the L.I.[18] I gave the audience a lesson on electricity with which all appeared highly satisfied. My lecture at the Royal[19] also seemed to give general satisfaction. I of course am the chief learner on such occasions. If I can get over a year or two I have no fears of ultimate success for I find my power of utterance becoming more and more unshackled, and I trust wholly to the inspiration of the occasion to furnish me with the requisite form of expression. By cultivating this trust I hope eventually to become a passable speaker. But I see my paper is drawing to a close and it would be preposterous to commence another sheet. It is also creeping up to 10 oC and my candles are sinking towards their jackets. I shall therefore turn my shins towards the fire and indulge in half an hour's rumination. Good night my dear boy and believe me as ever

Your affectionate John

> 'For pain, danger, difficulty, steady slaving toil and other highly disagreeable behests of destiny, shall in no wise be shirked by any hightest mortal that will approve himself loyal to his mission in this world.' Life of Sterling page 79.[20]

Ich bin zufrieden![21] | 11 oC.

RI MS JT/1/T/587
RI MS JT/1/HTYP/312

1. *[12 February 1853]*: Tyndall noted in his journal entry of Sunday, 12 February, 'Wrote to Tom in the evening' (Journal, JT/2/13b/659).
2. *a letter*: letter 0879.
3. *Anderton's*: see letter 0814, n. 5.
4. *a letter of yours*: not identified.

5. *blue devils*: 'a harmful or malignant demon, especially one that causes melancholy' (*OED*).
6. *Owen's lecture*: On Friday 10 February Richard Owen (1804–92), the English comparative anatomist and paleontologist, gave a lecture at the RI entitled 'On the Structure and Homologies of Teeth', *Roy. Inst. Proc.*, vol. 1, pp. 365–74.
7. *the Duke of Argyl*: George Douglas Campbell (1823–1900), the eighth Duke of Argyll, a scientist and politician who attempted to link social and scientific thought (*ODNB*). Tyndall recorded his impressions of Argyll on 10 February (Journal, JT/2/13b/658).
8. *some scientific bookseller*: not identified.
9. *The managers*: From May 1853–April 1854, the Managers of the RI were: William Wilberforce Bird (1784–1857), John Boileau (1794–1869), John Bate Cardale (1802–77), Henry John Codrington (1808–77), George Dodd (1808–81), Charles Fellows (1799–1860), John Peter Gassiot, Aaron Asher Goldsmid (1785–1860), Henry Bence Jones, George Macilwain (1797–1882), George Moore, James Parke (1782–1868), Frederick Pollock (1783–1870), Joseph William Thrupp (1799–1873), and Philip James Yorke (*Roy. Inst. Proc.*, vol. 1, p. 309).
10. *course of four lectures*: these would be on the theme of education. Eventually, the course would be expanded to seven lectures, including one by Tyndall (see letter 0900, n. 6). The other speakers included William Whewell, Faraday, Robert Latham, Charles Daubeny, James Paget, and W. B. Hodgson (see *Lectures on Education* [London: Parker, 1854]).
11. *Lyell*: Charles Lyell (1797–1875), English geologist, most notable for his work, *Principles of Geology: Being an Attempt to Explain the Former Changes of the Earth's Surface, by Reference to Causes now in Operation*, 3 vols (London: J. Murray, 1830–33) (*ODNB*). Lyell would not give one of the education lectures.
12. *Dawes*: Richard Dawes (see letter 0809, n. 7). Dawes would not give one of the education lectures.
13. *your friend John*: see n. 10.
14. *the Doctor sham*: see letters 0879, n. 5 and 0881 concerning Mummery's purchase of a German doctorate. See also letter 0893 regarding Edmondson's response to Tyndall's concern about mentioning Queenwood in his lecture on education.
15. *a page or two of your journal*: On 7 January, Hirst made note of a boy at Queenwood named W. Pilkington, who was 'about 12 or 13 years of age' with 'a pair of beautiful large soft blue eyes . . . As eloquent, affectionate eyes they are as I have ever seen'. Pilkington was mischievous and had a stammer, which impeded his learning, so one day Hirst took him into a field to teach him trigonometry (Hirst, 'Journals', 7 January 1854).
16. *Euclid*: Euclid (around 325–265 BCE), the Greek father of modern geometry (C. Hayhurst, *Euclid: The Great Geometer* [New York: Rosen Publishing Group, 2006], p. 13).
17. *Dr. Fox*: see letter 0800, n. 29.
18. *last Thursday at the L.I.*: Tyndall gave a lecture on Electricity and Magnetism at the London Institution on 9 February (see letter 0881, n. 1). On 9 February, Tyndall noted that although this lecture fell short of his ideal, he was nevertheless satisfied with it (Journal, JT/2/13b/658).

19. *My lecture at the Royal*: this lecture on the subject of volcanoes was given at the RI on 7 February (Journal, JT/2/13b/657).
20. *'For pain, danger . . . Life of Sterling page 79*: see T. Carlyle, *The Life of John Sterling* (London: Chapman and Hall, 1851), p. 79, a biography of the Scottish writer and poet, John Sterling (1806–44) (*ODNB*).
21. *Ich bin zufrieden!*: German, 'I am happy!'

To Christian Ludwig Gerling 17 February 1854 0883

London 17th Feb. 1854

My dear Professor.

I was much gratified to receive your long kind letter,[1] and shall feel the greatest pleasure in complying with the request which it conveys. I have consulted with my friend Francis who is accustomed to receive papers from the continent for the Philosophical Magazine, and he recommends you to send the lithographs[2] through a bookseller. If you send them through the post they are almost sure to be crushed and spoiled and there is also danger of difficulties arising with the post office in England. We have a foreign Bookseller here, Mr Nutt[3] who often receives papers from Giessen[4] directed to the Philosophical Magazine, and I think your safest plan would be to send your papers[5] as the Giessen people send theirs. Perhaps Herr Elwert[6] is in connexion with Winter of Frankfurt[7] or Marcus of Bonn;[8] either of these would forward the parcel to Nutt in the Strand.[9] When the lithographs reach me I shall at once gum the cards upon them and have them forwarded to the proper persons. I am much obliged to you for including my name in the list.

I am at present much occupied with my lectures and were it not for this I would send you a long German letter in return for your long English one. Your letter brought to my mind the occasion of my first seeing you in Marburg[10]—I believe you are truly one of my very oldest friends there, certainly you were the first that ever spoke a word of English with me. I have passed through a variety of places and scenes since I used to be a visitor to your lecture room but the remembrance of those days are still fresh in my recollection, and I can never think of Marburg save as a place where during upwards of two years I was constantly treated with kindness and friendship.

With kindest remembrances to Mrs. Gerling[11] and to the young ladies[12]

Believe me ever | Yours sincerely | John Tyndall

A gentleman named Guthrie[13] has written to me from Marburg, he is evidently an Englishman and probably not living at Marburg, this thought has prevented me from writing to him there, but I have arranged that his letter[14] (a scientific one) shall be published in the Philosophical Magazine. If you know him perhaps you would be kind enough to communicate this to him.

UB Mbg 319:738

1. *your long kind letter*: letter missing.
2. *the lithographs*: not identified. Lithographs are prints produced through an engraving on stone or metal. Ink is applied to the stone or metal to make impressions (*OED*).
3. *Mr Nutt*: see letter 0820, n. 5.
4. *papers from Giessen*: not idenified.
5. *your papers*: not identified.
6. *Herr Elwert*: Universitätsbuchhandlung Elwert, a well-known publisher and bookshop in Marburg, Germany.
7. *Winter of Frankfurt*: reference to the publisher and politician, Christian Friedrich Winter (1773–1858), who owned the C. F. Winter university bookshop in Heidelberg, located approximately fifty miles south of Frankfurt, Germany.
8. *Marcus of Bonn*: likely A. Marcus, a publisher located in Bonn, Germany, a city located three hundred miles southeast of Berlin.
9. *the Strand*: a major street in the city of Westminster, which ran along the bank of the north Thames River.
10. *the occasion of my first seeing you in Marburg*: they met on 27 October 1848, two days after Tyndall arrived in Marburg, when Tyndall called on Gerling (Journal, 27 October 1848, JT/2/13b/396).
11. *Mrs. Gerling*: see letter 0871, n. 4.
12. *the young ladies*: see letter 0871, n. 5.
13. *Guthrie*: Frederick Guthrie (1833–86), an English chemist and physicist. In 1854, Guthrie was working on his doctorate at the University of Marburg, which he would complete in 1855 (*ODNB*).
14. *his letter*: not identified. The earliest paper by Guthrie to appear in *Phil. Mag.* was published in 1857 (F. Guthrie, 'On Iodide of Acetyle', *Phil. Mag.* 14 [1857], pp. 183–86).

To Thomas Archer Hirst 18 February 1854 0884

Feb. 18th. 1854.

My dear Tom

I have written to Ruhmkorff,[1] but you must not calculate on moving the instrument[2] in less than a month. I think it well to mention this lest you should build upon it for your lectures. The following are one or two experiments:—first to shew the bending of your compound bar[3] by unequal expansion.

one

one end of the bar rests on a plate of metal on the black B. By bending the middle touches the metal rod C D and thus closes a galvanic circuit[4] the moment the circuit is closed the platinum lamp[5] *[P]* which can be placed at some distance from the table glows, G is the battery.

The same lamp may be used to shew expansion two stands A B are set so that a bar C nearly fills the space between them. By heating it does fill the space. Make A B close a galvanic circuit in which the little lamp is introduced.

To shew expansion of gases I have made use of current also. By sending a current through the bulb coil is heated the air expands and is transferred to the tube T which was previously filled by a coloured liquid. To shew currents of air bend a thin glass tube say an inch in diameter heat with a spirit lamp[6] at A—the current ascends—a smoky taper[7] placed at B has its smoke drawn upwards at C it is sucked downwards and traverses the entire tube.

The bursting of iron bottles by freezing is also a very good experiment. I burst two in my lecture on the subject.[8]

A balloon of heated air is also a popular experiment.

A Smeaton's pyrometer[9] shews the expansion of bodies for a popular audience very well.

Find out for me wherein Bulloughs patent[10] for throwing will work out of gear when a thread breaks consists. Edmondson knows all about it—I did know but have forgotten it—let me have this intelligence if you can by Tuesday morning.

Affectionately | John

RI MS JT/1/HTYP/314
LT Typescript Only

---

1. *I have written to Ruhmkorff*: letter missing.
2. *the instrument*: possibly Ruhmkorff's induction coil (see letter 0864, n. 1).
3. *compound bar*: a bar composed of two types of metals, which can be used to show the unequal expansion of differing metals when heated.
4. *a galvanic circuit*: 'a current generated by chemical action and coming directly from the cell, pile, or battery in which it is generated' (G. M. Beard and A. D. Rockwell, *On the Medical and Surgical Uses of Electricity*, 8th edn [New York: William Wood & Company, 1892], p. 777).
5. *platinum lamp*: 'an electric lamp in which the filament is made of platinum' (*OED*).
6. *spirit lamp*: 'a lamp fed by methylated or other spirits, and used esp. for heating, boiling, or cooking' (*OED*).
7. *taper*: a candle made of wax (*OED*).
8. *my lecture on the subject*: not identified.
9. *Smeaton's pyrometer*: An instrument developed by the English civil engineer John Smeaton (1724–92), which was used to measure the expansion of materials when subject to heat and cold (J. Smeaton, 'A Description of a New Pyrometer, with a Table of Experiments Made Therewith', *Phil. Trans.* 48 [January 1753], pp. 598–613).

10. *Bulloughs patent*: In 1841, John Bullough (1836–91), a manufacturer of textile machinery, patented the weft fork; the invention prevented cloth damage by stopping the loom when the thread broke (*ODNB*).

From Edward Frankland 19 February 1854 0885

Hanse House | Richmond Grove | Feb. 19/54

My dear Wat[1]

I am heartily glad to find you bubbling up again in that semicircular fountain of eloquence behind the R. I. lecture table, for I really had begun to think that you were gone spark out.

I have read your abstract[2] with much interest & think you have clearly demonstrated your point but do you think that the two solids ever come in contact at all, does not the rocker rather vibrate in free air, being repelled from the heated point below it and again falling back as the heat is abstracted from that point. I have not expressed myself very clearly but you know what I mean.

You are a decent fellow for providing 2 beds & if I come to London alone I shall certainly avail myself of your kind invitation; it is however my present intention to bring with me my sister-in-law Miss Fick[3] & therefore unless you turn Benedict[4] in the mean time I cannot have the pleasure of staying with you. My Sister-in-law whom we have expected since Friday comes to England with Frl.[5] Ballin.[6] A propos, I understand that this latter pretty little girl is an importation of yours, ostensibly to make a Governess of her, but in reality I know not what!

I was in London on the 1st, 2nd, 3rd & 4th of Jan. last & intended seeing you at the Royal Inst. on the latter day but it being the snow blockade I could get no cab and was obliged to make the best of my way from Waterloo[7] to Euston Station[8] with my luggage on my back. Had I known what was to follow I would quietly have taken up my quarters with you for 2 days as I had the pleasure of spending that night in a snowdrift between Birmingham & Stafford[9] & the 2 following days in Birmingham.

Thank you Mrs F[10] is quite well and desires to be very especially remembered to you. The little dot[11] is also thriving famously & will soon loose its title of 'baby'.

Will you not take a run down to Manchester at Easter or at all events in June when we shall be delighted to see you.

very sincerely yours | E. Frankland

RI MS JT/1/F/41

1. *Wat*: Wat Ripton, one of Tyndall's most common pseudonyms, under which he published a number of articles, poems, and translations (see, especially, Volume 1).
2. *your abstract*: J. Tyndall, 'On the Vibrations and Tones produced by the contact of bodies having different Temperatures', *Roy. Soc. Proc.*, vol. 6, pp. 392–93. See also letter 0876, n. 14.
3. *my sister-in-law Miss Fick*: sister of Sophie Frankland (see n. 11). Not further identified.
4. *Benedict*: a man newly married, especially one who initially appeared to be a confirmed bachelor (*OED*).
5. *Frl.*: abbreviation of *fraulein*, German 'miss'.
6. *Ballin*: see letter 0880, n. 6.
7. *Waterloo*: Waterloo Station (see letter 0705, n. 2).
8. *Euston Station*: a train station in London, England located north of the Thames River.
9. *Stafford*: a town in the West Midlands, England.
10. *Mrs F*: Sophie C. Frankland, née Fick (1821–74). Edward Frankland and Sophie Fick first met in Marburg in 1847 and married in 1851.
11. *The little dot*: Margaret Nannie Frankland (1853–1941), the first child of Edward and Sophie Frankland (Russell, *Edward Frankland*, p. 344).

From Thomas Archer Hirst 19 February 1854 0886

Queenwood College. | Feb. 19th 1854

My dear John.

When I said that I would write occasionally to you for medicinal purposes I thought I was generous in resolving upon a letter every Sunday. You seem to have anticipated more for you complain although I have acted up to my determination thus far. I am perfectly satisfied that 'through pure love of liberty you choose me as a butt against which to discharge your epistles'[1] for they interest me exceedingly and they serve to disentangle your thoughts which to you and me is most important You are quite right too in saying that 'yourself' is too often your theme at present and that your friends are partly to blame for it. To avoid this I have often talked about other things when your health was uppermost in my thoughts. At one time I thought I ought to avoid the subject & divert the conversation into other channels but I found it always interested me too much to do so successfully. And I consoled myself with the notion that it was perhaps less injurious to talk about yourself than to war against yourself without being talked to. The best thing of all to do would be to talk as little as possible about yourself, but to listen attentively to other people's talk on the same subject and to act accordingly. Unadvised you would in the end kill yourself, advised you talk about yourself too much and perhaps dwell a little too much on your weak health. Here are the two horns of a dilemma and of the two I begin to think the latter is most harmless. The Spiral[2] has arrived safely, is well made and acts well. What will it cost? Tomorrow I shall lecture at Salisbury.[3] A successful lecture depends upon many unforeseen things otherwise I should predict that tomorrow's lecture will be the best I have yet delivered. I have taken considerable pains & am in good health and spirits. You can never talk too much for me about Faraday every little incident interests me for all appear to proceed from a most kind and generous purpose towards you. Depend upon it, it will be an interesting chapter in the history of Physics & Physicists which records the good understanding and friendship that exists between you and Faraday. An occurrence took place yesterday which is new to Queenwood two little boys[4] ran away. After sending Wright after them Edmondson mentioned the matter to me and from what he said I thought he was certain that they had gone to Winchester[5] and I rested quiet all day. Wright returned at 12 P.M alone. He had been unable to trace them, at the Winchester Station they had not been seen. Edmondson had gone to bed, Mrs E.[6] had not; after delivering his report he came to me at the Lodge.[7] I saw that the affair ought to be more energetically sifted, that perhaps Edmondson had blundered. Late as it was (1 A.M) John Haas and I got into the College,[8] went quietly upstairs to see if they have come back again unseen, no: we then examined their clothes for letters &c examined their boxes, broke open their writing desks for any indications as to their intentions & from certain letters it seemed evident that Mr E.[9] was right. We then walked to Dunbridge[10] called up the Porter, got to know that they had taken tickets to Winchester; it was 3 AM when we returned; we sent for Emm's carriage[11] and early this morning set Wright off again to Winchester. All this was done quietly and without M^r^ E's knowledge.

When he heard of it this morning he opened his eyes and became as determined as a poor lion. Just as Wright was starting a letter from Barber's father[12] came saying that his son[13] together with the other Boy Orton[14] had arrived at Winchester and he would send them back on Monday. Well, Mr E. bustled up more still. 'no Richard Wright' says he, 'thou must go and bring Orton back with thee in any case, and if Mr Barber will not give up his son he must keep him altogether we will have him here no more. I think it is our duty to be prompt' Poor fellow he is like a bladder, in general all flabby and shriveled up, you may dilate him by blowing him up with courage and he sounds well, but prick him and he collapses again into flabbiness.

But it is just post time | Good Bye. | Thine as ever | T. A. Hirst

I know nothing about Bulloughs patent[15] and cannot well understand M^r^. E. I will see that he writes to you tomorrow morning.

RI MS JT/1/H/192
RI MS JT/1/HTYP/315–315a

1. *'through pure love . . . discharge your epistles'*: a quotation from letter 0882.
2. *The Spiral*: see letter 0876, n. 7.
3. *Tomorrow I shall lecture at Salisbury*: see letter 0874, n. 5.
4. two little boys: Barber and Orton (see nn. 13 and 14).
5. *Winchester*: a city in the county of Hampshire, England.
6. *Mrs E.*: Anne Edmondson (see letter 0818, n. 2).
7. *the Lodge*: see letter 0848, n. 12.
8. *the College*: Queenwood College.
9. *Mr. E.*: George Edmondson.
10. *Dunbridge*: see letter 0784, n. 25.
11. *Emm's carriage*: not identified.
12. *Barber's father*: not identified.
13. *his son*: Barber, a student at Queenwood. Not further identified.
14. *Orton*: R. Orton, a student at Queenwood. Not further identified.
15. *Bulloughs patent*: see letter 0884, n. 10.

From Tommy Grieve 22 February 1854 0887

Queenwood College | Feb. 22nd. 1854.

My dear Dr. Tyndall,

I hope you are quite well. Bellamy[1] told me to ask if you will come down to see his and George William's[2] Panorama of Italy,[3] it is on the 4th of March. There are five classes just now and above eighty boys, but they have not all returned yet. I have enjoyed myself very much since I returned to Queenwood. With kind to you

I remain | Your affectionate Friend | T. W. Grieve.[4]

RI MS JT/1/TYP/1/461
LT Typescript Only

1. *Bellamy*: H. Bellamy, a pupil at Queenwood College.
2. *George William's*: George William Edmondson (see letter 0797, n. 5).
3. *Panorama of Italy*: H. Bellamy and G. W. Edmondson had painted a panorama, a picture composed of a series of images depicting a passing scene (*OED*). Grieve misquoted the date, as the date of the panorama was noted in *The Queenwood Observer* as 2 March ('Miscellaneous', *The Queenwood Observer*, 2:2 [31 March 1854], pp. 9–16, on p. 10).
4. *T. W. Grieve*: in the postscript LT notes: 'all beautifully written with a flourish at the end—'

From James David Forbes 24 February 1854 0888

Clifton | Bristol | 24 Feb 1854

My Dear Sir

I have just received a sheet of the Royal Society's Proceedings which contains just enough of the commencement of a paper of yours on the Vibration of Metals,[1] to show that you have obtained results in some respects very different from those which I published long ago.[2] I never was well satisfied with that paper of mine. Though the experiments were made with due deliberation, the composition of the paper took place I recollect under the pressure of a very arduous engagement which prevented me from resuming the subject as I hoped to do.

I shall be particularly obliged if you will give me an early opportunity of becoming acquainted with your results.

I remain dear Sir | Yours faithfully | James D Forbes

Dr Tyndall

RI MS JT/1/F/30
RI MS JT/1/TYP/12/3940

1. *Royal Society's Proceedings . . . the Vibration of Metals*: see letter 0885, n. 2.
2. *those which I published long ago*: see letter 0814, n. 7.

From Thomas Archer Hirst [24–]26 February 1854[1] 0889

Queenwood College. | Feb.^ry 26th 1854

My dear John,

Last Tuesday morning on entering my cottage I found Miss Edmondson[2] there, she appeared to be doing something at my table, and on entering she hastily crushed up some paper in her hand and tried to look surprised and confused; but her attempt did not succeed, she had no business there and I knew she had come for a purpose: I wanted to see what that purpose was. Poor Miss E. she little thought as she chatted about trivialities that I saw through her so well, she moved to the door as if she was going I knew she was not; she pointed, as she stood in the porch to your picture as if by accident, whereas I knew we had got to her purpose now, and she had done so deliberately to introduce a conversation about you and your letter to M^r E on the Queenwood Mummery.[3] I was very brief in what I said indeed my talk consisted for the most part in replies for I wished her to feel throughout that she was meddling with things that did not belong to her. Well from the letter which you have sent me this morning[4] I see I was quite right, she had come to me in order to get an excuse for writing to you. After this her commencement is significant 'Circumstances, and my own strong feeling' says she 'led unexpectedly to a conversation with Dr H.' As far as I can remember the conversation was something like this.

Miss E. John Tyndall writes very warmly about the sham doctor,[5] but I fear he did not know what had been done in the matter. His motives are often very noble but his actions often give unnecessary pain. He has even threatened Father.[6]

Tom. As to the threat, as you call it, I delivered that simply because I knew it was necessary to remind Mr E. continually of his duty.

Miss E. Yes but he might be reminded without causing so much pain

Tom. (silent but thinking 'that the weak & imbecile of this world ought to be thankful if they can purchase the counsel of the wise at the expense of pain even' but the pain felt by Mr E. is scarcely comparable to the prick of a pin, is contemptible in fact)

Miss E. John does not seem to be aware that Father took active measures long ago.

Tom. In what way Miss Edmondson.

Miss E. All letters directed to Dr Mummery are enclosed in an envelope by Father.

Tom. I am glad to hear it for not long ago I happened to be present when such a letter came and Mr E. saw nothing he could do in the matter but to 'let things take their course'

Miss E. At the beginning of the half year Father told Mr M. he could not recognize his degree at which Mr M. was angry.

Tom. This was wise too; whether done by Mr E. himself or at the bidding of somebody else does not matter now.

Miss E. In fact I don't see that Father can do more. If his degree is not recognized here we have done our duty: we could not then send him away.

Tom. I think that Mr E. not only could but ought to send him away.

Miss E. But why?

Tom. Because the best men at Queenwood have lost all respect for him. It prevents our hearty co-operation & that is the secret of Queenwood's success.

Miss E. He will leave at Midsummer he does not like to go now for the sake of his own character.

Tom. Do you know this to be a fact? Has he told you or M^{r} E this?

Miss E. No, I scarcely like to say how I got to know it.

Tom. Good morning Miss Edmondson.

I should have liked to have told her that it was unbecoming in her to know so much from sources she dare not explain. That I should begin to suspect she had come to my cottage to get information. This system of espionage annoys me greatly I lose all respect for a woman when I detect her in such practices. The atmosphere about me becomes choking These women at Queenwood[7] know every trivial occurrence about the place they delight in it and boast about their secret sources of information. They shew this to you when talking of others, and little think that they are proving to you that you are similarly watched. I confess it they have willfully aroused my suspicions, I avoid leaving my letters about, as I did at first, because I believe Miss E or her Father would read them if they could; read them out of pure curiosity.

At present there is a cloud over Queenwood although in point of numbers we continue to increase. These two run-aways[8] have bothered us. I took the examination of them out of Edmondson's hands, much to his comfort. I investigated the matter very carefully. Orton[9] I found to be a little villain and Barber[10] a poor dupe. I advised Edmondson at once to expel Orton & release Barber, and without a word of objection he has done so. Then again there are thieves about the place, some of the boys' boxes have been broken open & money stolen. Wright has lost money and the thieves remain as yet undetected. I cannot rest until they are detected. One morning Mr E. told me that we must stir in the matter and that he was going to examine all the boys that morning, make them give up their money, so as to see if any of it was marked (Wright marked his money). I was astonished and asked him by all means not to spoil matters by acting prematurely. 'It appears strange to me,' I said 'that you should at once suspect your lads, sooner than your servants. For instance you know Waterman[11] is or has been dishonest; why not convince yourself first that he is not guilty and then examine your boys.' This stopped him & nothing further has been done.

I am suffering from head-ache too, all last Monday up to 7 ½ PM I was ill but when I entered the lecture room at Salisbury it vanished instantaneously and my lecture[12] was successful. Yesterday evening I was forced to go to bed by 9 o'clock and to-day I am but poorly. Hence I must close my letter.

Yours affectionately, | T.A.Hirst.

<Hereafter LT Transcript Only>

Sunday night: I have just returned from Wandsworth:[13] Bevington desires his kind remembrances: Hill is progressing in a very satisfactory manner,[14] he had the pony for a drive into Kent[15] to day.

RI MS JT/1/H/193
RI MS JT/1/HTYP/317–319

1. *[24–]26 February 1854*: presumably this letter was written over several days and only sent after a postscript was written on 'Sunday night', 26 February.
2. *Miss Edmondson*: Jane Edmondson.
3. *your letter to Mr E on the Queenwood Mummery*: letter missing, although Tyndall noted in his journal entry of 7 February that he 'wrote to him in a manner that will shake his nerves, told him flatly that he was the abettor of a fraud, that his duty was to see that no immoral trick of the kind should be transacted at Queenwood' (Journal, JT/2/13b/ 657–58). Tyndall also described his letter to Edmondson in letter 0881, n. 5. For other references to the 'Mummery affair' see letter 0817, n. 4.
4. *the letter which you have sent me this morning*: presumably the letter Jane Edmondson sent to Tyndall (see letter 0890, n. 2), which Tyndall forwarded to Hirst.
5. *John Tyndall writes very warmly*: letter missing.
6. *Father*: George Edmondson.
7. *These women at Queenwood*: see letter 0890, n. 16.
8. *two run-aways*: see letter 0886.
9. *Orton*: see letter 0886, n. 14.
10. *Barber*: see letter 0886, n. 13.
11. *Waterman*: likely a servant at Queenwood. Not further identified.
12. *Salisbury . . . lecture*: see letters 0874, n. 5 and 0898, n. 5
13. *Wandsworth*: a borough of London, England where Bevington lived located at the confluence of the Thames River and the River Wandle in the county of Surrey.
14. *Hill . . . manner*: see letter 0868, n. 7.
15. *Kent*: see letter 0861, n. 1.

To Jane Edmondson 25[–27] February 1854[1] 0890

Saturday night 25th. Feb. 1854

My willful sister

I received your letter[2] to day, read it twice over and then sent it down to Queenwood to shew my friend 'Tom'[3] how energetically you can write when you are aroused. My present writing is therefore from memory and perhaps this is all the better as a distance of 10 hours renders a man's feelings mellow and will prevent me from being energetic in my turn. I can truly say that I would not willingly give you a moments pain, I can just as truly say that I would not willingly give your father[4] a moments pain, and whenever I have done so there has been pain inflicted on both of us. I remember when in Germany your father wrote to me saying 'John Tyndall they say we are going asleep, come and rouse us.'[5] I took your father at his word and have often frankly told him that he had been too long accustomed to the soft lotions which you deem so commendable. I did rouse him, and no doubt have often roused your ire, but the result has been sufficient to console me for even your disapprobation. Besides in the case under consideration[6] I did first try the soft lotion and spared all allusion to the imprudence of conniving at the purchase of a spurious title. I heard it mentioned with universal applause at Queenwood during my visit[7] there that you had expressed yourself very decidedly against the imposture, but I heard it added at the same time that you did not know how far your father had committed himself. When I mentioned the subject to your father I was aware of this, still I forbore all allusion to it, actuated purely by the wish to see matters set right without wounding his feelings, for I believed, and still believe, that he countenanced the thing, without reflecting on its immorality. I believe that neither you nor he saw the true magnitude of the evil, for if you did you would have extinguished the notion at the outset and thus prevented the individual in question[8] from expending his money in the purchase of his own dishonour.[9] The best you can do at present is to cover up a weed which ought never to have been permitted to strike root. There ought to be an odour of sincerity about Queenwood, the people there ought to be so honest and highminded as to prevent the thought of such a transaction from being entertained by any body connected with the place. You have spoken to me frankly and I return the compliment, but I assure you I do so without the slightest feeling of anger. What I state is my cold intellectual conviction. You speak of my losing power, this may be a grave matter in your estimation but in mine it is one of extremely small moment. Certainly I should not like to lose power with the wise and good, for this would be indicative of my own error or perverseness. But I am extremely indifferent as to that power the retention of which depends on the withholding of an unpalatable truth. Just so far as I am false and incompetent do I lose power in reality, and just so far as I am true and competent do I retain it. This is my interpretation of the law, which I doubt not is widely different from yours. Surely I must have had truth on my side to render my presence at Queenwood at all tolerable: surely there must be some latent voice within your own soul pleading my cause and inducing you not withstanding all my roughness to your father to call me brother! I say it is the rectitude of my position alone which has been able to disarm those natural feelings of resentment which my necessary roughness was calculated to evoke; and when this transaction has receded among the memories of the past, I shall not blush to look upon the part I have acted in it, nor shall I fear your verdict concerning me although the nature of that verdict never entered into my calculations.

I could say more if I wished. Three years ago your father was a careworn man. I remember scanning his countenance sometimes shortly after my return to Queenwood[10] and seeing the cloud of anxiety upon his brow and the deep depression of heart depicted on his countenance; when I left Queenwood[11] he was a plump jolly faced man and compelled to increase the periphery of his waistcoat—what does this prove? Why that my pills though unpleasant were of a tonic character and good for both body and soul. Instead of quarrelling with me you ought like Socrates[12] to pray to the gods to grant your father men honest enough to act by him as I did. This is your reasonable service; and your 'strong feelings' are so many clouds and tinted spectra which distort and discolor and prevent you from discovering the veritable facts of the case. Now I think I have given you a Rowland for your Oliver.[13] One other remark in your letter crosses my mind—you say something about your father not doing a certain thing lest it should be considered he was bribing me to speak well of the place. This is a dark passage to me: I cannot unravel it. The truth is I rarely speak of the place at all, for I conceive the place must speak for itself, and here I see very clearly where your creed and mine begin to diverge from each other. I now leave the question to the judges who overlook earthly things, and leave it to them also to punish a young lady who instead of writing me a Valentine or something of the kind in which she excels has misapplied her time and talents in sprinkling me with gall and bitterness.

John Tyndall

I would strongly advise you to let this matter rest and to talk no more about it, and I trust you will accept this letter as the completion of all I have to say upon the subject.

I suppose you mean to say something hard when you speak of my being bribed to speak well of the place, or that your father thought I might consider something a bribe. Happily I do not understand the expression and therefore cannot feel angry regarding it. My belief is that Queenwood must speak for itself, and that its appearance must not be maimed by affording shelter to what you so justly call a 'sham'.[14] The other point which strikes me is an allusion to a 'threat' which, if I understand aright, was made use of by Dr. Hirst.[15] If I be correct in my interpretation you must be in great error regarding Dr. Hirst's character; for depend upon it he does not deal in threats. What he says I believe he will with the most coldblooded sincerity make good. I have no notion as to what the nature of his expressions may have been, but I am confident he does not mean to threaten any body. Do not let your feelings get you in a false position here, and cause you to imagine that Dr. Hirst is a mere blusterer who wishes to frighten you. If the governing party at Queenwood seeks to make truth its fulcrum it can well afford to smile at threats—but if the censure of an honorable man is really called for in the vindication of truth, than so much the worse for Queenwood if such censure be disregarded as *[only a]* threat. This is the third or fourth protest which it has been my lot to receive from the ladies of Queenwood.[16] But I cannot forget that when the 'ladies' of Queenwood had the entire control of matters, and soft lotions to order, the result was such as not to increase my estimation of their wisdom in a sphere which really does not belong to them.

RI MS JT/1/HTYP/320–321
LT Typescript Only

1. *25[–27] February 1854*: while the original letter is missing, Tyndall copied this out in his journal dating it 25 February. He presumably added the postscript, however, after receiving Hirst's letter (0889) dated 26 February, and therefore did not send this letter until the morning of 27 February (see letter 0893, n. 4).
2. *your letter*: letter missing.
3. *'Tom'*: Thomas Archer Hirst.
4. *your father*: George Edmondson.
5. *when in Germany your . . . come and rouse us*: letter not identified. Tyndall began teaching at Queenwood in August 1847, but left to pursue a doctorate in Germany at the University of Marburg in 1848. By 1851, student numbers at Queenwood had sunk tremendously and Edmondson asked Tyndall to return to teaching. Tyndall returned to Queenwood in September 1851, with his completed doctorate (Brock, 'Queenwood', p. 13).
6. *the case under consideration*: reference to John Mummery's purchase of a German doctoral diploma (see letters 0817, n. 4; 0833, n. 10; 0836, n. 7; 0874, n. 2; and 0876, n. 6). Tyndall and Hirst were disappointed with Edmondson's apparent indifference and advised him to take swift action against Mummery's fake diploma (see letters 0879, n. 5; 0881; 0882; and 0889).
7. *my visit*: see letter 0857, n. 3.
8. *the individual in question*: John Mummery.
9. *purchase of his own dishonor*: see n. 5.
10. *my return to Queenwood*: see n. 4.
11. *when I left Queenwood*: see n. 4.
12. *Socrates*: Socrates (470–399 BCE), the ancient Athenian philosopher and moralist known for the Socratic method—a dialectical form of questioning and inquiry. He was condemned to death by the city of Athens for impiety and the corruption of young minds (Plato and B. Jowett, 'Apology', in *Five Great Dialogues of Plato* [Claremont: Coyote Canyon Press, 2009], pp. 19–43).
13. *a Rowland for your Oliver*: tit for tat or to give as good as one gets. The line is referenced from the Medieval epic poem, *La Chanson de Roland*, in which the warriors Roland and Oliver could not beat one another in combat and were considered unbeatable as a team (*OED*).
14. *justly call a 'sham'*: see n. 5.
15. *a 'threat' which . . . by Dr. Hirst*: there was some confusion on Tyndall's part about the nature of Hirst's 'threat' concerning the Mummery affair (see letter 0893, n. 4).
16. *the ladies of Queenwood*: likely a reference to George Edmondson's wife Anne Edmondson, and his daughter Jane Edmondson.

To Thomas Archer Hirst [27 February 1854][1] 0891

Monday morning

Cheer up dear Tom. Things are only transiently black. You will learn to battle with the clouds[2] by and by. The enclosed[3] will perhaps amuse you. I have added a note since I read your letter[4] entreating that young lady[5] to meddle no further in school matters. Remember a portion of your vocation is to battle with insincerity and shallowness. I had a note from Forbes[6] this morning requesting *[me]* for a statement of my results when they differ from his. He says he was never fully satisfied with his own paper.[7] His tone of writing is very kindly.

Affectionately | John.

RI MS JT/1/HTYP/317–319
LT Typescript Only

---

1. *[27 February 1854]*: Tyndall recorded in his journal that he received Hirst's letter (presumably letter 0889) on 27 February, which was a Monday (Journal, 27 February, JT/2/13b/663–64).
2. *learn to battle with the clouds*: in letter 0889 Hirst referred to 'a cloud over Queenwood'.
3. *The enclosed*: presumably letter 0890, which included a postscript that suggested that the management of Queenwood was a sphere that 'does not belong to them', meaning the 'ladies' of Queenwood. It is, however, possible that the 'note' added here was the draft letter to Jane Edmondson that Tyndall never sent (see letter 0893, n. 4).
4. *your letter*: see letter 0889. The added note can be found after Tyndall's closing salutation in letter 0890.
5. *that young lady*: Jane Edmondson.
6. *a note from Forbes*: see letter 0888.
7. *his own paper*: Forbes's paper on the vibration of heated metals (see letter 0814, n. 7); Tyndall published a paper in January 1854 challenging a number of Forbes's claims (see letter 0876, n. 14).

To Thomas Archer Hirst [27 February 1854][1] 0892

My dear Tom

I hope your headache is gone, don't let the state of *[matters]* at Queenwood[2] depress you. Keep your heart cheerful boy and your hand ready to strike the right stroke whenever it is needed. Don't let that money affair depress you.[3] Let not vigilance be relaxed, but don't let it weigh down your spirits. I don't think you need fear your letters in the least: there is a kind of honour among them depend upon it: I had a note[4] from your brother[5] this morning he is coming to London next week. How old was Orton?[6] It was probably wise to send him away as an example, but when the delinquent is young and if the process do not occupy too much time some other correction may perhaps be applied. I am glad to see Edmondson acting so promptly on your advice.[7] Don't let your heart get too full of contempt Tom—mine often did—Remember I suffered all you suffer and it is right that you should suffer all I have suffered so that we may continue the same in soul and spirit. Take the matter as heroically as you can, and let a smile if possible play round your sternest resolution. This is the battle of life Tom. Cheer up boy. I am as merry as a lark, although I have not tasted brandy and water for a week. I would have given a crown to hear your lecture at Salisbury.[8] I preach tomorrow Good night Tom. Does Mummery[9] shew any signs of penitence?

Exit John.

RI MS JT/1/HTYP/323
LT Typescript Only

---

1. *[27 February 1854]*: both letters 0891 and 0892 were responses to letter 0889. Letter 0891 was written 'Monday morning', 27 February, and this one was written in the evening.
2. *the state of [matters] at Queenwood*: Three key matters were occurring at Queenwood, the 'Mummery affair' (see letter 0817, n. 4), two boys had run away (see letters 0886 and 0889, n. 8), and the theft of money (see letter 0889).
3. *that money affair*: see n. 2.
4. *a note*: letter missing.
5. *your brother*: John Hirst (see letter 0809, n. 9).
6. *Orton*: see letter 0886, n. 14.
7. *Edmondson acting . . . your advice*: see letter 0889.
8. *your lecture at Salisbury*: see letter 0889, n. 12.
9. *Mummery*: see n. 2.

From Thomas Archer Hirst [27 or 28 February 1854][1] 0893

Queenwood. | Tuesday Morn[g]

My dear John

Your letter[2] to Miss E.[3] is indeed very refreshing to me. It is the answer I knew you would write. If she feels the rebuke as keenly as she ought to do she will be a better girl from this time forth. As to the 'threat'[4] I think you are a little mistaken as yet. You wrote to me[5] saying that you were about to deliver a lecture on Education after Easter & might possibly mention Queenwood therein. but you could not do so if the Doctor sham[6] were permitted. Now I knew this was something which M^{r} E[7] would feel rather keenly and hence to keep the matter alive I told him it. I said 'John intends to draw upon his experience at Queenwood and if appropriate he may mention Queenwood in his lecture but he tells me he could not, of course, do so if the Doctor "sham" were tolerated' *[Well]* Edmondson straightened himself like a Bantam Cock[8] & cried 'It shall not be tolerated, it never has been, and if I see the least symptom of its being so I shall put my hand upon it. &c' You know his talk and what it is worth. My purpose was served I saw he felt the matter keenly. He took it to N° 13 of course and there the ladies interpreted it *[into]* a 'threat' a 'bribe' & so forth I am still very unwell & preparing my Romsey & Southampton lectures[9] in low spirits. Thine

T. A. Hirst

RI MS JT/1/H/483

---

1. *[27 or 28 February 1854]*: Hirst dated this letter Tuesday morning, which would make it 28 February. However, Tyndall wrote a draft of a letter to Jane Edmondson that was never sent (n. 4) that was presumably written after this letter. Tyndall claimed in his journal that he wrote that draft on 'Monday evening', or 27 February.
2. *Your letter*: letter 0890.
3. *Miss E*: Jane Edmondson
4. *As to the 'threat'*: Tyndall referred to a threat in his letter to Jane Edmondson (see letter 0890, n. 15). After receiving these comments from Hirst, Tyndall drafted another letter to Jane in his journal that was not sent (27 February 1854, JT/2/13b/663–64). It reads as follows:

   Dear Miss E.

   Today's post brought your letter back to me. The threat I find is my own and not Dr. Hirst's: I have only to remark that the term is a very unhappy one. Your assertion of your father's dignity is deserving of all praise, provided he did not countenance what both you and I consider to be a fraud. But if he did countenance this, then you know the assumption of dignity would be sadly out of place and it would be sinful to talk of 'self-respect' where the feeling ought to be self-condemnation. However I trust whatever his views were originally that there is now no difference of opinion between us as to the wrongness of the transaction.

   At the quiet of my own fireside I would earnestly repeat the advice which I urged upon you this morning—that is to keep yourself wholly unconnected with school matters. You cannot imagine what a repugnance some men feel to the interference of ladies; and moreover, these are the very men who will offer to ladies the most loyal service as long as they remain in their own sphere. Your wisest plan would be not to meddle directly or indirectly in school matters; a year's faithful adherence to this line of conduct will do much to expel the suspicion and corroding discontent which your interference has unhappily generated. Were your father surrounded by men of doubtful character my advice would be precisely the opposite, but this is not the case, and the more trustful and less mysterious you are the better. Now be a philosopher and conquer yourself so far as to act upon my advice. One act of self-renunciation of this kind would be a real victory although no eye might see it. Indeed I believe our greatest victories are those which are never seen. When you have accomplished this I may with truth subscribe myself

Your brother | J.T.

5. *wrote to me*: letter 0882, n. 14.
6. *the Doctor Sham*: this refer's to Mummery's purchase of a German doctorate (see letter 0817, n. 4).
7. *Mr E*: George Edmondson
8. *Bantam Cock*: A type of fowl, used as a comparison when someone is displaying 'cockiness' (*OED*).
9. *Romsey & Southampton lectures*: see letter 0898, nn. 3 and 4.

From Margaret Allen 28 February 1854 0894

Stanley Terrace, Feb 28th/54.

My dear John,

You will be surprised to hear the intelligence I am about to communicate. My beloved sister Penny[1] is about sailing for Australia, I think on the tenth of March or before, she goes in company of a clergyman & his family[2] who are our intimate friends. I assure you I feel her loss very greatly, much more so than you can conceive—will you then at this particular time permit 'me to use you as a brother', will you remit me a little cash?[3]—I feel much in asking you, but if a grateful heart & unbounded sentiments of respect will repay you in any degree, I can assure you you have them to the fullest extent. Would then I could have accompanied her. I could then have endeavored to accomplish some little myself, as it is, however large my desires upon this *[head]*, I am cramped, even in the very thought, by my connections—you cannot imagine how I feel this dependency, & Oh! to think that it may be for a life-time—is it not galling. I can truly say that my every moment is an anxious one—& yet all for the sake of the forms of society, I am to be thus fettered. In any case permit me to hear from you soon & *[Ever]* I shall continue,

Yours very sincerely | Gretchen[4]

Dr. John Tyndall.

RI MS JT/1/A/75

1. *Penny*: see letter 0699, n. 4.
2. *a clergyman & his family*: not identified.
3. *Remit me a little cash?*: see letter 0896, n. 1.
4. *Gretchen*: see letter 0872, n. 12.

Christian Ludwig Gerling 2 March 1854 0895

Marburg d. 2 März 1854.

Verehrter Herr und Freund!

Ich habe zwei Briefe von Ihnen zu beantworten. Der Brief vom 17 Februar kam am 21 F. Morgens bei mir an. Am Nachmittage desselbigen Tages wurde sodann Ihr Brief vom 16 Januar von den Herrn Bennett bei mir abgegeben, welche mich nicht zu Hause getroffen hatten. Diese beiden Herrn hospitirten darauf in meiner physikalischen Vorlesung am 22 Febr. und besuchten mich nachher freundlichst; sie waren aber sehr eilig indem sie schon 1½ Stunden nachher mit der Eisenbahn abreisen, und inzwischen noch die Elisabeth-Kirche in Augenschein nehmen wollten. Von diesen Herrn Bennett erfuhr ich denn auch, was ich Tages vorher schon von einem meiner Zuhörer auf Befragen vernommen, daß der in Ihrem Briefe vom 17 Febr. erwähnte Herr Guthrie ein hiesiges studiosus der Chemie war, welche hier bei Ihrem ehemaligen Hauswirth, dem Weißbinder Baum, wohnt. Darauf habe ich gleich die ihn betreffende Stelle Ihres Briefes abgeschrieben, und sie ihm zugesandt. Herr Guthrie hat mich darauf an einem der folgenden Tage besucht, das Missverständniss entschuldigt welches er dadurch veranlaßt hatte, daß er seinen hiesigen Wohnort nicht als solchen deutlicher bezeichnete, und mich gebeten Ihnen seinen Dank zu sagen, für die gefällige Besorgung seines Briefes zum Abdruck.

Was nun den weiteren Inhalt Ihres Briefes vom 17 Febr. betrifft; so habe ich, da ich Herrn Ewert gerade nicht treffen konnte, den anderen jüngeren Buchhändler Oscar Ehrhardt, (Nachfolger von Creuzer und Bayrhoffer) deshalb befragt und er hat mir versprochen die Rolle mit Lithographien durch den von Ihnen genannten Buchhändler Herrn Nutt in the Strad an Sie befördern zu wollen. Ich habe darauf heute Morgen ihm die Rolle geschickt in welcher Sie ausser den 5 Lithographischen Heften, auch noch einen Abdruck meines kleinen Aufsatzes über Briefwaagen, und einen Aufsatz von Kohlrausch über elektrische Residuen finden werden, worüber ich Ihnen auch noch einen Brief von Kohlrausch hier beifüge.—Herr Ehrhardt sagt mir nun daß er vorerst noch nicht angeben könne, wie lange es dauern würde bis Sie die Rolle bekämen, da er dieselbe über Leipzig schicken müsse, und nicht wisse in welchen Zeiträumen Herr Nutt sich von dort aus Sendungen machen lasse. Ich glaube deshalb daß es zweckmäßig seyn würde, wenn Sie

Herrn Nutt benachrichtigten daß diese Rolle für ihn von hier über Leipzig unterwegs ist.

Die freundlichen Grüsse welche Sie an meine Familie senden, erwiedert dieselbe unter dem besten Danke, und ich selbst, sehr erfreut über die in Ihren Briefen ausgedrückten freundschaftlichen Gesinnungen, bitte sehr daß Sie mir dieselben auch fernerhin erhalten wollen.

Von Herzen | der Ihrige | Gerling

Marburg 2nd March 1854

Dear Sir and friend!

I have two letters from you to answer. Your letter of 17 February[1] arrived with me in the morning of 21 F. In the afternoon of the same day your letter of 16 January[2] was delivered to me by the Messrs. Bennett[3] who had not found me at home. These two gentlemen attended my physics lecture[4] on 22 Febr. and visited me afterwards most kindly; however, they were very much in a hurry since they wanted to leave 1½ hours later by train, and in the meantime wanted to inspect St. Elizabeth's Church.[5] From these Messrs. Bennett I also discovered, as I had already heard the previous day by asking one of my listeners, that the Herr Guthrie[6] mentioned in your letter of 17 February was a student of chemistry here[7] and lived with your former landlord, Weißbinder Baum.[8] So I immediately copied the part of your letter concerning him and sent it to him. Herr Guthrie visited me thereafter on one of the following days, apologised for the misunderstanding that he had caused by not indicating his local residence more clearly as such, and asked me to thank you for kindly arranging the publication of his letter.[9]

Now, as to the further content of your letter of 17 Febr.; since I was unable to meet Herr Ewert,[10] I have asked the other young bookseller Oscar Ehrhardt,[11] (successor of Creuzer and Bayrhoffer[12]) about the matter, and he has promised me to forward the roll with lithographs[13] to you through the bookseller identified by you, Herr Nutt in the Strand.[14] Thus, I have this morning sent him the roll, in which in addition to the 5 lithographical volumes you will also find a reprint of my small article on letter scales,[15] and an article by Kohlrausch on electrical residues,[16] on which topic I also enclose herewith a letter from Kohlrausch.[17]—Herr Ehrhardt now tells me that he cannot yet predict how long it will be before you receive the roll, since he has to send it via Leipzig, and does not know in what intervals Herr Nutt has shipments made from there. I therefore think that it would be advisable if you informed Herr Nutt that this roll is on the way to him via Leipzig.

The friendly greetings which you send to my family is reciprocated by the same with the best thanks, and I myself, very gratified by the friendly feelings expressed in your letters, hope that you will continue maintain these towards us.

From my heart | your | Gerling

RI MS JT/1/G/2

1. *Your letter of 17 February*: letter 0883.
2. *your letter of 16 January*: letter 0871. Note that the letter is actually dated 15 January.
3. *Messrs. Bennett*: see letter 0871, n. 1.
4. *physics lecture*: Gerling was Professor of Mathematics at the University of Marburg.
5. *St. Elizabeth's Church*: a church in Marburg that was built in the thirteenth century by the Teutonic Knights in honor of St. Elizabeth of Hungary.
6. *Herr Guthrie*: Frederick Guthrie (see letter 0883, n. 13).
7. *student of chemistry here*: the University of Marburg.
8. *Former landlord, Weißbinder Baum*: Weißbinder Baum and his wife ran a student boarding house in Marburg house.
9. *the publication of his letter*: see letter 0883, n. 14.
10. *Herr Ewert*: misspelling of Elwert (see letter 0883, n. 6).
11. *Oscar Ehrhardt*: not identified.
12. *Creuzer and Bayrhoffer*: not identified.
13. *the roll with lithographs*: see letter 0883, n. 2.
14. *Herr Nutt in the Strand*: see letter 0820, n. 5.
15. *article on letter scales*: not identified.
16. *Kohlrausch on electrical residues*: R. von Kohlrausch, 'Theorie des elektrischen Rückstandes in der Leidner Flasche,' *Poggend. Annal.* 91 (1854), pp. 56–81; translated as R. Kolrausch, 'Theory of the Electric Residue in the Leyden Jar', *Phil. Mag.* 46 (May 1854), pp. 305–20, 412–26.
17. *a letter from Kohlrausch*: letter missing.

From Margaret Allen 4 March 1854 0896

Stanley Terrace, Preston | March 4th/54.

My dear John,

I can indeed just imagine you smiling & saying 'now my dear girl what do you want the money for'—but first I feel rather annoyed that I have permitted myself to trespass on your kindness to me, & under the circumstances wh. you have related,[1] I could not feel myself justified in receiving a portion of that which I know you have earned by the greatest industry. However be warned that even if you had said 'Gretchen,[2] I will not let you have my money!'—my opinion would not have changed one iota, it would take much, infinitely more, to eradicate my sentiments concerning you—& I trust in whatever I write to you, I shall always receive your best counsel.—Now if I relate a few of the many thoughts wh. have & do still occupy my mind you must judge leniently wh. I am confident you will. Now for instance you know my position[3] even with those nearest to me is peculiarly trying, certainly my dependency is very 'natural'; as every woman I suppose must be dependent in one sense or another—but could you not imagine, that I had a spirit, & a will, to raise myself from this I was thinking, degradation wh. I may say I have plunged myself into!—Could you not imagine my mind filled with the feeling of working for & supplying myself with suitable means. My meaning you thus see, concerning the fetters of society wh. are thrown around me. You would not wish me to banish every independent feeling, would you?—I am sure not; then, consider the difference, between a daughter before her marriage at home, & one returning to her fathers house when it might be considered she had secured a home;—it is easy to write, 'tis easy to talk; but the feelings are worked upon & it is not easy to endure. Thus, I have thought if I could study my French or German a little more—I might be enabled to take a private situation—here again many things intervene. However it may not always be thus. If you think I should banish every such thought I shall obey you, though against my inclination. I can scarcely write to you without making myself a very woman, for the tears will not keep back. You see if I almost want a penny I have to ask for anything I require I am told to ask Papa[4]—& though I am fully assured I am far from what may be called extravagant in my desires, I feel my position, you would not like it, now I am sure you could not endure it. I never was calmer than when I wrote that last letter,[5] to you, you judge me wrongly though I am proud of your having given me a knowledge of your thoughts concerning me.—I am still willing to suffer—but I shall study & drown every other feeling as *[soon]* as I possibly can, & though that which I shall thus gain may never be of any but private note—I think I shall have the gratification of feeling that I live above the low associations & mental *[mischief]* of those with whom I am surrounded—Your striving & resignation to your lot teaches me an impressive lesson—even write to me thus, I shall be repaid for all I have passed through.—Be careful of yourself for a letter's *[sake]*—Write to me soon, & be assured you will ever receive the devotion of a Sister from

Yours Gretchen[6]

RI MS JT/1/A/76

---

1. *circumstances wh. you have related*: Allen was responding to a hostile letter she received from Tyndall, which is missing. After receiving Allen's earlier letter (0894) asking for 'a little cash', Tyndall wrote the following in his journal: 'Received a letter two days ago from Mrs Allen which did not quite please me. . . . I thought it would be just to myself and of possible service to her to tell her plainly that I did not like it. I told her of my position: of my poverty: nevertheless that I should joyfully, even were I far poorer, share my pittance

with the writer of the letters which bore her name, for in them I discovered a suffering, striving, and superior nature: but that here she appeared to start aside a little from this character: asked her to sit down and write to me calmly and thus assured me that I am dealing with the same longsuffering Mrs Allen and she should have the money by return of post' (Journal, 4 March, JT/2/13b/664–65). After receiving this letter (0896) Tyndall wrote: 'A letter from Mrs Allen I am almost sorry for having written the entry on Saturday, for I had no right to let a single note disturb the estimate of a character founded upon years of friendship.... I should not have written to Mrs Allen as I did' (Journal, 6 March, JT/2/13b/665).

2. *Gretchen*: see letter 0872, n. 12.
3. *my position*: see letter 0872, n. 10.
4. *Papa*: Samuel Smith, a merchant in Preston, Lancashire ('Marriages at St John in the Town of Preston', http://www.lan-opc.org.uk/Preston/Preston/stjohn/marriages_1845-1846.html [accessed 26 March 2016]; Register: *Marriages 1845–1846*, p. 2, entry 3; Source: LDS Film 93993).
5. *that last letter*: see letter 0894.
6. *Gretchen*: see n. 2.

From Heinrich Debus 4 March 1854 0897

Queenwood College | Stockbridge | Hants | 4 March 54

Mein lieber John!

Dein Brief hat mir ausserordentlich viel Freude gemacht. Der Geist in dem derselbe geschrieben war bezeugte mehr als Alles andere dass du heile und munter und dass Deine Anstrengungen den verdienten und für Dich befriedigenden Erfolg haben. Ich habe das Stück Shell von dem wir sprachen noch nicht bekommen. Wenn du Zeit haben solltest dann sei so gut und schicke mir dasselbe im Laufe der nächsten Woche oder wenn es nicht eilt dann will ich es in London selbst bei Dir abholen. Heute erhielt ich die Chemical Gazette von Francis. In derselben findet sich eine kleine Abhandlung von zwei Americanern über eine angeblich neue Methode der Analyse von Kohlenarten und organischen Materien überhaupt. Diese Methode ist eine alte Sache die man schon in alten Büchern beschrieben findet und es scheint dass die Autoren der erwähnten Abhandlung unbekannt mit diesem Umstand waren; unverständlich ist es mir aber wie Francis der die Sache gekannt haben muss die Beschreibung einer alten Methode als etwas Neues drucken konnte.

Diese Woche hatten wir eine sehr unangenehme Sache hier wegen Geld stehlen aus Wrights Office; Tom wird dir die Sache erzählt haben besser als ich es thun kann und so will ich davon schweigen. Am letzten Donnerstag hatten wir das Panorama wo G. William wie gewöhnlich seine Talente entwickelte. T. Grieve wurde veranlasst Dich einzuladen jedoch es scheint dass Du Deinen Zuhörern nicht sagen wolltest dass Du wegen einem Panorama zu Queenswood Deine Vorlesung an einem andern Tag halten wolltest. Der Abend selbst verstrich zur vollkommenen Befriedigung der Anwesenden.

Wenn du Punch von dieser Woche zur Hand nehmen und die letzte Seite übersehn willst dann wirst Du eine Anzeige von Dr. Mussprat finden die eine der feinsten Proben von Aufschneiderei darstellt die mir bis jetzt vorgekommen ist.

Ich kann es mir lebhaft vorstellen dass Du jetzt mit vielen Schwierigkeiten zu kämpfen haben musst. Meine Vorlesungen hier in Queenwood machen mir oft viele Mühe und geben zu manchen Fragen Veranlassung. Wenn man keine blosse Maschine sondern ein Lehrer sein will dann muss die Wissenschaft die man andern Menschen mittheilt nicht wie aus zweiter Hand nur aus dem Munde des Vortragenden kommen, sondern die Rede muss aus dem Herzen fliessen, der Lehrer muss die Urquelle des Wissen selbst sein das aus ihm mit magischer Gewalt anderer Geister Eigenthum wird. Wenige sind es nur die dies erreichen aber jeder muss die Sache so gut zu machen suchen als er kann. Ich wollte daher ich könnte die grosse Last welche in diesem Augenblick auf Deinen Schultern reitet Dir in etwas erleichtern jedoch ich glaube nicht dass ich bei meinen geringen Kenntnissen in der Physik Dir etwas sagen kann das Du nicht schon längst weist. Du hast jetzt eine grosse und schwere aber auch eine höchst ehrenvolle Pflicht zu erfüllen die wenn Du Deine Kraft und Deine Talente dagegen setzest leicht von Dir erfüllt, einstmals Dir der Dank der Welt mit einem noch grössern und erhabern Lohn, dem Lohn den das Gewissen gewährt wenn ein Mensch nach vollbrachter Arbeit auf das gethanene Werk zurückblickt. Ich will nicht weiter gehn in dieser Weise weil ich sonst sentimental werden müsste wenn ich es nicht schon bin und es leben denn wohl

Dein | H. Debus

Queenwood College | Stockbridge | Hants | 4 March 54

My dear John!

I was extraordinarily pleased by your letter.[1] The spirit in which it was written proved more than anything else that you are happy and well and that you are satisfied with the well-earned success of your exertions. I have not yet received that piece of shell that we spoke of.[2] When you have the time be a good fellow and send it to me in the course of the next week or, if there is no rush, I will pick it up directly from you in London. Today I received the Chemical Gazette[3] from Francis. In it there is a brief essay from two Americans[4] about a supposedly new method of analyzing types of coal and organic matter in general. This is an old method that you can already find described in old books and it seems that the authors of the aforementioned essay were unfamiliar with this circumstance; it is incomprehensible to me, however, how Francis could have printed the description of an old method as something new, since he must have known about it.

Something very unpleasant happened here this week in regard to money being stolen out of Wright's office; Tom will have told you about the matter[5] better than I can and so I will not speak of it. Last Thursday we had the panorama[6] where G. William[7] displayed his talents as usual. T. Grieve would have seen to it that you were invited, but it seems that you did not want to tell your audience that you wanted to give your lecture[8] on another day because of a panorama at Queenwood. The evening itself progressed to the complete satisfaction of all present.

If you want to have a look at this week's Punch[9] and to throw a glance at the last page, you will find an advertisement from Dr. Mussprat[10] that represents one of the finest examples of self-aggrandizement that I have yet to see.

I can well imagine that you are battling many difficulties right now. My lectures here in Queenwood often cause me much consternation and give rise to some serious questions. If one does not want to be simply a machine but a teacher then the science that one communicates to other people cannot come out of the mouth of the lecturer second hand, instead your speech must flow from the heart, teaching must be the source of knowledge itself which emerges from it with a magical power to become the property of other minds. Few achieve this but each must try to do the best he can. I wish thus that I could lighten the great load that rests on your shoulders at the moment but I do not believe that with my limited knowledge of physics I would be able to tell you anything that you did not already well know. You now have a great and heavy but also a most honourable duty to fulfill, which you can easily fulfill if you direct your power and your talent toward it since the thanks of the world will be an even greater and more sublime reward. The reward that the conscience receives when after finishing a job a man looks back on the completed work. I do not want to continue in this vein, since I will become maudlin if I am not already and bid adieu

Yours | H. Debus

RI MS JT/1/D/16

---

1. *your letter*: letter missing.
2. *that piece of shell . . . of*: not identified.
3. *the Chemical Gazette*: *Chemical Gazette, Or, Journal of Practical Chemistry*, a chemistry journal edited by William Francis, founded in 1842.
4. *a brief essay from two Americans*: J. H. Alexander and C. Morfit, 'Improved Apparatus for the Analysis of Coal, and for Organic Analysis generally', *Chemical Gazette, Or, Journal of Practical Chemistry* 273 (March 1854), pp. 94–99.

5. *Tom will have told you about the matter*: see letter 0889.
6. *the panorama*: a panorama created by the Queenwood students H. Bellamy and G. W. Edmondson and displayed on 2 March (see letter 0887, n. 3).
7. *G. William*: George William Edmondson (see letter 0797, n. 5).
8. *your lecture*: Tyndall noted the deliverance of this lecture in his journal entry of 2 March, but did not indicate the topic or where it was given (Journal, JT/ 2/13b/664). It is likely, though, that Tyndall gave this lecture at the LI, as this is where he lectured on Thursdays.
9. *this week's Punch*: *Punch, Or the London Charivari*, a weekly satirical magazine running from 1841 until 2002. Debus was likely making reference to Issue 660, which was published on 4 March 1854.
10. *an advertisement from Dr. Mussprat*: not identified.

From Thomas Archer Hirst 12 March 1854 0898

Queenwood College. | March. 12th 1854

My dear John,

I have just received a printed form from the 'People's Assurance Society'[1] which I as an 'intimate friend' of Geo. S. Phillips am to fill up by stating what I know of his health and habits &c. I have done so as favourably as I conscientiously could. I am rather surprised he should refer them to me seeing that I have seen him so seldom during the last few years. It is a wise step however and I was glad to assist him. The Hampshire Independent[2] contains an absurd notice of my lecture at Southampton.[3] These fellows hide their ignorance in fulsome flattery and what is worse they do so with a kind of pretension. They are not able to give a simple report of the lecture and its subjects, but they administer a few hackneyed praises so worded as to tickle the lecturer and to bamboozle the reader into the notion that the Editor was able to criticize the lecture if he liked. He concludes his bare announcement of the lecture with the following high-sounding sentence. 'The learned lecturer, as might have been expected from his high scientific acquirements, evinced a perfect mastery of his subject, and at the conclusion of his lecture he received a hearty vote of thanks from the audience'. The report of the Romsey lecture[4] was modest and satisfactory. 'The experiments were beautiful and successful and they were explained in a lucid manner'.

At Salisbury[5] several people wished I had made a few remarks about Electro Biology[6] and such sciences, which are at present far too popular in this neighbourhood. On Tuesday next I intend doing so in my introduction at Lymington[7] and I will give the Secretary[8] this introduction to insert in the Hampshire Independent. In my last letter[9] I just mentioned that Mr Heathfield[10] had been down to see Alex. Cumming. He appears to be a kind hearted man and to have had a praiseworthy purpose in his visit down to Queenwood.

He is a Chemist half scientific & half manufacturing. He hoped to find Alex a chemist for then he thought he could have been of some service to him in the branch he himself followed. He seemed too much of a one-idea man, he appeared to think he could have cut out a little business for Alex, and plant him down in it to make money & prosper. I told him I thought any such notion would be premature at present, that in Alex, we could notice a tendency only or rather a capability to follow any intellectual pursuit with perseverance. That this tendency which at present takes a mathematical direction should be carefully developed now and in so doing it would find its own applications, a profession would in good time present itself. He agreed with me perfectly. At present Alex would like to be an Engineer; for my own part I think I would send him to a German University for a year or two after leaving Queenwood, and then think about a Profession. From our own experience we know that without some university education he would find it difficult to acquire that sound & broad theoretical knowledge which distinguishes the present from former professional men. There is plenty of time yet to consider these things, and in the interval I shall not neglect to study his best interests, and to cultivate an interest for such knowledge in him. Who knows he may turn to pure Science in the end, and for my part I should be best pleased to see him do so. Henry Hill writes still in good spirits, he has sent me copies of all Booths letters.[11] We have had beautiful weather for the last fortnight and internally also Queenwood is brighter. I hope you have used the fine weather too and that you have kept yourself at any rate as strong as you were if not stronger. This week you must not neglect to write to me and tell me all the news concerning yourself.

Yours affectionately | T.A. Hirst.

P.S. Mrs Simpson has got a little boy.[12] Miss Martin has written me one or two excellent letters lately.[13] I have some notion of paying a [pop] visit to Dublin either at Easter or Midsummer. You might be inclined to suspect the juxtaposition of the last two sentences did you not know by this time that I am a wary fellow. | T.A.H.

RI MS JT/1/H/194
RI MS JT/1/HTYP/325–325a

---

1. *'People's Assurance Society'*: an insurance company in London, that presented itself as working with the Equitable Provident Institution to 'offer the most sound and complete system of Life and Health Assurance that modern science and experience have hitherto been able to produce' (*The Register Insurance Directory and Almanac. 1854* [London: William Tweedie, 1854], p. 22).
2. *The Hampshire Independent*: a weekly Liberal newspaper founded in Southampton,

England, which ran from 1835 to 1923 (R. Preston, 'Thomas Leader Harman: A Gentleman of Fortune in Mid-nineteenth century Southampton', *Southampton Local History Forum Journal,* 16 [2010], pp. 3–27, on p. 8).

3. *notice of my lecture at Southampton*: not identified. As indicated in Hirst's journal, the lecture was delivered at the Victoria Rooms, Southampton, Hampshire, England on 8 March, on the subject of Heat (Hirst, 'Journals', 12 March 1854).
4. *The report of the Romsey lecture*: not identified. Hirst noted in his journal that he had delivered a lecture 'on Heat to a large audience at Romsey' on Tuesday, 7 March (Hirst, 'Journals', 12 March 1854). Romsey, Hampshire, England, a small town northwest of Southampton.
5. *At Salisbury*: reference to Hirst's lecture at Salisbury (see letters 0874, n. 5 and 0889, n. 12).
6. *Electro Biology*: 'a branch of biology which deals with electrical phenomena in living organisms' (*OED*).
7. *my introduction at Lymington*: a port town in Hampshire where Hirst was to give a lecture on 14 March.
8. *the Secretary*: not identified.
9. *my last letter*: letter missing.
10. *Mr Heathfield*: possibly W. E. Heathfield, Fellow of the Royal Society of Edinburgh and Fellow of the Chemical Society in London.
11. *Henry Hill . . . copies of all Booths letters*: letters not identified.
12. *a little boy*: not identified.
13. *one or two excellent letters lately*: not identified.

From Juliet Pollock[1] [14 March 1854][2] 0899

My dear M^r^. Tyndall,

If you will come and join us at our quiet dinner at 7 either Thursday or Friday next, you will see my little boy[3] & we shall be very glad to have the pleasure of an evening with you. We have sent away our man servant and are <*1 word missing*> pretending to ourselves that we are in the country, & our dinner is an *[*union*]* of dinner and tea, which I hope would not be disagreeable to you. It is most kind of you to think of my little man. I shall be obliged to you to *[*send*]* *[*me*]* a *[*line*]* to say on which of the days named we may hope to see you.

Yours very truly | <u>Juliet</u> Pollock | 59 Montagu Square | <u>Tuesday</u>

RI MS JT/1/P/210

1. *Juliet Pollock*: Juliet Pollock, née Creed (1819–99), writer and wife of William Frederick Pollock (1815–88), 2nd Baronet.

2. *[14 March 1854]*: Pollock dated her letter as 'Tuesday', and Tyndall noted in his journal entry of 17 March 1854 that 'Mrs. Pollock in passing asked me kindly how I did, and said that she was sorry I could not join them last night' (Journal, JT/2/13b/667). As such, this letter was likely written on 14 March 1854.
3. *my little boy*: the Pollocks had three sons, Frederick Pollock (1845–1937), 3rd Baronet, Walter Herries Pollock (1850–1926), and a third unknown son (*ODNB*). It is unclear which son is being referred to here, although 'little man' may refer to Walter Herries Pollock, who would have been three or four years of age at the time of writing.

To Thomas Archer Hirst 15 March 1854 0900

Wednesday 15th. March 1854.

My dear Tom,

I find I have not a sheet of note paper in my possession so you must be content with a leaf of foolscap. Now that I have commenced I find that I have got nothing to say—you desire me to write to you however and I am willing to gratify that desire by even making letters though they should overlap no idea. My health is just as usual. I have had very little relaxation lately and find that I want some. I will labour lightly for the remainder of this week, as I can afford to do it for there happens to be a gap in my course at the London Institution[1] which leaves me free to attend the Red Lions[2] on Thursday. On the 7th I lectured at the R.I.[3] on Vaporization. I had a mass of stuff in my head but not sufficient strength to manage it, and the consequence was that lecture, judged by my feelings was a miserable failure. Well on Thursday I lectured at the London.[4] I felt fresh—used no notes and had them all clapping uproariously before I ended. Yesterday I lectured again at the R.I.[5] my head was muddled but fortunately I had a good number of experiments which spared my logic to a considerable extent. I got on very well and the audience I believe thought so. Thus I oscillate—when in good health a lecture is mere amusement to me, when muddled and afflicted with headache it is a dreadful task.

Our educational lectures[6] are almost determined upon the course will be opened by Whewell Master of Trinity:[7] I should willingly back out of it myself, for it will cost me a great deal of labour. Besides the little experience I have had hardly furnished me with sufficient material. However like as a true portrait of any thing the features of a child, or even a pet lamb or a lapdog has something pleasing in it, I trust by adhering to the strict truth, and holding close on to the portraiture of my own mind that the picture thus drawn will on account of its fidelity possess interest for some. If the lectures are published, of which there is some probability, I know one at least who will feel an interest in what I have to say, and that one is yourself.

I have lately received several letters[8] from a gentleman in Greenwich[9] who is getting up a number of reading books for schools and who seems very anxious that I should be a contributor of his. He was recommended to apply to me by Playfair. The quanty he requires would be about equal to 80 pages of the Phil Mag. and for this he offers me £50. The money seems easily earned and I think I will undertake the work.[10]

An idea struck me a day or two ago which pleases me much, and the carrying out of which would be a source of profit and amusement to me—that is to write a series of letters on Natural Philosophy to boys.[11] When I have more leisure the thing would be a pleasant pastime to me, and I know the letters might be written so as to make them widely popular. However this is as yet in the region of conjecture—don't mention it to any body.

Your affectionate | John

Herr Hansen whom you saw in Gottingen is here at present. He has some notion of running down to see you. From Magnus's note of introduction[12] I judge him to be a superior young fellow.

RI MS JT/1/T/588
RI MS JT/1/HTYP/326

---

1. *my course at the London Institution*: the course of six lectures on Electricity and Magnetism Tyndall gave at the London Institution in 1854 (see letter 0865, n. 6).
2. *the Red Lions*: the Red Lions Club, founded by Edward Forbes, was an informal dinner club that met annually at the BAAS meetings beginning with the 1839 meeting in Birmingham. It was meant to provide a much less expensive and extravagant social space to the formal BAAS dinners (see letter 0975, n. 16), and would include only young men in their 20s and 30s as opposed to the mixed-sex and formal gatherings associated with the BAAS. After Forbes moved to London in 1844, the Red Lions club met regularly in London as well. Tyndall joined the club in 1850 and began dining with the club regularly after he moved to London in 1853 (see Journal, JT/2/13b/629, 654, 660, 692, 699, 705, and 714). It has been argued that the Red Lions Club was an important precursor to the influential X-Club (H. Ellis, *Masculinity and Science in Britain, 1831–1918* [London: Palgrave Macmillan, 2017], pp. 126–29). See also letters 0569, n. 11; 0622, n. 7; 0975, n. 16; and 1006, n. 5.
3. *I lectured at the R.I.*: this lecture was part of Tyndall's course on Heat at the RI (see letter 0865, n. 6)
4. *on Thursday I lectured at the London*: part of Tyndall's lectures on Electricity and Magnetism (see n. 1).
5. *Yesterday I lectured again at the R.I.*: see n. 3.
6. *Our educational lectures*: in the spring of 1854, the RI held a course of seven lectures on education. Tyndall delivered his lecture on 27 May, which was published as J. Tyndall, 'On

the Importance of the Study of Physics as a branch of Education for all Classes', *Lectures on Education* (London: J. W. Parker & Son, 1854), pp. 171–211. See especially letter 0933 for Tyndall's description of the event but also see letters 0906, n. 4; 0913, n. 4; 0924, n. 1; 0931; and 0932, n. 4.

7. *Whewell Master of Trinity*: William Whewell.
8. *several letters*: letters missing.
9. *a gentleman in Greenwich*: Edward Hughes (1819–59), an Irish teacher and Head Master of the Royal Naval Lower School at the Greenwich Hospital who 'edited and compiled several elementary educational books, principally on geography and arithmetic.' His publishing activities caused him much financial and physical distress ('Mr. Edward Hughes', *Minutes of Proceedings of the Institution of Civil Engineers* 19 [1860], pp., 190–91).
10. *I will undertake the work*: Tyndall's series of lessons on natural philosophy appeared as J. Tyndall, 'Natural Philosophy', in E. Hughes (ed.), *Reading Lessons: First Book* (London: Longman, 1855), pp. 289–338; 'Natural Philosophy', in Hughes (ed.), *Reading Lessons: Second Book* (London: Longman, 1855), pp. 296–322; 'Natural Philosophy', in Hughes (ed.), *Reading Lessons: Third Book* (London: Longman, 1856), pp. 238–66; and 'Natural Philosophy', in Hughes (ed.), *Reading Lessons: Fourth Book* (1858), pp. 229–76.
11. *Natural Philosophy to boys*: see n. 10.
12. *Magnus's note of introduction*: letter 0877.

To Thomas Archer Hirst [16 March 1854][1] 0901

Doggrel[2]

Dear Tom, the sky is gray
Today:
And like a vapour dense and dull
Something within my scull
Weighs heavily upon my aching brain.——
Would that the mind could rain!
And thus discharge in one full shower the weight
Which cumbers me of late.
Would that thou hadst a horse
To bound o'er grass and gorse
And make a breeze, though none should stir in heaven.
I'd start before eleven
Next Sunday with the bridle in my hand,
And gallop o'er the land.
The wind should stir the hair
Around my temples fair
And the bright crimson to my pale cheek bring
All maids should long to kiss so fair a thing.
And then the whisker's spare
Sublime appendix! which thou canst but scan

On cheeks like mine,
Denied alas! To Debus's and thine
Or half ashamed to be espied
Pays Haas a furtive visit at one side!
finis.

RI MS JT/1/HTYP/324
LT Typescript Only

1. *[16 March 1854]*: Hirst responded to this letter on 19 March (letter 0904), stating that he had some time to spare before dinner with the Lions, suggesting that it was most likely written on 16 March.
2. *Doggrel*: misspelling of doggerel, verse that is comic or burlesque and 'usually composed in irregular rhythm' (*OED*).

To Thomas Archer Hirst [16–17] March 1854[1] 0902

March 1854.

My dear Tom

I am very glad to hear that you have accomplished your lectures[2] satisfactorily. The magnet[3] shall be sent away to day or tomorrow as you desire.

Yours as ever | John

Friday

You had no right to expect a letter from me—for you are deeply in my debt.

I would strongly recommend Debus to make his lecture[4] as experimental as his subjects will admit of. It is far too soon to ask him to engage in a thing of the kind. It is no matter what the Secretary[5] 'seems to think.' The Secretary has no deep interest in the matter.

RI MS JT/1/HTYP/324
LT Typescript Only

1. *[16–17] March 1854*: the dating of this letter is highly speculative. Even though Tyndall already responded to letter 0898 with letter 0900, he did not comment on Hirst's lectures as he does here. The 'Friday' referred to here could therefore be 17 March.
2. *your lectures*: see letter 0898, nn. 3–5.
3. *The magnet*: presumably a reference to an electro-magnet that Hirst was borrowing from Tyndall (see letters 0874; 0879; and 0965).
4. *his lecture*: reference to Debus's lectures at Queenwood (see letter 0897).
5. *the Secretary*: not identified.

From Arthur Herbert Church[1] 18 March 1854 0903

To Dr. Tyndall, F.R.S., &c. | Royal College of Chemistry,[2] | March 18, 1854.

Dear Sir,

The successful method by which, in your last lecture,[3] the existence of a space between water in the spheroidal state[4] and the containing vessel was proved by the complete interruption that space offered to the passage of a galvanic current,[5] has induced me to devise a few experiments on the subject.

I have to describe in the present communication, in the first place, some experiments I have just performed for the purpose of obtaining decisive evidence of the isolation of all bodies in the spheroidal state from the surfaces on which they roll; and in the second place, to offer a few suggestions as to the probable causes of the phaenomena under consideration.

It was found by Boutigny,[6] that if into a clean, red-hot platinum capsule acids and alkalies be placed, the acid and alkaline liquids will roll about, repelling one another violently. This, though an interesting example of the suspension of chemical affinity, does not prove the existence of a space between the platinum vessel and the spheroids.

The first experiment I have to mention was this:—I took a copper basin, three inches in diameter and rather more than half an inch deep, polished its concave surface, and covered it with a thin film of silver by the galvanic process.[7] The plated basin was not brought to a very high temperature, and while thus heated, a few drops of a slightly alkaline solution of sulphide of sodium were poured into it. These drops instantly assumed the spheroidal form and rolled about, making, however, no mark or track upon the silver. The source of heat was now withdrawn: the temperature was soon so far reduced that the liquid exhibited its normal properties, the space between it and the silver no longer existed, and a black stain of sulphide of silver covered the dish.

Another instance of the assumption of the spheroidal state has been often noticed. It occurs when aether[8] is placed on the surface of boiling water. Now, if a fixed inorganic acid be dissolved in the aether, and the water be coloured with litmus,[9] no reddening of the latter will take place as long as the aether remains in the spheroidal state. The acidulated aether and the tinted water cannot, therefore, be in communication; they are separated by a film of air or of vapour.

I pass on to notice in as few words as possible the remainder of my experiments. I have remarked that in certain circumstances spheroidal globules form upon the surface of liquids during the processes of filtration and distillation. The phaenomenon to which I refer is exhibited by many liquids, more frequently and conspicuously perhaps by those that are the more volatile.[10] I have observed it with alcohol, water, aqueous and alcoholic solutions, syrup, with essential oils and many other organic substances. I have sometimes, however, found considerable difficulty in its reproduction, and will therefore describe in detail a method which is applicable in most cases for obtaining in this manner an example of the spheroidal state. We will employ a particular instance. If we take cymole, a hydrocarbon belonging to the benzole series, and half fill a bottle two or more inches in diameter with it, placing in the neck of the bottle a perforated cork through which passes a funnel-tube filled with cymole, and having a piece of sheet Indian-rubber[11] stretched over its mouth, we shall find that on adjusting the funnel-tube till its lower extremity is rather less than half an inch from the surface of the liquid in the bottle, and on letting fall a drop of cymole from it, beautiful spheroidal globules will be formed and roll about for some time, scarcely diminishing in size. This experiment may be performed with great advantage if the cymole be warmed first. An ordinary funnel with a filter may be substituted for the funnel-tube, and will answer well if the lower aperture of the funnel has a diameter of about Ÿ13 of an inch. The experiment may be repeated with other materials with similar effects. That the spheroids are not in contact with the surfaces on which they roll, may be proved by saturating the liquid in the tube with something that shall have a visible effect upon a substance dissolved in the liquid in the bottle. A beautiful illustration of this occurs when, under conditions similar to those before mentioned, we employ in the funnel-tube a solution of sugar containing sulphocyanide of potassium, and in the bottle a solution of sugar containing sesquichloride of iron; no red coloration takes place until the coalescence of the spheroids with the liquid beneath them. Many other chemical reactions may be made use of with similar results. It is very curious to see a solution of ferrocyanide of potassium floating upon a solution of sesquichloride of iron, while not a trace of Prussian blue[12] is formed. These experiments must of course be recommended whenever any union of the liquids employed has taken place. The cork spoken of above should have two perforations, one to admit the funnel-tube, and the other to allow the escape of air.

Is the employment of a volatile substance essential to the production of these phaenomena? I imagined that this question might be answered by the following experiment. A dish of platinum might be heated strongly, and a drop of melted lead then placed upon it; now if the production of vapour from the substances employed were essential to the formation of the spheroidal state, the lead should at once dissolve and perforate the platinum; if, however, the spheroidal state occurs when two non-volatile substances are employed, the platinum vessel should not be perforated until its temperature has been considerably reduced. The experiment was tried with every precaution to prevent the oxidation of the lead and ensure an accurate result; a perforation of the platinum ensued the moment of placing the melted lead upon it. This proves that the production of vapour is essential to the occurrence of the spheroid state; for it cannot be urged that that condition is never manifested when metals only are employed; for a drop of a volatile metal, mercury (melted mercury, we may say, in order to render its relationship to melted lead the more apparent), placed on an intensely heated surface of platinum instantly assumes the spheroidal form, and evaporating slowly, dances about in the vessel with peculiar movements. Upon thin, sonorous vessels of copper, &c., this movement takes place with such rapidity as to produce a musical tone of high pitch. I have sometimes seen globules of mercury and water rise to the height of six inches from the capsules in which they had been formed.

A word or two in conclusion, as an attempt at an explanation of the phaenomena observed may not be out of place.

Since a space always exists between the lower body and that in the spheroidal state, and since that state is not manifested by non-volatile substances, it seems reasonable to conclude that the vapour proceeding from those parts of the liquids nearest to the containing vessel or subjacent fluid tends to assist the internal molecular cohesive force of the drops in assuming and maintaining their spheroidal form. Other forces originated by the temperature may also be in operation.

I should premise that the solution of sulphide of sodium employed in my first experiment made a dark stain upon silver even in the cold. This renders the nullity of its action at a high temperature the more remarkable. It will be scarcely necessary to mention, that, for the successful performance of most of these experiments, considerable manipulative care is required.

If, in the further prosecution of my inquiries in this direction, I should discover anything worthy of notice, I hope to have the pleasure of communicating it to you.

I remain, dear Sir, | Truly yours, | Arthur H. Church.

A. H. Church, 'On the spheroidal state of bodies', *Phil. Mag.* 7:45 (April 1854), pp. 275–78.

---

1. *Arthur Herbert Church*: Arthur Herbert Church (1834–1915), a chemist, was at the time a student at the Royal College of Chemistry. This letter was published in the *Phil. Mag.*
2. *Royal College of Chemistry*: see letter 0717, n. 6.
3. *your last lecture*: likely a reference to Tyndall's 14 March lecture at the RI (see letter 0900, n. 5). In his journal entry of 13 March, Tyndall indicated he was preparing experiments that proved that the 'space between a spheroidal drop and the hot surface underneath does not conduct electricity' (Journal, JT/2/13b/666).

4. *spheroidal state*: in the approximate shape of a sphere (*OED*); refers to a drop (of water).
5. *a galvanic current*: see letter 0884, n. 4.
6. *Boutigny*: Pierre Hippolyte Boutigny (1798–1884), a French chemist and pharmacist.
7. *the galvanic process*: galvanize, 'to cover with a coating of metal by means of galvanic electricity' (*OED*).
8. *aether*: 'a sweet-smelling, volatile, flammable liquid made by distilling ethanol with sulphuric acid and used as a solvent, as an intermediate in chemical synthesis, and as a general anaesthetic' (*OED*).
9. *coloured with litmus*: a blue pigment obtained from lichens used for testing acidity or alkalinity; 'when reddened by an acid, it serves as a test for alkalis' (*OED*).
10. *volatile*: 'volatile' liquids have boiling points close to room temperature.
11. *Indian-rubber*: natural rubber, 'a polymerization product that is the coagulated latex of certain trees and other plants of South America, Africa, Indonesia, etc., which forms a highly elastic and flexible substance' (*OED*).
12. *Prussian blue*: a blue pigment consisting primarily of ferric ferrocyanide (*OED*).

From Thomas Archer Hirst 19 March 1854 0904

Queenwood College. | March 19th 1854

My dear John.

The scrap of verse you sent me[1] if not worth much in itself indicated that you had penned it for medicinal purposes. That you had been weary and were seeking some relaxation; or perhaps preparing yourself for a Red Lion dinner:[2] as such it was welcome. I shewed it to John Haas and he evidently enjoyed it Since my last letter[3] I have given the last of my lectures out of Queenwood at Lymington.[4] The audience was very respectable indeed, a 'select one' as it is called and it appeared pretty well satisfied It is a great satisfaction to be free once more and able to devote myself entirely to Queenwood duties I have had another eloquent letter from R. Carter Esq. accompanied by a Halifax Guardian which contains a report of his last lecture on the Geology of Halifax[5] He is now in London I believe on Parliamentary business and may probably visit you. Herr Hansen has not been though I should be glad to see him. He is an intelligent young fellow as far as science is concerned, that is, he is pretty expert at experiment and possesses considerable information, more than he has digested I think. He is young yet and his 'pretension' is perhaps excusable. He was very kind to me in Göttingen and on the whole I rather liked him

I have had a letter from du Bois[6] telling me that my Sirene[7] will be sent to you very shortly. In it are two passages which refer to you and which I will copy. The letter is written in English.

'I had a letter from Tyndall[8] lately concerning a paper which I read at

Belfast[9] and did not want to have printed, and which nevertheless has been printed in the report—if you write him tell him Er soll sich darum keine graue Haare waschen lassen.[10] I suppose I can bear it and nobody will prevent me in pointing out what that paper contains of erroneous statements.'

'You ask for my opinion upon Tyndall's paper on Animal Electricity[11] I have not yet had an opportunity of perusing it carefully, having only seen it once at the Royal Library.[12] So I would rather abstain from passing any judgment upon it, yet I must confess that I did not expect Tyndall to express himself so over skeptically with regard to my experimental results. He had perhaps better not volunteered his own opinion on certain points of controversy.'

When translated into honester English I take this to mean 'I have read Tyndalls article and been somewhat surprised at his skepticism. But I have not studied his objections sufficiently to be able to refute them or to satisfy either myself or him that they are unfounded. I am consequently not very well pleased with the article.'

Du Bois is a good fellow personally considered a healthy, hearty man. But when talking to him I never could admire him as much as a scientific man. With regard to science I always missed that simple honesty of action, that undivided love of Truth which in his private character was more observable. I am afraid his investigations are somewhat tarnished with selfishness My ideal of a scientific man is that he should seek truth for its own sake and that the finding of it should be his sufficient reward. He should feel that scientific Truths are no man's property, to be bought, sold & fenced in like acres of ground, but that like the landscape they belong to him who has an eye to see them. He should count himself fortunate in being permitted to discover such truths without seeking to found a personal claim to such discoveries

Judged by this ideal I am afraid du Bois and many others would be found wanting.

I like your scheme about Natural Philosophy for children[13] I hope you will think more of it. Your reliance on simple statements of experience in teaching will not mislead you.[14] In the end it is the most forcible of all arguments What will you do at Easter?

Yours affectionately | T.A. Hirst.

In last month's 'Household Words'[15] is an article on School teaching[16] which is worth reading.

RI MS JT/1/H/195
RI MS JT/1/HTYP/327–327a

---

1. *The scrap of verse you sent me*: letter 0901.
2. *a Red Lion dinner*: see letter 0900, n. 2.
3. *my last letter*: letter 0898.

4. *my lectures out of Queenwood at Lymington*: see letter 0898, n. 7.
5. *a Halifax . . . the Geology of Halifax*: not identified.
6. *du Bois*: Emil du Bois-Reymond.
7. *Sirene*: see letter 0733, n. 8.
8. *a letter from Tyndall*: letter missing.
9. *a paper which I read at Belfast*: see E. Du Bois-Reymond, 'On a New Effect produced on Muscles by the Electrical Current', *Brit. Assoc. Rep. 1853*, pp. 78–80.
10. *Er soll sich darum keine graue Haare waschen lassen*: colloquial German phrase meaning, 'He shouldn't lose any sleep over it'.
11. *Tyndall's paper on Animal Electricity*: see letters 0754, n. 9 and 0780, n. 5.
12. *the Royal Library*: The Prussian State Library, located in Berlin, Germany. From 1701 to 1918 the library was known as the Royal Library (*Encyclopedia of Library History*, ed. Wayne A. Wiegand and Donald G. Davis, Jr. [New York: Routledge, 2013], p. 517).
13. *your scheme . . . children*: Tyndall had mentioned an idea he had for producing a series of letters on natural philosophy for boys (see letter 0900).
14. *statements of experience . . . you*: Tyndall was concerned about his contribution to the educational lecture course held at the RI (see letter 0900, n. 6)
15. *'Household Words'*: a magazine that ran from 1850 to 1859, which was edited by the well-known Victorian novelist, Charles Dickens (1812–70).
16. *an article on School teaching*: the article Hirst was referring to was actually published in the 21 January issue of *Household Words*. See 'School Keeping', *Household Words* 8 (1854), pp. 499–504.

From A. Normandy[1] 20 March 1854 0905

To Dr. Tyndall, F.R.S. &c. | 67 Judd Street, Brunswick Square, | March 20, 1854.

DEAR SIR,

CONVINCED as I am that water frequently assumes the spheroidal state in boilers, in consequence of which these most important vessels, on which the very existence of the steam engine, that is, of most of the comforts of civilized life, depend, are too often converted by mismanagement or ignorance into frightful engines of death and destruction, I venture now to trouble you with a few observations in addition to those which you were kind enough to listen to the other day,[2] in the hope that, should the facts which I am about to relate appear to you as conclusive as they do to me, you may be induced, in your lectures on heat, to call once more the attention of the audience, should you deem fit to do so, to this, I believe, fruitful but almost unsuspected, or at any rate scarcely credited source of explosion.

I have already had the honour to relate to you that I had seen the plates of a Cornish boiler,[3] a quarter of an inch thick, become red-hot in the flue, although at the time the boiler contained its due quantity of water. It was a new boiler, about 18 feet long; and that the water had assumed therein the spheroidal state is proved by the fact, that a lead rivet, 1 inch in diameter, put in for safety was actually melted. It is, in fact, this rivet which led to the detection of the cause of the occurrence, and an investigation of the subject brought out the following details. The fireman or stoker was an ignorant mulatto, who having noticed the lead rivet in the boiler, had given it as his opinion that it would soon be melted by the fire; the possibility of such an accident having been denied by one of his fellow-workmen, a bet had taken place between them. The wager having thus been laid, the fireman set to keeping up a brisk, bright fire; in consequence of which, steam being generated exceedingly fast, *priming*, that is, a production of steam under the superincumbent water,[4] took place, and, as a matter of course, the water, or a portion thereof, not being any longer in contact with the surface of the iron plate, the latter in a very short time became red-hot, and *the rivet melted*; for some time afterwards, however, the water being in the spheroidal state, the boiler remained apparently sound, that is, it did not leak, although it had an opening of an inch diameter left in consequence of the melting of the rivet; it is only shortly afterwards I was told, that the heat having diminished, torrents of water poured through the hole with almost explosive violence, emitting volumes of scalding steam as it came in contact with the fire on the grate.

This took place at the Wenlock Timber Saw-mills;[5] and I can bear testimony to the above fact, and likewise to this circumstance, that when filled as usual with its normal quantity of water, the bottom of the boiler over the fire could, in the course of five or six minutes, be made red-hot at pleasure by brisk firing. In fact, having called at the above works, according to appointment, to meet the engineer there, my inquiry as to the then state of the boiler was answered by the foreman, in presence of his master, in these words:—'*The boiler is not red-hot now, but if you will step this way it will be made so immediately;*' and in spite of all remonstrance as to the danger of such an experiment, he came shortly afterwards to announce that the boiler was red-hot, and *I saw it* in that state in presence of the engineer, the master, his nephew, the foreman, and the fireman or stoker. The production of steam had been thereby spontaneously reduced, the bolt or lead rivet eventually melted, but not in my presence; for seeing the state of things, I speedily retreated with the engineer and the master, but almost reluctantly followed by the foreman, in whose somewhat jeering looks I could plainly read that he entertained but a poor idea of our pluck.

In addition to this, I may say that in 'waggon boilers,'[6] a shape than which none can be more unsafe, but which is now fortunately much less in use than in former years, bulging bumps are very often observed, which are certainly due to the pressure of the steam in the boiler upon the iron plates of the bottom, when by extravagant or incautious firing they have been made red-hot. Frequently, however, instead of bulgings or bumps being thus produced, a disruption of the boiler takes place, attended with more or less fatal consequences. Many such accidents have thus happened at Manchester, and particularly on Monday mornings.

The following occurrence, related to me by my friend Mr. A. M. Perkins[7] as having taken place *in his presence* in a boiler of a peculiar construction in which steam was generated by means of his hot-water pipes, is also well worth mentioning, as I think we may gather from it a correct explanation of the facts under consideration; and the rough sketch below, which represents a cross section of the boiler, will perhaps explain the matter better than I could do by writing.

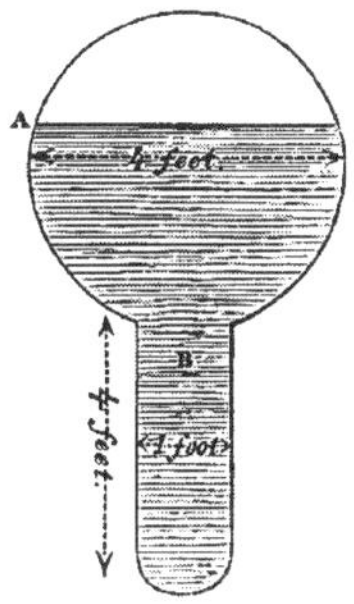

The boiler had the dimensions indicated, and was 8 feet in length; it was filled with water to about A as usual, the safety-valve had an area of 4 inches. From such a boiler the whole of the water was completely driven off through the valve by violent *priming*; that is to say, steam was generated so fast by a hot water pipe in the limb B, that it completely lifted up the water and violently squirted it out of the boiler, as just related. In such a case, the water being heated by pipes, and not by contact with the heated surfaces of the boiler, no other mischief could result than that of priming: but suppose, on the contrary, such a boiler to have been heated in the usual manner; there is no doubt that the steam generated in the limb B would, by lifting the superincumbent water, have soon permitted that part of the boiler to become red-hot, and thus have produced the curious phaenomenon *revived* and *studied* by Boutigny[8] d'Evreux,[9] and named by him a fourth or spheroidal condition of matter, but to which, so far as refers to water, the late Jacob Perkins[10] had, to my certain knowledge, called the attention of engineers and of savants more than twenty-five years ago (in fact, as far back as 1824).

Allow me to relate also the curious experiment[11] of a distinguished civil engineer, Mr. Alexander Gordon,[12] made several years ago, as related to me by Mr. A. M. Perkins.[13]

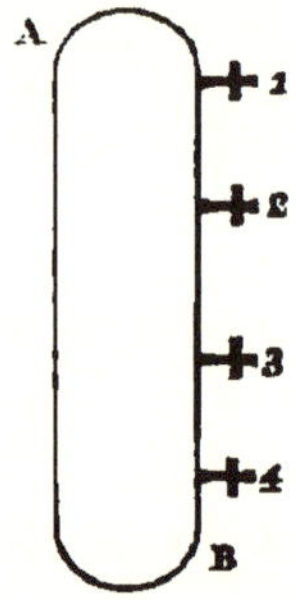

AB is a cylinder provided with four try-cocks, 1, 2, 3, 4; water being introduced into the cylinder and heated therein, the cocks on being successively turned gave the following results; namely, steam issuing from 1, 2, 3, and water from 4; but after urging the fire, the above order was found to be completely reversed; that is to say, cock No. 1 emitted water, and cocks Nos. 2, 3, 4 steam.

These facts, in my humble judgement, appear to me to prove in a decided manner,—

1. That the heating of surfaces previous to the introduction of water is not necessary to produce the spheroidal state.
2. That many boiler explosions may be referable to that condition.
3. That all boilers which offer an extensive surface to the heat, that is to say, all boilers with internal flues, are pre-eminently liable to explosions from this cause.

Earthy deposits in all kinds of boilers are favourable to the production of this dangerous phaenomenon.

I cannot terminate this informal letter without apologising for having so long trespassed on your valuable time and patience; but if a collection of facts be essential to the progress of science, I hope that you will indulgently receive this communication, and believe me,

Yours very faithfully, | A. NORMANDY.

A. Normandy, 'On the Spheroidal State of Water in Steam-boilers', *Phil. Mag.* 7:45 (1854), pp. 283–86.

1. *A. Normandy*: Alphonse René le Mire de Normandy (1809–64), a French-born chemist who lived and worked in England where he wrote a number of textbooks on analytical chemistry and invented an apparatus used to distil seawater into drinking water (*ODNB*). He was admitted as a member of the RI on 6 April 1857 ('General Monthly Meeting', *Roy. Inst. Proc.*, vol 2, pp. 413–14). This letter was published in the *Phil. Mag.*

2. *I venture now to trouble you . . . the other day*: not identified. However, Normandy may have attended the weekly meeting at the RI held on Friday, 17 March 1854.
3. *Cornish boiler*: a cylindrical flued boiler, also called a flue-boiler (*OED*).
4. *superincumbent*: overhanging (*OED*).
5. *Wenlock Timber Saw-mills*: Wenlock, Shropshire, England, a town now comprised of Much Wenlock and Little Wenlock, located close to the Wales border. While the mills are unidentified, the region is known historically to have housed a number of mills (A. P. Baggs, G. C. Baugh, D. C. Cox, J. McFall, and P. A. Stamper, 'Much Wenlock', in C. J. R. Currie [ed.], *A History of the County of Shropshire*, Vol. 10: *Munslow Hundred (Part), the Liberty and Borough of Wenlock* [London: Victoria County History, 1998], pp. 399–447, at *British History Online* http://www.british-history.ac.uk/vch/salop/vol10/pp399–447 [accessed 28 October 2016]).
6. *waggon boilers*: a steam boiler shaped with a semi-cylindrical top resembling a waggon. Used to power land engines, by 1875 they had all but disappeared from use due to their inefficiency (W. T. Brande, and G. W. Coxe, 'Steam Boiler', in Brande and Coxe [eds.], *A Dictionary of Science, Literature, & Art* [London: Longmans, Green, & Co., 1875], vol. 3 pp. 571–75, on p. 571.).
7. *Mr. A. M. Perkins*: Angier March Perkins (1799–1881), an American heating and steam engineer who became successful as an inventor and designer of heating systems, tools, and methods used in manufacturing and industry (*ODNB*).
8. *Boutigny*: see letter 0903, n. 6.
9. *d'Evreux*: the commune in northern France, Evreux, where Boutigny worked (see n. 8)
10. *Jacob Perkins*: Jacob Perkins (1766–1849), an American engineer and physicist, who worked on inventions used in printing coins and banknotes before he turned his attention to steam power and its advantages for use in locomotives (*ODNB*).
11. *the curious experiment*: Gordon's experiment is described in J. Perkins, 'Observations on the Circumstances Attending the Explosion and Bursting of Steam Boilers; With Animadversions on the Experiments of the Franklin Institute on that Subject', *Mechanics' Magazine, Museum, Register, Journal and Gazette* 727 (15 June 1839), pp. 185–88, on p. 188.
12. *Mr. Alexander Gordon*: Alexander Gordon (1802–68), an American–born civil engineer who worked in England and published an important treatise on locomotion (A. Gordon, *An Historical and Practical Treatise upon Elemental Locomotion by means of Steam Carriages on Common Roads* [London: B. Steuart, 1832]); he was also one of the founders of the Polytechnic Institution ('Obituary, Alexander Gordon 1802–68', *Minutes of the Proceedings of the Institution of Civil Engineers* 30:1870 [1870], pp. 435–36).
13. *Mr. A. M. Perkins*: see n. 7.

To Thomas Archer Hirst 21 March 1854 0906

21st. March 1854.

My dear Tom

To day I finished my 10th lecture,[1] one of the easiest and the most pleasant that I have given during the season. I was pale and hazy in the brain however and Mrs Faraday seeing this had me up to tea with them where I spent a couple of hours pleasantly enough. One good sign I see in my lectures, the audience instead of diminishing is steadily increasing and to day I had a very fair house full. I am now at home sipping a glass of Scotch ale and writing to you by way of cheating my thoughts into quietness. You certainly give me no encouragement to poetize, which I humbly suggest betrays a certain absence of poetic appreciation on your part—for I hold the little scrap I sent you[2] was well worth reading although you think proper to say it was not worth much in itself. You ask me what I am going to do with myself at Easter. I have uphill work before me I can assure you, for the next two months. I have a Friday evening to give,[3] one of those educational lectures,[4] and my course of seven besides.[5] Nevertheless I should not mind taking two or three days holiday at Easter. When do you break up? I think I should like to see you all together before this consummation takes place. I should not mind making a foot journey with you somewhere. But much time I cannot spare as I have not yet made a single experiment for my Friday evening nor do I yet know the experiments which the course I intend to give will admit of. Both of these things I must contrive to attend to during the holidays.

I wish you would contrive to carry the following request from me to DuBois.[6] To read that article to which he refers[7] from beginning to end and point out to me where I have cast a single doubt upon his experimental results. It has been said to me in London by people who have read that article—'well Tyndall you endorse all DuBois facts'—and DuBois himself on the contrary states that he did not expect me to speak so sceptically of his experimental results. I never doubted his experimental results. He must not have read the article when he wrote to you upon the subject. Since I received your note[8] I have read it over myself and I cannot for the life of me see where I have expressed myself sceptically. No man could gather from that article what my private convictions on the subject are, and I would take leave to say that DuBois himself does not at the present moment know what these convictions are. My opinion as to his experiments I unhesitatingly avow—that they are strictly correct and executed in a masterly manner. But my opinion as to his theoretic views I have carefully abstained from expressing publicly. Of course I claim the right of a free man to form an opinion upon these and all other matters which are within the scope of my intellect, but I dare affirm that what I have published on the subject does not furnish any reason to conclude that I entertain a trace of that scepticism to which DuBois alludes. If you could manage to say so much to him it would perhaps tend to set matters right between us. By what I have written I am prepared to abide until further enlightened; the article was composed with the fullest appreciation of DuBois merits and the most sincere desire to render those merits the fullest justice.

And now having finished my ale—finished all I had to say to you and nearly finished my paper, all three conclusions coincide harmoniously in the intimation that it is time for me to wish you good night and to call myself

Your affectionate | John.

RI MS JT/1/HTYP/328
LT Typescript Only

1. *my 10th lecture*: The tenth lecture of Tyndall's course on Heat, delivered at the RI (see letter 0865, n. 6).
2. *the little scrap I sent you*: letter 0901.
3. *Friday evening to give*: this would have been Tyndall's 19 May lecture at the RI 'On some Phenomena connected with the Motions of Liquids', *Roy. Inst. Proc.*, vol. 1, pp. 446–49.
4. *one of those educational lectures*: see letter 0900, n. 6.
5. *my course of seven besides*: reference to the second course of lectures Tyndall delivered at the RI in 1854, on 'Ebullition, Combustion, and other Phenomena of Heat'. This course of lectures commenced on 25 April (*Roy. Inst. Proc.*, vol. I, p. 420).
6. *DuBois*: Emil du Bois-Reymond.
7. *that article to which he refers*: see letter 0904, n. 11.
8. *your note*: letter 0904.

From Heinrich Gustav Magnus 22 March 1854 0907

Berlin 22 März 54

Werther Freund

Entschuldigen Sie daß ich Ihre freundlichen Zeilen vom 10tn. d. erst heut beantworte. Die Windfahnen und der Cylinder sind übrigens für Colonel Yorke bestellt. Der Apparat Fig 2 meiner Abhandlung, den Sie mir in Ihrem Briefe gezeichnet haben, ist deßhalb nicht in der Liste, die ich Ihnen sandte, erwähnt, weil ich mich statt deßelben einer des Apparats Fig. 6 bedient habe. Verbindet man nämlich die 3 Ringe mit einander so daß sie einen Ring bilden, so kann der cylindrische Theil des in den Ringen befestigten Körpers für den in Fig 2 abgebildeten Versuch benutzt werden. Es scheint mir als ob Colonel

Yorke nur die Versuche für die Abweichung der kugelförmigen Geschosse anstellen will, dann gebraucht er den Apparat Fig 6 nicht. Der Apparat Fig 2 (ohne den langen Balken) wird 5-6 Rthl kosten.

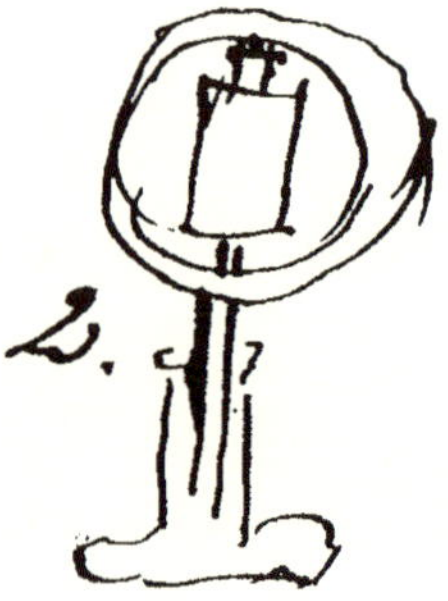

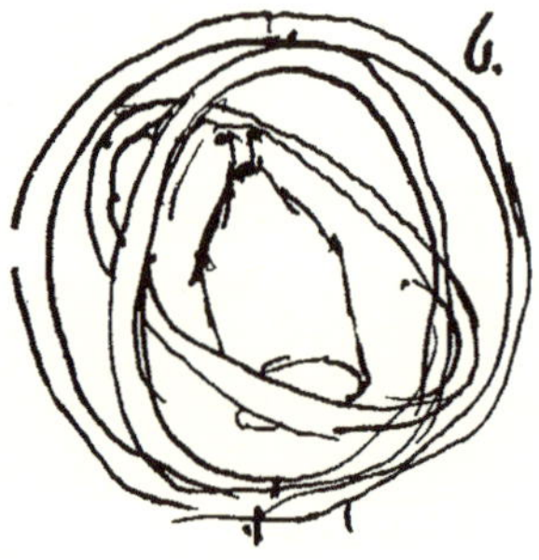

Barreaud kann erst in 14 Tagen anfangen ihn zu machen, will daher Colonel Yorke statt seiner den Apparat Fig 6 haben, so ist noch Zeit wenn Sie mir vor dem 1 April schreiben. Erhalte ich keine Nachricht so laße ich Fig 2 ausführen. Bei der Anfertigung des Cylinders ist noch eine kleine Schwierigkeit. Derselbe muß auf der Rotationsmaschine passen. Da aber die Rotationsmaschine dort gemacht ist, so bleibt nichts anderes übrig als den Cylinder fertig zu machen ohne die Hülse mit welcher er auf der Rotationsmaschine befestigt wird, und diese Hülse dort anzusetzen.

Die Windfahnen und der Cylinder sollen in den ersten Tagen des April fertig sein, und bitte ich mir zu schreiben ob ich sie durch Asher an Nutt soll befördern lassen. Ob sie abgehn sollen wenn auch der Apparat Fig 2. noch nicht fertig ist, und ob ich sie an Sie addressieren darf, oder welche die Addresse ist.

Der Grund weßhalb die Windfahnen früher fertig werden als das übrige, ist daß Barreaud selbst, für seine Person jetzt sehr beschäftigt ist. Sie werden vielleicht gehört haben daß Dr. Wiedemann einen Ruf als ordentlicher Professor an die Universität Basel erhalten hat. In der Mitte des nächsten Monats reist er mit all seinen Instrumenten ab, und da sind noch manche Veränderungen, Reparaturen u.s.w. erforderlich die Barreaud zu machen hat. Die Windfahnen kann er durch seine Leute fertig machen lassen.

Die schöne Ändrung die Wheatstone mit der Fesselschen Maschine vorgenommen, möchte ich wohl kennen. Mein Polytrop ist nach demselben Princip construirt oder richtiger Fessel hat seinen Apparat beschrieben nachdem ich den Fig 7 meiner Abhandl über die Geschosse beschrieben hatte, von dem das Polytrop nur eine Verbesserung ist. Doch sind die Bewegungen dieses letzteren sehr hübsch mitanzusehn.

Die kleine Note über die Gaslampe habe ich Ihnen nur gesendet weil

Hoffmann seinen Apparat in dem Journal der Chemical Soc. bekannt gemacht hat. Eignet sich die Sache nicht für das Phil. Mag so seien Sie so gut sie in der Chemical Soc. vortragen zu lassen. GM.

Berlin 22 March 54

Dear Friend

Excuse me for the fact that I am only today answering your kind lines[1] of the 10th of this month. The wind vanes and cylinder have, incidentally, been ordered for Colonel Yorke. The apparatus in Fig. 2 of my paper,[2] which you drew for me in your letter, is therefore not mentioned in the list that I sent you because I used one of the apparatus in Fig. 6 instead. If, namely, one ties the three rings together so that they form one ring, then the cylindrical part of the body fastened in the rings can be used for the experiment illustrated in Fig. 2. It seems to me as if Colonel Yorke only wants to do the experiments for the deviation of spherical projectiles, in that case he does not need the apparatus in Fig. 6. The apparatus in Fig 2 (without the long bar) will cost 5-6 Rthl.

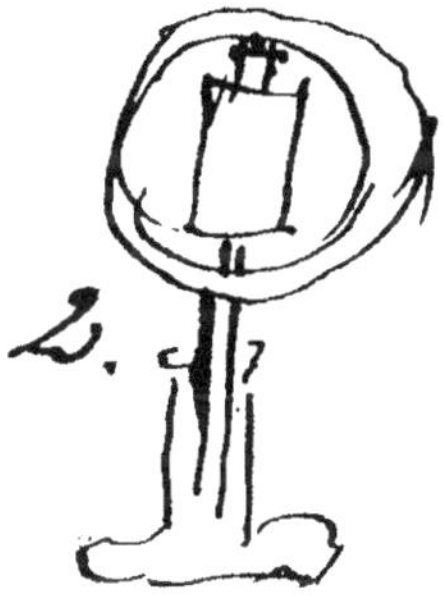

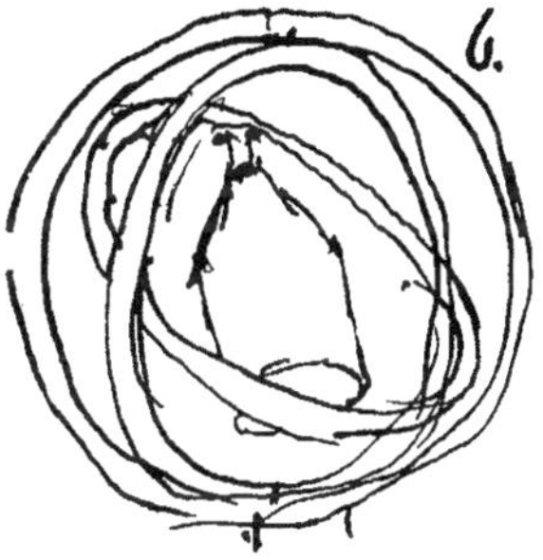

Barreaud can only start making it in a fortnight's time, so if Colonel Yorke wants the apparatus in Fig. 6 instead, then there is still time if you write to me before 1 April. If I do not receive any confirmation, then I will have Fig. 2 completed. There is still a minor difficulty with the making of the cylinder. It must fit on the rotation machine. As, however, the rotation machine is made there, there is nothing else left but to finish the cylinder without the case with which it is fastened to the rotation machine, and to attach this case there.

The wind vanes and the cylinder should be finished in the first few days of April, and I ask you to write to me whether I should have them sent to Nutt[3] by Asher, whether they should be dispatched even though the apparatus Fig. 2 is not yet ready, and whether I may address them to you, and what the address is.

The reason why the wind vanes will be finished earlier than the rest is that Barreaud himself is personally very busy now. You will perhaps have heard that

Dr. Wiedemann has received the offer of full Professor at the University of Basel. In the middle of next month he will be departing with all his instruments, and so there are still some modifications, repairs, etc. necessary that Barreaud has to do. He can have the wind vanes finished by his people.

I would very much like to know of the nice alteration that Wheatstone did on Fessel's machine.[4] My polytrope[5] is constructed on the same principle or, more correctly, Fessel described his apparatus after I had described that in Fig. 7 of my paper on projectiles,[6] of which the polytrope is only an improvement. However, the movements of this latter one are very pretty to watch.

I only sent you the short note about the gas lamp[7] because Hoffmann announced his apparatus[8] in the Journal of the Chemical Soc.[9] If the thing is not suitable for the Phil. Mag., then please be so good as to have it presented in the Chemical Soc. GM.

RI MS JT/1/M/9

---

1. *your kind lines*: letter missing.
2. *my paper*: see letter 0820, n. 12.
3. *Nutt*: see letter 0820, n. 5.
4. *modification that Wheatstone made to Fessel's machine*: Wheatstone's adaptation of Friedrich Fessel's gyroscope, an 'instrument designed to illustrate the dynamics of rotating bodies, and consisting essentially of a solid rotating wheel mounted in a ring, and having its axis free to turn in any direction' (*OED*). German instrument maker Friedrich Fessel's (1821–61) own machine was especially useful in exhibiting 'the mechanical laws of light' (*Proceedings of the American Academy of Arts and Sciences* 3 [1975], p. 207).
5. *polytrope*: see letter 0870, n. 7.
6. *Fig. 7 of my paper on projectiles*: see n. 2.
7. *my short note about the gas lamp*: note missing.
8. *Hoffmann announced his apparatus*: likely reference to A. W. Hofmann, 'Compound testing jet', *Journal of the Chemical Society* 4 (1852), pp. 39–40.
9. *Journal of the Chemical Soc*: *Journal of the Chemical Society*, founded in 1849 in London, England.

From François-Napoléon-Marie Moigno 22 March 1854 0908

Mon cher M[r]. Tyndal

Je vous envoie le numéro du Cosmos dans le quel j'ai reproduit votre leçon si intéressante sur les vibrations des corps chauds; si vous avez la bonté comme le fait M[r]. Faraday de m'envoyer de bonne heure à l'adresse du Cosmos les rédactions scientifiques que vous publiez, je les traduirais sur le champ;

Ma lettre a encore un autre but; j'ai vu avec plaisir que votre nom figurait parmi ceux des directeurs et des rédacteurs du Philosophical Magazine, et j'ai pensé qu'en m'adressant à vous je pouvais obtenir l'échange du Cosmos avec le Philosophical; j'en serais bien satisfait, et je crois avoir d'autant plus de droits à cet échange que j'envoie gratuitement le Cosmos à Sir David Brewster, depuis sa fondation. Plaidez, je vous en prie, ma cause auprès des propriétaires de votre journal, et daignez me transmettre une réponse favorable.

Je suis dans les sentiments de la considération la plus distinguée, | Votre très humble serviteur | l'abbé F. Moigno | 22 mars 1854 | 18 rue de l'ancienne comédie

My dear M^{r}. Tyndal

I am sending you the Cosmos[1] issue in which I reproduced your extremely interesting lesson on the vibrations of hot bodies.[2] If you would be so kind, as M^{r}. Faraday does, to send me to the address of Cosmos, early on, the scientific essays that you publish so that I would be able to translate them right away.

My letter also has another purpose. I witnessed with much delight that your name appeared among the directors and editors of the Philosophical Magazine,[3] and I thought that I could ask you about setting up an exchange of Cosmos with the Philosophical. I would be very satisfied with it, and I think that I may have some rights to this exchange since I am sending Sir David Brewster[4] the Cosmos free of charge, ever since its founding. Please plead my case with the owners of your journal, and please send me a favourable answer.

I am sending you my best wishes | Your very humble servant | Father F. Moigno | 22 March 1854 | 18 rue de l'ancienne comédie

RI MS JT/1/M/104

---

1. *Cosmos*: *Cosmos: Revue Encyclopédique Hebdomadaire des Progrès des Sciences*, a French science encyclopedia founded and published by M. B. R. (Benito R.) de Monfort and compiled by Moigno in 1852.
2. *your extremely interesting lesson on the vibrations of hot bodies*: see letter 0876, n. 14.
3. *your name . . . editors of the Philosophical Magazine*: see letter 0867, n. 5.
4. *Sir David Brewster*: David Brewster (1781–1868), a Scottish physicist, natural philosopher, and academic administrator noted for his contributions to optics and the popularization of science (*ODNB*). He was also a 'conductor' of the *Phil. Mag.*

To William Ginty 25 March 1854 0909

Saturday night | 25th. March 1854.

My dear Ginty,

I suppose long before this the Insurance agents[1] have received my answers to their questions. I made no delay in replying.

It is a selfish question to trouble you with on the eve of your departure[2]—but could you contrive an experiment by which I could shew the manufacture of gas at the Royal Institution after Easter? If any thing strikes you which might be made use of as an instructive experiment I wish you would let me know.

I may possibly be down in Hampshire at the date of your departure and if so will give you my benediction before you sail. This is a strange world—I was astonished the other day when I had to say I knew you for upwards of 15 years—it seems like yesterday when I turn my eyes backward. Well thus far you have fought your way bravely through the world, and I hope you will return from your tropical sojourn with sufficient spoils to enable you to live without anxiety for the remainder of your days.

most faithfully yours | J. Tyndall.

RI MS JT/1/TYP/11/3641
LT Typescript Only

1. *the Insurance agents*: not identified.
2. *your departure*: William Gilbert Ginty, a civil engineer, had been selected to take on the role of general manager of the Imperial Gas Works in Rio de Janeiro ('Mr. William Gilbert Ginty', *Minutes of Proceedings of the Institution of Civil Engineers* 27 [appendix to vol. 27] [1868], pp. 582–602, on pp. 589–90).

From Thomas Archer Hirst 26 March 1854 0910

Queenwood College, | March 26th 1854

My dear John,

Yesterday Ruhmkorff's Apparatus[1] arrived with safety—and yesterday afternoon I made a few experiments with it and found it to be a beautiful instrument. Although I was very cautious and used glass handles when I touched the connecting wires of the induced current I got one terrible shock. I want now to get a look at any papers containing experiments that have been made with the machine, (I have read Gassiot's in the Phil. Mag.[2] only) or containing investigations on secondary currents and their effects, that is such as are capable of being best illustrated by this apparatus. If you can refer me to any or procure me any do so.

On what subject will your Friday evening lecture[3] be? And on what subject and to what audience will those six after Easter[4] be delivered? I shall write to du Bois[5] and convey to him what you say in your last,[6] but between ourselves I think it will be useless. When he says that you[7] have expressed yourself sceptically about the results of his investigations, he means of course that among other things you have not deduced the same theoretical opinions. You may easily prove to him from your article[8] that you have not thus expressed yourself, but it will avert little, for on the other hand you have not defended him, as his partizan,[9] against Matteucci.[10] You have not expressed yourself as a believer in du Bois' expressed opinions and you have not seen reason to insist upon his superiority over Matteucci all of which du Bois probably expected. And again you mentioned once a certain concluding paragraph which Bence Jones declined to have published. May not du Bois be aware of the scepticism there expressed? And knowing this he would naturally interpret your printed article differently. At all events time will rectify all. Your article was clearly and honestly written and in comparison to du Bois' rather selfish disappointment these are lasting virtues and will in due time command the respect they deserve, even from du Bois himself.

To come to home matters Queenwood is filling finely. There are upwards of 107 on the books for Midsummer and only 6 who will certainly leave, leaving a balance of 101 already. The only drawback is that we are threatened with Scarletina.[11] Four have already had it and two are at present ill. Mrs Edmondson's motherly care of them is praiseworthy indeed. In such matters her true character is best seen. For a week and more she has been down at the firm[12] with the invalids, has attended to them herself and never been to the college so as to avoid all infection. For hopes to stop its spreading so that we all keep silent about the matter. As to Mummery nothing new has transpired.[13] Edmondson bullies him rather vulgarly occasionally. A few days ago I told him the Quarter's end was close upon us and I asked him if Mummery had given notice. "No," he said. "Then what do you propose to do?" I asked. "I'm at a loss" said he hesitatingly. "What would'st thou advise?" "I know what I should do in the case." "Well, what's that?" he asked. "Why I should give him notice."

Of course he was undecided and hinted that Mummery had obeyed orders and kept quiet. As an indication to the contrary I mentioned a case when at Fox's Mummery had claimed his title "But" said I "that is a secondary matter, for which there may be excuses, the principal thing is we have lost all respect for Mr M. If he kept quiet even, it would be more satisfactory to every body here to have a more upright man to work with us. In our prosperity it is all we have to look too, to gather men round us with whom we can heartily co-operate" This is one of those sentiments he could applaud. 'Oh yes' he replied 'he must go; indeed I expect he will give me notice, we will give him that chance; but if not and all comes to the worst he can go at Midsummer a quarters salary is no object" "That would be money thrown away" I returned "as long as there is yet time to give him notice. I see no necessity to consult his feelings so much, we have seen that on certain points they are not very sensitive. Depend upon it he will not give you notice until a better situation presents itself. Mr M. is prudent enough to consult his interests sooner than his feelings of honour" Here our conversation ended without a decision, but he shall be noticed before the month's end.

I have business in Yorkshire again and I ought to see Dickinson. I think I must do both at Easter. I might postpone these until Midsummer and have a walk with you but it would be more inconvenient then than now for I have other things in view. I propose therefore that you come down to Queenwood a day or two before our holidays see us all *[again]* (as you say) and renovate yourself the break up on the 12th I believe (Wednesday). On Saturday the 8th there will be a Mutual Imp. meeting on Sunday we can have a good stroll together. On Tuesday the 11th there will be a party and concert if you like to remain, but judging from your last visit you will not. At any rate you must come on the Saturday and return when you like.

Yours affectionately | T. A. Hirst

RI MS JT/1/H/196
RI MS JT/1/HTYP/329

1. *Ruhmkorff's Apparatus*: Ruhmkorff's induction coil (see letter 0864, n. 1).
2. *Gassiot's in the Phil. Mag.*: J. P. Gassiot, 'On Some Experiments Made with Ruhmkorff's Induction Coil', *Phil. Mag.* 7: 43 (1854), pp. 97–99.
3. *your Friday evening lecture*: the lecture entitled, 'Phenomena Connected with the Motion of Liquid' was delivered on 19 May ('Weekly Evening Meeting', *Roy. Inst. Proc.*, vol. 1, pp. 446–48).
4. *six after Easter*: Hirst was likely referring to the seven lectures Tyndall would give at the RI on 'Ebullition, Combustion, and other Phenomena of Heat' (see letter 0906, n. 5).
5. *du Bois*: Emil du Bois-Reymond.
6. *what you say in your last*: letter 0906.
7. *you*: prior to this point, the typescript has been relied upon, as the first page of the handwritten letter is missing.
8. *your article*: see letter 0904, n. 11.
9. *partizan*: 'An adherent or proponent of a party, cause, person, etc.; esp. a devoted or zealous supporter' (*OED*).
10. *against Matteucci*: see letters 0724; 0730; 0732; 0737; 0744; 0754; and 0780.
11. *Scarletina*: 'scarlet fever in a mild or uncomplicated form' (*OED*).

12. *firm*: either a reference to a farm or to the infirmary, or 'the sick-quarters in a religious establishment, a school, a workhouse, or other institution' (*OED*).
13. *As to Mummery nothing new has transpired*: see letter 0817, n. 4 for references to the 'Mummery affair'.

To Thomas Archer Hirst 28 March 1854 0911

Tuesday evening | 28th. March 1854.

My dear Tom

I have just reached home after the completion of my 11th. lecture.[1] I am tired, tired with the walk and tired with the talk, but it is a wholesome kind of weariness, a weariness to which rest is sweet and salutary. I have had my tea and slice of brown loaf the weather is wild without and my fire sparkles within—I have ink and paper and a friend to write to and have therefore reason to thank the gods for their bounty to me. I quite agree with your remark concerning DuBois,[2] it would be perfectly useless to write to him as I suggested, and if you have not done so, do not do it. It is only an indication of haste or weakness on my part when I evince any anxiety to set such matters right. It is not by logic that I ever expect to prove to DuBois or any body else the rectitude of my conduct. Surely there are rewards open to him who acts an upright part irrespective of the approbation of his fellows. Simply there are whispers sometimes audible to the human soul which reveal secret sources of support and encouragement of which the world seldom dreams. I am at peace when I trust these whispers fully, and am at war with myself when I try to gain the praise of men. Not that I contemn[3] man's good will; but there ought to be a power in the soul which shall enable it to hold things precious without longing for them—a great sacrificial capability to renounce the desired object without loss of serenity. But where am I wandering? DuBois has set me off. Now my feeling towards DuBois is twofold I wrote to him once for a testimonial[4] and he sent me a capital one.[5] We have spent many pleasant hours together in Berlin; when I think of this my heart softens and I deprecate the idea of a coolness establishing itself between us. Again I look upon circumstances which reveal to me his enormous avidity for scientific fame; and the utter absence of that /unhasting/ trust which in my opinion is the first qualification of a great mind. When I dwell upon this side of the picture I am disposed to be indifferent. Or perhaps a dash of anger rises in my own heart and I say with a feeling which has a slight tincture of the diabolical in it 'I will shew you some day that I had a right to volunteer an opinion upon your animal electricity.' But such feelings are transitory, and as the spring advances and the trees begin to sprout into verdure[6] my heart opens in sympathy with that of nature and all the gall[7] runs away. Unfortunately

I have not your letter[8] beside me and I am not sure that I remember all its contents. The plan you propose seems to me a very good one. I think you said you break up on the 12th. and that you have a mutual improvement Society[9] on the 8th. Well this is very tempting and then the stroll on Sunday would be worth gold to me. I am almost inclined to close with the plan. And then you have a concert[10] on Tuesday. It would be harsh to judge of me from the last concert[11] I did not appear to enjoy it, but I did so until my head gave way in a rare and singular manner. A dull sleepiness beset my brain which must have been evident to every body. I have rarely before or since felt any thing in my head so peculiar and so sudden in its appearance as what I experienced that night. The probability therefore is if I go to Queenwood at all that I shall be only too glad to join your concert and sing you a song myself if such an honour be permitted to me. With regard to Ruhmkorff's apparatus[12] I have myself only read a scrap or two in the Comptes Rendus[13] from which you could derive no benefit. As regards induced currents generally I would recommend you to study the original investigation of Faraday upon the subject. It is the first paper in the first volume of his 'Experimental Researches.'[14] You must help yourself to experiments. If I go down I will shew you a number of the principal ones, and I doubt not you will readily contrive others. I don't think you could get a shock through a glass handle unless it were wet. Perhaps you had the glass handle in one hand and touched the induction wire with the other; if so the glass is no protection to you: it is not necessary to lay hold with both hands to get the shock. If a single hand touch it is sufficient. For many experiments a good air pump is indispensable. I hope your pump is in good order.

Ginty is going to south America on the 9th. of April. He sails from Southampton[15] in the Great Western. We might perhaps take a trip over and see him set sail.

The sirene[16] reached me to day—it was addressed to the Revd. Dr. Hirst care of the Revd. Dr. Tyndall. I redirected it and probably it is sent to the carriers by this time.

I am now wandering into twaddle, forsaking by slow degrees the sentiment which actuated me at the outset of the letter. It is like turning from the song of a seraph[17] to listen to the gossip of old women. But here my philosophy may be at fault. However I wont analyze it lest my head should ache. If I go to Queenwood Heinrich[18] must be prepared to go through a number of those combustion experiments with me. I hope he will not forget this, but in his leisure moments put a little material aside so that it may be ready if I come. But a truce to science; it has invaded the sanctity of my thoughts while writing this letter and I will now expel it from my brain perforce.

Goodbye | Your affectionate | John

RI MS JT/1/HTYP/331–332
LT Typescript Only

---

1. *my 11th. lecture*: reference to the eleventh lecture of Tyndall's 1854 course on Heat at the RI. Speaking about the lecture in his journal entry of 28 March, Tyndall noted that even though he had a headache and 'lectured with the feeling of a man who was deliberately spilling out his life', he 'Got through it well' (Journal, JT/2/13b/669).
2. *your remark concerning DuBois*: Emil du Bois-Reymond. See letter 0910 for Hirst's suggestion that a response from Tyndall would be useless (see also letters 0904 and 0906).
3. *contemn*: 'to treat of small value, treat or view with contempt' (*OED*).
4. *wrote to him for a testimonial*: see letters 0512.
5. *he sent me a capital one*: see letter 0515.
6. *verdure*: 'the fresh green colour characteristic of flourishing vegetation; greenness' (*OED*).
7. *gall*: 'bitterness of spirit, asperity, rancour' (*OED*).
8. *your letter*: letter 0910.
9. *mutual improvement Society*: see letter 0819, n. 16.
10. *a concert*: this concert at Queenwood took place on 10 April (P. Grubb, 'Miscellaneous', *Queenwood Observer* 2:3 [29 April 1854], pp. 17–24, on p. 17).
11. *the last concert*: likely a reference to Queenwood's Christmas concert (see letter 0848, n. 6).
12. *Ruhmkorff's apparatus*: see letter 0910, n. 1.
13. *the Comptes Rendus*: see letter 0737, n. 12.
14. *the original . . . his '<u>Experimental Researches.</u>'*: reference to Series I of Faraday's work, which starts with a section entitled 'Induction of electric currents'; see Faraday, Experimental Researches in Electricity, vol. 1 (London: R & J. E. Taylor, 1839), pp. 1–41.
15. *Ginty . . . Southampton*: see letter 0909, n. 2.
16. *The sirene*: see letters 0904, n. 7 and 0733, n. 8.
17. *seraph*: commonly associated with angels. In the Bible, the seraphim were 'living creatures with six wings, hands and feet, and a (presumably) human voice' (*OED*).
18. *Heinrich*: Heinrich Debus.

From Tommy Grieve 29 March 1854 0912

Queenwood College | March 29th. 1854

My dear Dr. Tyndall,

Dr. Hirst told me to write to you. I will do so now. There are about ninety boys here now.[1] Papa[2] is coming at Easter, you will have a chance of seeing him.

George William[3] is going to take my portrait to-day to send to you

I remain | Your affectionate Friend | T. W. Grieve

RI MS JT/1/TYP/1/462
LT Typescript Only

---

1. *There are about ninety boys here now*: students at Queenwood College.
2. *Papa*: possibly H. M. Grieve (see letter 0839); not further identified.
3. *George William*: George William Edmondson.

To Thomas Archer Hirst [31] March 1854[1] 0913

My dear Tom

Be kind enough to give me credit for the following:—

Cost of Ruhmkorff's coil[2]	£2.2.6
Duty carriage &c.	1.1.6
Carriage of Syren[3]	0.7.2

I think I may safely say that I shall see you at the end of next week. I believe it will be a clear gain to me to get out of London for a couple of days, and I may learn something at Queenwood which can be turned to account in my lecture on education.[4]

RI MS JT/1/HTYP/333
LT Typescript Only

---

1. *[31] March 1854*: as Tyndall left for Queenwood on Wednesday, 5 April (Journal, 5 April, JT/2/13b/670), he likely wrote this at the end of the previous week, namely Friday, 31 March.
2. *Ruhmkorff's coil*: see letter 0910, n. 1.
3. *Syren*: misspelling of siren. See letter 0911, n. 16.
4. *my lecture on education*: see letters 0900, n. 6; 0906, n.4; and 0933, n. 1.

To Thomas Archer Hirst 1 April 1854 0914

Saturday April 1st. 1854.

My dear Tom

My lectures[1] end on Tuesday next, and I may be tempted to come down to Queenwood on Wednesday. I hope George Bell[2] will be able to give me some light work to do digging, picking grass, raking, or something of the kind. It would I am afraid be mere loss of time and strength (of which there is not much to lose) to remain here after the lectures have been finished. Possibly my brief sojourn in Queenwood may even aid me in the preparation of my educational lecture.[3] But this is a secondary point with me at present. The first thing is to get these absurd nerves and muscles into a normal condition. I go out to Bennetts[4] tomorrow. So you see that I am allowing myself much holiday time. Not more than I need I am sorry to say. I saw Ginty yesterday. He will probably be at Queenwood on Saturday night the 8th. to inspect some sick gas apparatus.

In a few days therefore I shall probably

Breathe the breath
Of the cowslip and primrose sweet
With the sky above my head
And the grass beneath my feet.

The thought is comforting.
Good bye | John

RI MS JT/1/HTYP/333
LT Typescript Only

---

1. *My lectures*: likely a reference to Tyndall's lectures on Heat at the RI (see letter 0865, n. 6). The last of this course of lectures was delivered on 4 April.
2. *George Bell*: not identified.
3. *my educational lecture*: see letter 0900, n. 6; 0906, n. 4; 0913, n. 4; and 0933, n. 1.
4. *Bennetts*: possibly Edward Bennett and his brother (see letter 0871, n. 1); not further identified.

To Thomas Archer Hirst 3 April 1854 0915

Monday April 3rd. 1854.

My dear Tom

I like any arrangement which will preserve me from putting them to any trouble in the house. I hope you will insist on my sleeping at the Lodge.[1] Can I not sleep in the same little room as before. There is not the slightest fear of my taking cold.

Yours affectionately | John.

RI MS JT/1/HTYP/333
LT Typescript Only

---

1. *the Lodge*: see letter 0848, n. 12.

From Hector Tyndale 4 April 1854 0916

Staffordshire Potteries[1] | Stoke-upon-Trent April 4th 1854

Dear Cousin

Being in England—on business matters—for a short time and as I shall pass through London in a week or two on my way to Paris, I write to you to ask your exact address in the great wilderness of London.

I confess myself very desirous to meet you for many—obvious—reasons, among them the necessity I am under of apologising to you for my failure to acknowledge your last kind letter[2] and the several papers[3] all duly rec'd by me.

I write now because I may remain, perhaps, a very short time in London, and if possible I must see you to satisfy a very natural and proper wish of my own, and of all my family, a wish to meet my—our—nearest Atlantic relative.

Hoping to receive your reply within a week and with my regards to your Mother[4] and to your Sister[5] I am

very truly Yours | Hector Tyndale

Please address to me to the care of Samuel Asbury Esq:[6] | Shelton | Staffordshire.

RI MS JT/1/T/44
RI MS JT/1/TYP/5/1586

---

1. *Staffordshire Potteries*: a well-known industrial area for ceramics in Staffordshire, England. It consisted of potteries in the towns of Tunstall, Burslem, Stoke, Fenton, Longton, and Hanley. As is indicated, Tyndale was writing from Stoke, or Stoke-upon-Trent, the largest city in Staffordshire.
2. *your last kind letter*: letter missing, though see letter 0605, which is the last extant letter in their correspondence.
3. *the several papers*: not identified.
4. *your Mother*: Sarah Tyndall, née McAssey, (1793–1867), Tyndall's mother and wife of John Tyndall Senior (1792–1847). Sarah descended from a prosperous farming family with lands in Ballybrommell, County Carlow. She was known to have encouraged the education of her children.
5. *your Sister*: Emma Tyndall (1817–1904).
6. *Samuel Asbury Esq*: a relative of John Astbury (around 1688–1743), a potter who worked in the Staffordshire Potteries (*ODNB*).

To Hector Tyndale 5 April 1854 0917

5th April 1854.

My dear Cousin

It gratifies me exceedingly to learn that you are in England and that I am likely to see you. My address is 9 Chadwell Street Myddelton Square Islington—but you will be more likely to find me at the Institution (in Albemarle Street) than at my lodgings. I have often thought of you during the last half year and waited only for a peaceful opportunity to write to you again. But I am rejoiced to find that the necessity of writing is, for the present at least, to be superseded by the sound of your voice and your own visible presence before me. One thing I cannot help regretting and that is my inability to offer you a bed, being prevented from doing so by the family arrangement of the people[1] in whose house I live. I am going into the Country today or tomorrow[2]—as my head is in rather a weak condition. On next Wednesday (this day week) I shall be back again and hope to find a note informing me when I may expect you—till then

Believe me | most truly yours | John Tyndall

RI MS JT/1/T/1441

1. *the people*: not identified.
2. *I am going into the Country today or tomorrow*: Tyndall indicated in his journal that he left for Queenwood on 5 April (Journal, JT/2/13b/670).

From Margaret Allen 7 April 1854 0918

Blackburn,[1] April. 7th /54.

My dear Tyndall,

I trust now that your lectures[2] are concluded, & that the Easter recess[3] is at hand that you will not disappoint me but that I shall receive one of your highly prized letters. I have looked at the syllabus of those lectures[4] every week, feeling almost anxious that they were completed, for, I know how it is with you; you 'work ever', but 'rest never', however if you do not feel that anxiety, & possess that care for your health as you ought, I can assure you there are those who do remember you—ever since I received those letters[5] (previous to your last visit to the continent) respecting the state of your health at that period, I have been doubly concerned regarding your safety.

I have been here nearly three weeks, but how very short the time appears, I can assure you I feel as if I was the inhabitant of another sphere—it must arise from the associations I have; & a congeniality with the disposition of my mind; can you have an idea of the refreshing feeling I experience, the true enjoyment which arises in my mind from the intercourse I have with one or two who entertain similar thoughts, and who are pursuing the same path, though alas! I have not as yet gained any great height;—however, 'Onward', ever being my motto I believe I shall not be disappointed—how the thoughts, and the mind fructify behind these genial and almost imperceptible influences, as in the affairs of general life, so in the highest aim which fills the heart, are we not instructed and warmed into life by kindred spirits?

Last evening I made acquaintance with your esteemed friend Dr. Hirst—As you say, how 'how strangely matters sometimes come about—& moreover, drunk your health in true German style, in sundry glasses of good old port; you must not lay, fie! fie!—So you see I have forestalled you, I now know something of his appearance, and shall be most happy if ever I have the pleasure of meeting his acquaintance—John Charles Dickenson,[6] possesses his portrait.—Oh—what a delightful ramble I had through his books, & now at last I have got the 'Faust'[7] in the original—now I am to be repaid—for what I will not say—but it is a pleasure to know that I shall be able to read it, having a little help from my dictionary—I also read in a volume of Emerson's Essays 'J.T. Gasthof züm Ritter, Cassel, 1851.'[8] What with the conversation, and these little mementoes I could have almost imagined you were present—the wish was heartily expressed, that we might have been together in that comfortable library—Ah! Well—I hope to see you some day before very long—I am infused by a new impulse very often when I think of you, and your ever cherished Friendship—but silence on this subject will probably be more gratifying to you, for I know that you are fully aware of my thoughts concerning you without my making a full declaration. Do not send me one of those very tiny notes of yours—if you knew how much they are prized & how deeply your every sentiment sinks into my mind you would endeavour to mould your pupils thoughts after your own standard.

Believe me ever | Yours most sincerely | Gretchen[9]

Be sure & seal your letter if you direct to Stanley Terrace,[10] in case I should not have returned—. if you find you have time to write me a few days after you receive this, address

Care of Chas. Boardman Esq.[11] | King Street | Blackburn

RI MS JT/1/A/1441

1. *Blackburn*: Blackburn, Lancashire, England, a town northwest of Manchester where Allen's father, Samuel Smith was a merchant (see letter 0896, n. 4). Allen had been having financial and marital difficulties and had mentioned that she may have to relocate to her father's home (see letter 0896).
2. *your lectures*: presumably the lectures Tyndall had been delivering at the RI and LI (see letter 0865, n. 6).
3. *the Easter recess*: the Easter break at the RI.
4. *the syllabus of those lectures*: see letter 0865, n. 6.
5. *letters*: letters missing. Tyndall had travelled to the continent on 11 July 1853 before returning to London in September (Journal, JT/2/13b/610–11; 624–25). Allen also refers to Tyndall's health in letter 0779.
6. *John Charles Dickenson*: John Charles Dickinson (b. 1835), a machine maker from Blackburn, Lancaster ('Notices of Intention to Proceed Dated November 30, 1855', *The Mechanics' Magazine* 1689 [22 December 1855], pp. 577–600, on p. 598).
7. *'Faust'*: Allen is most likely referring to the original German edition of Goethe's *Faust: Eine Tragedie* (Tübingen: Cotta, 1808). She quotes a line from *Faust* in letter 0699, n. 5 (see also n. 9).
8. *a volume of Emerson's Essays 'J.T. Gasthof züm Ritter, Cassel, 1851'*: Allen is likely quoting the owner's inscription that appears in her copy of Emerson, suggesting that it once belonged to Tyndall. Such inscriptions often denoted the location where it was first read. Tyndall would have inscribed it at a 'gasthof', or inn, at Cassell, Germany in 1851. Not further identified.
9. *Gretchen*: see letter 0872, n. 12.
10. *Stanley Terrace*: see letter 0699, n. 1.
11. *Chas. Boardman Esq*: possibly Charles Boardman (1823/24–56), a painter in the town of Blackburn (W. Durham, *Chronological Notes of Prominent Historical Events in the Town and Parish of Blackburn: From A.D. 448 to A.D. 1860. With the Census of 1851 and 1861, and a Table of Distances* [Blackburn: Charles Tiplady, 1861], p. 26).

To May Ann Coxe 12 April 1854 0919

Queenwood 12th April. 1854.

My dear Mrs. Coxe

Most heartily do I thank you for your friendly note[1] and Dr. Coxe[2] for his valuable present.[3] I shall undoubtedly study the book[4] and hope to have sufficient pace to enact to some extent the sound practical wisdom which it contains. I came to Queenwood on Thursday and for two days did nothing but amuse myself. I ran, jumped, climbed and played cricket and was astonished to find the change wrought in the state of my health. A vigour of new life seemed to run through my veins and I was surprised at the elasticity of my constitution. I commenced however to work a little—only a little—and today I find myself the worse for it—So I suppose I must give up work for a day or two longer. I am sometimes discontented to find what a trifle affects my health at present, but I dare say my duty is to accept such things as admonitions and shape my conduct accordingly. Be assured of it both your letter and your book will have due effect in promoting this resolution upon my part.

Your reference to the words of Dr. Playfair made me laugh.[5] at all events it proves how willing you are to remember any thing that may be said in my favour. I have often thought that there must be a kind of moral gravitation in this world which draws towards me those people whose friendship I prize most highly. My last accession in this line is yourself, and the thought of your kindness will often furnish me with a pleasant Subject for Contemplation. I am very glad you saw Dr. Playfair—He is a man of much practical knowledge and of great influence in England at present—He may possibly furnish efficient help in the solution of the problem regarding Alick.[6]

With all kind wishes | Believe me dear Mrs. Coxe | Most truly yours | John Tyndall

RI MS JT/1/T/246

1. *your friendly note*: letter missing.
2. *Dr. Coxe*: see letter 0781, n. 10.
3. *his valuable present*: see n. 4.
4. *the book*: not identified.
5. *reference to . . . Playfair made me laugh*: see letter 0961, n. 2 for an excerpt from Coxe's letter about Playfair.
6. *the problem regarding Alick*: a reference to the concern regarding the vocational path of Alexander Cumming, May Anne Coxe's nephew (see letter 0898).

From Peter Riess 13 April 1854 0920

Geehrtester Herr

Sie würden mir eine große Gefälligkeit erweisen, wenn Sie die untenstehenden Zeilen in einer getreuen Übersetzung in das philosophical magazine baldigst einrücken laßen wollten. Daß ein Physiker sich sein mühsam erworbener, unbestreitbarer Eigenthum nicht entreißen laßen will, bedarf bei Ihnen und Ihren Landsleuten keiner Entschuldigung. Sollte es Ihnen nicht thunlich erscheinen, meinen Wunsch zu erfüllen, so bittet um ein paar Zeilen

Ihr | ganz ergebenster | Dr. P. Rieß | (Spandauerstraße 81)

Die elektr. Wärmeerregung betreffend.

Hr. W. Thomson hat im philos. magazine (March 1854 p. 193) angegeben, daß die durch Entladung einer leydener Flasche erregte Wärmemenge proportional dem Quadrate der entladenen Elektricitätsmenge sei, was zuerst Hr. Joule 1840 bewiesen habe, im Widerspruche mit den von Sir W. Snow Harris früher angestellten Versuchen. Diese Angabe ist in jeder Beziehung unrichtig. Hr. Joule hat das angeführte Gesetz niemals bewiesen, sondern nur in einer Anmerkung zu seiner Arbeit über die Wärme des voltaischen Stromes die Vermuthung (presumption) geäußert, daß unter den nöthigen Beschränkungen, die elektr. Erwärmung dem Quadrate der Ladung einer gegebenen Batterie proportional sei. (phil. magaz. october 1841 p. 265). Als diese vage (vague) Vermuthung geäußert wurde, lag bereits seit vier Jahren eine Abhandlung vor, in welcher ich Harris's Angabe widerlegt und das Gesetz des Quadrats durch Versuche bewiesen hatte (Poggend. Annal. März 1837 p. 341) Ich hatte bereits in zwei ausgedehnten Arbeiten die vollständige Formel für die Erwärmung durch die Batterieentladung gegeben, nachdem ich die merkwürdige Beziehung Relation der in einem Drathe erregten Wärmemenge zu dem Widerstand des Drathes aufgefunden hatte (Poggend. Annalen Januar 1838 p.47. September 1838 p.1)

Solche Reclamationen, die dem, der sie veranlaßt, nicht peinlicher sein können, als sie dem sind, der sie nothgedrungen geben muß, würden vermieden werden, wenn überall die billige Forderung anerkannt würde, daß eine Behauptung so allgemeiner Art, wie Hr. Thomson gewagt hat, erst die Folge einer sorgfältigen Prüfung der, nicht nur im eigenen Lande, sondern auch in andern Ländern über den Gegenstand publicierter Arbeiten sein darf.

Berlin 13 April 1854. | P. Rieß

Most honoured Sir

You would do me a great favour, if you would have the lines below inserted in a faithful translation in the Philosophical Magazine[1] as soon as possible. That a physicist does not want to let his hard won and indisputable property to be snatched away from him requires no excuse with you and your fellow countrymen. Should it not seem possible to you to carry out my wish, then a few lines are requested by

Your | most humble | Dr. P. Riess | (Spandauerstrasse 81)

Concerning the Electr. Generation of Heat.[2]

Herr W. Thomson has stated in the Philos. Magazine (March 1854 p. 193),[3] that the quantity of heat generated by the discharge of a Leyden jar[4] is proportional to the square of the quantity of electricity discharged, which Herr Joule supposedly first proved in 1840, in contradiction to the experiments done earlier by Sir W. Snow Harris.[5] This statement is incorrect in every regard. Herr Joule never proved the law referred to, but only expressed the Vermuthung (presumption), in a footnote to his work on the heat of Voltaic current, that, under the necessary constraints, electr. heating was proportional to the square of the charge of a given battery (Phil. Magaz. October 1841 p. 265).[6] At the time when this vague presumption was expressed, a paper had already been in existence for four years, in which I had refuted Harris's statement and proved the law of squares[7] by experiments (Poggend. Annal. March 1837 p. 341)[8] I had already given the complete formula for heating by battery discharge in two extended papers, after I had discovered the remarkable relation[ship] of the quantity of heat generated in a filament to the resistance of the filament (Poggend. Annalen January 1838 p.47. September 1838 p.1)[9]

Such complaints, which cannot be more embarrassing to the man who prompts them than they are to the man who is forced to make them, would be avoided, if everywhere the proper demand were acknowledged that an assertion of so general a nature as Herr Thomson has ventured, may be only the result of a careful examination of works published on the subject, not only in one's own country but also in other countries.

Berlin 13 April 1854. | P. Riess

RI MS JT/1/R/26

---

1. *the Philosophical Magazine*: Tyndall became co-editor in January (see letter 0867, n. 5).
2. *Concerning the Electr. Generation of Heat*: this letter was translated and published in the May issue as P. Riess, 'On the Generation of Heat by Electricity', *Phil. Mag.* 7:46 (May 1854), p. 348.
3. *the philos. magazine (March 1854 p. 193)*: see W. Thomson, 'On the Mechanical Values of Distributions of Electricity, Magnetism and Galvanism', *Phil. Mag.* 7:44 (March 1854), pp. 192–97, on pp. 193–94.
4. *Leyden jar*: 'an electrical condenser consisting of a glass bottle coated inside and outside with tinfoil to within a certain distance of its mouth, and having a brass rod surmounted by a knob passing through the cork, and communicated with the internal armature.' The jar is significant for its ability to store electrical charges (*OED*).
5. *Sir W. Snow Harris*: Sir William Snow Harris (1791–1867), an English natural philosopher notable for his work on electricity and magnetism (*ODNB*).
6. *(Phil. Magaz. October 1841 p. 265)*: see J. P. Joule, 'On the Heat evolved by Metallic Conductors of Electricity, and in the Cells of a Battery during Electrolysis', *Phil. Mag.* 19:124 (October 1841), pp. 260–77, on pp. 264–65.
7. *law of squares*: see letter 0708, n. 2.
8. *a treatise . . . (Poggend. Annal. March 1837 p. 341)*: The page number indicated by Riess was incorrect; see P. Riess, 'Ueber gebundene Electricität', *Poggend. Annal.* 40 (1837), pp. 321–54.
9. *two extended papers . . . (Poggend. Annalen January 1838 p.47. September 1838 p.1)*: see P. Riess, 'Ueber die Erwärmung im Schliessungsbogen der electrischen Batterie', *Poggend. Annal.* 43 (1838), pp. 47–48; and P. Riess, 'Ueber die electrische Verzögerungskraft und das electrische Erwärmungsvermögen', *Poggend. Annal.* 45 (1838), pp. 1–24.

From William Wright[1] 17 April 1854 0921

Springfield Clark, Co. Ohio.[2] | April 17th 1854.

My dear John,

Your welcome favour of Oct. 16th[3] was received in due season and met with a hearty welcome although my silence since discredits my assertion. Nevertheless it was so, I will not attempt to appologize for my apparent neglect. I was much pleased to hear from you and learn that although wrapt in your profession that you occasionally gave some thoughts to the days long past and gone. The days of our happy boyhood, when the world seemed all a pleasant scene and anxious for the completion of our manhood. But alas how little did we anticipate the cares and toils of that manhood, the disappointments and sorrows which ever follow in its train. To some of us at least it has brought unrequited toil so far as position and a name extends. Our mutual friend, poor Joe,[4] was I fear far from that station which his talents and opportunities would have entitled him to occupy had they been properly applied. Had he lived I have no doubt he would have done well here as I had secured for him an engagement with the Dargan of our Western country.[5] But the all-wise fiat of Omnipotence had ordained it otherwise and we must bow in humble submission to his will. Joe's decease was very melancholy I recollect no rational answer he gave to the many enquiries I made of him on his arrival here. One of my first was to ask for you and your prospects in life. One of the greatest men living was his reply. On further enquiries of a like nature I ascertained his mind was unsettled and so continued until his death. To me the pursuits of life has not been so ungenerous. You are well aware that my ambition was never of a lofty nature. I well recollect a severe rebuke you gave me many years ago when I informed you that I would like to be a guard on a Mail coach and intended applying for it although I doubt whether I had then sufficient interest to obtain so lofty a position. Your ideas had some weight in preventing me from the trial and has often occurred to me since. By some fortunate circumstance I got engaged in the Railroad business in 1845 and have continued in it and with the same company ever since, so that if I have gained nothing more by the connexion we at least ought to have an intimate knowledge of each other by this time. I hold an office which requires a personal attention of at least 10 hours per diem and requited by a salary which enables me to keep my family very comfortable, a large one now, and prevents me from acquiring much hold on this world's goods as is generally expected of emigrants to this part of the world. I am however making some little progress for myself and in a few instances been of some benefit to my fellow man. How pleasant it is to turn from the preceding incidents to look upon your career when by your own unaided industry and talent to have so soon acquired a position as to enable you to be the colleague of one of the first men of the age whose fame neither ocean or mountain barrier can confine. The name of Faraday is as familiar here as household words, when scientific research is mentioned, and I have no doubt it is so on the slopes of the Rocky Mountains[6] and the, used to be far off shores of the Pacific. To see your name coupled with his was a proud day for me and a happy one to learn that you sometimes could lose sight of your station to allow your thoughts to ramble back to the time when we dabbled in the crystal waters of Madlin[7] when our highest prize was a small Eel or speckled trout and our most formidable weapon of attack a three [pronged][8] fork whose usefulness in the culinary department was often missed by our respected mothers.[9] Well, those days are past and gone and the dark side scarce remembered whilst the sunshine of them often flashes before our vision. I was indeed sorry to hear your health was not good, take care that your too close application to your studies does not prevent you from accomplishing much that you would otherwise complete were you to pay a proper regard to your health. I do not know but what the advice or practise of it would be beneficial to myself and the effects of close attention to my business is very visible. Parties who knew me seven or eight years ago will scarcely recognize me as the same individual, I have got so completely emaciated in flesh, having lost some 50lbs., in weight since my arrival in the States. Notwithstanding, my health is tolerably good but I lack strength to perform the same amount of labour I did a few years ago. I had some idea last winter of dropping into your rooms this Spring and given you a surprise. But a bargain in a house and lot offering improvements made this Spring, made me apply the funds I had set apart for my trip. Perhaps in another season I may accomplish the object with less disadvantage to my pecuniary means. It is an object I have long had in view just to spend a few days in the vicinity of the old home to visit Ballyknockan Moat,[10] Maudlin Lug[11] and Burgage,[12] to greet the few of my old acquaintances who still reside there, and for-ever bid adieu to scenes around which my heart will fondly linger to my latest breath. How happy would I be if I could have you for an associate in that visit. I am well aware of the vast difference which exist between our minds and modes of thought still I cannot but cherish the belief that were we seated on the old familiar Moat or the Grave Stones in Ballynockan for a short period our thoughts and feelings would flow in unison. Perhaps that day may yet arrive, at least I shall always look forward to accomplishing it. I was happy to see notwithstanding your entire devotion to science you did not allow it to crush out all thoughts of a matrimonial nature. You cannot in justice to mankind allow all your talents and acquirements to pass away with yourself but should endeavour to transmit them to a host of Tyndalls. There is plenty of time yet and I know enough of good looks left to secure for you a helpmate in so righteous a cause. So let me hear from you soon that you are thinking seriously of committing yourself in the cause of matrimony. It would I have no doubt add much to your happiness. You know from my experience I ought to be capable of judging in this matter. When writing home to Ireland please give my best respects to Miss Emma[13] and tell her I am much indebted to her for the privilege of once more addressing you, for I feared my former neglect would have deprived me of it, and I trust my present delay will not prove an obstacle in future. The fact I had two letters written but they did not suit my views, nor does this, but if I wait until I write one that would I should never send it, and so have concluded to Mail this for what it is worth. If perused by you in some of your leisure moments and provokes a reply it will afford me pleasure.

I have taken the liberty to promise a letter of introduction to you to a friend of mine who will visit London this Summer, J.B. Morris Esq.,[14] Agent of the Mad River and Lake Erie Railroad Com.,[15] with whom I have transacted business as the agent of a connecting *[firm]*,[16] for the last three years. You will find him a gentleman well acquainted with the Eastern Seaboard and Western country. He has been extensively engaged in commercial transactions the last few years and been very successful in his operations which has enabled him to visit Europe, more in the hope of restoring his health, than the eclat[17] of a European tour. Any attentions which your engagements will admit of during his stay in London will be extremely gratifying to me. I will be happy to reciprocate the favour if ever in my power. I have promised him much from the interview he will have with you in London, and some of my friends in Dublin.

I will now close with my best wishes for your health and future prosperity in which my Mother[18] and Wife[19] join. Let me hear from you soon and oblige

Truly yours | William Wright.

RI MS JT/1/TYP/5/1828–1831
LT Typescript Only

---

1. *William Wright*: William Wright (1818/9–92), a childhood friend of Tyndall's. The two had a long-term correspondence (see Volumes 1, p. 391 and 2, p. 448) and Wright often urged Tyndall to join him in America.
2. *Springfield Clark, Co. Ohio.*: a city located in Clark County in southwest Ohio, USA.
3. *Your welcome favour of Oct. 16th*: letter missing.
4. *Joe*: likely Joseph Payne, who was a civil assistant in the 5th Division, C District of the Irish Ordnance Survey. He joined the Survey on 1 April 1839, the same day as Tyndall, and

worked closely with him. On 8 April 1842, he was transferred to the English Survey. He is possibly the Joseph Payne, son of William and Jane Payne, who was born on 5 November 1821 in Dunleckny, County Carlow (see vol. 1, p. 441).

5. *Dargan of our Western country*: reference to the western United States, though Dargan is specifically unidentified.
6. *the Rocky Mountains*: a major mountain range in North America, located on the western side of the continent.
7. *the crystal waters of Madlin*: reference to the Madlin River in County Carlow, Ireland, where Tyndall was born and grew up.
8. [*pronged*]: in the typescript LT was not sure what this word might be and suggested 'pronged (?)' in the margin.
9. *our respected mothers*: Sarah Tyndall (see letter 0916, n. 4). William Wright's mother is not identified.
10. *Ballyknockan Moat*: Ballyknockan was a village located in County Wicklow, in northern Ireland. Likely a reference to Maudlin Moat, near Ballyknockan, which was a rath or 'enclosure of roughly circular form made with a strong earthen wall, and originally serving as a fort and place of residence' (S. Lewis, *A Topographical Dictionary of Ireland, Comprising the Several Counties, Cities, Boroughs, Corporate, Market, and Post Towns. Parishes, and Villages, with Historical and Statistical Descriptions*, 2 vols [London: S. Lewis & Co., 1837], vol. 2, p. 695).
11. *Maudlin Lug*: misspelling of lough, or lake. Maudlin Lough was likely a lake or water feature near Ballyknockan that later became part of the Poulaphouca Reservoir, or Blessington Lakes.
12. *Burgage*: Burgage Castle, located in County Wicklow, in northern Ireland.
13. *Miss Emma*: Emma Tyndall (1817–1904), Tyndall's sister, whose education he encouraged. She does not appear to have married and later in life suffered from a mental illness, which was manifested as an extreme form of religiosity.
14. *J.B. Morris Esq.*: John B. Morris.
15. *the Mad River and Lake Erie Railroad Com.*: The Mad River and Lake Erie Railroad Company was a major railway company in Ohio from 1832 until 1881; it was renamed the Sandusky, Dayton and Cincinnati Railroad Company in 1858 (*Annual Report of the Commissioner of Railroads & Telegraphs of Ohio, For the Year Ending June 30, 1880* [Columbus: G. J. Brand & Co., 1881], p. 496–99).
16. [*firm*]: in the typescript LT inserted 'firm(?)' here.
17. *eclat*: reputation, social distinction, or renown; often used in a disparaging manner during the nineteenth century (*OED*).
18. *Mother*: see n. 9.
19. *Wife*: Sarah Wright, née Neale/Neill (1815–88), who was from a prosperous family in Roscrea, County Tipperary, Ireland. She and William Wright married on 7 February 1841 and emigrated to America in early March 1841.

To Peter Riess 17 April 1854 0922

London. | 17th April 1854.

Dear Professor Riess,

I received your note[1] this morning and shall feel the greatest pleasure in complying with your desire. I translated a letter from Dr. Clausius[2] upon the same subject which was published in the Magazine[3] of last month. It is certainly very unpleasant to be compelled to claim your undoubted right in this manner, but your note will be of service and will teach some of our scientific men nor to be hasty in assigning the credit of a discovery concerning which they have not given themselves the trouble of acquiring accurate information. I may truly say that it will always give me pleasure to see the honour of a discovery secured to him who deserves it, and I trust the Philosophical Magazine will be ever ready to open its columns to the claims of justice. Might I beg of you to present my kind remembrance to Mrs. Riess[4] and to your family.

Believe me | most sincerely yours | John Tyndall.

You would oblige me by sending the accompanying note[5] to Prof. Magnus.

Staatsbibliothek zu Berlin (Stabi) Slg Darmstaedter 1855 Tyndall Be. 1–58, 42–43

1. *your note*: letter 0920.
2. *a letter from Dr. Clausius*: R. Clausius, 'On the Heat Produced by an Electric Discharge', *Phil. Mag.* 7:45 (April 1854), p. 297.
3. *the Magazine*: the Phil. Mag.
4. *Mrs. Riess*: Fanny Riess (née Güterbock) (1810–89), the daughter of the merchant Levin Isaac Güterbock (1745–1815), married Peter Riess in 1835. They had three children, including daughter Rebecca Riess (1836–1924), who married the physicist Georg Quincke (1834–1924).
5. *accompanying note*: letter missing.

To Hector Tyndale [20 April 1854][1] 0923

My dear Hector,

I should feel much obliged if you would slip into one of the photographic establishments and get your portrait taken for me. I write this in the hope of it reaching you sufficiently early to enable you to get it done today—If not I hope you can contrive to do it tomorrow. I have seen some capital things on the left hand side as you go up Regent Street.[2] The artist is a Frenchman named Mayer[3] I think. It is not far from Regent's Circus Piccadilly.[4]—If you cant spare the time then put in a claim against the blue sky of Paris during your stay there.

ever yours | John Tyndall | Thursday

RI MS JT/1/T/1478
RI MS JT/1/TYP/4/1588

---

1. *[20 April 1854]*: at the top of the letter LT noted details that were presumably from the envelope: 'Hector Tyndale *[Esq]* | Mr. Chapman's | 142 Strand | Royal Institution | *[April]* 20'. This suggests that the date of 'April 20' was from the postmark.
2. *Regent Street*: a major commercial street in London, England.
3. *a Frenchman named Mayer*: not identified.
4. *Regent's Circus Piccadilly*: located in Westminster, London a junction constructed in 1819 to connect Piccadilly and Regent Street, which functioned as an important public space. It is now known as Piccadilly Circus.

From Thomas Archer Hirst 23 April 1854 0924

Queenwood College | April 23rd 1854

My dear John

I have just read your lecture[1] which arrived this morning and as I shall never have a better opportunity I will try to say a word about it. As I expected it contains many excellent remarks which I am glad to think will be listened to by your fashionable audience. There is much high-toned thought in it and it possesses that rare property of being simple in its adherence to facts and in its interpretation of them without being hampered by insisting too much on crotchets of minor importance. I like the way you commence exceedingly well, that distinction between a <u>means</u> and a <u>branch</u> far from being trivial is of vast importance in my opinion and deserves to be dwelt upon even more than you have done. You perhaps might improve upon the manner of illustrating the subject by means of the two images of a stomach and a seed and I would do so by expanding your remarks on the subject It appears to me that the objectors to science and scientific enquiry are misled often by considering it as a <u>branch</u> instead of a <u>means</u> of education. Regarding the mind as a store-house or vessel for containing knowledge they naturally wish to seek materials of the best quality and thus they presume to judge the comparative value of different kinds of knowledge and in thus judging they often embarrass themselves with crude notions of utility and so forth. John Haas will hold that a knowledge of atmospheric pressure is of no use to him, he performed the experiment of respiration and inspiration just as well before he knew how it was effected. That he sucked milk from his mothers breast and got fat upon it without knowing either how it flowed into his mouth or what was its chemical constitution. But if we take the higher position and regard the development of the seed as the primary object of education then all facts fall into their proper places and the different branches of knowledge take rank according to the amount of nutriment they are capable of furnishing to this precious seedling. It would be well worth the trouble to expand this first portion of your lecture if by so doing you could induce your hearers to regard the subject of Physics from this point of view in fact they ought never to lose sight of this object of education throughout the whole lecture, for then what you have said about Physics in the latter portion, and you have said it well, will be seen in the best light the whole lecture would appear well knit together and instead of being rather fragmentary as some might at first deem it, it would be found to be all to the point and all illustrative of its subject the Importance of Physics as a branch of education. Then the word branch itself would have a new and higher meaning. The study of Physics tends to develop the mind in a certain direction. As you have very well expressed it it teaches man to traverse with safety that line between cause and effect. Your remarks on the moral tendencies of Physical investigations are very good Here again I am glad that your audience will hear what you have to say on this point. The petty jealousies of scientific men are very distressing indeed and many will listen to you who might profit by what you have here said for it is just what wanted saying It is true that it is not so immediately connected with your subject as other parts of the lecture otherwise I should ask you to expand this part too. But I want to send you the extract from my journal[2] which you asked for so I will conclude my letter.

Yours affectionately | T.A. Hirst.

N.B. Please return M[r] Pilkington's[3] letter.[4]

Pilkington[5] stammers distressingly and this impeded his progress very much for, teaching his class on the simultaneous principle,[6] I often passed him by knowing that I could not get an intelligible answer to any question I might ask him, indeed had it not been for his eloquent eyes which often said 'I know it Sir if I could but speak' I might have mistaken him for a dunce and thus have done him great injustice. What with his love of mischief, however, and his inability to cope with his schoolfellows, owing to his unfortunate stutter, it was evident that he was losing interest in his work, or rather that he had never felt much interest in it, and it became necessary to try and awaken it in him. One day after he had been more noisy and mischievous than usual, and had been made to stand out of his seat, I walked up to him at the end of the hour and told him to get permission to go with me next hour instead of to M[r] Wright. On his return I told him rather sternly to get his cap and follow me. He did so and I walked on towards the lodge[7] he at my heels in a state of anxious suspense. After some moments silence I asked 'Do you know what I am about to do with you Pilkington?'. 'Ne—ne—ne... no Sir' he stuttered 'Well' I replied 'I will tell you. You will not work with me, I have spoken often enough to you and now I am about to make you do better for the future' He began to stutter again, no doubt it would have been an excuse or a prayer for mercy if I had turned to look at his eyes, but I would not, and I stopped his stuttering with a stern 'follow me Sir'. Without speaking a word we walked past the lodge, through the gates and up the road to the Avenue his anxiety increasing every step. At length, putting my arm quietly round his neck I broke silence 'Can you tell me what an angle is my boy?' 'Ye—ye—ye—yes Sir an angle's a—a—a—a' he could get no further and turned his eyes up to me imploringly, as much as to say 'Don't be cross Sir, I'm sure I know it but I cannot tell you'

'Well', I replied to his silent appeal, 'go and pull two long grass stalks and shew me what an angle is.' This he did and with the grass stalks he continued to answer my questions about the geometrical definitions. Thus continuing our subject we turned into a stubble field 'What is a right angled triangle?' I asked. By this time he had lost all fear and spoke often quite distinctly 'It has all its angles right angles Sir' 'Indeed' I replied taking my arm from about his neck 'it has three right angles has it? Will you kneel down a few minutes' He saw he was wrong, stuttered 'two', looked at me pitifully, and hesitated. Again I cried 'go on your knees Sir' which he did, and then dropping on my knees too I said 'now boy pull up some stubble and make me a triangle with either two or three right angles' He began doing so, at once saw his error, and the absurdity of our situation as we knelt together there making geometrical figures of stubble, and springing to his feet he shook with laughter and cried 'It has only one right angle Sir, only one of course.' I replied 'of course' too and with my arm round his neck we turned homewards and continued our lesson successfully. 'This is the punishment I had in store for you' I said as we reached home, 'now go and transgress no more' to which his eyes answered 'I will Sir'.

(Extract from my journal of Oct 7th 1853)[8]

We had several such walks afterwards and I have since found that although he has forgotten much that he learnt in school with other boys, he has never forgotten what we learnt together in that stubble field.

T.A.H. April 23rd / 54

RI MS JT/1/H/197
RI MS JT/1/HTYP/334–335

---

1. *your lecture*: Tyndall's lecture on education (see letters 0900, n. 6; 0906, n. 4; 0913, n. 4; and 0933, n. 1).
2. *extract from my journal*: see letter 0882, n. 15.
3. *Mr Pilkington's*: refers to the father of William Pilkington, a student at Queenwood College. Although the younger Pilkington was quite a troublemaker, Hirst cared substantially for the boy.

4. *letter*: Hirst indicated in his journal that he had received 'a very pleasing letter from Pilkington's father' on 21 April (Hirst, 'Journals', 23 April 1854); not further identified.
5. *Pilkington*: the following part of this letter is a rewritten portion of Hirst's journal entry of 7 January 1854.
6. *the simultaneous principle*: reference to a method of education in which an entire class is treated as a single unit. In this method, when the teacher asks a question students are asked to respond all at once (J. S. Laurie, 'General Report on the British Schools, and the Wesleyan and other Denominational Schools, Inspected in the Year 1857', in *Minutes of the Committee of Council on Education; Correspondence, Financial Statements, &c. and Reports by her Majesty's Inspectors of Schools. 1857–8* [London: Her Majesty's Stationery Office, 1858], pp. 554–73, on p. 560).
7. *the lodge*: see letter 0848, n. 12.
8. *(Extract from my journal of Oct 7th 1853)*: in the handwritten letter, it appears that someone changed the date from 'Jan 7th 1854' to 'Oct 7th 1853'; see also n. 5.

From Heinrich Gustav Magnus 26 April 1854 0925

Berlin d. 26 April 54.

Theurer Freund.

Zwei von Ihren Briefen liegen unbeantwortet vor mir, der letzte vom 15 d. M.; der frühere vom 29 März. Zunächst will ich Ihnen berichten daß alle Aufträge ausgeführt sind. Die Instrumente für Colonel Yorke sind fertig. Ich lasse sie verpacken, und bitte mich wissen zu lassen: 1. Was für eine Adresse auf die Kiste gesetzt werden soll? 2. Wie ich die Kiste schicken soll? Ob ich sie dem Buchhändler Asher wieder übergeben soll?—Die Auslagen betragen

für 2 Windfahnen	3 [Rthl.] 20 [Sgr.]
" ein Cylinder im Ring	7.
" ein dito auf die Rotationsmaschinen zu setzen	4
	14.20
Emballage	1
	15.20

Das Polytrop für Mr. Gassiot ist bestellt, es werden aber mehrere Wochen vergehn bis es fertig wird. Ich werde es so schnell als möglich zu schaffen suchen.

Colonel Yorke wird sich müssen zu dem Cylinder im Ringe noch eine Art von Drehwage machen lassen, wie in Fig 2 meiner Abhandlung abgebildet ist.—Die Schwierigkeiten welche Sie erwähnen wegen der Erklärung der Abweichung der Kugeln sind nicht schwer zu beseitigen: Wenn eine Kugel durch die Gase, welche sich aus dem Pulver entwickeln, in dem Geschütz vorwärts getrieben wird, so ist der Druck den sie erfährt überall derselbe, d.h wenn man einen grössten Kreis der Kugel senkrecht gegen die Axe des Geschützes sich vorstellt, so erfährt dieser in jedem Punkte denselben Druck. Liegt nun der Schwerpunkt der Kugel in diesem Kreise, aber nicht im Mittelpunkt desselben sondern z.B. mehr nach rechts in Bezug auf einen hinter dem Geschütz stehenden Beobachter;

so ist links von dem Schwerpunkte zwischen as eine größere Fläche des Kreises als rechts zwischen sb; die Kugel wird also auf der rechten Seite stärker gedrückt als auf der linken, und folglich rotirt sie von links nach rechts

sowohl im Geschütz als wenn sie dasselbe verlassen hat. Wiewohl auf beiden Seiten des Schwerpunkts gleichviel Materie ist, so ist doch die Quantität der Bewegung verschieden weil dieselbe sich nach der Fläche richtet.

Wirft man einen Körper mit der Hand, so erhält jeder Punkt der Masse gleichviel Bewegung, unter der Voraussetzung daß man nicht mittelst der Hand den einen Theil stärker gestossen hat als den andern. Wenn bei einem mit der Hand geworfenen Körper der schwerere Theil voran eilt, so rührt das davon her daß derselbe im Verhältniß zu seiner Bewegung einen geringeren Widerstand erfährt als der andere, weil er im Verhältniß zur Masse eine geringere Fläche darbietet.—M^r^. Baden Powell's Vortrag, oder richtiger der Auszug desselben in dem Phil Mag., ist nicht sehr klar. Er wirft mir Unklarheit der Darstellung vor, aber ich kann nicht finden daß er überhaupt etwas Neues in der Sache vorgebracht hat. Sehr neugierig wäre ich etwas von Wheatstone's Apparat zu erfahren, namentlich ob er im Princip etwas Neues enthält. Ich hoffe Sie werden sich an dem Polytrop erfreuen, und ich fühle mich sehr dankbar gegen M^r^. Gassiot daß er eins bestellt hat.

Von Herzen bedaure ich daß Sie unwohl gewesen sind, hoffentlich ist alles vorüber, und Sie sind wieder ganz wohl und frisch. Prof. Faraday bitte bestens zu grüßen, ich sende ihm nächstens etwas rothen und schwarzen Schwefel, über die ich neulich in der Academie gelesen. Viele Empfehlungen an D^r^. B. Jones, M^r^. Gassiot, und Alle die sich meiner erinnern. Warrington bitte zu sagen daß ich dies Jahr schwerlich nach England komme. Wir wissen ohnehin nicht ob wir nicht in Krieg gerathen, und das ist für uns ernster als für die Engländer, die keinen Feind im Lande sehn werden, denen also der Krieg nur Geld kostet, während uns die Russen leicht mit ihrer Knute beglücken könnten. Leben Sie wohl und schreiben Sie bald

Ihrem | treuen | GMagnus.

Meine Frau und die Kinder, so wie mein Bruder der jetzt aus Paris zurück gekehrt ist, grüßen vielmals.

P.S. Die Reclamation von P. Rieß haben Sie wohl erhalten?!

Berlin, 26 April 54.

Dear friend.

Two of your letters[1] lie before me unanswered; the last one from the 15th of this month, the earlier from 29 March. I want to tell you, first of all, that all the jobs have been completed. The instruments for Colonel Yorke are finished. I am having them packed, and ask you to let me know: 1. What kind of address should be put on the case? 2. How should I send the case? Whether I should hand it over to the bookseller Asher again?—The expenses amount to

for	2	wind vanes are	3 [Rthl.] 20 [Sgr.]
"	one	cylinder in the ring	7.
"	one	ditto to be put on the rotation machines	4
			14.20
Packaging			1
			15.20

The polytrope[2] for Mr. Gassiot has been ordered, but it will take several weeks until it is finished. I will try to do it as quickly as possible.

Colonel Yorke will still need to have a kind of torsion scale[3] made for him for the cylinder in the ring, as it is illustrated in Fig. 2 of my paper.[4]—The difficulties which you mention about the explanation the deviation of bullets are not difficult to eliminate: If a bullet is driven forwards in the gun by the gases which are produced from the gunpowder, then the pressure that it experiences is the same everywhere, i.e., if one imagines a greatest circle of the bullet perpendicular to the axis of the gun, then this will experience the same pressure at every point. If, now, the centre of gravity of the bullet lies within this circle, though not at its centre but rather e.g. more to the right relative to an observer standing behind the gun,

then, on the left of the centre of gravity between as, there will be a larger area of the circle than on the right between sb; the bullet will therefore be pushed more powerfully on the right side than on the left and consequently it will rotate from left to right,

both within the gun as well as when it has exited it. Although there is the same amount of matter on both sides of the centre of gravity, the quantity of motion will nevertheless be different because it depends on the area.

If one throws a body by hand, then every point of the mass receives the same motion, assuming that one has not pushed the one part more powerfully by means of one's hand than the another. If, with a body thrown by hand, the heavier part rushes forwards, then that is due to the fact that it experiences less resistance in proportion to its motion than the other part, because it presents less area in proportion to the mass.—Mr. Baden Powell's lecture, or more correctly, the excerpt of it in the Phil Mag.,[5] is not very clear. He accuses me of a lack of clarity in my presentation, but I cannot find that he has put forward anything new at all in the matter. I should be very curious to learn anything about Wheatstone's apparatus,[6] in particular if it contains anything new in principle. I hope that you will take pleasure in the polytrope and I feel very grateful to Mr. Gassiot for ordering one.

I sincerely regret that you have been unwell, hopefully everything has passed, and you are quite well and bright again. I ask you to give my best regards to Prof. Faraday, I shall shortly send him some red and black sulphur, about which I recently read at the Academy.[7] Many compliments to Dr. B. Jones,[8] Mr. Gassiot and everyone who remembers me. I ask you to tell Warrington[9] that I shall hardly make it to England this year. We do not know anyway if we shall not get into a war,[10] and that is more serious for us than for the English, who will not see an enemy in their country and for whom war will therefore only cost money, while the Russians could easily favour us with their lash. Farewell and write soon to

Your | loyal | GMagnus.

My wife[11] and the children,[12] as well as my brother,[13] who has now returned from Paris, give you their best regards.

P.S. You have no doubt received the complaint from P. Riess?![14]

RI MS JT/1/M/10

1. *Two of your letters*: letters missing.
2. *polytrope*: see letter 0870, n. 7.

3. *torsion scale*: a scale used for weighing in which the levers function through torsion, a twisting motion.
4. *my paper*: see letter 0870, n. 3.
5. *the excerpt of it in the Phil Mag*: see B. Powell, 'On Certain Phenomena of Rotary Motion', *Phil. Mag.* 7:45 (April 1854), pp. 291–96.
6. *Wheatstone's apparatus*: see letter 0907, n. 5.
7. *the Academy*: see letter 0852, n. 13.
8. *Dr. B. Jones*: Henry Bence Jones.
9. *Warrington*: likely Sir Warington Wilkinson Smyth (1817–90), a British geologist and lecturer at the Government School of Mines.
10. *a war*: the Crimean War (see letter 0873, n. 30). Although Prussia initially resisted joining the war, they eventually followed Austria in trying to limit Russia's expansion of power (G. Arnold, *Historical Dictionary of the Crimean War* [Lanham: Scarecrow Press, 2002], pp. xviii, 113).
11. *wife*: Bertha Magnus (née Humblot) (1820–1910), whose family were French Huguenots.
12. *the children*: Heinrich and Bertha Magnus had three children: Anna Magnus (1841–75), Christine Magnus (1842–1936), and Paul Gustave Magnus (1845–1930).
13. *my brother*: in 1854, Magnus had five living brothers: Friedrich Martin Magnus (1796–1869), Eduard Leopold Samuel Magnus (1799–1972), Ludwig Immanuel Magnus (1790–1861), Ferdinand Ludwig Magnus (1804–65), and Albert Magnus (1809–59).
14. *the complaint from P. Riess?!*: see letter 0920. Tyndall had already responded to Riess's letter (see letter 0922).

From Thomas Archer Hirst 30 April 1854 0926

Queenwood College. | April 30th 1854

My dear John.

On looking at Wertheims paper[1] of which I am to give a report in the next Phil. Mag.[2] I find that before I can well understand it I shall have to make myself conversant with a good deal of optics, especially of the portion which has reference to the Polarization of Light. This will involve considerable labour which I am very willing to give though it interferes very much with a Mathematical investigation I had in hand, for I find that I must put the latter aside in order to give this report and prepare my lectures on Acoustics,[3] which will commence immediately. I mention this because you asked me to finish your translation of Kohlrausch's paper[4] and I cannot promise to do both before the end of the month. Wertheim alludes to an apparatus of Ruhmkorff's[5] connected with the Polarization of Light and having two electro-magnetic coils. Can you tell me in what this instrument consists and for what purpose it is used. It must be somewhat similar to that which shews Faradays experiment on the solution of Sugar.[6]

More new boys have come since Easter, we have now 95 or 96, only 4 short of our maximum! From Mr Iliff I learnt that Mr Mummery has given Mr Edmondson notice[7] that he will leave at Midsummer. How this has been brought about I do not know. Mr Iliff thinks of leaving also. Why I cannot exactly tell but I suspect it is because of M^r^ E's treatment. Iliff is a shrewd, good fellow and was not long before he discovered what our principal was. He says that from M^r^ E's professions and from the Prospectus (!)[8] he was induced to think that Queenwood was quite a model school and he finds it quite the contrary He expected to learn something and says he is disappointed, so that he thinks he might as well be working for his father as for Mr Edmondson. Indeed, said he, there is no satisfaction whatever in working for the latter gentleman. I replied that he was quite right, that Queenwood was no place for anyone who did not find satisfaction enough in working for the lads and not for the principal. 'But I can do that at home' said he. 'Most assuredly,' I replied, 'you can and in your place I should do so. You came to Queenwood expecting to find in Mr Edmondson a man with whom you could work well & thus see your own principles carried out into practice. In this sense of course you will be disappointed. But I do not think that your time at Queenwood will be without profit. You came to learn how a school should be conducted, you expected to find here a system which you could transplant home and experience has shewn you that there are here 7 different systems corresponding to 7 different masters That it is fruitless to go seek a system for yourself but that you must stop at home and from experiment make your own system. If you leave with this conviction the six months will have been well spent. And now you have mentioned the prospectus which you say mislead you I may tell you that it was the production of a different man from Mr Edmondson. Not Mr Edmondsons but Dr Tyndalls system is there recorded for the most part. It was what he aimed at and induced Mr E. to aim at and I may add that I am working here with a similar purpose.' With all this he agreed indeed I have since heard that it is by no means certain that he will leave us. Debus has had similar conversations with him and told him much the same. It is possible that the little fellow feels more interest in the place now that he has learnt to know Debus and myself better. He is a hard working, conscientious and sensible little fellow at bottom and if he can stop and work with us I shall be glad.

Henry Hill will be married at Whitsuntide.[9] He told me of another characteristic act of Bevington's which nearly made me cry when I read it, it was so beautiful. One morning he gave Henry a parcel containing some silver spoons marked H. and R.H. Henry & Rebecca Hill.[10] In his quiet way he told him that he hoped some day to visit him and drink a cup of tea with one of them.

Yours affectionately | T.A. Hirst.

Mr Haas wishes me to ask you where you left the 'Question Book'[11] it is wanted. Send me Mrs Allan's address[12] please.

RI MS JT/1/H/198
RI MS JT/1/HTYP/337–337a

---

1. *Wertheims paper*: G. Wertheim, 'Sur la double refraction temporairement produite dans les corps isotropes, et sur la relation entre l'élasticité mécanique et entre l'élasticité optique', *Annal. Chim. et Phys.* 3:30 (1854), pp. 156–221.
2. *a report in the next Phil. Mag.*: the report was eventually published in two parts as G. Wertheim, 'On the Double Refraction Temporarily Produced in Isotropic Bodies', *Phil. Mag.* 8:52 (October 1854), pp. 241–61; *Phil. Mag.* 8:53 (November 1854), pp. 342–57.
3. *my lectures on Acoustics*: in May 1854, Hirst delivered a lecture at Queenwood on 'the nature of sound, the conditions on which it depends, and the powers of conducting it, as possessed by various gases, liquids and solids' ('Lectures', *Queenwood Observer* 2:4 [31 May 1854], pp. 25–32, on p. 25).
4. *translation of Kohlrausch's paper*: See R. Kohlrauch, 'Theory of the Electric Residue in the Leyden Jar', *Phil. Mag.* 8: 46 (May 1854), pp. 305–20; *Phil. Mag.* 8: 47 (June 1854), pp. 412–26; *Phil. Mag.* 8: 48 (June 1854), pp. 476–89.
5. *an apparatus of Ruhmkorff's*: the apparatus is described on p. 242 of Wertheim's article (see n. 2).
6. *Faradays experiment on the solution of Sugar*: refers to what is known as the Faraday effect, 'the rotation of the plane of polarization of light or other electromagnetic waves when transmitted through certain substances in a magnetic field that has a component parallel to the direction of transmission' (*OED*). Faraday discovered the effect and published experiments using various solutions, including sugar, thus revealing a relationship between light and electromagnetism (Faraday, *'Experimental Researches in Electricity'*, *Phil. Trans.* 136 [1846], pp. 1–20, on p. 8).
7. *Mr Mummery has given Mr Edmondson notice*: Tyndall and Hirst had much discussed their wish to see Mummery tender his resignation after his purchase of a spurious doctoral degree (see letter 0817, n. 4 for full references).
8. *the Prospectus (!)*: a document on the philosophy of education followed at Queenwood and a description of how the school was ran.
9. *Whitsuntide*: the festive season that surrounds Whitsunday (see letter 0759, n. 2).
10. *Rebecca Hill*: Henry Hill's wife. Not further identified.
11. *the 'Question Book'*: likely a book of questions used for teaching students at Queenwood.
12. *Mrs Allan's address*: Hirst was likely asking for the permanent address of Margaret Allen, which was Stanley Terrace, in Preston, Lancashire, England. Hirst had recently met her while visiting Blackburn, England, where Allen was temporarily residing (see letter 0918, n. 10).

From Charles Cumming 1 May 1854 0927

Queenwood College | May 1st. /54.

Dear Dr. Tyndall,

Mr Taylor[1] requested me to ask you whether you would be an honourable Member a Patron or a Donor of the large cricket club which is now just commencing, and he says he will send you a copy of the rules and different things about the club when they are printed and that will be in two or three days. I am enjoying myself very much at Queenwood. It has been very wet weather lately. Those boys which like can have a little piece of ground for a garden and I have got one, my seeds are now a good height above the ground. I must now say good bye dear Dr. Tyndall as I have no more to say at present

With kind love to you | Believe me | Ever your affect. friend | C. Cumming.

RI MS JT/1/TYP/1/1/260
LT Typescript Only

1. *Mr Taylor*: either a staff member at Queenwood College who worked as a treasurer, or a student who took on the role of treasurer for the Queenwood Cricket Club (see letter 0933, n. 11). Not further identified.

To Thomas Archer Hirst 2 May 1854 0928

2nd. May 1854.

My dear Tom

It would be very unreasonable in us to expect you to give a translation of Kohlrausch[1] and a report of Wertheim[2] this *[month]*. If you send us Kohlrausch we shall be quite satisfied. I have found such obstacles as you are now encountering[3] of the greatest service to me, for they have driven me to study subjects of importance which I should have otherwise passed over in a superficial manner. As soon as I get a little time I will write more reports though I have scarcely any other inducement than the instruction to be derived from writing them. Faraday found on submitting heavy glass and other bodies to the action of a magnet or of an electric current that the plane of polarization was turned as in the case of the sugar solution, the direction of rotation depending on the direction of the current and changing with the latter.[4] The apparatus to which Wertheim refers[5] is a magnet, but the poles are perforated so that a ray of light can be sent in an axial direction through any transparent body placed between the poles, the said ray being observed by a person placed behind the other pole. Suppose you send a ray already polarized through such a body and conceive a plate of tourmaline[6] placed before the observer's eye; he will see a maximum of light in a certain position of the plate. Now excite the magnet. The light is quenched or diminished, for this position of the tourmaline, and in order to find the place of maximum strength you must turn the plate round: the angle through which you must turn the plate expresses the amount of the rotation of the plane of polarization.

I read your remarks about Iliff[7] with great interest and great concern. I think you acted the very wisest part to state your plain feelings to him, and this mode of acting will tell far more upon a sincere man, and tend more to secure him to Queenwood than if you had endeavoured to bolster up a case for E[8]— he is the most miserable solecism[9] I ever encountered. But I trust you will be able to smelt Iliff into your nucleus and induce him to work for a higher ideal than that of pleasing Mrs. E.[10] Iliff I suppose has been much in Mummery's company if so I am not surprised at his discontent as you may be assured the former, to assign a reason for his leaving has not spared Queenwood. But tell me Tom, do you not think the school in a healthy condition? It appeared to me to be so when I was there. There was a spirit of cheerfulness and activity apparent everywhere which I think must be accepted as a sign of health. Perhaps the 7 different systems might have much said in their favour, perhaps this apparent irregularity tends to strengthen both teachers and taught. If you thought well of it, I will write a note to Iliff—it is a pity to lose such a man if the loss could be avoided. If this be not the case, if he is determined to go, why boy let him.

We had Whewell last Saturday. He gave us a good lecture[11] which gave general satisfaction. He did not by any means fill the measure which would have satisfied either your thirst or mine upon the subject. He talked a good deal of Socrates[12] and Plato[13] which interested me very much. Prince Albert[14] was in the Chair and there was a good full house. Faraday comes on next Saturday[15]

I went out to Bevington on Sunday evening and spent a few hours in a truly cordial and pleasant manner with him. I did not know of that little matter to which you allude,[16] nor did he tell me. He is a very choice good soul and I am glad we found him out.

Affectionately yours | John

It behoves Edmondson to look out for a classic at once enough to have his man marked before school breaks up. But if not drawn to this he will neglect it.

RI MS JT/1/HTYP/338
LT Typescript Only

1. *a translation of Kohlrausch*: see letter 0926, n. 4.
2. *a report of Wertheim*: see letter 0926, n. 2.
3. *such obstacles as you are now encountering*: in letter 0926, Hirst noted that in order to give a report on Wertheim's paper (see n. 2), he had to learn a great deal about optics. This interfered with his mathematical investigation.
4. *Faraday found . . . the latter*: see letter 0926, n. 6.
5. *The apparatus to which Wertheim refers*: see letter 0926, n. 5.
6. *tourmaline*: 'a brittle pyro-electric mineral, occurring in crystals, also massive, compact, and columnar, originally obtained from Ceylon' (*OED*).
7. *your remarks about Iliff*: Iliff, a teacher at Queenwood, was considering leaving his position at the school (see letter 0926).
8. *E*: George Edmondson.
9. *solecism*: 'a breach or violation of good manners or etiquette; a blunder or impropriety in manners' (*OED*).
10. *Mrs. E.*: Anne Edmondson.
11. *a good lecture*: reference to Whewell's lecture of 29 April, which was the first of the RI's Lectures on Education (see W. Whewell, 'Influence of the History of Science Upon Intellectual Education', *Lectures on Education* [London: J. W. Parker & Son, 1854], pp. 3–36).
12. *Socrates*: see letter 0890, n. 12.
13. *Plato*: see letter 0855, n. 5.
14. *Prince Albert*: Franz Karl August Albrecht Immanuel (1819–61), prince consort of Queen Victoria, who played a major role in developing the Great Exhibition of 1851 (*ODNB*).
15. *Faraday . . . next Saturday*: reference to Faraday's lecture of 6 May, which was the second of the RI's Lectures on Education (see M. Faraday, 'Observations on Mental Education', *Lectures on Education* [London: J. W. Parker & Son, 1854], pp. 39–90)
16. *that little matter to which you allude*: in letter 0926, Hirst noted that Bevington had given Henry Hill a set of engraved spoons as a gift in celebration of his and Rebecca Hill's marriage.

From William Hopkins 6 May [1854] 0929

Cambridge | May, 6 '55.[1]

My dear Sir,

You were kind enough some time ago to say that you would undertake some experiments which I suggested on the conductive powers of certain compressed substances.[2] I have had specimens formed of several substances, three of each under different pressures up to 12 or 14,000 lbs. on the square inch. My object is to ascertain the effect of these different pressures on the conductive power. I have wished you to undertake these experiments knowing that you have a delicate apparatus[3] adapted for these, and conceiving also that you have established a sort of claim to add these new substances to those you have already experimented on, supposing you to have a wish to extend these researches. If therefore other engagements will allow you to undertake them, I again say how glad I shall be to place the experiments in your hands.

The specimens which have been formed for me of spermaceti,[4] wax, tin, zinc, lead, clay &c are cylinders 4 or 5 inches long and a bar of about 3 inches diameter. From these I will have smaller specimens out of any form and size which may be most convenient to you if you will have the kindness to let me know what they shall be.

It is also of the first importance in certain geological speculations to ascertain the conductive powers of certain hard crystalline rocks as compared with those of mere earthy structure. Of these also I will have specimens formed for you should you undertake the experiments.

Believe me | Yours very truly | W. Hopkins.

RI MS JT/1/TYP/2/606
LT Typescript Only

---

1. *May, 6 '55.*: This letter was either dated incorrectly or reflects a typographical error. LT noted '54?' in the margin, which is probably correct. Hopkins wrote here that 'some time ago' he had requested Tyndall to undertake some experiments, which likely referred to experiments discussed in letters 0753 and 0862. See also letter 0938.
2. *You were kind enough . . . certain compressed substances*: not identified. Tyndall had presumably responded to letter 0753. See also letter 0862.
3. *a delicate apparatus*: see letter 0753, n. 5.
4. *spermaceti*: see letter 0753, n. 2.

From Peter Riess 11 May 1854 0930

Geehrtester Herr

Empfangen Sie meinen verbindlichsten Dank für die Mühe, die Ihnen die Übersetzung meiner Notiz verursacht hat, die so getreu meine Meinung wiedergibt, daß ich die fortgefallenen beiden Citate leicht verschmerzen kann. Leider muß ich Sie mit einer neuen Bitte belästigen, die untenstehenden Zeilen zu übersetzen und aufzunehmen doch thue ich dies zuversichtlicher, als früher, weil ich Ihren Namen unter den Herausgebern der philos. Journ. bemerkt habe. Es kann Ihnen nicht unerwünscht sein, daß eine Taktik, die in Ihrem Journale geübt ist und nach Jesuitismus schmeckt, gerügt werde. Mit der Bitte dem folgenden Schreiben bei der Übersetzung die üblichen Höflichkeitsformeln am Anfange und Schluß zu geben, verbleibe ich

Ihr | ganz ergebenster | P. Rieß.

An die Herren Herausgeber des philosophical Magazine

Es thut mir wahrhaft leid, daß Hr. Thomson im phil. magaz. May p. 347, indem er einen Irrthum zugesteht, einen neuen und schlimmeren Irrthum begangen hat, den ich nicht unerwähnt laßen darf. Hr. Thomson nimmt die Entdeckung des Gesetzes, daß die durch die leydener Flasche erregte Wärmemenge dem Quadrate der Elektricitätsmenge proportional ist, nicht mehr für Hr. Joule, sondern für Cuthbertson in Anspruch, nach deßen Versuchen, wie er behauptet, Hr. Becquerel das Gesetz schon 1835 "quite explicitly enunciated". Ich bedauere, daß die vollständige Mittheilung der Bemerkung Becquerels unterblieben ist, die mir diese Gegenbemerkung erspart haben würde. Die Stelle lautet im Originale, wie folgt: Cuthbertson et autres physiciens qui ont mesuré avec des électromètres l'action calorifique produite par la décharge d'une batterie, ont trouvé qu'elle croissait à peu près comme le carré des charges des batteries pour certaines longueurs des fils. Cette loi varie en outre suivant l'épaisseur des jarres qui composent la batterie. Celles qui sont épaisses ont une puissance de fusion moins grande. Man braucht die Versuche Cuthbertson's nicht zu kennen, um aus dem Zusammenhange zu sehn, daß Hr. Becquerel von Schmelzung, und allein von Schmelzung spricht. Daß aber durch die bekannten Schmelzversuche Cuthbertson's und Anderer, die ich in einer Abhandlung über Schmelzung discutiert habe (Pogg. Annal. 65. 497. übers in Scientific memoirs) das in Rede stehende Gesetz bewiesen worden sei, wird schwerlich Jemand behaupten wollen, selbst wenn ihm unbekannt geblieben wäre, daß bei der elektr. Schmelzung neben der Erwärmung eine andere elektrische Wirkung thätig ist. Ich habe die Ehre

Berlin 11 Mai 1854. | P. Rieß

Most honoured Sir

Please accept my most profound thanks for the trouble that the translation of my note[1] has caused you; it conveys my opinion so faithfully that I can easily get over the two quotations which were omitted. Unfortunately I must pester you with a new request to translate and include the lines below, but I do this with more confidence than earlier, because I have noticed your name among the editors of the Philos. Journ.[2] It cannot be unwelcome to you that a tactic that is practices in your journal and smacks of Jesuitism[3] may be admonished. With the request that you will give the following communication the usual polite phrases at the beginning and end when you translate it, I remain

Your | most humble | P. Riess.

To the publishers of the philosophical Magazine[4]

I am truly sorry that in admitting an error in the Phil. Magaz. of May p. 347,[5] Herr Thomson has committed a new and worse error that I may not leave unmentioned. Herr Thomson no longer arrogates the discovery of the law that the quantity of heat generated by the Leyden jar[6] is proportional to the square of the quantity of electricity to Herr Joule, but rather to Cuthbertson,[7] following whose experiments, as he claims, Herr Becquerel[8] had already 'quite explicitly enunciated' the law in 1835. I regret that the complete communication of Becquerel's observation was omitted, as it would have spared me these contradictory remarks. The passage in the original is as follows: Cuthbertson and other physicists, who have measured with electrometers the calorific action produced by the discharge of a battery have discovered that it increased approximately as the square of the charges of the batteries for certain lengths of wires. Moreover, this law varies according to the thickness of the jars which comprise the battery. Those which are thick have a lower power of fusion.[9] One does not need to know the experiments of Cuthbertson to see from the context that Herr Becquerel is speaking of fusion alone. But scarcely anyone is going to want to claim that the law in question has been proven through the well-known fusion experiments of Cuthbertson and others that I have discussed in a paper on fusion (Pogg. Annal. 65. 497. transl. in Scientific memoirs),[10] even if it had remained unknown to him that, that with electr. fusion, another electrical effect is active in addition to heating. I have the honour

Berlin 11 May 1854. | P. Riess

RI MS JT/1/R/27

---

1. *the translation of my note*: see letter 0920, n. 2.
2. *name among the editors of the Philos. Journ.*: this is a reference to the *Phil. Mag.* Tyndall became a co-editor in January (see letter 0867, n. 5).
3. *smacks of Jesuitism*: in this instance Jesuitism referred to a casuistic approach to morality.
4. *To the publishers of the philosophical Magazine*: a translation of the original German version of this letter was published as P. Riess, 'On the Generation of Heat by Electricity', *Phil. Mag.* 7:47 (June 1854), pp. 428–29.
5. *the Phil. Magaz. of May p. 347*: W. Thomson, 'On the Heat Produced by an Electric Discharge', *Phil. Mag.* 7:46 (May 1854), pp. 347–48.
6. *Leyden jar*: see letter 0920, n. 4.
7. *Cuthbertson*: John Cuthbertson (c. 1743–1821), an instrument maker who wrote numerous papers and was notable for his glass-plate frictional electrical machines (*ODNB*).
8. *Herr Becquerel*: Antoine César Becquerel (1788–1878), a French scientist and pioneer of the field of electrochemistry.
9. *Cuthbertson and other physicists . . . a lower power of fusion*: A. C. Becquerel, *Traité Expérimental de L'Électricité et du Magnétisme, et de leurs Rapports avec les Phénomènes Naturels*, 7 vols (Paris: Firmin Didot Frères, 1835), vol. 3, p. 150.
10. *a paper on fusion (Pogg. Annal. 65. 497. transl. in Scientific memoirs)*: P. Riess, 'Ueber das Glüben und Schmelzen von Metalldrähten durch Elektricität', *Poggend. Annal.* 65 (1845), pp. 481–537, translated as P. Riess, 'On the Incandescence and Fusion of Metallic Wires by Electricity', in Tyndall and Francis (eds.), *Scientif. Mem.*, pp. 432–75.

From Heinrich Debus 13 May [1854][1] 0931

Queenwood am 13 Mai.

Mein lieber John!

Ich habe Deine Vorlesung aufmerksam durchgelesen und habe dieselbe vortrefflich gefunden. In der That es würde schade sein viel daran zu ändern, sondern es erscheint mir am besten wenn Du dieselbe im wesentlichen wie sie jetzt ist in die Welt schickst.—Einige Puncte sind darin über die ich vielleicht mich in anderer Weise ausgedrückt haben würde, jedoch diese Puncte sind von untergeordneter Bedeutung, sie sind klein im Verhältniss zum Ganzen. Ich will kurz über die mir aufgefallenen Stellen weggehn.—

1. Ist die Verbesserung der menschlichen Stimme von einer Veränderung der Membranen abhängig?

   Ich glaube nicht. Ich glaube dass bei der Erziehung der Stimme unser Geist zum Bewustsein kommt „wie die Membranen zu gebrauchen sind" um den angenehmsten Ton hervorzubringen.

2. „considering the education of the human mind to be the exercise and development of the human mind"

   Dies ist richtig so weit wie es geht aber ich glaube dass es über der Entwicklung des Verstandes noch eine andere und höhere Erziehung giebt. Erziehung zerfällt meiner Ansicht nach in drei Theile. 1. Moralische (religiöse) 2. Verstandes && & 3 physische Erziehung. Mache die Seele „edel", den Verstandt „scharf und schnell", und den Arm „stark", und Du wirst einen guten vortrefflichen Menschen geschaffen haben. Nun es scheint mir als wenn die Verstandeserziehung zu ausschliesslich vor Deinem Auge gewesen wäre.

   Wenn ein junger Mensch von 15 oder 16 Jahren, ich studierte jeden Tag einige Stunden. Jede neue Thatsache die ich in mich aufnahm und die dunkle Sachen mehr oder weniger klar machen konnte war für mich in jener Zeit ein grosser Schatz. Ich war stolz und glücklich wie ein König, mein Geist war mehr erhaben und ich hätte in solchen Augenblicken „Nichts Schlechtes thun können".

   Wenn „mind" das geistige Princip im Gegensatz zu der todten Körpermasse bezeichnet denn ist meiner Ansicht nach die obige Definition der „education of the human mind" zu eng, weil am Ende der Teufel auch ein Stück davon ist.

   Diese Bemerkungen beziehn sich nur auf die obige Definition.

3 „the body is fed by oxygen"

   Ich würde nicht so sagen weil Oxygen mehr ein Zerstörer des Körpers ist. Sauerstoff brennt Fleisch und Blut.

4. Physics is the midway between Astronomy and Chemistry &&& Es scheint mir besser zu sein hier einen andern Ausdruck zu gebrauchen. Der Physiker ebensogut wie der Chemiker, beide sind gezwungen immer zu Atomen ihre Zuflucht zu nehmen. Ich glaube nicht dass man sagen kann dass Physik angewandt auf Atome Chemie giebt. Wir arbeiten in der Regel mit grossen Massen und erkllfe der Atome. heinungen n sagen kann dass Physik angewandt auf Atome Chemie giebt. zu nehmen. der Teufel ären unsere Erscheinungen mit Hülfe der Atome. Dasselbe thut der Physiker. Er muss Atome haben um die Phänomene des Lichtes, der Electricität und der Wärme, der Molecular *[aggregation]* &&& zu erklären. Physik & Chemie sind hier gleich.

Nun geht die Vorlesung weiter und spricht über die Erregung unserer geistigen Thätigkeiten, unserer Wissbegierde, unserer Einbildungskraft && durch das Studium der Physik. Die Untersuchungsmethode dieser Wissenschaft und den grossen Einfluss den dieselbe auf die Ausbildung unseres Verstandes hat, erklärst Du auf den folgenden Seiten. In allen diesen Dingen stimme ich ganz vollkommen mit Dir überein und ich habe daher Nichts darüber zu sagen.

Noch ein Wort über die Erziehung der arbeitenden Classen.

Wird es möglich sein den Arbeitern eine so gute Ausbildung in der Physik zu geben dass die Wissenschaft Aussichten bekommt 1000 mal schneller entwickelt zu werden als es bisher der Fall war? Neue Ideen kann ich kaum hinzufügen. Soll ich Dir aber sagen was ich von dem Nutzen des Studiums der Physik für die Erziehung denke so ist es etwa folgendes: Was macht den Menschen edler, was erhebt uns mehr über die drückenden Sorgen des gewöhnlichen Lebens als ein ruhige *[Beschauung]* der Natur, sei es wenn die Elemente friedlich nebeneinander walten oder wenn dieselben im wilden Kampfe einander zu vernichten drohen. Wir sagen in unserer deutschen Sprache „Lasse uns in das Freie gehen" tief und bedeutungsvoll, denn in der Natur herrscht Freiheit, Gesetz und Ordnung und wie durch Zauber macht sie den Menschen zum Menschen. Kleine Sorgen verschwinden, die Seele fühlt sich erhaben. Es ist das Gefühl welches milder gestimmt, höher erregt wird, durch den Anblick des Sternenhimmels, des Weltmeeres oder einer üppigen Vegetation.

Wo gehen Liebende hin wenn sie seelig sein wollen? Was beschreibt David wenn er von der Grösse und Allmacht Gottes spricht? Wer kennt nicht die lieblichen Bilder welche eine ruhige heitere Landschaft in unserer Seele hervorruft. Die Einbildungskraft fängt an zu spielen, und wir vergessen uns selbst. Wenn ich meine Apparate, wenn ich Zahlen und Bücher bei Seite lege, und Hoffnungen habe ein kleines Korn Wahrheit zu finden, ein bischen mehr den Schleier der Natur zu lüften, denn bringt meine Einbildungskraft tausend angenehme Dinge zum Vorschein und ich phantasire wie ein Dichter. Freilich nur in Stunden der Erholung; sobald wie Zahlen, Bücher und Apparate wieder in meine Hand kommen tritt kalte Berechnung, ruhige Überlegung, strenges und besonnenes Forschen an die Stelle des Spieles der Phantasie. In der Natur herrscht strenges Gesetz, eiserne Consequenz, unveränderliche Ordnung. Der Geist der dies studirt gewöhnt sich daran die Dinge mit natürlichen Augen, mit Logischen Gedanken zu verfolgen. Kurz „das Studium der Physik" gewöhnt unseren Verstandt an eine logische Denkungsweise, es macht denselben schärfer und schneller, Gefühl und Einbildungskraft werden lebhaft erregt, der Mensch wird menschlicher, er wird edler & besser; er lernt Gott durch Gottes Werke kennen.

Wo ist ein edlerer Gegenstandt für unsere Wissbegierde als die Werke einer höhern Hand? Dies ist der geistige Nutzen der Physik. Gross aber ist der praktische Vortheil den sie uns bringt. In jedem Moment unseres Lebens brauchen wir die Physik. Eisenbahnen, Telegraphen, Regen, Thau, Licht, Wärme, Compass &&&

Dieser Brief wird von wenig Nutzen für Dich sein weil ich so wenig über den Gegenstandt gedacht habe. Soll ich meine Meinung über die Vorlesung „als ein Ganzes" sagen denn muss ich mit Tom übereinstimmen dass dieselbe sehr gut ist.

Dein H Debus

Queenwood, 13 May

My dear John!

I have read your lecture[2] through attentively and have found it excellent. In fact, it would be a pity to change much in it; rather it appears best to me if you send it off into the world essentially as it is now.—There are some points in it about which I would perhaps have expressed myself in a different way, but these points are of secondary importance, they are minor in relation to the thing as a whole. I shall go quickly over the parts that caught my attention.—

1. Is the improvement of the human voice dependent on a change of the membranes?

   I think not. I think that when educating the voice, our mind comes to the awareness of 'how the membranes are to be used'[3] in order to produce the most pleasant sound.

2. 'considering the education of the human mind to be the exercise and development of the human mind'

   This is correct as far as it goes, but I think that there is another different and higher education over and above the development of the mental faculties. In my view, education breaks down into three parts. 1. Moral (religious) 2. that of the mental faculties, etc. & 3. physical education. Make the soul 'noble', the mental faculties 'sharp and quick', and the arm 'strong' and you will have created a good, excellent human being. Now, it seems to me as if you had the education of the mental faculties too exclusively in mind.

When I was a young person of 15 or 16 years, I used to study for a few hours every day. Every new fact which I took in and which made the obscure things more or less clear was a great treasure for me at that time. I was proud and happy as a king, my mind was more uplifted and, in such moments, I would have 'been able to do nothing wrong'

If 'mind' denotes the intellectual principle, in contrast to the dead corporeal mass, then in my view, the above definition of the 'education of the human mind' is too narrow because in the end, the devil is a part of it too. These comments refer only to the above definition.

3 'the body is fed by oxygen'

I would not say so because oxygen is more a destroyer of the body. Oxygen burns flesh and blood.

4. Physics is the midway between Astronomy and Chemistry etc. && It seems better to me to use a different expression here. The physicist as well as the chemist, are both compelled to always recourse to atoms. I do not think that one can say that physics applied to atoms gives chemistry. We work as a rule with large masses and explain our phenomena with the help of atoms. The physicist does the same. He must have atoms in order to explain the phenomena of light, electricity and heat, of Molecular *[aggregation]* etc. && Physics & chemistry are identical here.

The lecture now goes on and speaks of the excitement, through the study of physics, of our intellectual activities, of our thirst for knowledge, of our imagination, etc. & You explain the method of investigation of this science and the great influence that it has on the training of our mental faculties on the following pages. In all these things I agree quite completely with you and I therefore have nothing to say about it.

One more word about the education of the working classes.

Will it be possible to give workers such a good training in physics that the science has prospects of being developed 1000 times faster than was the case until now? I can hardly add any new ideas. But if I should tell you what I think about the utility of the study of physics for education, then it is something like the following: What makes a human being nobler, what raises us more above the pressing worries of everyday life than a calm *[contemplation]* of nature, be it when the elements work peacefully side by side, or when they threaten to destroy each other in a wild. In our German tongue we say 'let us go outside',[4] deeply and meaningfully, for in nature freedom, law and order prevail, and as if by magic it makes a human being a human being. Small worries disappear, the soul feels uplifted. It is the emotions, which are put in gentler mood or are more highly excited by the sight of the starry sky, the ocean, or luxuious vegetation.

Where do lovers go when they want to feel bliss? What is David describing when he speaks of the greatness and omnipotence of God?[5] Who does not know the lovely images that a calm and serene landscape evokes in our soul. Imagination begins to play and we forget ourselves. When I put aside my instruments, when I put aside numbers and books, and have hopes of finding a grain of truth, of lifting nature's veil a little bit more, then my imagination produces a myriad pleasant things and I fantasize like a poet. Of course, only at times of relaxation; as soon as numbers, books, and instruments come to hand again, cold calculation, calm deliberation, disciplined and considered research take the place of games of the imagination. In nature, strict law, resolute consistency, immutable order prevail. The mind that studies this accustoms itself to follow things with natural eyes and with logical thoughts. In short, 'the study of physics' accustoms our mental faculties to a logical way of thinking, it makes it sharper and quicker, the emotions and the imagination are actively excited, a human being becomes more human, he becomes nobler & better; he gets to know God through God's works.

Where is a nobler subject for our thirst for knowledge than the works of a higher hand? That is the intellectual utility of physics. The practical advantage which it brings us, though, is great. We use physics in every aspect of our lives. Railways, telegraphs, rain, dew, light, heat, compass, etc. &&

This letter will be of little use to you because I have thought so little about the subject. Should I give you my opinion about the lecture as 'a whole', then I have to agree with Tom[6] that it is very good.

Your H Debus

RI MS JT/1/D/85
RI MS JT/1/TYP/1/2403–2405

---

1. *13 May 1854*: since this letter is about Tyndall's lecture on education (see n. 2), which he delivered on 27 May 1854, it must have been written in 1854.
2. *your lecture*: Tyndall's lecture on education (see letters 0900, n. 6 and 0933).
3. *'how the membranes are used'*: Tyndall would later discuss this more in depth in J. Tyndall, *Sound: A Course of eight lectures delivered at the Royal Institution of Great Britain* (London: Longmans, 1867), on pp. 48–84.
4. *'let us go outside'*: 'das Freie', German for 'the outdoors'; it could also be translated as 'the open air' or 'open space'.
5. *What is David . . . of God?*: Psalm 19 in the Bible, in which David speaks about revelation and the ability that studying nature has to reveal God's glory to man.
6. *Tom*: Thomas Archer Hirst. Hirst also gave feedback to a draft of Tyndall's lecture on education in letter 0924.

To Thomas Archer Hirst 20 May 1854 0932

Royal Institution May 20th 1854.

My dear Tom

You have been looking at the subject of billiard balls and will probably be able to enlighten me on a point or two.

Suppose the ball A struck by the cue BC at the point C, the force being thus applied oblique[1] to a radius what kind of motion will follow?

Supposing the same ball to be withdrawn from the table and suspended in air, what will be its motion when the same shock as before is imparted to it?

You have seen the ball dancing on a jet of water in the filter maker's shops[2] in London I will give you a sketch of the thing and you will perhaps state to me the precise mechanical forces concerned in the production of the phenomenon.

The ball appears to rest upon the jet in a manner which leaves its centre of gravity unsupported and a quick motion of rotation is imparted to it by the jet. It casts rays of drops on all sides as I have sketched them but a portion of the water actually encircles the ball like a noose. If you place a glass cylinder in the path of a jet the latter is deflected and may be caused to twirl round the cylinder several times, owing to its adhesion to the latter. If the cylinder were moveable it would move—and this in fact is the case with our dancing ball.

Suppose the force acting upon the ball in the direction of the jet AB to be decomposed into two others, one acting in the direction of the radius BC and the other in the direction of the tangent. This latter owing to the adhesion of the water will make the ball revolve, the former will I imagine tend to drive the ball from the jet. Why does it not accomplish this?

My Friday evening lecture[3] was given last night and was very successful. The experiments were truly beautiful; and the applause rather exceeded the bounds permitted in the Royal Institution.

I gave my lecture[4] to Mrs Faraday to read, and requested Faraday also to read it. He has done so and expressed himself greatly pleased with it. In fact he said he had no expectation of any thing of the kind and that it caused me to stand higher in his estimation than ever I did before. His remarks about means and branch were quite the same as those made by yourself[5] and he said it was an error to introduce the word branch into the titles.

Now I have said enough about myself. We are to have Daubeny of Oxford[6] to day, and I come on next week.[7] My chief troubles for the session will be over at the end of next week and I shall begin to look forward to the time when I can have another game upon the Queenwood playground.

Yours affectionately | John

RI MS JT/1/HTYP/340
LT Typescript Only

1. *oblique*: at an angle that is not a right angle (*OED*).
2. *the filter maker's shops*: not identified.
3. *My Friday evening lecture*: the lecture took place on 19 May at the RI (see letter 0910, n. 3).
4. *my lecture*: Tyndall's lecture on education (see letters 0900, n. 6 and 0933).
5. *His remarks about . . . those made by yourself*: Hirst had provided feedback on Tyndall's lecture in letter 0924.
6. *Daubeny of Oxford*: reference to Daubeny's lecture for the RI series on education published as C. Daubeny, 'On the Importance of the Study of Chemistry as a Branch of Education for all Classes', *Lectures on Education* (London: J. W. Parker & Son, 1854), pp. 117–70.
7. *I come on next week*: reference to Tyndall's lecture for the RI series on education (see n. 4).

To Thomas Archer Hirst 27 May 1854 0933

London Saturday evg | May 27/54

My dear Tom.

I should not wonder if you thought me remiss in writing to you but the fact is I have been compelled to shut out the world, yourself included, for the last three or four days while I ran my lecture[1] once more through the crucible of thought—I added some and changed some, and think you will approve of both changes and additions when you see the printed lecture. Today it was given. I had been in the house for 3 days previously, permitting my beard to grow and altogether when I had the matter complete I looked very much like a scoundrel with my unwashed and unshaven face.—I felt clear in head and not cowardly at heart when I entered the theatre. I acquitted myself to the satisfaction of every body—the bursts of applause were frequent throughout, and at the end the applause was long and loud—A kind of procession walked past me afterwards and both gentleman and ladies, quite unknown to me shook me by the hand. The audience was a fine one, included many men of eminence—Many Oxford men were present and Faraday told me that they frowned and looked gruff at the commencement but that every man of them melted before the lecture was concluded. Edmondson would be proud if he heard the people enquiring about Queenwood afterwards, I was glad to see this, for indeed my experience at Queenwood was a great help to me—I had the boys and the playground for some time before my minds eye and spoke, I doubt not, with the feeling which my affection for the little fellows prompted. It is a comfort to me to feel that I did not disappoint my friends.

Some however in shaking hands with me said I was two centuries in advance of my time and I replied that Gallileo[2] was the same. It is wonderful to me how things seem to have been arranged for me since I came to London—Here for example was an opportunity which never occurred before and may never occur again and which enabled me to to write my own testimonial in a manner somewhat congenial to me—Coupling it with the success of my Friday evenings[3] I am sure if I had the arrangement of the circumstances of my life I could not have managed matters half so well. In the printed lecture I will try and introduce the extract from your journal[4] in the form of a note, although I did not read the extract, but I gave Faraday the lecture to read beforehand and although your portion was crossed out he managed to read it and it was he suggested that it ought to be appended as a note. He evidently was highly pleased with it, and will probably shake your hand with increased cordiality the time he sees you next. Lyell[5] was present and other men of similar calibre. On the whole, though physically very weak from fasting and confinement, I got through my task with vigour. I had studied Emerson's lecture on Plato[6] in the forenoon of the day, and entered the room in some measure imbued with the spirit of the man, who of all that I have read of comes nearest to the expression of my constitution. I had a cup of tea with Faraday afterwards, and the affectionate light of his clear eye was as grateful to me as anything else. I know all this will interest you for otherwise I would not write it.

You sent me a letter from Phillips[7] on Sunday, and with a scrap from yourself—thinking I suppose that January's letter[8] would be an equivalent for the weekly epistle which I have a right to expect from you. I am not disposed to accept any such equivalent, but I suppose you were busy like myself and compelled to make your letter brief. With regard to January I dont know what to say—I believe he is improving but the letter is so very different from what I should write myself that I hardly know what to think of it—On the whole I am of opinion that it would be a greater service to him to leave him to his resources than to send him the money—I fear he has got money in this way too easily and that it has done him far more harm than good. If I were convinced of his stern probity in such matters I should not hesitate an instant to send him the money myself, but as yet he has given no sufficient proof of his amendment and until he does I would not lend him money. He must pay the penalty of his own folly, and learn, if need be by bitter experience, that his friends require honourable conduct at his hands—I have been forced before now to cast away friends for whom I entertained a warm affection purely because their want of probity destroyed all affinity between me and them. I hate that expression of his that you are to consider the matters he mentions to you 'sacred'—He said the same to me when he wanted to borrow ten pounds from me. I hate it because it is unmanly and shows a disposition to live under false appearances. This is all I have to say about January, and I know that in the main your thoughts will coincide with mine. I cannot conclude however without one word of admiration for little Smith,[9] who although he knows his man has been willing to incur such risque to serve him—this last sentence may seem inconsistent with what I have said before, but I know you will not deem it necessary for me to reconcile both, as you will do it yourself without my aid.

If you would be kind enough to hand over ten shillings as my subscription to the Queenwood Cricket Club[10] I will pay you when I come down—give the money to Charles Cumming and he will hand it to the Treasurer.[11]

Your affectionate | John

RI MS JT/1/T/589
RI MS JT/1/HTYP/341–342

1. *my lecture*: Tyndall's Lecture on Education (see letters 0900, n. 6; 0924; and 0933).
2. *Gallileo*: Galileo Galilei (1564–1642), the Italian astronomer, physicist, and mathematician, who was famous for his telescopic observations of sunspots and the phases of Venus and Jupiter's moons. Tyndall would later argue in his 1874 'Belfast Address' as the President of the BAAS that Galileo's forced renunciation of the Copernican system represented just one of many historical confrontations between scientific progress and theological dogmatism.
3. *my Friday evenings*: reference to Tyndall's Friday evening lectures at the RI.
4. *the extract from your journal*: Hirst included an extract from his journal in his letter of 23 April (see letter 0924). Tyndall's lecture was published later in 1854 (see letter 0900, n. 6).
5. *Lyell*: see letter 0882, n. 11.
6. *Emerson's lecture on Plato*: see letter 0855, n. 5.
7. *a letter from Phillips*: letter missing.
8. *January's letter*: 'January' was the pen name of George Searle Phillips (see also letter 0759, n. 15). The letter mentioned here is the same as n. 7.
9. *little Smith*: John Stores Smith (1828–1892/3), a young literary man and disciple of Carlyle. He was introduced to Hirst by George Searle Phillips (see n. 8). Smith was educated at the Manchester Grammar School and was the author of *Mirabeau: A Life History*.
10. *my subscription to the Queenwood Cricket Club*: Charles Cumming had asked Tyndall whether he would join as an honourable member of the Queenwood Cricket Club by becoming a patron or donor (see letter 0927).
11. *the Treasurer*: likely Mr. Taylor (see letter 0927, n. 1).

From Charles Buxton 30 May 1854 0934

Brewery[1] | Spitalfields[2] | May 30/54

Dear Sir.

I had the pleasure of hearing your admirable lecture[3] on Sat[urda]y & was much struck by the account you gave of your mode of teaching the boys. We have been trying in vain to find a schoolmaster here[4] for our workmen's sons, of just the kind you described—a man who would excite feelings of 'wonder & delight' in their minds, by stimulating them to reflect upon the laws of nature, & men. Marvelous work: & it occurred to me that possibly at the agricultural college you were at,[5] or elsewhere, you might have met with some young man, (or perhaps taught one) who would engage in the work of teaching in that spirit. I enclose our prospectus to show the kind of school. The salary <u>to begin with</u> would be £80: but for a first rate teacher we would raise it. Pray forgive my troubling you—the cause *[1–2 words illeg]* my excuse.

Your faithful listener | Charles Buxton[6]

RI MS JT/1/B/169

1. *Brewery*: the well-known brewery of Truman, Hanbury, Buxton & Co., of which Buxton was a partner. The brewery was located on Brick Lane, Spitalfields, London (*ODNB*).
2. *Spitalfields*: a parish located partly in Central London and partly in the East End of London.
3. *your admirable lecture*: Tyndall's lecture on education delivered on 27 May (see letters 0900, n. 6; 0924; and 0933).
4. *schoolmaster here*: not identified, but presumed to be a local school at or near the Brewery.
5. *the agricultural college you were at*: Queenwood College.
6. *Charles Buxton*: Charles Buxton (1822–71), a politician with an interest in educational reform. Politically, he was considered to be an independent Liberal (*ODNB*). Buxton was the son of Sir Thomas Fowell Buxton (1786–1845), first baronet, a philanthropist, and Member of Parliament.

To Thomas Archer Hirst [5 June 1854][1] 0935

Monday night.

My dear Tom,

I ought to have written to you today. You would then receive my letter tomorrow morning, but I have been hard at work in the laboratory till 7 ½ oC[lock] preparing for my last lecture[2]—which thank the gods will be given tomorrow. The said gods are good to me—for two days ago I found a pretty thing Connected with the subject of tomorrows lecture which I know will interest the audience and enable me to bid them a graceful goodbye. You have I suppose by this time received the tube for Savart's experiments[3]—at least I gave directions last week to have it sent to you—I must endeavour to run down to you in time for the Concert[4]—For I should feel a kind of blank in my brain if I permitted the boys to scatter without seeing them. I must if need be come up again to Hodgson's lecture[5]—We had an excellent lecture on Saturday last on physiology.[6] The man talked without notes for an hour and a quarter and I believe delighted his audience. The reporter of the morning Chronicle[7] made use in his report[8] of my lecture[9] attack Latham the ethnologist,[10] and Latham writes to me today sending me his lecture[11] in order to permit me to modify my statement—My disclaimer however fortunately appears in the Chronicle on Thursday last.[12] I thought you would have seen the Chronicle for otherwise I should have been tempted to send it to you—though the matter is really not worth the reading. Whewells and Faraday's lectures[13] are already published. I received copies of them today from the publishers—they appear singly and are well worth reading. I spend tomorrow evening with the eldest son[14] of Chief Baron Pollock[15]—a man I like well and who was drawn towards me by a kind of natural gravitation. Now I have discharged all my gossip—I will go down to you on Saturday in all probability. Frankland gave us a lecture on Friday last[16]—goodbye. How is Derichlet's[17] name spelt?

Affectionatly John

RI MS JT/1/T/590

---

1. *5 June 1854*: although Hirst wrote 'June 6' at the top of this letter, suggesting when he had received it, the letter must have been written on Monday, 5 June. Tyndall indicated that he wrote the letter on Monday evening, the evening before his lecture. Tyndall delivered his final lecture on 6 June (Journal, 6 June 1854, JT/2/13b/677).
2. *my last lecture*: Tyndall had commenced a series of seven lectures on 'Ebullition, Combustion, and other Phenomena of Heat' at the RI on 25 April 1854 ('General Monthly Meeting', *Roy. Inst. Proc.*, vol. 1, p. 420). This was the concluding lecture (see n. 1).
3. *tube for Savart's experiments*: likely refers to an experiment used to demonstrate the Biot-Savart law relating to the magnetic field generated by an electric current, or 'that the magnetic flux density near a long straight conductor is inversely proportional to distance from the conductor and directly proportional to the current in the conductor' (*OED*). This law was first identified by Félix Savart (1791–1841) and Jean-Baptiste Biot (1774–1862).
4. *the Concert*: a concert at Queenwood College that took place on 12 June (H. Smithers, 'Miscellaneous', *Queenwood Observer* 2:4 [31 May 1854], pp. 25–32, on p. 26). Tyndall noted his attendance in his journal stating, 'a large number of people were present, it was very successful' (Journal, JT/2/13b/677).

5. *Hodgson's lecture*: W. B. Hodgson gave the seventh and final RI Lecture on Education on 10 June, 'On the Importance of the Study of Economic Science', in *Lectures on Education* (London: Parker, 1854), pp. 261–316.
6. *an excellent lecture . . . on physiology*: reference to the sixth RI Lecture on Education on 3 June by James Paget, 'On the Importance of the Study of Physiology', in *Lectures on Education* (London: Parker, 1854), pp. 215–60. Sir James Paget (1814–99), first baronet, was a surgeon notable as a founder of medical pathology (*ODNB*). At this time, he was assistant surgeon at St. Bartholomew's Hospital and had recently published *Lectures in Surgical Pathology, Delivered at the Royal College of Surgeons of England* (London: Longman, Brown, Green, and Longmans, 1853). He was made a baronet in 1871 (*ODNB*).
7. *The reporter of the morning Chronicle*: not identified. *The Morning Chronicle* was founded in 1770 by William Woodfall (1739–1805) under the name *The Morning Chronicle and Daily Advertiser* ('The Morning Chronicle', in L. Brake and M. Demoor [eds.], *Dictionary of Nineteenth-century Journalism in Great Britain and Ireland* [Gent and London: Academia Press and the British Library', pp. 426–27).
8. *his report*: see 'Royal Institution', *The Morning Chronicle*, 29 May 1854, p. 5. For Tyndall's response to the report see n. 12.
9. *my lecture*: Tyndall's installment in the series of lectures on education presented at the RI, although Latham was not explicitly mentioned in the lecture (see letters 0900, n. 6 and 0933); also see n. 13.
10. *attack Latham the ethnologist*: Robert Gordon Latham (1812–88), an English ethnologist, physician, and philologist notable for his works *The English Language* (1841) and *The Natural History of the Varieties of Man* (1850) (*ODNB*). Tyndall's response to the report included a quote from the initial report stating, 'In conclusion, Dr. Tyndall expressed himself warmly relative to the slighting manner in which physical science had been spoken of by Dr. Latham, in the third lecture of the course' ('Mr. Tyndall's Lecture At The Royal Institution', *The Morning Chronicle*, 1 June 1854, p. 3).
11. *his lecture*: Latham's lecture on education, which was part of the series on education presented at the RI (see n. 10). See R. G. Latham, 'On the Importance of the Study of Language as a branch of Education for all Classes', *Lectures on Education* (London: J. W. Parker & Son, 1854), pp. 91–113.
12. *My disclaimer . . . Thursday last*: in his letter to *The Morning Chronicle*, Tyndall emphasized that he had neither knowledge of, or thought about, Latham while conducting his own lecture. Nor had Tyndall referred to Latham in his own lecture and he pointed readers to the soon to be published copy of his own lecture for further clarification on his position on the physical sciences ('Mr. Tyndall's Lecture At The Royal Institution', *The Morning Chronicle*, 1 June 1854, p. 3).
13. *Whewells and Faraday's lectures*: Whewell's and Faraday's lectures on education. See W. Whewell, 'Influence of the History of Science Upon Intellectual Education' and M. Faraday, 'Observations on Mental Education', *Lectures on Education* (London: J. W. Parker & Son, 1854), pp. 3–36; 39–88. The lectures also must have been published as single pamphlets.

14. *the eldest son*: Sir Frederick Pollock (1845–1937), third baronet. Pollock was an English jurist and well-known legal author (*OED*).
15. *Chief Baron Pollock*: Sir William Frederick Pollock (1815–88), second baronet and an English lawyer and author.
16. *Frankland gave us a lecture on Friday last*: see E. Frankland, 'On the dependence of the Chemical properties of Compounds upon the Electrical character of their constituents', *Roy. Inst. Proc.*, vol. 1, pp. 451–54.
17. *Derichlet's*: misspelling of Peter Gustav Lejeune Dirichlet.

To Thomas Archer Hirst 7 June 1854 0936

Royal Institution | 7th. June 1854.

My dear Tom

Lest I should forget it I will draw your attention to a very common and therefore very serious misconception under which Smithers appears to have laboured when he reported your lecture upon Barker's mill.[1] He speaks of the issuing water striking against the air and driving the arms round. Now the air you know has nothing to do in the matter and the instrument would rotate if it were placed in a vacuum. Of course the force of the issuing jet, whatever that may be, is exerted in the opposite direction to the jet's motion and this it is which produces the motion.

Thine | John

RI MS JT/1/HTYP/343
LT Typescript Only

---

1. *your lecture upon Barker's mill*: H. Smithers, 'Local Information', *The Queenwood Observer*, 2:4 (31 May 1854), p. 25. Barker's mill was 'a mechanical contrivance for producing rotary motion, consisting of a hollow vertical revolving axis with two (or more) horizontal arms fitted into its lower end' invented during the late seventeenth century (*OED*).

To Thomas Archer Hirst 10 June 1854 0937

Royal Institution Friday | June 10th. 1854

My dear Tom,

I have gone to Newman's[1] and have been assured that the tube has been sent away.[2]

I found it good to put a little grease round the rim of the disk so as to ensure good contact with the tube—not sufficient however to permit any of the grease to protrude itself at the bottom of the tube

I. Stop the orifice with a little peg of wood—fill your tube—let it rest a little and then withdraw the peg.
II. Let the water fall upon an inclined board or two so as to extinguish the noise of the fall. If you allow it to fall into a vessel of water the bubbling interferes greatly with the sounds and almost masks it.
III. Let the jet issue into water. Place a beaker of water underneath the tube end—so that the latter just dips beneath the surface of the water in the beaker. The sound—slightly lowered in pitch is heard as before.

Your note[3] has rather unsettled me and therefore I give these practical hints just to save your time for it is just possible that I may not be present.[4] With the idea of going I had connected a number of other circumstances. These your note informs me are to be changed and I am inert and find the sudden transition difficult.[5]

I am very much obliged to both Mr and Mrs Wright.[6]

Your affectionate | John

RI MS JT/1/HTYP/344
LT Typescript Only

---

1. *Newman's*: see letter 0717, n. 3.
2. *assured that the tube has been sent away*: Tyndall had arranged for a Savart tube to be sent to Hirst to use in his experiments; he must not have received it yet (see letter 0935, n. 3).
3. *your note*: letter missing.
4. *may not be present*: on 6 June, Tyndall stated that he would visit Hirst at Queenwood in time for a concert on 12 June (see letter 0935, n. 4).
5. *transition difficult*: below this line appear 12–15 illegible words handwritten by LT that have been crossed out.
6. *Mr and Mrs Wright*: Tyndall would stay with the Wrights during his stay at Queenwood College (see letter 0939, n. 1).

From William Hopkins [12] June [1854][1] 0938

Cambridge | June

My dear Sir,

To-morrow I set out for Manchester[2] whence I hope to bring back a supply of specimens[3] for you to try their conductive powers whenever you may feel at leisure to make the experiments. I shall return in about ten days or a fortnight.

You ask me whether Prof. Thompson's[4] results on the effect of pressure in resisting solidification of water are not opposed to those which I have obtained with other substances in which the solidification is promoted by pressure.[5] In the one case the temperature of fusion[6] is lowered in the other it is elevated. According to the explanation which the dynamical theory of heat[7] affords of the phenomena, such ought to be the case, because water expands in freezing, whereas the substances I have experimented on contract in becoming solid. The general explanation appears to be something of this kind. Suppose a substance to expand in passing from the solid into the fluid state. Conceive it to be enclosed in a given space and subjected on all sides to a uniform pressure; and suppose moreover that its temperature is such that it is just on the point of liquefaction. Conceive the temperature to be elevated by a small quantity. The substance will begin to melt, and if a sufficient quantity of heat were constantly supplied to it to maintain this increased temperature, the liquefaction would go on till the whole mass became liquid, supposing always the process to be affected by nothing but the temperature. But if we suppose the envelope enclosing the substance to be perfectly rigid so that the space occupied by the substance remains invariable the pressure on all sides will be increased by the tendency of the substances to expand in liquifying. Now the theory tells us that if the volume of any substance be kept constant while its temperature is increased by a small amount, the corresponding increase of pressure cannot exceed a small corresponding amount. Consequently if the process of liquefaction under the conditions above mentioned tends to increase the pressure more than this small amount assigned by the theory, the process will be arrested, while it would necessarily have proceeded in consequence of the supposed increased temperature, had not the pressure been altered. This shows that the temperature of fusion must be increased by pressure when the substance expands in becoming fluid; and similar reasoning would show that that temperature will be diminished by pressure when the substance contracts in becoming fluid. Accurate investigation gives the amounts of the effect thus produced by any assigned pressure in a prepared substance. One object of my experiments is to compare the results of theory and experiment in such substances as are sufficiently tractable to be experimented on.[8]

Believe me | Yours very truly | W. Hopkins.

RI MS JT/1/TYP/2/607–608
Typescript Only

1. *[12] June [1854]*: it looks as if the original typescript stated 'June 5th '56', but '5th 56' was crossed out and replaced, in pencil with '12, 56'. And then, possibly later, '54?' was added as a possibility. The date 12 June 1854 seems the most likely as this letter follows from a previous letter concerning the conductive power of certain specimens that was dated 6 May (see letter 0929).
2. *set out for Manchester*: Hopkins travelled to Manchester where he was assisted by James

Prescott Joule (1818–89) and William Fairbairn (1789–1874) to conduct experiments on the effects of pressure on the melting point of substances (*ODNB*) (see also letter 0862).

3. *supply of specimens*: these are likely the 'hard crystalline rocks' that Hopkins refers to on 6 May (see letter 0929).
4. *Prof. Thompson's*: misspelling of William Thomson, Lord Kelvin.
5. *those which I have obtained . . . promoted by pressure*: Hopkins was involved in research on the effects of pressure on the melting points and conductivities of solids (see n. 2). For further explanation see C. Smith, 'William Hopkins and the Shaping of Dynamical Geology: 1830–1860', *British Society for the History of Science* 22:1 (March 1989), pp. 27–52, on pp. 39–40.
6. *temperature of fusion*: this refers to the temperature at which a solid becomes a liquid or a liquid becomes a solid.
7. *dynamical theory of heat*: this is a foundational theory of thermodynamics advanced by William Thomson, based largely on the experiments of James Joule concerning the mechanical action of heat (W. Thomson, 'On the Dynamical Theory of Heat, with numerical results deduced from Mr. Joule's equivalent of a Thermal Unit, and M. Regnault's *Observations on Steam*', *Phil. Mag.* 22 [1852], pp. 8–21) (see also letter 0808, n. 5).
8. *One object of my experiments . . . to be experimented on*: see n. 5 and, for example, W. Hopkins 'An account of some experiments on the effect of pressure on the temperature of fusion of different substances', *Brit. Assoc. Rep.* 24 (1854), pp. 57–58. For more detail see C. Smith and M. Norton Wise, *Energy and Empire: A Biographical Study of Lord Kelvin* (Cambridge: Cambridge University Press, 1989), pp. 552–59.

To Thomas Archer Hirst 17 June 1854 0939

Friday June 17th. 1854[1]

My dear Tom

I know Mr Tidmarsh's[2] family will be glad to see you. They live I think on Merchants Quay Dublin. The number I forget. Should you call remember me kindly to them. This dismal weather has checked my advance into the New Forest.[3]

Yesterday I absolutely smoked three cigars to lift the weight of time from my shoulders. It is not yet noon, the rain drizzles drearily outside and I have already completed my first cigar. As long as I remain idle I can bear a large amount of fumigation. Wright is off this morning to Lincoln; he received your letter before he went: and no doubt will commence his journey all the more gladly on account of it. Though I believe he will not find it necessary to avail himself of your offer. Debus is pottering amid his bottles, and the ladies are all in bed.

Your affectionate | John

RI MS JT/1/HTYP/344
LT Typescript Only

1. *My dear Tom*: LT notes to the left of the date: 'to Hirst in Ireland written at Q[ueen]wood prob[abl]y staying with the Wrights'.
2. *Mr Tidmarsh's*: John Tidmarsh (see letter 0873, n. 26).
3. *New Forest*: New Forest, Hampshire, an area of southern England that includes a large tract of unenclosed pastureland and heavily wooded forest. It was designated a royal forest and hunting ground by William the Conqueror in 1079, became a protected area under the New Forest Act of 1877, and became a National Park in 2005.

From Heinrich Gustav Magnus 20 June 1854 0940

Berlin 20 Juni 54.

Theurer Freund

Leider ist das Polytrop für M^r^. Gassiot noch nicht fertig und es werden noch 3 Wochen darüber hingehen bis ich es erhalte. Brauchen Sie den Treppensteiger und die /Mäuse/ früher so sende ich Ihnen solche durch Asher direkt. Ich werde sie jedenfalls von Reichel kommen lassen, damit ich sie sogleich kann verpacken laßen sobald ich Ihre Antwort erhalte.—/Sicher/ wissen Sie wie Sie sich Rath schaffen können. Dabei wird das Unternehmen gewiß vortheilhaft in pecuniaerer Beziehung, und erhält Sie zugleich in Vertrautheit mit der ganzen Physik.

Lassen Sie mich bald hören ob Sie sich zu dem Unternehmen entschloßen. Nur rathe ich Ihnen es nicht zu weitläufig zu machen.

Von Herzen | Ihr | GMagnus.

P.S. Entschuldigen Sie bei M^r^. Gassiot das lange Ausbleiben des Polytrop ich habe mich so darüber mit Barreaud entzweit daß ich schwerlich wieder etwas werde bei ihm arbeiten lassen. Doch weiß ich es nicht fertig zu schaffen.

Berlin 20 June 54.

Dear friend,

I regret that the polytrope[1] for Mr. Gassiot is not finished yet and another 3 weeks will pass until I receive it.[2] If you need the stair climber and the /mice/ sooner, then I shall send them to you directly through Asher. I shall send for them from Reichel[3] in any case, so that I can have them packed at once as soon as I receive your reply.—You will surely know how you can be shown what is to be done. The undertaking[4] will certainly become advantageous in a pecuniary respect as well, and will keep you familiar with the whole of physics at the same time.

Let me hear soon whether you have resolved upon the undertaking. I shall only advise you not to make it too extensive.

Sincerely | Yours | GMagnus

P.S. Apologise to Mr. Gassiot for the long absence of the polytrope, I have so fallen out with Barreaud about it, that I shall hardly have any work done by him again. But I do not know how to get it finished.

RI MS JT/1/M/11

---

1. *polytrope*: see letter 0870, n. 7.
2. *until I receive it*: Magnus discussed ordering the polytrope for Gassiot in letter 0925, n. 2.
3. *Reichel*: possibly refers to C. Reichel, a mechanic located at 65 Alte Jakobstr in Berlin ('1412. Reichel, C.', in Commissioners of the Zollverein-Governments [ed.], London Exhibition 1862, *Special Catalogue of the Zollverein-Department* [Berlin: R. Decker, 1862], p. 80).
4. *the undertaking*: not identified.

From John Herschel 24 June 1854 0941

Sir J. Herschel presents his compliments to Prof. Tyndall and begs to thank him for his interesting Lecture on the Study of Physics as a branch of Education[1]

32 Harley Street. | June 24/54

RI MS JT/1/H/91
RI MS JT/1/TYP/2/498

---

1. *branch of Education*: see letters 0900, n. 6 and 0933.

From Thomas Archer Hirst 24 June 1854 0942

Limerick | Saturday, June 24th/54

My dear John,

Immediately after posting my last letter[1] at Cork I received yours.[2] After a hard weeks work I have now arrived at Limerick and in an hour or two shall start by train to Dublin. We have had rain everyday since I started I am sorry to say, for through it I am now suffering from a severe cold which forces me to hasten to Dublin and remain quiet for a day or two. On Thursday morning I started from my Inn to ascend Mangerton[3] one of the highest mountains about Killarney.[4] It was rather hard work and I perspired very much; as usual on the top it was very cold and foggy for the mountain was covered with clouds. The cold drizzling rain wet me through. The toil of descending, however, again warmed me and dried my clothes so that I did not change them and anticipated no evil results, for in Switzerland I have often done the same thing. Next day I rode in a jaunting car from Killarney to Tarbert[5] through Tralee;[6] it rained all day, wet me through and through the result of all which is that today I am worth nothing. I got up about noon after a miserable night's unrest, during which I did nothing but sneeze and grow alternately very hot & very cold. With the exception of this furious cold, however, I have profited much from my journey. My face is no longer of a pale yellow colour, but a healthy brown, and my coat is already more difficult to button than it was. The country between Cork and Killarney fully satisfied my expectations it is wild and in many places grand whilst near Killarney there are some of the sweetest little spots in the world. I did not go direct from Cork to Killarney but went round by Inchegula,[7] Gougane Barra,[8] Glengariff[9] and Kenmare. I stopped a night at Glengariff and was delighted with it. It stands at the head of a quiet bay which opens into the celebrated Bantry Bay.[10] From whatever side you approach it you have to pass through wild barren scenery which reminds one of Grimsel,[11] high hills and deep valleys where nothing but bare rocks and bog will grow. Whereas Glengariff is rich and full of beautiful trees. One great disadvantage of Irish scenery is that you are never left for many minutes in quiet contemplation of it before you are disturbed by some poor beggar. They will follow you for miles in order to wring a halfpenny from you by making themselves nuisances. If a penny is thrown to them they tear one anothers eyes out to get it. When you descend from a coach or steamer you are nearly devoured by carmen.[12] The only way to escape is to seize your own trunk and with it to knock the first carman down. If you talk to any of them you find they are saving money to go to America. Fellows who scream for a penny have changed me half a crown when necessary. But good Bye write soon to the Post Office Dublin.

Yours affectionately | T.A. Hirst

RI MS JT/1/H/199
RI MS JT/1/HTYP/345

1. *last letter*: letter missing.
2. *I received yours*: letter 0939.
3. *Mangerton*: a mountain in County Kerry, southwest Ireland.
4. *Killarney*: a popular tourist destination in County Kerry, Ireland.
5. *Tarbert*: a coastal town in County Kerry, Ireland.

6. *Tralee*: a town in County Kerry, Ireland located nearly halfway between Tralee and Tarbert.
7. *Inchegula*: misspelling of Inchigeelagh, a small village in County Cork, Ireland near the border of County Kerry, Ireland.
8. *Gougane Barra*: a settlement in County Cork, Ireland known for its monastic ruins.
9. *Glengariff*: a village in County Cork, Ireland situated on Bantry Bay.
10. *Bantry Bay*: a bay with a long inlet in southwest Ireland.
11. *Grimsel*: the Grimsel Pass is a high mountain pass in Switzerland, crossing the Bernese Alps that connect Bern to the north and Valais to the south.
12. *carmen*: carmen were often employed by railway companies to make local deliveries of goods and parcels with horse-drawn vehicles.

To Michael Faraday 25 June 1854 0943

25th, June, 1854.

My dear Prof Faraday

I was very sorry to find on coming to the Institution on Monday last that you had departed an hour before. I had been in town for two or three days previous, and once called at the Institution: But I had got wet in the country, sat in a draught on my way to town and done sundry other imprudent things: these brought on a cold and confined me to the house for two whole days.[1] When I called at the Institution it was late and I was ill, for otherwise I should have intruded my bodily presence upon you. I am now at work, and as usual sadly bewildered—I know nothing of magnetism—The experiments which every body seems to understand are those which puzzle me most—At least I find the accepted theories of magnetic action no refuge at present. Well patience is sure to bring something out of the bewilderment at least. Magnus informs me that he has sent you a paper on black sulphur[2] and that some specimens of the substance are also on the way to you. He obtains the black sulphur by melting the substance over and over again—perhaps 20 times—and cooling it very quickly each time. But I dare say you will be glad to forget science for the present so will not trouble you with it further.

Kind remembrance to Mrs Faraday

Believe me | Most sincerely Yours | John Tyndall

RI MS JT/TYP/12/4025
Faraday Correspondence 4:2858

---

1. *got wet in the country... two whole days*: in his journal Tyndall noted the bad weather and his catching of a cold 'which kept me in the house all day' (Journal, 10 June, JT/2/13b/678).

2. *paper on black sulphur*: G. Magnus, 'Ueber rothen un schwarzen Schwefel', *Poggend. Annal.* 92 (June 1853), pp. 308–24. It was translated by Hirst (see letters 0948, n. 4 and 0961, n. 7) and published in English as 'On Red and Black Sulphur', *Phil. Mag.* 8:51 (September 1854), pp. 177–86.

From James B. Bevington 26 June 1854 0944

Wandsworth Common | 26 June 1854

Dear John.

I received my accustomed & most welcome invitation to join the party that meet annually at Birch Wood.[1] The father of the Club my old friend Edward Newman[2] expressed a hope that you would join the party and, I can assure you it will give me increased pleasure if I can count upon you as companion for the day.

I have forgot, the manner of your disposing yourself about the month of July as the meeting is to take place on the 12th, will you take the matter into consideration & let me know. If you can join me I will arrange the mode of travelling.

Henry Hill tells me that Hirst is wandering through Ireland & Scotland[3]—Do you think he will return in time.[4] I think it a good practice that we should have some annual meeting & if this does not fit all parties we must seek some other opportunity.

My Mary & boys[5] are all well for which I thank God. I am getting an old kermudgeon & want some chastisement.

Commend me to the kind Hirst when you write & believe me very affectionately

James B Bevington

Dr. John Tyndall | Royal Institution

RI MS JT/1/B/94

---

1. *Birch Wood*: the Entomological Club held an annual field day and social dinner at Birch Wood in Kent County in July (F. Smith, 'Captures on the Birch-Wood Day', in E. Newman [ed.], *The Entomologist*, vol. 2 [Bishopsgate: Edward Newman, 1964–65], p. 210). Not further identified.
2. *Newman*: Edward Newman (1801–76), leading London naturalist who specialized in entomology and botany. Together with four others in 1826, Newman established the Entomological Club. In 1840 he established and edited *The Entomologist* and its successor *The Zoologist* until 1863. He was a notable opponent of Darwinism and used his journals as vehicles for spreading anti-evolutionism (*ODNB*).

3. *Ireland & Scotland*: for Hirst's accounts of his travels see letters 0942; 0949; and 0953.
4. *return in time*: Hirst suggested that 12 July may be too early for such a meeting (see letter 0949).
5. *My Mary & boys*: Bevington's wife and sons. Not further identified.

To Thomas Archer Hirst [26 June 1854][1] 0945

Royal Institution Monday

My dear Tom,

I wrote to you to Kilkenny as you suggested but you will probably never get my letter.[2] I am very sorry to hear of your being ill. We have had glorious summer weather here for the last week. *[But]* the week I spent at Queenwood was dismal. I also got cold which kept me at home for 2 days. I had a letter from Mrs Coxe,[3] Alexander Cummings' aunt, a few days ago, containing a beautiful and expensive purse which she had worked for me herself.[4] If you see her you must warn her that there is some danger of my being made a spoiled child. I gave you a description of the scene of my boyhood[5] but I should be sorry if you went there in bad weather for the rain spoils all the poetry.[6] We will make the tour together at a future day. Indeed I have <u>often looked forward</u> to this. You know nothing of what the carmen of Ireland used to be.[7] They are now sucking doves[8] in comparison. Time was when you and I had no voice in the choice of your own carman. The fellows would toss up for it and you must abide by the result. I hope when you write to me again it will be to tell me that the cold is vanished. I am working in my den before Faraday's big magnet. I shall soon set off for Wales with Francis, and there I trust to become a little broader in my transverse section also.

Present my kind remembrances to Mr and Mrs Simpson and to Miss Martin if you see her. I feel half inclined to run over to Dublin—intercept you there and go to the Causeway along with you. I was yesterday rambling through the groves of Richmond and Kew, and feel all the better for it to day. Paget[9]—one of the lecturers in the educational course,[10] came to me on Friday and asked me to accept a lectureship[11] at St. Bartholmew's Hospital.[12] I scarcely think they will make it worth accepting. As yet however I have said neither yes nor no. Faraday is in the Isle of Wight[13] and I must write to him first. I am strongly of opinion that Alick Cumming might to be off to Germany soon. What say you? If you think the same I will write to his aunt about it.[14]

Affectionately yours | John

RI MS JT/1/HTYP/346
LT Typescript Only

1. *[26 June 1854]*: as this was written on a 'Monday' and refers to the previous Friday, which was 23 June, the letter was presumably written on 26 June.
2. *never get my letter*: letter missing. Hirst did receive it according to his letter to Tyndall on 29 June (letter 0949).
3. *letter from Mrs Coxe*: letter missing.
4. *purse . . . herself*: Tyndall referred to the 'beautiful purse' that May Anne Coxe sent to him in his journal entry of 25 June (Journal, JT/2/13b/678).
5. *description of my boyhood*: see n. 2.
6. *spoil all the poetry*: LT noted an alternate phrasing for the end of this sentence in the margin: 'scene of my boyhood rain w spoil all the poetry'.
7. *carmen of Ireland used to be*: Hirst referred to being 'nearly devoured by carmen' on 24 June (see letter 0942, n. 12).
8. *sucking doves*: refers to a bird still with its mother, but was used figuratively to refer to someone gentle and soft-spoken such as in Carlyle's *French Revolution*: 'Some loud as the lion; some small as the sucking dove' (*OED*).
9. *Paget*: see letter 0935, n. 6.
10. *educational course*: Paget gave one of the RI's education lectures (see letter 0900, n. 6) that was published as 'On the Importance of the Study of Physiology as a Branch of Education for All Classes', *Lectures on Education* (London: J. W. Parker & Son, 1854), pp. 213–60.
11. *lectureship*: Tyndall referred to Paget's offer of a lectureship 'On Friday last the 23rd' of June in his journal stating that 'I think they want too much work and offer too little money, it is not at all likely that I shall accept it' (Journal, 10 June 1854, JT/2/13b/678).
12. *St. Bartholmew's Hospital*: misspelling of St. Bartholomew's Hospital (often called Barts), the oldest hospital in Britain. Founded in 1123, it became a medical school in the eighteenth century and was formally recognized in 1822 ('Our History', *St. Bartholomew's History Timeline*, 2016, at http://bartshealth.nhs.uk/our-hospitals/st-bartholomew's-hospital/our-history/ [accessed 26 August 2016]).
13. *Isle of Wight*: England's largest island, which is located in the English Channel just off the coast of Hampshire. The Isle of Wight became a fashionable resort area after Queen Victoria and Prince Albert built Osborne House there, which became their private family home in 1851.
14. *write to his aunt about it*: Tyndall would write to James Coxe (letter 0993) about sending Cumming to Germany so that he could get a better education than he would receive at Queenwood.

To Thomas Archer Hirst 27 June 1854 0946

Royal Institution | 27th June 1854

My dear Tom,

Here is a note[1] from our good friend Bevington. Just say whether it is likely that you will be home in time to go to Birchwood[2] *[dinner]*. The meeting is on the 12th of July.[3] I fear you will be away among the Orangemen of the north[4] at the time.

Affectionately John

RI MS JT/1/HTYP/346
LT Typescript Only

---

1. *Here is a note*: letter 0944. It was either enclosed or copied and sent to Hirst along with this letter.
2. *Birchwood*: see letter 0944, n. 1.
3. *The meeting 12th of July*: the meeting for the Entomological Club was scheduled for 12 July 1854 (see letter 0944).
4. *Orangemen of the north*: the Orange Order was a political organization that promoted Protestant and Loyalist ideas; it was particularly active in Northern Ireland and parts of Scotland (*OED*).

From Margaret Allen 27 June 1854 0947

Charnley House[1] | Freehold Park E. Fulwood[2] | June 27th / 54

My dear Tyndall,

I am afraid that you will consider me extremely neglectful in not acknowledging your kindness in sending me the lecture on Education[3]—but you must excuse me, I cannot tell you how much instructed and delighted I have been in its perusal, & tho as yet its contents have under gone but one reading be assured it will not be laid upon the shelf as one of those everyday pamphlets with which the public are daily supplied.

I suppose you are now enjoying some little recreation after your laborious duties, I trust you will employ your leisure in the recovery & strengthening of your health. All your attainments will be of little value to yourself & the world if you do not take care of yourself & why shorten your existence. Why have the world more ignorant than it might be by not using proper precautions for the preserving of your bodily strength, thus enabling you to become a great blessing to the world. I am hoping to see you in September,[4] by all means you must remain two or three days with us, longer if you can. Mamma[5] says I am to tell you that you are not to stay at M^r^. Hebden's;[6] now you will remember. I think you will be much pleased with our new residence, it is a very agreeable change from Stanley Terrace & that place was not to be despised. I had a delightful note from Dr. Hirst[7] a short time ago what a fine noble fellow he is, proud indeed you may be of your 'son'.[8] I hope all your adopted children may prove so worthy of such a Father.

I am | Yours ever sincerely | Gretchen[9] | Dr. Tyndall.

RI MS JT/1/A/78

1. *Charnley House*: a townhome located at 13 Winckley Square, Preston, Lancashire, England ('Charnley House', *Charnley House*, https://historicengland.org.uk/listing/the-list/list-entry/1279781 [accessed 15 October 2016]).
2. *E. Fulwood*: a district of Preston, Lancashire, England where Margaret Allen lived.
3. *sending me the lecture on Education*: letter missing. The lecture Allen refers to was Tyndall's 'On the Importance of the Study of Physics as a Branch of Education' delivered at the RI on 27 May (see letters 0900, n. 6 and 0933).
4. *see you in September*: Tyndall visited Allen on 24 September, finding himself 'cordially received by her parents and sisters' (Journal, 24 September 1854, JT/2/13b/692). See also letters 0967 and 0972.
5. *Mamma*: Margaret Allen's mother; not otherwise identified.
6. *not to stay at Mr. Hebden's*: see letter 0872, n. 8. Tyndall must have stayed at Mr. Hebdon's residence in the past. Allen had previously requested that Tyndall not stay at Hebdon's when he visited her in Preston (see letter 0872).
7. *delightful note from Dr. Hirst*: Hirst had asked Tyndall for Allen's address on 30 April (see letter 0926, n. 12).
8. *your 'son'*: Tyndall used to refer to Hirst as his son. See Volumes 2 and 3.
9. *Gretchen*: see letter 0872, n. 12.

From Michael Faraday 28 June 1854 0948

Ventnor[1] | Isle of Wight[2] | 28 June 1854

My dear Tyndall

You see by the top of this letter how much habit prevails over me. I have just read yours from thence[3] and yet I think myself there. However I have left its science in very good keeping & am glad to hear you are at Experiment: but how is the health? not well I fear. I wish you would get yourself strong first and work afterward:—as for the fruits I am sure they will be good, for though I despond as for myself I do not for you. You are young I am old, you are fresh in thought I am exhausted in thoughts and little more than a thing of habits—but then our subjects are so glorious that to work at them rejoices & encourages the feeblest delights & enchants the strongest.

I have not yet heard any thing from Magnus—thoughts of him always delight me—we shall look at his black sulphur together.[4] I heard from Schoenbein[5] the other day who tells me Liebig[6] is full of ozone i.e of allotropic oxygen

Good bye for the present. We are moderately well—my wife weak.

Ever My dear Tyndall | Yours Truly | M. Faraday

APS Mss.Ms.Coll.200
Faraday Correspondence 4:2859

1. *Ventnor*: 'Royal Institution' is crossed through above 'Ventnor'. Ventnor, a town on the Isle of Wight, was built along the southern cliffs.
2. *Isle of Wight*: see letter 0945, n. 13.
3. *read yours from thence*: Faraday refers to Tyndall's letter of 25 June (see letter 0943), when Tyndall was in Ventnor.
4. *from Magnus . . . black sulphur together*: in a letter of 25 June, Tyndall informed Faraday to expect a paper and specimens of black sulphur from Magnus (see letter 0943, n. 2).
5. *Schoenbein*: Christian Friedrich Schönbein (1799–1868), a German-Swiss Professor of Physics and Chemistry at the University of Basel, Switzerland. He was known for his discovery of the fuel cell, guncotton, and ozone. He also corresponded extensively with Faraday; see *The Letters of Faraday and Schoenbein, 1836–1862*, ed. G. W. A. Kahlbaum and F. V. Darbishire (London: Williams & Norgate, 1899).
6. *Liebig*: Justus Liebig (see letter 0852, n. 7).

From Thomas Archer Hirst 29 June 1854 0949

Dublin | June 29th/1854

My dear John

The weather is still very changeable and my cold still clings to me and annoys me very much. It has attacked all my weak points. first it threatened my chest, then my throat and now it unfortunately resides in my jaws and keeps me awake very often during most part of the night. But I shall shake it off soon. Thanks to Simpson & family I have been well taken care of in Dublin and now I only wait for finer weather to continue my journey. Miss Martin is here too assisting Mrs S who has been in a delicate state of health.[1] I delivered your compliments to all and I assure you they were kindly received and returned. Miss Martin is much improved and it is just possible that you may some day have to accept her as a daughter[2] instead of a grand daughter I have said this to you but to no other soul as yet indeed it is but the expression of some vague notions that have floated about in my head for the last day or two and I feel half inclined to run my pen through it as being premature, the writing it has almost startled one but no matter you will know how to read it. I am in the midst of an experiment and you know I am a cautious fellow. Whatever may be the result you will hear of it and need not question me.[3] I have received the letter you sent to Kilkenny[4] and my first trip as soon as fair weather returns will be to Leighlin Bridge[5] I will sit on the moat and think of you and your boyhood of our friendship and of many other things. This morning Bevingtons letter arrived.[6] In answer to it I sincerely hope we may all three meet this year and spend a day together infact we will do so, but the 12th of July will be too early. I am afraid I do not think I shall return so soon, it will depend upon my purse I shall travel about until it is empty.

I have had a letter from Steiner this morning which has pleased me and which I will shew you when I come.

You will hear from me again very soon

Yours affectionately | T A Hirst

RI MS JT/1/H/200
RI MS JT/1/HTYP/348

1. *Mrs S . . . state of health*: Mary Simpson had birthed a son in March, so her condition may have been related to complications during childbirth (see letter 0898, n. 12).
2. *as a daughter*: see letter 0816, n. 11.
3. *need not question me*: Hirst described the result of this 'experiment' to Tyndall in letter 0953.
4. *letter you sent to Kilkenny*: letter missing, though Tyndall referred to it in letter 0945.
5. *Leighlin Bridge*: see letter 0784, n. 26.
6. *Bevingtons letter arrived*: letter 0944, which was enclosed in letter 0945.

To Michael Faraday 30 June 1854 0950

June 30th, 1854

My dear Prof Faraday

It is just three quarters of an hour to bed time and I think you will not call me imprudent if, as a sedative to my thoughts, I employ this interval in writing to you. You have kindly asked after my health[1]—Though not robust it is still in a sound condition—My face is pale as usual, but it would be hasty to infer from this that there is any thing radically wrong with me. Dr Bence Jones once hinted to me that he thought I might be consumptive. I smiled inwardly at the time, for it brought to my mind a similar remark made by others who when they knew me better came round to the opposite opinion. The fact is, there is greater toughness in these attenuated muscles of mine than many give me credit for, and it has often been my lot in mere physical exertion to weary out men of far greater promise. I do intend however to give myself three weeks holiday. Francis and I have arranged a trip into Wales together and thence to Liverpool to the British Association.[2] This will endow me with an amount of vigour sufficient to cope with the requirements of our next campaign. I have been for the last week endeavouring to decide a point or two in magnetism and have got myself into a labyrinth of difficulties; sometimes in the deepest intellectual darkness, relieved now and then by a gleam which cheers a man on to renewed effort. I often think that the qualities which go to constitute a good christian are those most essential to a man of science, and that above all things it is necessary for him to become as 'a little child.'[3] But how apt is a man to forget this docile spirit:—how apt to rise disaffected and unhappy from his task when he has failed to confirm some foregone conclusion on which he has set his heart:—And then again in the midst of all his discontent a still small voice[4] seems to reason with him and like the harp of David acting upon Saul drives away the evil spirit from his heart and makes him once more fresh and hopeful.[5] I have found myself, even recently, converted from a miserable, complaining, rebellious wretch, into a loyal and happy worker in less time than it has taken me to write this sentence. A thought has rifted and scattered the cloud of discontent as the wind disperses the mist upon the hills.

I wish I could hear you say that you were more than "moderately well" and that Mrs Faraday was more than moderately strong,[6] and I do trust that the fresh breezes of the cozy little spot at which you now reside will increase the strength of both of you. Your last letter[7] is to me in many respects more precious than 'much fine gold', but I have found difficulty in interpreting one phrase—Though I despond as for myself"—I can only account for it on the supposition that nothing short of infinitude can satisfy the soul of man. When I look at the intellectual conquest which lies at your feet, and which, as I heard Verdet say last autumn in Paris,[8] stands unequalled since the time of Newton, the attitude which presents itself to my mind as the natural one is that of a warrior who, after the day's successful conflict, wipes his iron bow and looks with tranquil satisfaction upon the spoils of victory. With regard to my own feelings in the matter I can only say that the ejaculation of Baalam[9] would express the asperation of my soul "Let me die the death of such a worker"![10]

They are still plying me with missives from the Society of Arts.[11] Were I to undertake all they request of me my whole time would be scarcely sufficient for them. Dr Lotham has written[12] requesting me to give a lecture pointing out the connexion of physics with chemistry, on the one side, and with mathematics on the other. Of course I have been compelled to decline it. A Mr Michall wrote[13] requesting me to come and help him on Friday and Saturday to arrange some philosophical apparatus. He writes as if he thought he had nothing to do but to ask me; but notwithstanding this, lest they should consider my persistent refusals illnatured, I have consented to go down to them for a couple of hours tomorrow. A few days ago Dr Bence Jones came into the laboratory and told me that the time for voting the grants by the Royal Society had just arrived. I applied on Friday last for 50 or 100 pounds[14]—On Wednesday evening I met Sabine at Col Yorke's and he informed me that they had given me the 100—It is very kind of them.

My candles are now burnt out and wishing you and Mrs Faraday good night, I end my scroll by subscribing myself

Most faithfully Yours | John Tyndall

RI MS JT/TYP/12/4026–4028
Faraday Correspondence 4:2861

---

1. *asked after my health*: letter 0948.
2. *Liverpool to the British Association*: the Twenty-Fourth Meeting of the BAAS was held in Liverpool, Wednesday, 20 September–Saturday, 23 September 1854. For Tyndall's account of the meeting see letter 0975.
3. *'a little child'*: see, for example, Matthew 18:2–3 (*Faraday Correspondence* 4:2861).
4. *still small voice*: 1 Kings 19:12 (*Faraday Correspondence* 4:2861).
5. *harp of David . . . fresh and hopeful*: 1 Samuel 16:23 (*Faraday Correspondence* 4:2861).
6. *moderately well . . . moderately strong*: see letter 0948.
7. *your last letter to me*: letter 0948.
8. *Verdet say last autumn in Paris*: Tyndall noted in his journal that on 5 September he 'went to the Academy Met Verdet there' (Journal, JT/2/13b/622).
9. *Baalam*: misspelling of Balaam, a prophet in the Torah.
10. *death of such a worker*: see Numbers 23:10 (*Faraday Correspondence* 4:2861).
11. *Society of Arts*: founded in 1754, the Society for the Encouragement of Arts, Manufactures and Commerce was given a Royal Charter in 1847. Its mission was to encourage 'innovation and excellence through six areas—Agriculture, Manufacture, Chemistry, Mechanics, Polite Arts, Colonies and Trade' ('Archive & History', *RSA*, 2016, at https://www.thersa.org/about-us/archive-and-history [accessed 26 August 2016]).
12. *Dr. Lotham has written*: letter missing. 'Lotham' is possibly a misspelling of Robert

Graham Latham (see letter 0935, n. 10), who had been in contact with Tyndall since the confusion regarding reports on their lectures on education in *The Morning Chronicle* (see letter 0935, nn. 10–12).

13. *A Mr Michall wrote*: not identified. Letter missing.
14. *I applied . . . 20 or 100 pounds*: in his journal entry of 2 July, Tyndall noted that he applied for these grants on Tuesday, which would have been 27 June not Friday, 23 June, as mentioned here. He also noted his receipt of the £100 grant (Journal, JT/2/13b/678).

To Thomas Archer Hirst [1 July 1854][1] 0951

My dear Tom,

Although you complain of illness the hand in which you directed your last note[2] to me is a bold and vigourous one. I sincerely trust that your cold after having made so many migrations through your body will eventually communicate itself to the air. It has evidently an upward tendency—first the chest then the throat, then the jaws—let us pray that the next move may be right through the top of your head into the atmosphere above. Considering my relation to you I did not receive that intelligence or rather that hint at a distant possibility with any palpitation of heart—for in the first place I have had thoughts of the kind myself which came to nothing in the end and in the next, and by far the more important place, I have firm confidence that whenever you take the step it will not be unwisely taken—it will not be done in haste. I believe that whatever golden mists the affections may generate around you that your reason with eagle eye will still see through them and be the guide which you will follow. You give me credit for knowing how to read your remarks[3]—and I give you the credit of being able to read mine and to distinguish what I have said from the advocacy of a small worldly prudence. It would be worth the world to me if I could in such a matter throw my heart and understanding into the same scale. For I hold that a man who cannot love is an incomplete man, but I also hold that to place an alliance of the kind upon a firm basis the understanding ought to be satisfied as well as the heart. And indeed I believe if these notions of yours ever ripen into realities it will have been a healthy growth from manly esteem and manly appreciation into manly love. I therefore do not fear, and at any given moment I am ready to give you my fatherly blessing.

Did I not believe it to be my duty to stand by my magnet at the present moment I would go over to you.[4] I am bewildered again. A week ago I thought I knew something of magnetism, I now feel that I know nothing, and that nobody knows any thing about it. Perhaps this conviction of ignorance is the commencement of knowledge, as the Christians say the conviction of sin is the first step towards purification.[5] At all events if I fail to get out of my difficulties it will not be through neglecting to make the attempt.

Faraday writes to me from the Isle of Wight.[6] The following is a quotation from his letter.[7] 'I have left its[8] science in very good keeping, and I am glad to hear you are at experiment: But how is the health? Not well I fear. I wish you would get yourself strong first and work afterwards. As for the fruits I am sure they will be good; for though I despond as for myself I do not for you. You are young, I am old. You are fresh in thought, I am exhausted in thought and little more than a thing of habits—but our subjects are so glorious that to work at them rejoices and encourages the feeblest—delights and enchants the strongest.'

To which I reply as follows:[9]—'Your last letter is to me in many respects more precious than "much fine gold", but I have found difficulty in interpreting one phrase:—"though I despond as for myself." I can only account for it on the supposition that nothing short of infinitude can satisfy the soul of man. When I look at the intellectual conquest which lies at your feet, and which, as I heard Verdet say last autumn in Paris, stands unequalled since the time of Newton, the attitude which presents itself to my mind as the natural one is that of a warrior who, after the day's successful conflict, wipes his iron brow and looks with tranquil satisfaction upon the spoils of victory. With regard to my own feelings on the matter, I can only say that the ejaculation of Baalam would express the aspiration of my soul—

'Let me die the death of such a worker!'

Goodbye Tom | Affectionately yours | John

The Royal Society has given me a hundred pounds to help me in my experiments.[10]

RI MS JT/1/HTYP/349
LT Typescript Only

1. *[1 July 1854]*: Tyndall did not date this letter but LT's suggestion that it was 'probly July 1' is likely correct as it was written in response to letter 0949 from Hirst, dated 29 June.
2. *your last note*: letter 0949.
3. *how to read your remarks*: in letter 0949, Hirst suggested that Tyndall would 'know how to read' his cryptic remarks about Anna Martin.
4. *over to you*: Hirst was travelling throughout Ireland and was making his way to Tyndall's birthplace of Leighlin Bridge (see letters 0949 and 0953).
5. *as the Christians . . . first step towards purification*: see Romans 3:20–22, on the idea of righteousness through faith.

6. *Isle of Wight*: Faraday and his wife were on holiday in the Isle of Wight (see letters 0945, n. 13; 0948; and 0950)
7. *quotation from his letter*: see letter 0948.
8. *I have left its*: LT notes: '(the Royal Institution)'.
9. *I reply as follows*: see letter 0950.
10. *The Royal Society . . . in my experiments*: see letter 0950, n. 14.

From Thomas Charles Higginson 6 July 1854 0952

Ipswich[1] | July 6th. 54

My dear old Friend,

I was truly delighted to see your well remembered hand writing again[2] and not a little flattered to find that you still desired to remember your Youghal bed-fellow.[3] A few lines of poetry for which you were partial (Oh let that eye wild as the Gazelle's &c[4]) whenever I happened to remember them brought the 'light of our youthful days around me' and your features were brought so clearly before me that if we met in the King of Otahuti's country[5] I should recognise you. My present employment of Superintending the Recruiting in this part of England keeps me so busy,[6] especially at the beginning of the quarter, that I have little time for private correspondence therefore won't apologise for the shortness of this but if you let me know your address when I go to town (which I hope to do in 12 or 14 days) I shall pay you a visit and have a long chat about the strange things that have happened to both and with a first grasp of the hand renew our former friendship

Your old and sincere | Friend | T. C. Higginson | 16th Reg.

RI MS JT/1/TYP/2/596
LT Typescript Only

1. *Ipswich*: a port town in Suffolk, England.
2. *Delighted to see . . . hand writing again*: letter missing.
3. *Youghal bed-fellow*: both Higginson and Tyndall were civil assistants with the 5th Division of the Irish Ordnance Survey in the early 1840s. They must have lodged together when Tyndall was stationed in Youghal, County Cork. See Volume 1.
4. *Gazelle's &c*: from the fourth stanza of the dedication 'To Ianthe' in Lord Byron's *Childe Harold's Pilgrimage,* 'Oh! let that eye, which, wild as the Gazelle's' (*Childe Harold's Pilgrimage*, vol. 1 [Brussels: Du Jardin–Sailly Brothers, 1829], pp. 7–8, on p. 8).
5. *King of Otahuti's country*: misspelling of Otaheite, or Tahiti.
6. *Superintending the Recruiting . . . keeps me busy*: at this time Higginson was a Lieutenant in the Royal Artillery.

From Thomas Archer Hirst 8 July 1854 0953

Dublin | July 8th 1854

My dear John,

Day after day I have postponed writing in the hope that when I did so I could tell you that I had been to Leighlin[1] and day after day the weather prevented me from doing so. At last tired of waiting I set out on Thursday last in the Teeth of wind and rain. Weather like Fortune favours the brave.[2] I reached Bagnals town[3] at 8 P.M the rain had cleared away the moon had risen and I had a delightful stroll up the side of the Barrow to Leighlin Bridge. A more beautiful welcome to the birth place of my best friend I could not have wished for. It was like entering a fairy land. A boat was gliding down the river and from it proceeded the music of two flutes beautifully played. It was indeed a beautiful greeting. Unconsciously I fell into a happy tranquil train of thought. The memory of past events in which we had both been actors came to me and accompanied me as I strolled on towards Leighlin Next morning the weather was still fine I called upon your Mother and Sister,[4] got William Wright's letter[5] and strolled out to read it on the moat. Believe me I spent a delightful hour or two amongst those scenes of your boyhood than which I know none more beautiful. I returned and got a simple lunch with your Mother and Sister. We talked for a long time together, principally of course about yourself. Right glad I am that I came here that I have seen your Mother and Sister and the place where you were born. I seem to know and love you the better for it, our friendship seems more complete and intimate. I went to see your old Schoolmaster Conwill[6] too, and a hearty reception he gave me I can tell you Both he and your mother told me little anecdotes about you that I loved to hear and that I shall long remember. But I shall see you soon in London and we shall talk more of my visit to Leighlin. At present I have something else to tell you The experiment I hinted to you about[7] has proved even more successful than I anticipated Miss Martin is no longer my daughter she will some day be my wife.[8] I have not been quite candid with you on this matter partly because I felt that in choosing a wife I must decide for myself and partly on account of the vagueness & uncertainty that always attends such intentions The thought of making Miss Martin my wife is by no means a new one for two years it has gradually grown upon me, and the chief almost the sole object of my visit to Ireland was to decide the matter. It was on this account that I did not press you to accompany me which otherwise I should have done. I have told you that I admired her for many things in Marburg but that her fun and frolic hid from me much of her character about which I wished to know more. Her letters revealed more of this to me and in them I noticed a certain cordial friendship almost amounting to respect for me which I began soon to think might form a good foundation for a warmer feeling In this I was not mistaken, when after mature consideration I discovered my real sentiments towards her she was at first somewhat astonished but soon recovered herself and timidly half fearful that she was not good enough for me she has told me she can and will love me sincerely.

Ah me but I have been spending happy days. Between us a mutual trustfulness has gradually established itself which promises to ripen and chasten our whole lives.

Perhaps she has an exaggerated notion of my 'goodness' and of her own unfitness but I do not quarrel with it as long as I can see that it is an ideal after which she is earnestly striving herself. Whatever may be the result, and by God I do not fear it, I have this satisfaction that I have forced nothing but simply submitted myself to an instinct that never yet deceived me. I firmly believe she will make me a good and true little wife and that my own life will be ennobled and chastened by having her to love and cherish.

We have talked much about you and I believe she truly appreciates our friendship In talking of you we have learnt more of one another and now I believe she is as wishful as I am for the fatherly blessing which you have promised us.

Goodbye! If you feel desirous you may write to her at 1, Brougham Place Wellington Road, Dublin as for me I shall be with you soon, for I leave here in a day or two.

Yours affectionately | T.A. Hirst.

RI MS JT/1/H/201
RI MS JT/1/HTYP/350

---

1. *Leighlin*: Leighlin Bridge (see letter 0784, n. 26).
2. *fortune favours the brave*: a version of the Latin proverb '*fortes Fortuna adiuvat*' referred to as early as in Cicero, *Tusculanae diputationes*, 2.4.11.
3. *Bagnals town*: misspelling of Bagenalstown, County Carlow, Ireland.
4. *your Mother and Sister*: Sarah Tyndall (née McAssey) and Emma Tyndall, respectively (see Volume 1, p. xxxix).
5. *Wright's letter*: William Wright (see letter 0921, n. 1); letter missing.
6. *Old Schoolmaster Conwill*: John Conwill (1802–80) was Tyndall's schoolmaster at the National School at Ballinabranna from *c.* 1836. He was central in Tyndall's early education particularly with regard to mathematics (see Volume 1). Hirst recounts his meeting with Conwill in his journal entry of 8 July 1854 (Hirst, 'Journals').
7. *the experiment I hinted to you*: see letter 0949. See also letter 0951, n. 3.

8. *some day be my wife*: Tyndall wrote in his journal that he 'received intelligence from Tom that he has taken one important step and arranged his marriage with Miss Martin. I had long foreseen the probability of such a consummation and was not surprised by it. He has been to Leighlin and written me a sweet account of his visit to the dwelling place of his "best friend". The moon was on the Barrow as he walked from Badnalstown to Leighlin, and the sound of a flute came mellowed along the waters: next day he visited the moat and spent some hours among the scenes of my earliest boyhood. What wonderful things are in the womb of time! May God bless Tom and keep him strong and upright, and may his wife be a perennial treasure to him' (Journal, 5 August 1854, JT/2/13b/679–80).

To Anna Martin 13 July 1854 0954

July 13th. 1854.[1]

My dear Miss Martin

I had, as I thought, folded up my papers for the night, and in order to make me sleep, had been selfish and sinful enough to prepare for myself a dose of brandy and water, when the thought of a letter which I received from my friend Tom this morning[2] occurred to me and I could not resist the luxury of reading it once more. The night is so calm, so silent, so different from what I am accustomed to in London—for I am stopping in a cottage at the outskirts of a quiet Hampshire village,[3] I myself, if I except the flies and crickets, being the only living creature within its walls—that, added to the effects of Tom's letter, and perhaps in part to the brandy and water, the tranquility of my position sheds a kind of dewiness over my old frozen heart, and renders it more pleasant to me to write than to go to bed. You have doubtless read the story of Damon and Pythias[4] whose friendship, evidenced in the willingness of one of them to suffer death for the other, has been handed down to us as a sample of better ages than our own. In the chivalrous moods which used to come upon me, when I was younger than I am now, I have often thought that such a piece of friendship might be enacted even now. That the same fresh and faithful heart was still beating in the present world as in the world of yore, and that it might be perfectly possible for two young men of the 19th. century to entertain as noble a friendship and to act as brave a part as that which history has recorded of the two friends of Syracuse.[5] There have been times when I have felt it possible that Thomas Hirst and myself, if called upon to do so, might act in a similar manner towards each other. And though this foolish feeling is now tempered—withered perhaps by time and choaked by study—the memory of it at least still clings to me and keeps my affection for Tom alive. I read his letter[6] this morning with deep interest, and I reread it to night with undiminished pleasure. The boy (as I used to call him) cunningly commences with a description of a visit to my poor birthplace,[7] calling many things to my mind which he must have known would soften its parched soil and render it fit for the further intelligence he had to communicate, intelligence which I may say did not take me by surprise. The intelligence was that you were no longer his daughter[8] but occupied a relation to him entirely new. I have watched the growth of this consummation more closely than perhaps Tom himself imagines, or, to speak more correctly, I saw without watching the tendency and probable termination of the correspondence which had been established between you. Knowing this, there are people in the world who would doubtless deem it very strange that I never uttered a thought upon on the subject. But the friendship between Tom and me is not a meddling friendship, and I believe that our lives demonstrate that a union of no common strength is compatible with the perfect individual freedom of the friends united. I have left Tom to his own unbiassed choice, and now that that choice is made the fact of <u>his</u> having made it is a sufficient proof to me that it is a wise one. Most gladly however do I add to this conviction another derived from the perusal of some of your letters to Tom—letters so clever and so good, that my eyes were quickly opened to their effect upon a mind receptive of all cleverness and of all goodness. The fact is, that if Tom had not taken his present step I think I should have become your wooer myself, and thus instead of the athletic son, with the wine of life rushing through all his arteries, lending light to his eye and vigour to brain, you would have had for a suitor the dilapidated old father, possessing only half a heart and with a cheek and forehead furrowed by monitions from the world to come.

I will not say that Tom is perfect, nor am I even polite or poetical enough to tell a lady that she is so. But this I will say that he has the power of understanding the best moments of the best men, and the greatest moments of the greatest: and this power he could never have without having been, at least during some portion of his life, great and good himself. Now when I know this, and that to this he adds the wisdom of practical life, and a competent knowledge of the world; what am I to think of the little sorceress who has wound her spell around him and tied my Hercules[9] to her apronstrings? But that sorceress knows the residence of her power as well as I do, that it is based upon the affection and the truth of her own character. These are qualities the retention and exercise of which will preserve both Tom and her the same faithful lovers through long future years, as they feel themselves to day.

And now I have time to reflect for a moment on the probable increase to my own comfort which this alliance promises. I dream of tea trays and arm chairs by the cosy fireside, and kind looks and kind words spoken when my head aches. I am sure we shall all be very happy when we come together.

And now goodbye and excuse this badly written letter for my eyes are dim and weary.

And believe me | With a bit of Tom's love | Your friend | John Tyndall

RI MS JT/1/HTYP/351–352
LT Typescript Only

1. *July 13th, 1854*: LT notes: 'Broughton near Stockbridge (sic)'. According to his journal, Tyndall spent a week in Broughton, which is very close to Queenwood, in mid-July (Journal, 3 August 1854, JT/2/13b/679).
2. *letter which I received from . . . Tom*: letter 0953.
3. *outskirts of a quiet Hampshire village*: Broughton, Hampshire, England (see n. 1).
4. *story of Damon and Pythias*: a Greek legend concerning the Pythagorean ideal of friendship. As one version of the story has it, during the reign of Dionysius I (r. 405–367 BCE), Pythias was accused of plotting against Dionysius and was sentenced to death. Damon offers to take Pythias's place while the latter visits his home for one last time. Different versions of the story appear in Aristoxenus, Cicero (*De Offic.* 3.45), and Diodorus Siculus (10.4). The story also became a tragicomic play by the English poet Richard Edwardes in 1564 (*Damon and Pythias*) and was made into a ballad by Frederich Schiller in 1799 (*Die Bürgschaft*). By the Victorian period 'Damon and Pythias' had become idiomatic of true friendship.
5. *two noble men of Syracuse*: see n. 4. Refers to Syracuse (now known as Sicily), in southern Italy and home of the tyrant, Dionysius.
6. *read his letter*: letter 0953.
7. *my poor birthplace*: Leighlin Bridge (see letter 0784, n. 26).
8. *no longer his daughter*: see letter 0816, n. 11.
9. *Hercules*: Hercules is the Romanized name of the Greek divine hero Heracles, who was known for his courage, strength, and adventurous spirit.

From Unidentified 19 July 1854 0955

Manchester | July 19th 1854.

Sir,

At the request of my friend Mr Hopkins of Cambridge, I have forwarded specimens[1] of substances 3/10[2] of an inch square—compressed under pressures, some of them to the extent of nearly 80,000 lbs on the square inch. You will find two specimens of each kind and I trust of such dimensions as will enable you to ascertain with considerable accuracy the conducting powers of the different substances which Mr Hopkins desired me to submit to experiment.

It will afford me much gratification if you will kindly furnish me with a copy of your results,[3] and if not asking too much, I should at the same time be glad to have the specific gravity of each. I am desirous of having these results for other objects than those contemplated by Mr Hopkins, and I shall esteem it a favour if these objects can be accomplished without inconvenience.

I am, Dear Sir | yours faithfully[4]
Dr. Tyndall | &c. &c.

RI MS JT/1/TYP/2/609
LT Typescript Only

1. *specimens*: on 12 June Hopkins referred to a 'supply of specimens' that he was hoping to send to Tyndall from Manchester (see letter 0938, n. 3).
2. *3/10*: the number 10 is partially handwritten and covering an illegible, typed character, suggesting that LT corrected a typographical error.
3. *your results*: Hopkins wanted Tyndall to experiment on the conductive powers of the specimens (see letters 0929 and 0938).
4. *yours faithfully*: the typescript did not include a signature.

From Heinrich Gustav Magnus 21 July 1854 0956

Berlin d. 21 Juli 54.

Theurer Freund!

Endlich bin ich im Stande Ihnen anzeigen zu können daß das Polytrop mit den Spielereien an Ihre Adresse expedirt worden ist, und zwar wieder durch Asher. Meine Auslagen werden Sie aus den beifolgenden Rechnungen ersehen. Zugleich will ich nicht unterlassen anzuzeigen daß mir die früheren Auslagen für Col. Yorke sehr ordentlich durch Asher gezahlt worden sind. Mich freut sehr daß Col. Yorke mit der Sendung zufrieden ist und ich wünsche nur daß M^r^. Gassiot ebenso zufrieden sein möchte.—Von Wheatstone's Apparat ist Beschreibung und Abbildung schon hier, scheint mir aber nichts besonderes zu sein, und ich kann nicht begreifen wie Baden P. so viel Wesen davon hat machen können.

Hier ist ein Angriff gegen meine Erklärung von der Abweichung der Geschosse erschienen, der aber ganz miserabel und ohne Kenntniß ist. Er rührt von einem Artillerie Officier und ich denke es wird sich nächstens Gelegenheit finden ihm zu antworten.

M^r^. Barlow ist noch nicht hier gewesen, ich freue mich seine Bekanntschaft zu machen. Sie haben jetzt nun die beiden Herrn Schlagintweit in London und werden wohl eine Ausstellung ihrer Instrumente zu sehn bekommen. Wahrscheinlich werden Sie auch noch Dove in diesem Herbst dort sehen.

Hier geht alles in gewohnter Weise vorwärts; dazu gehört auch daß Du bois ein Töchterchen bekommen hat.

Daß Sie sich mit Magnetismus beschäftigen ist sehr gut, denn ich bin überzeugt Sie würden es nicht thun wenn Sie nicht schon irgend eine Idee hätten wie dem näher zu kommen sei.

Viele Grüße von meiner Frau und den Kindern. Lassen Sie mich recht bald von sich hören, und empfehlen Sie mich an M^{rs} und M^{r}. Faraday.

Was hat Brodie zu dem rothen und schwarzen Schwefel gesagt? Schreiben Sie mir doch darüber.

Von Herzen | Ihr | GMagnus.

Berlin, 21 July 54.

Dear friend!

I am finally in a position to be able to report to you that the polytrope,[1] along with the gadgets, has been sent to your address, and again through Asher. You will see my expenses in the enclosed from the enclosed invoices.[2] At the same time, I shall not omit to report that the earlier expenses for Col. Yorke have been paid to me in very orderly fashion by Asher. I am very pleased that Col. Yorke is satisfied with the consignment and I wish only that M^{r}. Gassiot might be just as satisfied.—The description and illustration of Wheatstone's apparatus[3] are already here, but it does not seem to me to be anything special, and I cannot comprehend how Baden P.[4] has been able to make so much fuss about it.

An attack on my explanation of the deviation of projectiles has appeared here[5] which, however, is quite wretched and lacking any knowledge. It comes from an artillery officer, and I think that there will be an opportunity shortly to respond to him.

M^{r}. Barlow has not been here yet; I shall be pleased to make his acquaintance. You currently have both the Messrs Schlagintweit[6] in London now and will probably get to see an exhibition of their instruments. You will probably see Dove there as well this coming autumn.

Everything is going forward here in the usual way; this also includes the fact that Du bois[7] has got a little daughter.

It is very good that you are busy with magnetism, for I am convinced that you would not do so, if you had not already some idea or other as to how the subject was to be approached.

Many greetings from my wife and the children. Let me hear from you very soon, and give my compliments to M^{rs} and M^{r}. Faraday.

What did Brodie[8] say about the red and black sulphur? Write me something about that.

Sincerely | Your | GMagnus.

RI MS JT/1/M/12

---

1. *polytrope*: see letter 0870, n. 7.
2. *enclosed invoices*: invoices missing.

3. *Wheatstone's apparatus*: see letter 0907, n. 4.
4. *Baden P.*: Baden Powell (see letter 0925, n. 5).
5. *An attack . . . appeared here*: not identified.
6. *Messrs Schlagintweit*: brothers Hermann Schlagintweit (1826–82) and Adolf Schlagintweit (1829–57) were German naturalists and explorers (see letter 0470, n. 14).
7. *Du bois*: Emil du Bois-Reymond.
8. *Brodie*: Benjamin Brodie (1783–1862), see letter 0816, n. 7. While there is no mention of any discussion of red and black sulphur with Brodie, Magnus had planned to send samples to Faraday (see letters 0925 and 0943, n. 2). As of 28 June, Faraday had not received them (see letter 0948, n. 4). Tyndall had also recently met Brodie at a dinner party held at Pollock's home on 6 June 1854 (Journal, JT//2/13b/677).

From Hector Tyndale 21 July 1854 0957

Paris 21st July 1854

My dear Cousin John

Whether it is your fault or mine I don't know, but it is somebody's or I should have rec'd some news of and from you before this. On my part I can only say that I have been working so hard to get through my business here and have had, of necessity, so much writing to do that I could think of, or do nothing else until I had finished all. Having nearly closed up everything here I shall remain but about a week longer to enjoy Paris under this blue Summer sky[1]—the which I shall do heartily as it ceased raining but the day before yesterday, having commenced the day I entered France, and with the exception of five or six days, has rained incessantly dampening me completely up to this moment.

And now let me ask you whether you will be ready and willing to make a trip of <u>about</u> a week[2] on my return to England? You see I have shortened the time we spoke of for travel[3] from three weeks to one—but recollect I am homeward bound and it is accounted for. If you feel like <u>tripping</u> it say so, and if you have not the inclination or the time why say so. Of course you will feel under no obligation as tho' fulfilling a bargain—at least I hope you won't. Now if you can spare the time and can raise the courage to sustain a week's ennui in my company, can you also think of any particular place, person or thing on which to bestow 'all our tedium your worship'?[4]

In plain English will or can you go & when and where can we go for one week?[5] Pray pardon this rambling note of mine, but I have been now nearly three months without hearing more than an occasional English word and have forgotten the little Yankee English I had learned before.

If you have the time to reply to this addressing me to the care of

E. Didier | Cité Trevise No 5 | Paris

I will be obliged. Anything in the way of comfort you can throw into your note will be very acceptable—also any stray bit of news you may have for I have heard nothing since I left London worth remembering. With my kind regards to Doctor Francis should you see him—I am Cousin John very faithfully & truly

Your friend | Hector Tyndale

I shall bring you a <u>beautiful</u> portrait[6] when I come!

RI MS JT/1/T/45
RI MS JT/1/TYP/5/1589

---

1. *in Paris . . . Summer sky*: Tyndall's American cousin was visiting England on business and had taken a trip to Paris (see letter 0916). The mention of the blue sky references Tyndall's previous letter to Tyndale (see letter 0923).
2. *trip of about a week*: Tyndall discussed their trip in his journal entry of 5 August 1854 (Journal, JT/2/13b/680) and in letter 0960.
3. *time we spoke of*: On 4 April, Tyndall noted that he and Tyndale had discussed taking 'a ramble together' once Tyndale returned from Paris (Journal, JT/2/13b/671).
4. *'all our tedium your worship'?*: this is paraphrased from Shakespeare: 'But truly, for my own part, if I were as tedious as a king, I could find it in my heart to bestow it all of your worship' (*Much Ado About Nothing*, 3.5).
5. *where can we go for one week?*: they determined to travel throughout the Isle of Wight (see letter 0960).
6. *beautiful portrait*: Tyndall had asked Hector to have a portrait taken (see letter 0923). And Tyndale would indeed give Tyndall the portrait. Tyndall wrote in his journal on 5 August that Tyndale 'left his portrait with me at my request, but this was not needed to keep my affections for him alive' (Journal, JT/2/13b/681).

From Wilhelm Weber 23 July 1854 0958

Hochgeehrter Herr Professor

Durch eine kleine Reise, die ich zu machen genöthigt war, ist meine Antwort auf Ihr Schreiben vom 3 Juli etwas verspätet worden. Ich habe die mir übersandte Bestellung eines Apparats zur Nachweisung der diamagnetischen Polarität Herrn Leyser in Leipzig überschickt und habe die Veränderungen des Instruments, die zu Ihren Zwecken erforderlich schienen, genau angegeben und hoffe, dass er darnach ein Ihren Wünschen ganz entsprechendes Instrument liefern kann.

Das Instrument, wie es in meiner Abhandlung beschrieben ist, war nicht bestimmt, zu feinen und genauen Beobachtungen zu dienen, wie Sie anstellen wollen, und Sie bemerken selbst in Ihrem Briefe mit Recht, dass eine grössere

Wirkung erhalten werden kann, wenn zwei astatisch verbundene Nadeln statt einer einfachen Nadel gebraucht werden, womit dann einige andere Veränderungen des Instruments zu verbinden waren, die ich Herrn Leyser genau angegeben habe.

Der Preis des Instruments wird dadurch beträchtlich erhöht werden und ich habe Herrn Leyser aufgefordert, Ihnen darüber sogleich näheren Bericht abzustatten.

Mit grösster Hochachtung stets | Ihr ganz ergebenster | Wilhelm Weber.

Göttingen, 1854 Jul. 23.

Esteemed Herr Professor

My reply to your letter of 3 July[1] has been somewhat delayed by a short trip that I was forced to make. I have posted the order sent to me[2] for an apparatus for the detection of diamagnetic polarity[3] to Herr Leyser[4] in Leipzig and have precisely specified the modifications to the instrument that seemed essential for your purposes,[5] and hope that he can accordingly supply an instrument that completely meets your wishes.

The instrument, as described in my paper,[6] was not meant to be used for fine and precise observations such as you wish to do, and you correctly remark yourself in your letter[7] that a greater effect can be obtained if two astatically connected needles[8] are used instead of one simple needle, whereby some other modifications of the instrument then needed to be connected, which I have precisely specified to Herr Leyser.

The cost of the instrument will therefore be considerably increased because of this and I have requested Herr Leyser to give you a more detailed report about this at once.

Always with the greatest respect always | Your most humble | Wilhelm Weber

Göttingen 1854 Jul. 23.

RI MS JT/1/W/13

1. *your letter of 3 July*: letter missing.
2. *the order sent to me*: not identified (but see letter 0959).
3. *diamagnetic polarity*: the polarity of diamagnetism was discussed in letters 0710 and 0821, n. 4. Tyndall later elaborated on his opinion of Weber's views on diamagnetic polarity in letter 0970. The specific apparatus referred to here is not identified.
4. *Herr Leyser*: Georg Moritz Leyser (1816–81), a German instrument maker attached to the University of Leipzig, whose electrodynamometer had been on display at the Great Exhibition of 1851 ('Saxony', *Official Descriptive and Illustrated Catalogue of the Great Exhibition of the Works of Industry of All Nations*, Part IV [London: W. Clowes & Sons, 1851], pp. 1104–13, on p. 1105). See also letter 0959.

5. *precisely specified the modifications . . . your purposes*: Leyser described the apparatus in letter 0959.
6. *my paper*: not identified. Weber published a number of papers on diagmagnetism (see letter 0820, n. 13).
7. *your letter*: letter missing.
8. *astatically connected needles*: 'one[s] so situated or arranged as to be unaffected by the earth's magnetism' (*OED*).

From G. M. Leyser 27 July 1854 0959

Leipzig d. 27 *[July]* | 1854.

Hochgeehrter Herr!

Durch die *[Güte]* des Herrn Prof. Wilh. Weber ist mir der ehrenvollen Auftrag von Ihnen geworden ein Instrument für den Nachweiß der diamagnetischen Polarität anzufertigen. Da das Instrument noch empfindlicher werden soll, als das vom Herrn Prof. Weber beschriebene, so hat Herr Prof. Weber an mich deshalb geschrieben und eine Zeichnung entworfen, wo 2 Magnete nebst Spiegeln und 2 Drathspiralen angewendet werden sollen. Das Instrument weicht somit von dem frühern bedeutend ab; ist aber freilich wie vorauszusehn auch weit empfindlicher.

Ein solcher Apparat würde 60 [Rthl.] kommen. —Das größere Fernrohr nebst Scale kostet 50 [Rthl.]—Zusammen würde ein solcher Apparat auf 110 [Rthl.] zu *[stehn]* kommen; Natürlich würde für die Brauchbarkeit garantirt, da das Instrument vom hiesigen Physiker oder auch vom Prof. Wilh. Weber selbst vor Absendung probirt werden müßte.

Ich sehe deshalb Ihrem Entschluße entgegen und zeichne mit Ergebenheit:

G. M. Leyser | Inspector a. d. Univ. | Verte

Leipzig, 27 *[July]*[1] | 1854.

Esteemed Sir!

I have received from Herr Prof. Wilh. Weber your honourable assignment[2] to construct an instrument for the demonstration of diamagnetic polarity.[3] Because the instrument is to be made more sensitive than the one described by Herr Prof. Weber, Herr Prof. Weber wrote to me and created a drawing according to which 2 magnets beside mirrors and 2 wire coils will be used. The instrument differs therefore considerably from the previous one; but it will be, as expected, significantly more sensitive.

Such an apparatus would amount to 60 [Rthl.]. —The larger telescope and scale cost 50 [Rthl.]. —Altogether, such an apparatus would cost 110 [Rthl.];

Naturally, its usability would be guaranteed because the instrument would be tested before dispatch by the in-house physicist or also by Prof. Wilh. Weber himself.

I will therefore look forward to your decision and sign respectfully:

J. M. Leyser[4] | Inspector at the Univ. | <u>Verte</u>[5]

RI MS JT/1/L/21

---

1. *[July]*: while part of the date is cut off, it must be July as it follows from, and references, letter 0958.
2. *I have received from . . . honourable assignment*: see letter 0958. Tyndall must have requested from Weber an apparatus for the detection of diamagnetic polarity similar to one described in one of Weber's papers (see letter 0958, n. 6).
3. *diamagnetic polarity*: see letter 0958, n. 3.
4. *J. M. Leyser*: see letter 0958, n. 4.
5. *<u>Verte</u>*: not identified. Possibly Latin to mean 'turn over', but there does not appear to be another page to the letter.

To Thomas Archer Hirst 6[–7] August 1854[1] 0960

Sunday 6th. Aug. 1854

My dear Tom

You desired me to let you hear from me after my return from the Isle of Wight,[2] and here I, like a magnanimous boy sit down to comply with that desire. I thought to finish the week writing to you last night. I thought to lie down in bed with the image of you freshest in my brain; I thought you might cling to me during the night and poke me up in good time in the morning: but I have to mourn the inefficacy of the means, for I lay in bed till nearly 8 o'clock. The fact is Tom I have slidden into bad habits. When I came to London[3] it was my righteous determination to rise at least at 6 o'clock every morning. Well, my aching head compelled me to relinquish this resolution, compelled me to indulge myself in bed, and now that the aching has passed away I find the habit around me like a chain weighing me down and preventing me from lifting myself beyond the mud. Instead of a bird soaring aloft on triumphant wing and with fearless eyes looking into the sun I am a mere reptilian waddling amid swamps and quagmires, and seeking my enjoyments in mere slush. How is the creature to be winged again? How are its broken and draggled pinions to be once more fitted to cleave the spiritual atmosphere? Not by <u>knowledge</u>. Knowledge is inefficacious, I know enough, my head assents to all your moral propositions—to every statement of duty. But where is the force which can convert the assent of my head into the practice of my life? This is the grand point. This is the great defect of myself and the world. Where are the fountains of living water at which the fainting soul may quicken its energies? Fool! You would eat the fruit and not earn it. You would feel the force and not toil for it. The patch of the desert lies between you and the palm-tree wells but you shrink upon the desert's edge. But I am wandering—The very asking of such questions as these tends to awaken me, tends to stir up the embers of the life within me, and here I see your possible value to me. I may make use of you as a purchase to regain the point from which I have slidden.

I left you on Monday last,[4] 2 o'clock saw us at Ryde[5] and at 4½ we were on the road to Sandown.[6] Thence along the beach to Shanklin,[7] and thence through mist and drifting clouds to Bonchurch.[8] We did not pause even here, but were tempted on to Ventnor[9] where we rested our weary limbs at the Marine Hotel.[10]

Next morning the sun shone gloriously from a heaven of the deepest blue. We had passed the old burial ground of Bonchurch[11] the evening before, without calling to see it. This morning we turned back along the cliffs and soon found ourselves within the little 'city of the dead.'[12] It is a sweet spot and the thoughts which I had then were able to cast a fragrance over many hours of my existence. Once while looking at John Sterling's grave[13] the thought occurred to me of stretching myself between the head and first stone, just to see whether the same grave would answer for me. I did so, looked upward to the ultramarine of the sky and to the green and golden elmtree leaves which twinkled against it. The lark was singing and the sea was sounding while the morning wind blew the perfume of a cluster of roses in my face. I had been warm when I lay down, and after some time I felt chilled and arose. Hector[14] and I faced the hill, and soon reached the summit of the Downs.[15] I rejoiced in my strength as I mounted the declivity when on the top we threw ourselves down with the Island spreading at one side and the ocean at the other and there he sang the 'Starspangled banner'[16] for me. Thence along the ridge towards Ventnor, but before reaching it lay down again to drink the glory of the scene. Reached our hotel, paid our bill, had a glass of beer which was manifestly <u>sour</u> and started on our way. In an hour or so I found myself invaded by symptoms like cholera.[17] I was forced to pause and rest. The strength in which I had gloried seemed to ooze from every pore, and half an hour's pain reduced me to a helpless prostrate mortal. I clung to Hector's arm and staggered feebly on to the Sandrock Hotel.[18] Here I threw myself upon a sofa while Hector prepared some hot brandy and water and ginger for me—drank it but found myself getting quickly weaker. Went to bed, and for two or three hours felt very uncertain whether I should ever leave the Island. It is well that

I have been strong previously, for had such a thing invaded me while feeble in health I think it must have carried me off. Towards evening the attacks of pain became less frequent and I had evidently got over the worst. Hector nursed me as if I had been his sweetheart, and having some knowledge of the thing, which I had not, his presence was of great service to me. Our journey was however completely interrupted I crawled with him as far as Blackgang Chine,[19] but the effort to get there merely demonstrated my own feebleness. I remained at the Sandrock two days, and my strength rapidly returned. On Thursday I was well enough to go any where, but Hector prevailed upon me to turn homewards So we left the Island on that day and reached London the same evening. I am now completely restored, thank Heaven, without pain or ache, and fit to accept a wager from you to run round the Island if you have the courage to make such a proposition.

I must now shut out the world and get to work. I think seriously of doing as Bonaparte did and seeing my letters[20] hence a fortnight. They distract me greatly. Dr. Simpson has called upon me, but where he is I know not—as he simply left his name, but not his address.

Hector is now gone—I wish you had known more of him. He is extremely clever and honest to the core. He possesses the greatest command of mental imagery which I have ever met in any man.[21] Fair winds attend him!

Just before writing my name I cast my eye upon my journal which lies open before me and there I read. 'This day Tom and I dined at Werda &c.'[22] But I must not open this vein.

Your affectionate | John

Monday morning up at 6.

Oh! I see the crescent promise of my spirit hath not set Ancient founts of inspiration well through all my fancy yet.[23]

RI MS JT/1/HTYP/353–354
LT Typescript Only

---

1. *6[–7] August*: while the typescript dates the letter '6 August', there is a postscript that was written on 'Monday', which was August 7.
2. *Isle of Wight*: see letter 0945, n. 13. Tyndall and his cousin Hector Tyndale spent a week travelling throughout the Isle of Wight (see letter 0957, n. 2).
3. *When I came to London*: they returned to London on Thursday, 3 August 1854 (Journal, 5 August 1854, JT/2/13b/680).
4. *Monday last*: Tyndall and his cousin had stopped at Queenwood before their travel to the Isle of Wight (Journal, 5 August 1854, JT/2/13b/680).
5. *Ryde*: Ryde is a town on the Isle of Wight, England.
6. *Sandown*: Sandown is a town on the Isle of Wight, England.

7. *Shanklin*: Shanklin is a town on the Isle of Wight, England.
8. *Bonchurch*: Bonchurch is a town on the Isle of Wight, England.
9. *Ventnor*: see letter 0948, n. 1.
10. *Marine Hotel*: A hotel in Ventnor owned by George Bush (n.d.) ('Ventnor', in *Post Office Directory of Hampshire, Wiltshire, and Dorsetshire, with Maps Engraved Expressly for the Work* [London: Kelly and Co., 1855], pp. 216–18, on p. 218).
11. *old burial ground of Bonchurch*: likely refers to the grave yard at St. Boniface Old Church, which was built by the Normans. There was also a new St. Boniface Church, founded in 1847 ('St. Boniface', *St. Boniface*, http://www.bonchurch.info/ [accessed 15 October 2016]).
12. *city of the dead*: see n. 11.
13. *John Sterling's grave*: John Sterling (1806–44), Scottish writer and poet. He was a member of the Cambridge Apostles and was closely associated with F. D. Maurice (1805–72), John Stuart Mill (1806–73), and Thomas Carlyle (1795–1881). The latter wrote a memorial of Sterling's life entitled *The Life of John Sterling* (London: Chapman & Hall, 1851). Sterling died of consumption in 1844 and was buried in a churchyard in Bonchurch (*ODNB*).
14. *Hector*: Hector Tyndale.
15. *the Downs*: refers to Ventnor Downs, the highest point on the Isle of Wight.
16. *'starspangled banner'*: 'The Star-Spangled Banner' has been the official national anthem of the United Sates of America since 1931. The lyrics, however, were penned by the attorney, Francis Scott Key (1779–1843), who wrote the song after witnessing the bombardment of Fort McHenry in 1814. Key borrowed the melody from John Stafford Smith's (1750–1836) popular English song of 1775, 'To Anacreon in Heaven' ('Star-Spangled Banner', *Smithsonian*: *National Museum of American History*, 2016, http://amhistory.si.edu/starspangledbanner/default.aspx [accessed 27 August 2016]). See also 'The Lyrics', *Smithsonian*: *National Museum of American History*, 2016, http://amhistory.si.edu/starspangledbanner/the-lyrics.aspx [accessed 27 August 2016] to hear the song as it would have been played in 1854.
17. *symptoms like cholera*: Tyndall described his symptoms in his journal entry of 5 August (Journal, JT/2/13b/680).
18. *Sandrock Hotel*: a hotel in the village of Niton, Isle of Wight, near Ventnor ('Rock Cottage History', *Rock Cottage, Niton, Isle of Wight,* http://www.rockcottageniton.co.uk/information/history-2/ [accessed 15 October 2016]).
19. *Blackgang Chine*: the United Kingdom's oldest amusement and curiosity park, founded in 1843 by Alexander Dabell (1808–98) and made popular by its display of a whale skeleton that had beached nearby in 1842 ('About Us', *Blackgang Chine*, http://www.blackgangchine.com/about-us [accessed 15 October 2016]).
20. *seeing my letters*: not identified.
21. *ever met in any man*: Tyndall noted his growing appreciation for Hector in his journal: 'A sound solid liking grew up in my heart towards Hector, and although we differed sometimes our differences were so slight and of such a nature as to be converted for each into indications of the character of the other' (Journal, 5 August 1854, JT/2/13b/680–81).

22. *dined at Werda &c*: this refers to an important entry that Tyndall made in his journal on 2 March 1851. After stating that 'This day Tom and I dined at Werda', Tyndall wrote the following about Tom: 'He thinks my thoughts and shares my creed, and knows almost as much about me as I do myself' (Journal, JT/2/13b/523).
23. *my fancy yet*: paraphrased from Tennyson's *Locksley Hall*: 'Oh, I see the crescent promise of my spirit hath not set: Ancient founts of inspiration well through all my fancy yet' (Alfred Tennyson, *Locksley Hall*, in *Poems* [Boston: W. D. Ticknor, 1842], lines 187–88). Tyndall wrote to Hirst in April 1853 that he 'bought Tennyson's poems the other day—they are wine to the intellect' (see letters 0743, n. 11 and 0793, nn. 8 and 9).

From Thomas Archer Hirst 11 August 1854 0961

Queenwood College | Augst 11th 1854

My dear John

Enclosed I return you Mrs Coxe's letter[1] which pleased me so well that I have copied a portion in my journal.[2] I admire the woman's frank unhesitating admiration as well as her unusual penetration but I am afraid on acquaintance we should find her too intellectual, a little too fond of defining attributes. I was rather startled at your attack of cholera on the Island[3] and am thankful you have re covered from it. I expect your Cousin[4] will have left you before now, I should have been glad too to have known more of him. We are exceedingly busy now our 110 boys having almost all arrived.[5] I have already given two lectures. The introductory one I send you,[6] you will understand it better than my hearers It is plain but sincere and although it did not, I believe, please some of the masters I feel convinced that I only said what required to be said. Send me it back when read and recollect that it is written for delivery and not for publication. I have received Magnus' memoir[7] and will set about translating it immediately I will have it finished as soon as my duties will permit. I am on the point of taking a house at Broughton it is close to Tomkins'[8] but a little further down the valley towards Mottisfont.[9] I had Simpson here one day last week he was sorry he did not see you. Every week I have cheering news from Ireland, my weekly letter[10] encourages me greatly it is a solace to me throughout the week. D^{r}. Atkinson will soon call on you he is paying a visit to England do not tell him of my prospects but tell him I want him to come to Queenwood for a day. He wishes to earn a little money by translating chemical articles; if Francis or yourself could put a little money in his pocket in this way I know you will for he is a deserving fellow and has not much money to spare.

Wrights prospects are favourable enough now. Next month he enters his house and will have his four boys. We are going to have another junior master immediately and in October I shall remove to Broughton in all probability.

I hope your health continues good and that you are progressing with your researches to your own satisfaction. As soon as I have translated Magnus' I will commence Beetz's[11] memoir though it has not yet arrived.

With best love believe me | Yours affectionately, | T.A. Hirst.

RI MS JT/1/H/202
RI MS JT/1/HTYP/355

---

1. *Mrs. Coxe's letter*: letter missing but see n. 2 for an excerpt.
2. *copied a portion in my journal*: from Hirst's journal entry of 11 August 1854: 'In a letter from Mrs Coxe (A. Cumming's aunt) to Tyndall occurs the following passage worth recording. It was written shortly after he had seen her for the first time:—"I will tell you now the very words of Dr Playfair in reply to my (womanlike!) question "Is he handsome?" "Not in the ordinary sense of that term, but it is a countenance full of genius—he is simple and playful as a child—and of most fascinating manners." One element of this "fascination", the existence of which I admit most fully, is in the genial spirit with which you seem to raise others to your own level, instead of making them feel a painful consciousness of inferiority. Above all, one feels thoroughly to "believe in" you, and I think unless you have the conviction that one is honest and true, all the qualities in the world are worth little or nothing"' (Hirst, 'Journals', 11 August 1854). Tyndall referred to Coxe's comments about Playfair in letter 0919.
3. *attack of cholera on the Island*: on his recent trip to the Isle of Wright, Tyndall told Hirst that he found himself 'invaded by symptoms like cholera' (see letter 0960, n. 17).
4. *your Cousin*: Hector Tyndale.
5. *110 boys . . . arrived*: Hirst was referring to the beginning of classes at Queenwood College.
6. *introductory one I send you*: not identified. However, Tyndall did respond with criticism to the lecture, as noted in his journal entry of 16 August 1854 (Journal, JT/2/13b/683).
7. *Magnus' memoir*: Tyndall must have sent Magnus's memoir on red and black sulphur to Hirst to be translated for the *Phil. Mag.* (see letters 0943, n. 2 and 0948, n. 4).
8. *Tomkins'*: may refer to the established residence of the Steele Tomkins family, longtime residents of Broughton near Stockbridge, Hampshire, England ('Anne Steele on Suffering and the Silence of God', in C. Y. Aalders, *To Express the Ineffable*: *The Hymns and Spirituality of Anne Steele* [Eugene: Wipf and Stock, 2008], p. 105, n. 9).
9. *Mottisfont*: refers to Mottisfont Abbey, a historic priory that was transformed by Sir John Barker-Mill (1803–60) into an estate known for its hunting and shooting ('The History of Mottisfont', *Mottisfont*, https://www.nationaltrust.org.uk/mottisfont/features/the-history-of-mottisfont [accessed 15 October 2016]).
10. *from Ireland, my weekly letter*: this refers to his correspondence with Anna Martin, who Hirst recently proposed to marry (see letter 0953). See also letters 0949; 0951, n. 3; and 0954.

11. *Beetz'*: Wilhelm von Beetz (1822–86), a German physicist who worked on magnetism and electrochemistry. Hirst was to translate Beetz's paper and address from 13 June to the Berlin Academy of Science, 'Ueber die Leitungsfähigkeit für Elektricität, welche Isolatoren durch Temperaturerhöhung annehmen', *Poggend. Annal.* 92 (July 1854), pp. 452–66. It was published in English as M. W. Beetz, 'On the power of conducting Electricity assumed by Insulators at High Temperatures', *Phil. Mag.* 8:51 (September 1854), pp. 191–201.

From Thomas Archer Hirst 15 August 1854 0962

Queenwood College | Aug^st^ 15th 1854

My dear John.

Your candour has lost you no nights lodging, if you were to become less candid towards me you would deprive me of one of the most valuable privileges of your friendship I sent you the lecture in order to receive from you the excellent letter you have sent me.[1] I anticipated most of it and there are but few of your remarks against which I am inclined to kick.[2] The verbal objections are all just in this respect your criticisms are always invaluable they have already been of great service to me. What you say about systems of education is perfectly correct to another audience I should have used no such strong expression, but I spoke to people who under system understand the treatment of boys and have but one blind method of treatment, they could not understand a system which should reserve free scope for all idiosyncrasies and I chose to make my sentence paradoxical by using the term 'system' as they virtually understand it. The paragraph about Natural & Scholastic Laws is parenthetical and perhaps misplaced. I hesitated whether to strike it out or expand it which I think I could then have done I never yet wrote a lecture without being conscious that there were sentences which were mere echoes of others I had written or thought of before. In which my own life blood may once have pulsed but did not then do so. Good writing consists in sticking close to the feelings which are uppermost in you when writing. But unfortunately there come periods when having dwelt upon these feelings sufficiently long you feel urged to pass to others before you have properly matured them. Perhaps we might say the heart possesses more inertia than the intellect and thus the transition from one subject to another is made easily by the latter then the former. Nevertheless I can justly say that a great part of this lecture was as sincerely felt when written as any other I ever attempted.

Magnus is translated[3] and will be sent to Francis by this post. I shall commence Beetz[4] immediately. I underlined Dr Atkinson merely as an easy manner of telling you he had made his examination I do not think you will offend him by calling him M^r^. But when new there is a certain allowable satisfaction to a student in the title (particularly when properly earned) of which I would not care to deprive him. If he is worth anything he will some day hold it as cheaply as you do.

Goodbye dear John Thine as ever | T. A. H.

RI MS JT/1/H/203
RI MS JT/1/HTYP/356

1. *letter you have sent me*: letter missing.
2. *remarks against which I am inclined to kick*: Tyndall was pleased with Hirst's response, recording in his journal entry of 16 August that he had received 'an excellent letter from Tom, on my criticism of his lecture' (Journal, JT/2/13b/683).
3. *Magnus is translated*: see letters 0943, n. 2; 0948, n. 4; and 0961, n. 7.
4. *Beetz*: see letter 0961, n. 11.

From Thomas Charles Higginson 21 August 1854 0963

Ipswich | August 21st. 1854.

My dear Tyndall,

What will you say when I tell you that I have been in London since the rect. of your letter[1] and that I have not been able to see you, the facts of the case will explain how it occurred; I went to meet my sister,[2] who volunteered to cheer my old-Bachelor home. We stopped with some friends out at Kentishtown[3] and as I could only spare from my duties two or 3 days I was obliged to take her at Railway speed to see all the sights of London that could be seen in the above time—Sydenham,[4] the Tower[5] &c—so that I could not leave her for an hour. I greatly regretted my inability to see you, but console myself with the thought that the pleasure of seeing such a sincere and old friend is only deferred for a short time as I hope to be in London again shortly. You did not tell me whether you had become a benedict.[6] Nearly all my contemporaries have left me to mourn over my single blessedness (if such it should be called) so I suppose you are to be included in the list. I hope your present appointment[7] is a lucrative one.[8] No one would be more delighted to hear such was the case than

your old and sincere | Friend | T. C. Higginson.

RI MS JT/1/TYP/2/597
LT Typescript Only

1. *your letter*: letter missing.

2. *my sister*: possibly Mina Higginson (see letter 0040, n. 11). Not further identified.
3. *Kentishtown*: Kentish Town is an area of northwest London in the Borough of Camden.
4. *Sydenham*: now an area of southeast London, Sydenham was located in Kent until 1889 when the County of London was created. In 1854 when Higginson would have visited, the Crystal Palace (see letter 0767, n. 7), which was from the Great Exhibition of 1851, had been relocated to Sydenham Hill, making the area an important tourist destination.
5. *Tower*: the Tower of London.
6. *benedict*: see letter 0885, n. 4.
7. *present appointment*: Tyndall was appointed Professor of Natural Philosophy at the RI on 4 July 1853. See letter 0762 for the offer of appointment.
8. *a lucrative one*: Tyndall received £200 a year (see letter 0762).

To Thomas Archer Hirst [25 August 1854][1] 0964

Friday Afternoon

My dear Tom

I was so awfully busy yesterday that I had not time to thank you for your valuable letter[2]—and I am so awfully busy to day that I can do nothing more than this. I have the spiral:[3] all I want is the magnet.[4]

Ever yours affectionately | John

Last Sunday's excursion[5] set me up for the week. I have got a great deal of work done, but the brain is now beginning to simmer again. So on next Sunday I will take a run to Gravesend[6] or somewhere else.

RI MS JT/1/HTYP/359
LT Typescript Only

1. *Friday*: Hirst dated this letter 'Aug. 25th. 1854', which was indeed a Friday, and appears to be accurate based on when Tyndall travelled to Gravesend (see n. 6).
2. *your valuable letter*: letter missing.
3. *the spiral*: Tyndall may have been borrowing the spiral that he had ordered for Hirst earlier in the year (see letters 0876, n. 7).
4. *all I want is the magnet*: see letters 0879 and 0965.
5. *last Sunday's excursion*: Tyndall had travelled by train to Winchester, Hampshire and then walked to Queenwood on Sunday, 20 August (Journal, JT/2/13b/684). Hirst recorded in his journal on 20 August 1854 that 'After dinner to-day John Tyndall walked unexpectedly into the Lodge' (Hirst, 'Journals').
6. *take a run to Gravesend*: in his journal, Tyndall recorded travelling to Gravesend, an ancient town in northwest Kent, England, on Sunday, 27 August (Journal, JT/2/13b/685).

From Thomas Archer Hirst [27 August 1854][1] 0965

Dear John,

The enclosed[2] will be sufficient for your purpose[3] I think. I go this minute to pack up the electro-magnet[4] for you and in it I shall place the 12 cigars which I stupidly forgot to give you at Winchester.[5] Our new master[6] arrived yesterday and commences work to-morrow. The question about the perpetual motion[7] I will consider soon.

Yours affectionately, | T.A. Hirst.

On looking at the electro magnet I find that the coils which fit around the soft iron poles are not here. Have you got them and is it the remaining part which you now want?

(Here follows Problem)[8]

RI MS JT/1/HTYP/316
LT Typescript Only

---

1. *[27 August 1854]*: LT noted at the top of the typescript that the letter 'must be between Feb 20 and March 2'. However 27 August 1854 seems more likely as Tyndall and Hirst were together at Winchester the previous week while Tyndall requested that Hirst send him back the electro-magnet in letter 0964 dated 25 August (see n. 4).
2. *enclosed*: in the typescript LT inserted '[a long calculation]' here. Not otherwise identified.
3. *your purpose*: not identified.
4. *electro-magnet*: Hirst was referring to the electro-magnet he borrowed from Tyndall (see letters 0833; 0874; 0879; 0902; 0965; and 0968).
5. *Winchester*: in his Hirst wrote that he and Wright 'took [Tyndall] back to Winchester' (Hirst, 'Journals', 20 August 1854). Tyndall also described going to Winchester with Hirst and Wright in his journal entry of 20 August (JT/2/13b/684).
6. *new master*: most likely Mr Blanche (see letter 0968, n. 10). In his journal, Hirst wrote: 'A new master, Mr Blanche, has arrived' (Hirst, 'Journals', 27 August 1854).
7. *The question . . . perpetual motion*: Craven had mentioned that he expected Tyndall would eventually make a major discovery, such as perpetual motion (see letter 0873). Not further identified.
8. *(Here follows the Problem)*: enclosure missing.

To Thomas Archer Hirst [31 August 1854][1] 0966

Thursday

My dear Tom.

Do you feel inclined to spend Sunday in /putting/ heath.[2] If so and if you name any place within excursion range I will meet you. If you like I will meet you on the horse monument[3]—or at Winchester or at Bishopstoke[4] at 10 oc. I am working very hard, and my head is sure to break down at the week's end. If you don't feel heartily inclined just say so—I will then run out to Windsor[5] and spend the evening with Francis at Richmond.[6]

John.

Let me hear from you by Saturday morning

Elton Hall FAM 124

---

1. *[31 August 1854]*: while Hirst noted 'John Tyndall | Sep. 30th 1854' and LT wrote '[Probly Aug 30]', the date of this letter is Thursday, 31 August 1854. Letter 0968, also written on 31 August, was a direct response to Tyndall's request contained herein to make an excursion with Hirst. In his journal entry of Sunday, 3 September 1854, Tyndall recounted his journey to Windsor where he 'Sauntered through the castle labyrtinths, heard a portion on the Church service. . . . Dined at the White Hart and came to Richmond, purposing to spend the evening with Francis' (Journal, JT/2/13b/688).
2. *heath*: possibly refers to Horton Heath, a village in Hampshire, England located approximately 10 miles southeast of Winchester.
3. *horse monument*: refers to the folly located at the top of Farley Mount, the highest hill in Hampshire, erected in remembrance of a race horse. The plaque reads, 'Underneath lies buried a horse, the property of Paulet St. John Esq., that in the month of September 1733 leaped into a chalk pit twenty-five feet deep afoxhuntiing with his master on his back and in October 1734 he won the Hunters Plate on Worthy Downs and was rode by his owner and was entered in the name of "Beware Chalk Pit"' ('The "Horse Monument" at Farley Mount', *Hampshire Country Council,* http://www3.hants.gov.uk/countryside/fmcp/monument.htm [accessed 14 December 2016]). Farley Mount is located 6 miles west of Winchester. Tyndall took a walk in Farley Mount and visited the Monument on Sunday, 20 August 1854 (Journal, JT/2/13b/684).
4. *Bishopstoke*: a village in Hampshire, England located 8 miles south of Winchester, near Horton Heath.
5. *Windsor*: refers to the town in Berkshire, England and site at which Windsor Castle was built by William the Conqueror beginning in 1070.
6. *Richmond*: Francis and Sarah Taylor lived in Alton Lodge, Richmond (Brock and Meadows, *Lamp of Learning*, p. 134). See also letter 0786, n. 10.

From Margaret Allen 31 August 1854 0967

Charnley House Fulwood | Aug 31st /54.

My dear Tyndall

I thank you for the letter[1] I lately received from you but the conclusion of the epistle gave me cause for some little regret—Now, can you not possibly contrive to visit us,[2] either a day or two before your services are required in Liverpool,[3] or after your duties are concluded? If not I must make my best endeavour to see you, by coming over to Liverpool myself—I have long had a great desire to be present at some of those scientific meetings—and if we are not to have the pleasure of having you eventually here, I ask you to have the kindness to direct my infant footsteps & name to me on what day I shall visit you. I should much like to hear my dearest Friend speak, how proud I should be to see you standing giving instruction[4] even to some of the most intelligent men in our land. Tell me on what day the Association commences[5] to hold their meetings. I have seen the date in a newspaper but the time has slipped my memory. —It certainly will be a disappointment I shall not readily overcome if I do not see you. —it is above a year & a half[6] since I had that gratification.

Do you remember a period in your life when you resided at Goosnargh[7] I was there the other day with Mamma.[8] I have never been there until this time, since the Sunday I once spent there with you—that visit ever remains in my mind the most pleasing remembrances, from thence I date the commencement of my firm attachment towards you—the house you resided in has passed into other hands, we were ushered into the large room on the right-hand-side of the house, but I said to Mamma 'I don't like this room'—So I had to assume a very pleasant countenance and request 'Will you have the kindness to show us into the small parlour beyond the bar?'—here I felt myself; although the appearance of the room is changed. yet I dispensed with material objects & I could fancy your spirit still traversing around the spot—here is the window where we chatted so pleasantly in that corner stood a small table which contained writing materials &c & from which, when you were absent, I so audaciously took up your journal & read. —but I must denote by telling you that I then *[knew]* a contrast, which even at that period was painful to my mind, but how fully has it been exhibited—I solve not this for you—You can easily guess to what I refer. —My dear Tyndall permit me to hear from you as soon as your engagements will permit

I am ever | Sincerely Yours | Gretchen[9]

RI MS JT/1/A/79

1. *the letter*: letter missing.
2. *visit us*: see letter 0947, n. 4.
3. *services required in Liverpool*: Tyndall served as secretary for Section A (Mathematical and Physical Science) of the Twenty-Fourth Meeting of the BAAS in Liverpool (see letters 0950, n. 2; 0972, n. 3; and 0975).
4. *giving instruction*: Tyndall presented the papers 'On the Peculiarities of the Magnetic Field' and 'On the Diamagnetic Force' during section A of the BAAS meeting. See 'Twenty-Fourth Meeting of the British Association for the Advancement of Science', *The Athenaeum* 1405 (30 September 1854), pp. 1172–83, on p. 1174; and 'Twenty-Fourth Meeting of the British Association for the Advancement of Science', *The Athenaeum* 1406 (7 October 1854), pp. 1202–12, on pp. 1203–4, for lengthy discussions of Tyndall's papers.
5. *Tell me . . . commences*: see n. 3.
6. *year & a half*: Tyndall had visited Allen in Preston in January 1853, when Allen's son was dying (see letter 0698, n. 5). Tyndall also recorded details about this visit in his journal (1 January 1853, JT/2/13b/594).
7. *when you resided at Goosnargh*: Tyndall reminisced about his time in Goosnargh with the Allens in his journal entry of 18 December 1849 (Journal, JT/2/13b/471–72).
8. *Mamma*: not identified.
9. *Gretchen*: see letter 0872, n. 12.

From Thomas Archer Hirst 31 August 1854 0968

Queenwood College | Aug[st] 31st 1854

My dear John.

Go to Richmond on Sunday next and let us postpone our meeting[1] for a week or two when I know you will require another journey. My reason for so doing is a good one Atkinson has been and stopped until Wednesday,[2] his visit was very pleasant to me but it has interfered with my work and made me a little behind with it. Tomorrow I shall have a clear days work before me and on Sunday I must write two or three important letters. Next Sunday week or fortnight as you like I will arrange to meet you wherever you please.[3]

The Phil. Mag. this month is very meagre.[4] It does not contain your report of Wertheims memoir[5] as I expected it would.

On Monday I sent the Electro-magnet[6] which I hope you received The other day I tried to recall your reasonings on the Diamagnetic and Magnetic problems[7] and I could not do so to my satisfaction I wish you would write me a short account of them[8] and I will give them my best consideration for you.

In my leisure time I am reading a little book just now which you once recommended to me a tract on Education by the Dean of Hereford.[9] There are many satisfactory hints and suggestions in it plainly shewing that the man has broad and liberal notions of education

Queenwood is in excellent condition at present. McDougall is gaining experience and I have some hopes that he will profit by it. He is a grim earnest fellow and not without common sense though his normal school training has unfitted him greatly for a position such as Queenwood. He appears to have been accustomed to coercing rough ill cultivated pupils, and he is only just beginning to see that his present pupils are really generous and gentlemanly little fellows. It is just possible that one or two collisions which have already occurred may teach him something. The fellow has really some character about him and on that fact all my hopes rest; he is dogged and blunt enough but you can push against him. His companion Blanche the last comer is far inferior he is an empty headed canting fellow, last Mutual improvement night he volunteered a lecture[10] on the fixed stars and gave them a twaddling Methodist parson's sermon which made every teacher present uneasy in his seat. Unfortunately or perhaps fortunately I was not there. If it had been a good lecture it would have been a great service and a good introduction for him, as it was it was a miserable failure and it will be his last lecture at Queenwood. Let me hear soon from you and believe me

Yours affectionately | T. A. Hirst

RI MS JT/1/H/204
RI MS JT/1/HTYP/358

---

1. *Go to Richmond . . . postpone our meeting*: see letter 0966, wherein Tyndall invited Hirst to meet him near Winchester. Hirst declined, and Tyndall instead travelled alone to Windsor and Richmond (see letter 0966, n. 1). See also letter 0786, n. 10.
2. *Atkinson . . . Wednesday*: Tyndall noted Atkinson's arrival in his journal entry of 26 August 1854 (Journal, JT/2/13b/685).
3. *meet you wherever you please*: Tyndall later suggested they meet in Winchester on 17 September (see letters 0970, n. 12 and 0971).
4. *Phil. Mag . . . meagre*: see *Phil. Mag.* 8:51 (September 1854), pp. 161–240.
5. *Wertheims memoir*: Hirst had been translating M. G. Wertheim, 'Mémoire Sur la double réfraction temporairement produite dans les corps isotropes, et sur la relation entre l'élasticité mécanique et entre l'élasticité optique', *Annal. Chim. et Phys.* 40 (1854), pp. 156–221. The English translation would appear in two parts: M. G. Wertheim, 'On the double Refraction temporarily produced in Isotropic Bodies', *Phil. Mag.* 8:52 (October 1854), pp. 241–61; and *Phil. Mag.* 8:53 (November 1854), pp. 342–57.
6. *I sent the Electro-magnet*: Tyndall requested the magnet in letter 0964, n. 4 and Hirst said that he was packing it up in letter 0965, n. 4.
7. *your reasonings on diamagnetic and magnetic problems*: see letters 0710, n. 7 and 0821, n. 4.
8. *write me a short account of them*: see letter 0970.

9. *tract on Education by the Dean of Hereford*: Richard Dawes (1793–1867), an educator and the Dean of Hereford from 1850 until his death in 1867. The 'tract on Education' to which Hirst referred was Dawes's *Suggesting Hints towards Improved Secular Instruction, Making it Bear upon Practical Life* (London: R. Groombridge & Sons, 1849).
10. *Blanche . . . volunteered a lecture*: the lecture is mentioned in a report of the 'Mutual Improvement Society', *The Queenwood Observer* 2:5 (31 August 1854), pp. 33–40, on p. 35.

From Tommy Grieve 3 September 1854 0969

Queenwood College | Sept. 3rd. 1854.

My dear Dr. Tyndall,

We have just come from church and it was very warm I go to east Titherly[1] to church because there is not room for us all at Broughton there are a hundred and ten boys.[2] We had a cricket match at stock bridge on Wednesday and we were beaten[3] but I think we are going to have another match soon. Last Tuesday Papa[4] was in London he was going to the continent I think he will come and see us. Yesterday I went with a good many other boys to bathe and Mr Iliff came with us I had a very nice bathe and on Wednesday we bathed in the cannal and Mr Haas came with us and before when he came to look for a place to bathe George William[5] got in and there was a boat coming along and he caut[6] hold of the boat and it carryed him along down the canal. It has been very warm for the last week and I hope it will continue this morning there was a thick fog but it soon cleared away.

I remain, | Your affectionate friend | Tommy Grieve.

RI MS JT/1/TYP/1/463
LT Typescript Only

1. *east Titherly*: misspelling of East Tytherley, a small village in Hampshire, England.
2. *a hundred and ten boys*: students at Queenwood College. Hirst had mentioned earlier that 110 boys arrived for the beginning of classes (see letter 0961, n. 5).
3. *We had a cricket match . . . were beaten*: an account of the match can be found in R. Dunville, 'Miscellaneous', *The Queenwood Observer* 2:5 (31 August 1854), pp. 33–40, on pp. 33–34.
4. *Papa*: possibly H. M. Grieve (see letter 0839); not further identified.
5. *George William*: George William Edmondson.
6. *caut*: LT notes: 'sic'. Misspelling of caught.

To Thomas Archer Hirst [11] September 1854[1] 0970

Sep. 1854.

My dear Tom

Although I am dreadfully pushed still I think I must write to you. Many thanks for the proof.[2] The paragraphs were numbered, not by me, but by Hughes himself, and your remark serves simply to strengthen my determination to alter them. I will talk about the 'information' you require[3] when we meet, five minutes conversation will save a great deal of writing. I have been engaged in a tolerably exhaustive comparison of magnetism and diamagnetism,[4] and have I trust established a compete antithesis throughout, and proved beyond a doubt that diamagnetic bodies possess polarity the reverse of that of iron. Weber's views and mine thus far coincide, but I am firmly convinced that Weber's hypothesis is false.[5] The experiments please me because they leave no doubt upon the mind and I have strengthened the action, and improved the mode of operation so as to permit me to detect the actions sought in masses of bismuth not a few grains in weight, but in cylinders 14 inches long and an inch in diameter. Feilitsch[6] has written to me[7] recently and sent me his recent memoirs[8] where he professes to prove that diamagnetic bodies and magnetic ones possess the same polarity. Many of the actions which he failed to obtain I have succeeded in obtaining, and I imagine his experiments cannot stand for an instant in opposition to the evidence which I can bring to bear against them.

I had a note from Col. Sabine[9] a few days ago saying that Dove had arrived and would be glad to see me—he is staying with Sabine, and I spent a very agreeable evening with him there.[10] On Saturday evening I had both him and Faraday here to tea. We spent two or three hours quite delightfully—an uninterrupted and almost ardent flow of talk. We were all pleased. Faraday told me on going that he had not spent such an evening for years. Dove I and Francis spent yesterday at Windsor.[11] I was suffering from a severe cold, but the excursion did me good and I am well to day.

What do you say if we meet at Winchester on Sunday[12] Name any other place you like

Your affectionate | John

It is needless for me to say that I rejoice to hear of the cheering news from Ireland. Faraday asked me on Saturday 'how is your young friend'—meaning you—I told him you were about to get married.[13]

RI MS JT/1/HTYP/359
LT Typescript Only

1. *[11] September 1854*: based on Tyndall's journal entries, and discussion of events of the week contained herein (see nn. 9–12), this letter was written on 11 September (Journal, JT/2/13b/688–90).
2. *the proof*: the proof was mentioned in Tyndall's journal entry of 23 August 1854 (Journal, JT/2/13b/685). Possibly refers to S. Hughes, 'Ueber den Ausfluss des Leuchtgases und Bewegung desselben in den Leitungsröhren', *Der Civilingenieur, 1* (1854), pp. 61–66. Samuel Hughes (n.d.). Not further identified.
3. *'information' you require*: Hirst had asked Tyndall to write a short account of his 'reasonings on the Diamagnetic and Magnetic problems' (see letter 0968).
4. *comparison of magnetism and diamagnetism*: Tyndall would present the results of these experiments at the BAAS meeting in Liverpool (see letter 0967, nn. 3 and 4).
5. *Weber's views . . . hypothesis is false*: presumably Tyndall he accepted Weber's results and conclusion (on polarity of bismuth being the opposite to iron) but not the theory/hypothesis underlying it. (See Jackson, 'Tyndall and Diamagnetism', pp. 467–68, and 472–73; W. Weber, 'Ueber die Erregung und Wirkung des Diamagnetismus nach den Gesetzes der inducierten Ströme', *Poggend. Annal.* 73 [1848], pp. 242–56; and also letter 0958, n. 6).
6. *Feilitsch*: misspelling of Fabian Carl Ottokar von Feilitzsch (1817–85), Professor of Physics at the University of Greifswald, Germany. See also letter 1001, n. 3.
7. *written to me*: letter missing, although Tyndall noted its receipt in his journal on 9 September (Journal, JT/2/13b/690).
8. *recent memoirs*: Feilitzsch likely sent Tyndall these memoirs: Feilitzsch, 'Erklärung der diamagnetischen Wirkungsweise durch die Ampère'sche Theorie', *Poggend. Annal.* 92 (1854), 366–401, 536–76; and Feilitzsch, 'Erklärung der diamagnetischen Wirkungsweise durch die Ampère'sche Theorie', *Annalen der Physik und Chemie* (1852), pp. 87, 206–26, 427–54. See Jackson, 'Tyndall and Diamagnetism'.
9. *note from Col. Sabine*: letter missing. However, Tyndall noted receipt of the letter in his journal on 4 September 1854 (Journal, JT/2/13b/688).
10. *spent a very agreeable evening with him there*: Tyndall dined at Col. Sabine's home with Dove on 6 September 1854 (Journal, JT/2/13b/689).
11. *spent yesterday at Windsor*: on 10 September, Tyndall noted in his journal that he had 'met Dove at Waterloo and took up Francis at Richmond. Walked from Windsor to Virginia Water where we dined—a long political discussion between Francis and Dove after dinner' (Journal, JT/2/13b/690). See also letter 0966, n. 5.
12. *Winchester on Sunday*: Tyndall would meet Hirst on 17 September, having noted in his journal, 'A glorious day on the Downs above Winchester with Tom' (JT/2/13b/690). See also letter 0971.
13. *about to get married*: Hirst had recently become betrothed to Anna Martin (see letters 0949; 0951, n. 3; 0953; and 0954).

To Thomas Archer Hirst [15 September 1854][1] 0971

Royal Institution Friday

My dear Tom

You ought to have written to me[2] thus:—

John be it hail rain or snow, thunder or lightning or any other elemental convulsion whatever, I shall expect you in Winchester on Sunday.[3] I must get out of London on Sunday and to Winchester I will go

Yours | John

RI MS JT/1/HTYP/360
LT Typescript Only

1. *[15 September 1854]*: Hirst dated this letter 'Sep. 17th 1854' but this was likely the date that he received it. 'Friday' was 15 September. Additionally, Tyndall and Hirst were together in Winchester on 17 September (Journal, JT/2/13b/691); see also n. 3.
2. *written to me*: this is presumably a response to a missing letter.
3. *Winchester on Sunday*: Tyndall had previously asked Hirst to meet him in Winchester on Sunday, 17 September (see letter 0970, n. 12).

From Margaret Allen 15 September 1854 0972

Charnley House, Fulwood | Sep $15^{\underline{th}}$ /54.

My dear Tyndall,

I suppose you will have received my last letter,[1] and as I have not had any reply I think I anger favourably, that I may expect to see you in Preston[2]—but if not I shall see you in Liverpool[3] if you will mention to me the most convenient day for visiting you. Now, I have a request to make & I believe you will not deny me—for if I thought it at all necessary I might urge the claim on some points, but I know your goodness of heart—well it is this—will you get for me—a striking facsimilie of yourself by means of a photograph—I have long wished to be in possession of this. —I think you will probably be able to procure it, both better taken & at a more reasonable charge than elsewhere. Dr. Hirst has your portrait, he told me how pleasant it was to look upon, and greet you each coming day. I frequently conjure your image & words, the one before me the other in reflection, how much more happy shall I be if able to contemplate the likeness of my absent friend, & almost hold a conversation with him.

You see I am a very tire some child, teasing you in this way & by my scrawls, well you can scold me if you are in the humour when I see you, till then and ever believe me

Yours Sincerely | Gretchen[4]

RI MS JT/1/A/80

1. *my last letter*: letter 0967.
2. *see you in Preston*: see letter 0947, n. 4.
3. *see you in Liverpool*: see letter 0967, n. 3.
4. *Gretchen*: see letter 0872, n. 12.

From Heinrich Debus 24 September 1854 0973

Queenwood College | Stockbridge | Hants | 24 Septbr 54

Lieber John!

Gern würde ich schon früher die Analyse des Stückes „Shell" welches Du mir übergabest vollendet und Dir die Resultate zugeschickt haben wenn ich nicht durch besondere Umstände verhindert gewesen wäre. Ich wollte nämlich dass Richards die Analyse mit mir machen sollte um denselben den Vortheil einer solchen Arbeit zuzuwenden. Da nun Richards wenig Zeit hat und ich wusste dass die Analyse nicht eilig war so erklärt sich hieraus dass ich jetzt erst im Stande bin Dir die gewünschten Resultate zu übersenden.

Alle Wägungen sind von mir selbst und alle Arbeiten unter meinen Augen und mit meiner Hülfe ausgeführt; ausserdem ist die Analyse zweimal gemacht um die erhaltenen Resultate zu controliren.

0.808 grm Substanz bei 120° im Luftbad in einem Strom Kohlensaure getrocknet, verloren 0.017 grm an Gewicht welches 2.10 per cent hygroscopisches Wasser anzeigt.

0.359 grm verloren beim Glühen an der Luft 0.050 grm welches entspricht 13.92 Wasser mit einer Spur Bitumen. Zieht man hiervon das hygroscopische Wasser ab so bleiben 11.82 per cent Verbindungswasser.

0.339 mit *[*$NaOCO_2$*]* aufgeschlossen gaben. 0.155 SiO_3; 0.034 Fe_2O_3. 0.067 Al_2O_3; 0.012 $CaOCO_2$; 0.031 *[1 word illeg]*.

Hieraus berechnet sich für 100 Theile

SiO_3 = 45.72 Kieselsaure
Fe_2O_3 = 10.03 Eisenoxyd
Al_2O_3 = 19.76 Thonerde

CaO = 1.99 Kalkerde
MgO = 3.35 Magnesia
2.10 Hygroscopisches Wasser
11.82 Verbindungswasser mit einer Spur Bitumen.
94.77

Es fehlen hier also noch 5.23 per cent; diese sind Kali, Natron, Schwefelsaure. Ich habe einen Apparat bestellt zur Bestimmung der Alkalien und sobald ich denselben erhalte werde ich sogleich die Menge des Kalis und Natrons ermitteln und Dir das Resultat schicken. Man muss nämlich das Mineral mit Flussäure aufschliessen und hierzu ist ein Bleigefäss nöthig das jetzt in Romsey gemacht wird.

Das Mineral enthält fast nur Eisenoxydul (FeO).—Wenn sich eine Substanz in Säuren auflöst dann kann man bestimmen wie viel FeO und wie viel Fe_2O_3 darin ist. Wenn aber eine Substanz in Säuren unlöslich ist und Wasser enthält das erst bei hoher Temperatur entweicht dann ist es fast unmöglich zu sagen wie viel von dem Oxyd und wie viel von dem Oxydul des Eisens darin ist. Wenn aber eine Substanz in Sauren unlöslich ist und Wasser und Bitumen enthält wie es mit dem „Shell“ der Fall ist dann sind die Schwierigkeiten noch grösser die sich der Bestimmung des Eisenoxydul's entgegen stellen. Darum ist in der obigen Analyse alles Eisen als Eisenoxyd aufgeführt. 10 per cent Eisenoxyd entsprechen 9 per cent Oxydul.

Aus denselben Gründen ist das Verbindungswasser das erst beim Glühen entweicht nicht ganz genau. Da die Substanz Eisenoxydul enthält und das letztere beim Glühen Sauerstoff aus der Luft, oder unter Umständen aus dem Wasser aufnimmt das die Substanz enthält, so muss der Gewichtsverlust den der „Shell“ beim glühen erleidet aus folgenden Grössen zusammengesetzt sein

Shell vor dem Glühen = Wasser + Spur Bitumen + SiO_3 + AlO_3 + FeO + Fe_2O_3 + CaO <etc. etc. etc>

Shell nach dem Glühen = SiO_3 + AlO_3 + FeO + Fe_2O_3 <etc. etc. etc> + einer kleinen Menge O die das Eisenoxydul absorbirt haben kann

Die Differenz zwischen beiden Grössen ist daher das Wasser + Spur Bitumen—dem Sauerstoff den das FeO absorbirt hat. Da nun aber unter der Voraussetzung dass Alles Eisen als Eisenoxydul vorhanden ist nicht mehr als 1 per cent O aufgenommen sein kann aber in der That viel weniger absorbirt ist so muss der wahre Werth des Verbindungswassers zwischen 11.82 und 12.82 per cent liegen. Demnach ist das Resultat

SiO_3 = 45.72
Fe_2O_3 = 10.03 oder FeO = 9.03
Al_2O_3 = 19.76
CaO = 1.99

MgO = 3.35
Hygroscopisches HO= 2.10
Verbindungs HO = 11.82 bis 12.82
Das Bitumen ist sehr gering nur eine Spur.

Indem ich wünsche dass diese Bestimmungen Deinen Zwecken entsprechen bin ich wie immer

Dein | aufrichtiger | H. Debus.

Queenwood College | Stockbridge | Hants | 24 Septbr 54

Dear John!

I would gladly have completed earlier the analysis of the piece of "Shell" which you forwarded to me[1] and sent you the results, if I had not been prevented by special circumstances. That is to say, I wanted Richards[2] to do the analysis with me, in order to give him the benefit of such a job. As Richards has little time right now and I knew that the analysis was not urgent, then this explains that I am only now in a position to send you the desired results.

All the weighings were done by myself and all work was carried out under my supervision and with my assistance; in addition, the analysis was done twice in order to check the obtained results.

0.808 g[3] of the substance dried at 120° in an air-bath under a current of carbonic acid lost 0.017 g of weight, which indicates 2.10 per cent of hygroscopic water.[4]

0.359 g lost 0.050 g during incandescence in air, which corresponds to 13.92 of water with a trace of bitumen. If one removes the hygroscopic water from this, then there remains 11.82 per cent of bonding water.

0.339 dissolved with *[*$NaOCO_2$*]* yielded 0.155 SiO_3; 0.034 Fe_2O_3. 0.067 Al_2O_3; 0.012 $CaOCO_2$; 0.031 *[1 word illeg].*

From this, one can calculate, for every 100 parts

SiO_3	=	45.72 silicic acid
Fe_2O_3	=	10.03 ferric oxide
Al_2O_3	=	19.76 aluminium oxide
CaO	=	1.99 calcium oxide
MgO	=	3.35 magnesia
		2.10 hygroscopic water
		11.82 bonding water with a trace of bitumen.
		94.77

There is therefore 5.23 per cent still missing here; this is potash, sodium carbonate, sulphuric acid. I have ordered an apparatus to determine the alkalis and, as soon as I receive it, I shall at once ascertain the quantity of potash and sodium carbonate and send you the result. One must, namely, dissolve the mineral with hydrofluoric acid, and for this a lead vessel is necessary, which is now being made in Romsey.

The mineral contains, almost exclusively, ferrous oxide (FeO).—If a substance dissolves in acids, one can then determine how much FeO and how much Fe_2O_3 is in it. If, however, a substance is insoluble in acids and contains water that escapes only at high temperature, then it is almost impossible to say how much of the higher oxide and how much of the lower oxide of the iron is in it. If, however, a substance is insoluble in acids and contains water and bitumen, as is the case with the "Shell", then the difficulties which stand in the way of determining the ferrous oxide are even greater. For this reason, in the above analysis all iron is listed as ferric oxide. 10 per cent of ferric oxide corresponds to 9 per cent of the lower oxide.

For the same reasons, the bonding water that escapes only during incandescence is not entirely precise. As the substance contains ferrous oxide and the latter, during incandescence, takes oxygen from the air or, possibly from the water that the substance contains, then the loss of weight that the "Shell" sustains during incandescence must be made up of the following variables

Shell before burning = water + trace of bitumen + SiO_3 + AlO_3 + FeO + Fe_2O_3 + CaO <etc. etc. etc>

Shell after incandescence = SiO_3 + AlO_3 + FeO + Fe_2O_3 <etc. etc. etc> + a small quantity of O that the ferrous oxide can have absorbed

The difference between the two variables is therefore the water + trace of bitumen—the oxygen that the ferrous oxide has absorbed. As, however, assuming that all the iron is present as ferrous oxide, [and] no more than 1 per cent of O can now be taken up, but in actual fact much less is absorbed, then the real value of the bonding water must lie between 11.82 and 12.82 per cent. Accordingly, the result is

SiO_3 = 45.72
Fe_2O_3 = 10.03 or FeO = 9.03
Al_2O_3 = 19.76
CaO = 1.99
MgO = 3.35
Hygroscopic HO = 2.10
Bonding HO = 11.82 to 12.82

The bitumen is very small, only a trace.

Wishing that these determinations meet your purposes, I am as always

Your | sincere | H. Debus.

RI MS JT/3/49, interleaved in Volume

---

1. *the results . . . over to me*: see letter 0897, n. 2.
2. *Richards*: W. Richards, a teacher at Queenwood College. Not further identified.
3. *g*: abbreviation of gram.
4. *hygroscopic*: 'said of bodies which readily absorb moisture from the air' (*OED*).

From James B. Bevington 25 September 1854 0974

Wandsworth Common | 25th. Sep. 1854.

My dear Tyndall

Two letters which I received to day remind me specially of you, add to this the conscience—the result a billet.

The one was from our ever charming friend Hirst, in answer to one I wrote to him about Harry Hill whose excellent conduct elicits my best praise—this informs me of your whereabouts and your proposed wanderings in North Wales.[1] The second was from my sister in law, Mrs Sarah Jenkins,[2] whom I think you once met at Wandsworth—at any rate she speaks of you—fancies you must be at Liverpool,[3] and would be delighted to see you at her house—Rock Park, Rock ferry, on the opposite bank of the Mersey[4]

In case you have the time and the inclination, I enclose you a note of introduction and hope you will give them a call—Mr Jenkins[5] is also to be found at his office—say 44 or there abouts—Castle Street Liverpool. So Hirst has had some attractive influence upon a distant comet whose path hereafter will circle near his own. Although his orbit may be some-what changed thereby yet it will give a beautiful halo of light to the constellation in which he moves and I hope thereby myself to gain a glimmer from this cometic wanderer.

This is a festive night with me—Good news at all hands—Hirst the Batchelor to be married;[6] my son that has been ill—well; George,[7] my brother, escaped from death's untimely clutch—nearly taken off by cholera, but now rapidly recovering. Myself, the curmudgeon, tempered with these trials and wiser made.

Take me with you at eventide as the sun descends behind the Welsh mountains and when you wander alone. Take me with you when you early rise and brush the fresh dew from the upland path. Leave me to play in the Rainbow of the Waterfall, then I will accompany you arm in arm when the night falls and we will enter the Hotel together and call for our drink.

'Twas thus that I sometimes beguiled my imagination—at the Rigi,[8] on the banks of Lac Lutzern[9] and on the mountains near Inter Laken.[10]

I remain | my dear friend | Yours affectionately | J.B.B.

RI MS JT/1/TYP/1/150
LT Typescript Only

1. *proposed wanderings in North Wales*: Tyndall travelled to North Wales from Liverpool after the BAAS meeting (see letters 0950, n. 2; 0967, n. 3; 0972, n. 3; and 0975). He recounted his journey in his journal entries of 27 September through 5 October 1854 (Journal, JT/2/13b/692–94).
2. *Mrs Sarah Jenkins*: presumed to be the wife of Mr. Jenkins (see n. 5).
3. *fancies you must be at Liverpool*: see n. 1.
4. *Rock ferry . . . the Mersey*: Rock Ferry in Birkenhead, Cheshire, England a town located on the west bank of the River Mersey across from Liverpool.
5. *Mr Jenkins*: a solicitor who resided at Rock Park, Rock Ferry, Cheshire ('Assignments for Benefit of Creditors', *The Solicitors' Journal & Reporter* [26 December 1857], pp. 121–40, on p. 140).
6. *Hirst the Batchelor to be married*: Hirst was engaged to Anna Martin (see letters 0949; 0951, n. 3; 0953; 0954; and 0970, n. 13).
7. *George*: George Bevington. Not further identified.
8. *Rigi*: Mount Rigi, also known as 'Queen of the Mountains', located in Central Switzerland in the Alps. It is almost completely surrounded by three lakes, Lake Lucerne, Lake Zug, and Lake Lauerz.
9. *Lac Lutzern*: refers to Lake Lucerne (see n. 8).
10. *Inter Laken*: the region located on an alluvial plain between Lake Thun and Lake Brienz in the Bernese Oberland.

To Thomas Archer Hirst 1[–2] October 1854[1] 0975

Foot of Snowdon[2] Oct. 1. 1854.

My dear Tom

My first act on this sunshiny Sunday morning after leaving my bed is to commence a letter to you. I thought of writing to you from Liverpool[3] but the confusion around me was so great, and the splashing and dashing of thoughts through my own cranium so bewildering that I was unable to settle down and do what I otherwise should have done. I met Faraday at the Euston Square station the morning on which I left London.[4] 'Now Tyndall' he said 'how do you travel? have you got your ticket?' I said no, and that before I saw him my intention was to travel second class. You shall do so, he said, I will travel second class too. He disappeared in a twinkling and purchased tickets for both of us: Murchison[5] and many of his friends were in a first class coupé[6] next door, and I endeavoured to persuade him to join them—No, it was useless, with me he would travel. Tennant[7] the mineralogist joined us afterwards and a most pleasant journey we made of it. I got a parcel of instruments unpacked the same evening.[8]

Next day I made my first communication,[9] it excited much discussion, in which Faraday Thomson and Whewell joined. A gentleman present asked me whether I had ever made experiments upon the human body with a galvanometer[10] on the production of electricity 'of animal magnetism in other words' he continued. My reply not appearing to satisfy him he looked at me enquiringly and demanded 'You have seen a galvanometer you know I suppose what a galvanometer is.' I assured him gravely that I had that felicity, and there was a general burst of laughter in the Section.

That evening there was a soiree in the magnificent new hall. Next day I made my second communication.[11] I was afraid to give it its proper dimensions and thus cribbed and curtailed it more than I ought to have done. Thomson completely backed out of the position which he had assumed in Belfast,[12] and completely disowned the interpretation of his views[13] as stated in Faraday's lecture. Thomson has in fact backed out of almost every position he has assumed in regard to the phenomena of diamagnetism and magnecrystallic action.[14] And he has done so leaving the public to suppose that he had been misconstrued or misapprehended—which tact may certainly increase his reputation with the general public, but in the private opinion of me at least does not add a whit to his nobleness.

Friday evening we had an address from Owen.[15] The room was crammed, in short the meeting as regards numbers has been a very successful one.

Saturday the Red Lions[16] dined together and we had a specimen of the tendency of human nature to relapse into the error which it once condemned. The Red Lions were instituted as a kind of concrete protest against the formality and expense of the dinner of the President. They were instituted for the purpose of enabling a few genial souls to dine genially and cheaply together, but alas the last dinner was more than double the expense of the President's, and the brilliancy of talent and eloquence though considerable was not in proportion. Foucault[17] was my guest and Dove was Francis's. Foucault and I have been much together. In some respects there was an affinity between us which brought us often into contact.

Wrightson and Atkinson have been at Liverpool. Wrightson enquired after you very warmly.

Sunday last I went to Preston to see Mrs Allen and met Schnackenberg[18] and Dickinson there. I spent a most pleasant day. I told you some thing of Mrs Allen's husband.[19] On Sunday I saw some letters from his father to her[20] in which, with bitterness of heart, he tells her that her life would not be safe in the society of his *[1 word illeg]* that he drinks and in his drunken fits is more like a fury than any thing human. She has a sweet intelligent little girl a daughter[21] too sweet and too intelligent, I fear, to live long in this world. We had plenty of music and I at least had plenty of laughter. The consequence was that I returned to Liverpool a far stronger man than I had departed. Dickinson desired to be remembered to his father, Schnackenberg's countenance and demeanour pleased me: there was a quiet earnestness in each which to me was pleasant to observe.

Beddgelert,[22] Monday morning

I must dash through the record of my doings and tell you at once that we crossed Snowdon yesterday. The morn was glorious—blue sky and bright sunshine. A couple of hours toil brought us above the region of the clouds which belted the hills and through which the black summits protruded, like volcanic islands through the sea. The sunlight upon these vapour masses was reflected back to us a dazzling white as if from fields of snow. Gradually as the day advanced the clouds melted and by the time we reached the summit the lakes at the bottom of the valleys were visible. We had many grand glimpses from the mountain's beak as we ascended, but the view from the summit and the peculiar structure of the summit itself are of the grandest description. The summit is not rounded as the Brocken, but scooped out here and there, leaving a knife edge at the top and a basin more than a thousand feet below. This adjacent horror was sustained by the black boldness of the crags and peaks around. The mechanical action which upheaved and tore asunder these immense ridges must have been a most extraordinary one, sea and subterranean force must have worked together to produce the rents and scoopings. We rested two hours at the summit pondering upon the sublime doings around us and then descended: the path part of the way lay along an edge at a sharp angle, the steep sides of which were nearly half a mile in depth. Deviating from the path, to which Francis with many warnings to me clung persistently, I clambered straight down the mountain side and reached the bottom half an hour before he did.

Festiniog,[23] Monday night.

I slept late last night at Beddgelert—one of my earliest remembrances is the reading of the poem on the death of Gelert,[24] and this morning I stood beside the hound's reputed grave. I remember weeping bitter tears over the dog's fate before I was 8 years old. We had a glorious day to day—the sun clear and cloudless over head and an earth beneath worthy of his light. It is not the height of these hills, though that is considerable, that I admire, but it is their peculiar contours and grouping. They set themselves at times like prostrate vultures with their claws turned defiantly against the sky. Some of wilder shape lift their conical summits up to the firmament as if they wished to kiss the evening star. Traces of vast glacial action were every where present the rock masses being rounded by the action of water and scratched by sliding icebergs. It is long since I felt such a vigour in my muscles as I do this evening.

I have gradually increased in strength since the first day, and do not know where this accumulation of force is to end. I said to Francis to day (he desires to be kindly remembered to you) I will advise Tom to come here when he gets married[25] instead of going to Paris: but then when I called to mind the wintry sky, the valleys wadded with mist and the sublime summits hidden by the grey clouds—when I thought of the biting air and the keen winter wind, I finally came to the conclusion that it would be better to let you carry out your own notion. Nevertheless when you have the time and money to spare you and Anna[26] would enjoy a trip in North Wales. Excuse this horrible rigmarole of a letter, it is written by bits and scraps whenever I can seize a pen. To night as the sun sunk amid the hills I thought of going without my dinner for the purpose of writing to you in a worthy manner, but the mountain air, a roast chicken and some choice Welsh mutton were too much for me. I dined like a hungry man and you must consequently be content with a sodden epistle. Francis is asleep on the sofa. We are perched upon the summit of a high hill at present, and tomorrow will see us facing the road to Dolgelley[27] and Llangollen.[28] Good bye Tom. I trust you are getting your house ready, it will add to my comfort when I can feel that there is a place in Broughton into which I can turn and give nobody any thanks

Affectionately yours | John

RI MS JT/1/HTYP/361–362
LT Typescript Only

1. *1[–2] October 1854*: while this was dated 1 October, which was a Sunday, sections of the letter were written on 'Monday morning' and 'Monday night'.
2. *Snowden*: the highest mountain in Wales. Tyndall noted in his journal entry of 1 October 1854 that he climbed Snowden from Llanberis, Wales, stating, 'Snowden is a magnificent hill: I have been on higher summits far but scarcely on one that pleased me more' (Journal, JT/2/13b/692).
3. *Liverpool*: Tyndall had been in Liverpool for the annual BAAS meeting held on 20–23 September 1854 (see letter 0950, n. 2; 0967, n. 3; 0972, n. 3; and 0975).
4. *morning on which I left London*: Tyndall met Faraday on 20 September 1854 (Journal, JT/2/13b/691).
5. *Murchison*: Roderick Impey Murchison (1792–1871), a Scottish geologist who was a key participant in the long controversy over the Silurian and Devonian Systems. He was elected President of the Geological Society of London and the Royal Geographical Society several times, and was the first to establish the geological sequence of Early Paleozoic strata (*ODNB*).
6. *coupé*: 'an end compartment in a railway carriage, seated on one side only' (*OED*).
7. *Tennant*: James Tennant (1808–81), an English mineralogist to Queen Victoria and Professor of Mineralogy and Geology at Kings College, London (*ODNB*).

8. *I met Faraday . . . the same evening*: Tyndall recounted this exchange and experience in detail in his journal entry of 20 September 1854 (Journal, JT/2/13b/691).
9. *my first communication*: Tyndall gave a paper 'On some Peculiarities of the Magnetic Field' on 21 September 1854 (see letter 0967, n. 4).
10. *galvanometer*: an instrument for measuring electrical current with a moving coil (*OED*).
11. *my second communication*: Tyndall gave a paper 'On the Diamagnetic Force' on 22 September 1854 (see letter 0967, n. 4).
12. *the position . . . in Belfast*: likely refers to papers given by Thomson at the 1852 meeting of the BAAS in Belfast, 'On certain Magnetic Curves; with applications to Problems in the Theories of Heat, Electricity, and Fluid Motion' and 'On the Equilibrium of elongated Masses of Ferromagnetic Substance in uniform and varied Fields of Force' (*Brit. Assoc. Rep. 1852*, pp. 18–20).
13. *his views*: his views on the polarity of Bismuth (Jackson, 'Tyndall and Diamagnetism', p. 31).
14. *phenomena of diamagnetism and magnecrystallic action*: see letter 0710, nn. 5–7.
15. *Owen*: see letter 0882, n. 6. On 22 September Owen lectured 'On the Anthropoid Apes' *Brit. Assoc. Rep. 1854*, pp. 111–13).
16. *the Red Lions*: see letter 0900, n. 2.
17. *Foucault*: Léon Foucault (1819–68), a French physicist who developed the 'Foucault pendulum', which was 'designed to demonstrate the rotation of the earth by the rotation of its plane of oscillation' (*OED*).
18. *Schnackenberg*: not identified.
19. *of Mrs Allen's husband*: Robert Copeland Allen, with whom Margaret Allen was having difficulties (see letter 0779, n. 2).
20. *letters from his father to her*: Allen wrote to Tyndall on 4 January 1854 (letter 0865) that her husband's father 'is ashamed of his Son'.
21. *daughter*: likely refers to Sophy Allen (see letter 0865, n. 2).
22. *Beddegelert*: a village in Gwynedd, Wales near to Snowden.
23. *Festiniog*: misspelling of Ffestiniog, Gwynedd, Wales.
24. *death of Gelert*: Gelert, the faithful canine of medieval Prince Llewelyn the Great, has been the subject of several poems. According to legend, having gone hunting without his hound, the Prince returned to find Gelert covered in blood and the bedclothes of his son empty and bloodstained. He promptly killed Gelert only to discover that his son was unharmed. The legend of his death is associated with Beddgelert, where Tyndall was staying. A tombstone at Beddelgelert recounts the legend ('Gelert', *Beddelgelert Tourism Association,* 2014–16, http://www.beddgelerttourism.com/gelert/ [accessed 21 December 2016]).
25. *when he gets married*: Hirst married Anna Martin in December 1854 (see letter 0983, n. 18).
26. *Anna*: Anna Martin, who was betrothed to Hirst.
27. *Dolgelley*: a town in Gwynedd, Wales also known as Dogellau.
28. *Llangollen*: a small town in Denbighshire, Wales.

From Margaret Allen 2 October 1854 0976

Charnley House Oct 2nd/54.

My dear Tyndall,

I was indeed gratified by the reception of your kind letter.[1] And I was also somewhat surprized at its contents, tho much pleased, for I had almost reconciled myself to wait nearly a twelvemonth before I should again have the subject mentioned wh[ich] your letter contains. I shall take the first presentable opportunity of framing the matter to Papa and Mamma & though I feel it my duty, as well as in accordance with your desire, I find I shall have to summon a considerable amount of courage to surmount the barrier wh[ich] has ever been placed between my Papa & his children.[2] but, 'nothing venture nothing have'[3] is and old proverb, therefore I trust if I venture in this somewhat troubled sea, I shall in the end by proper skill attain the desired post. I assure you I am most anxious, it will be such a luxury after all to be able to assist myself. If you have not written to the Friends you name perhaps you will do so, I wish to be guided by you in these affairs. But what shall I say to another portion of your letter, what can I say. Silence, you will perhaps say is the best return you can make. I feel this kindness deeply, a feeling touches my breast wh[ich] is not for utterance—accept my most heartfelt gratitude. I trust however I may not have to draw so largely upon you. I hope the plan will succeed. I should indeed like my little Sophie[4] to accompany me if possible I should feel certainly a little uncomfortable in leaving her. It would in part be an education for her, as she is an observant child. I shall hope to hear soon from you. I hope you will derive every benefit from your tour, & that you will gain strength to enable to to prosecute all your studies. Your daughter, says [s]he feels very proud of her relationship & desires to be remembered to you. She says I must give my love to her dear Father

I am ever sincerely yours | Gretchen[5]

Sophie also sends you a kiss though she would not give you one.

RI MS JT/1/A/82

1. *your kind letter*: letter missing.
2. *Papa & his children*: Allen was having difficulties with her husband (see letters 0975, n. 20 and 0699, n. 6).
3. *nothing venture nothing have*: variations of the proverb 'nothing ventured nothing gained' are first dated to Chaucer's 'Troilus and Criseyde' in the late-fourteenth century (*OED*).
4. *Sophie*: see letter 0865, n. 2 and 0975, n. 21.
5. *Gretchen*: see letter 0872, n. 12.

From Thomas Archer Hirst 5 October 1854 0977

Vine Cottage, | Broughton.[1] | Octr 5th 1854

My dear John

You will see from the above that I have already moved to my new residence and have commenced my initiation into the mysteries of household matters. I have already started in good earnest to make the place a neat and habitable one, by the time you next visit Queenwood I shall have a roof to shelter to you and you may dwell under it in peace as long as you wish and thank nobody for your lodging. The reports of the Liverpool meeting[2] which reached me through the papers sent by yourself[3] and Atkinson were all very meagre indeed and I began to have a very poor opinion of the association and its meeting. Your letter[4] partially removes this, but I wish you had told me a little more about the discussion in which Faraday, Thompson and Whewell joined. Whether they were able to shake any of your results or views as to the nature of Diamagnetism,[5] or whether they shewed any symptoms of changing their own views. As to Thompsons shiftiness[6] I can well conceive that it must be annoying It is, however, satisfactory to reflect that the time will come when Thompson will be judged not from what he said at this meeting or at that, or wrote in this memoir or the other but from what he has said and written altogether when the real merit of what he has done and misdone will be apparent to all. To me he has always appeared to be a man of greater brilliancy than solidity, he appears to be an acute but a too-hasty thinker and writer. The fame of such an one always increases most rapidly during his own lifetime and decreases almost as rapidly afterwards.

I am right glad to hear of your increasing strength, depend upon it you will be thankful for the store of health you have laid up when you return to London. I never was <u>in</u> North Whales[7] but I have seen the picturesque contours of the mountains about Snowden[8] from the sea and I fully agree with you as to their beauty. Some day when I have money enough I hope to take Anna[9] to see them. My brother[10] is with me at present he has come over professionally to give me some assistance in repairing my house. To day the lads have a whole holiday and this evening a concert. It is a beautiful day for them & I am sure they are in high spirits. I lecture now at 3 P.M at Queenwood and I think the arrangement will suit us all very well I lose very little time. My morning's walk to Queenwood has thus far agreed with me very well and at present I enjoy very good health I dine from 5 to 6 every day so that before dinner I can get a usual days work done At present of course I have my evenings very cosily to myself, and am less interrupted than when at Queenwood. The physical exercise which I so unwillingly took when health was the only object is now more easily attainable. I can at any time earn my dinner and give it a relish by an hours work in my garden. I expect much benefit from an hours digging per day.

Good-bye, John. Get strong and healthy and don't hurry yourself about returning to London. Where are you going to reside to write your treatise on Natural Philosophy.[11] Write me another line soon.

Yours affectionately | T. A. Hirst.

RI MS JT/1/H/205
RI MS JT/1/HTYP/364

1. *Broughton*: see letter 0872, n. 6.
2. *Liverpool meeting*: the BAAS meeting (see letter 0950, n. 2).
3. *reports of the Liverpool meeting*: not identified.
4. *your letter*: letter 0975.
5. *your results . . . Diamagnetism*: Tyndall presented two papers on magnetism and diamagnetism at the Liverpool meeting (see letter 0967, n. 4).
6. *Thompsons shiftiness*: refers to Thomson and their sense that he was changing his position on diamagnetic properties of Bismuth (see letter 0975, nn. 12 and 13).
7. *North Whales*: after the Liverpool meeting, Tyndall and Francis travelled throughout northern Wales from 27 September to 5 October 1854 (Journal, JT/2/13b/692–94).
8. *Snowden*: see letter 0975, n. 2.
9. *Anna*: Anna Martin.
10. *My brother*: John Henry Hirst (see letter 0809, n. 9).
11. *treatise on Natural Philosophy*: possibly a reference to Tyndall's 'lessons' on natural philosophy for boys (see letter 0900, n. 10).

From Tommy Grieve 7 October 1854 0978

Queenwood College | October 7th. 1854.

My dear Dr. Tyndall,

I hope you are quite well. I have just been up to the avenue to get wood for the bonfire last Thursday we had a hole holiday and we had the concert in the evening and chrades[1] after; we had tea down in the long room. There is not much time just now and so I have to write rather fast. The number of boys has encreased to a hundred and foreteen.[2] The boys play very much at shinty[3] they do not play with wooden balls but with Indian rubber ones. It was raining all yesterday and part of to day. I am not going home to Scotland for it is too cold for Mamma and she is coming to Devonshire or the Isle of Wight.[4] In the Midsummer holidays Mr Edmondson had built a new laboratory and he is making the old one into a class room it is either for the second or third class, he has finished the new one some time ago and the boys work in it, Dr. Debus has got his room up there and his little store room two by the side of it it is where the tanlk[5] is where no water was in some the shinty balls come near the windows so we have to take care not to hit the balls in that direction. The fifth class go into number nine because there is not room in the dinning room for us all sometimes the forth class come in.

I remain | Your affectionate Friend | Tommy Grieve Esqr.[6]

RI MS JT/1/TYP/1/464
LT Typescript Only

1. *chrades*: LT notes: [sic]. Misspelling of 'charades'.
2. *foreteen*: LT notes: [sic]. Misspelling of 'fourteen'.
3. *shinty*: see letter 0713, n. 11.
4. *Isle of Wight*: see letter 0945, n. 13.
5. *tanlk*: LT notes: [sic]. Misspelling of 'tank'.
6. *Esqr*: LT notes: [sic]. Misspelling of 'esquire'.

From Heinrich Gustav Magnus 12 October 1854 0979

Berlin 12 Octob 1854.

Theurer Freund

Sie werden sich gewundert haben, oder Sie werden sich vielleicht nicht gewundert haben, daß ich Ihren lieben Brief vom 11ten Aug. noch nicht beantwortet habe. Derselbe ist hier eingetroffen nachdem ich Berlin schon verlassen, und ich habe ihn erst bei meiner Rückkehr vorgefunden. Wenn ich nicht irre so habe ich Ihnen schon mitgetheilt daß ich wegen meiner Augen nach Carlsbad gegangen bin, und ich muß gestehen, daß der Gebrauch des Wassers, gegen mein Erwarten, mir gutgethan hat. Unterdeß ist Dove zurückgekehrt und hat mir Ihre Grüsse gebracht, er ist ganz voll von Entzücken über seinen Aufenthalt in England und über die Aufnahme die er dort gefunden. Er hat mir auch erzählt von Versuchen die Sie in Livpool gezeigt haben, indeß ist unsere Unterhaltung bis jetzt nur sehr kurz gewesen, so daß ich ihn nicht recht verstehen konnte. Hoffentlich bekommen wir Ihre Arbeit bald zu lesen.

Daß Mr. Gassiot mit dem Polytrop zufrieden ist, freut mich sehr. Aber Dove sagte mir auch in aller Eil, daß Foucault so schöne Drehungsapparate in Liv[er]pool gezeigt habe. Sagen Sie mir doch gelegentlich ein Wort darüber.—Von Asher ist die Zahlung sehr regelmäßig eingegangen.

Mr. Barlow ist zu meinem aufrichtig[en] Bedauren hier durch gekommen während ich noch abwesend war. Es thut mir leid daß er nicht Poggendorff oder Rieß aufgesucht hat die beide in Berlin waren Empfehlen Sie mich ihm und drücken Sie ihm mein Bedauern aus.

Für die Uebersetzung meiner Abhandl über schwarzen Schwefel meinen verbindlichsten Dank.

Sie fragen welche einzelne Gegenstände ich für die electrische Lampe ausgewählt habe, es würde Ihnen diese Mittheilung nichts helfen, weil ich nur das ausgewählt habe was ich nicht schon früher von Duboscq erhalten. Ubrigens bin ich kein Freund von der electrischen Lampe. Sie giebt ein sehr intensives Licht aber es ist nicht regelmäßig genug; man müßte sich für die Behandlung der Lampe erst jemand anlernen, denn man kann selbst, wenn man sie reguliren will, nichts anderes dabei thun, weder andre Versuche vorbereiten noch viel reden. Sie werden sich erinnern daß Dubosq einen jungen Mann hatte der nichts anderes that als den Apparat bedienen. Mit einer recht intensiven Lampe kommt man gewiß eben so weit, und ich habe vor, mir ein Gebläse zu construiren aus Leuchtgas und atm. Luft die durch einen Blasebalg zugeführt wird, und dies auf Kalk oder Magnesia wirken zu laßen. Ich bin überzeugt daß dies wenigstens in Bezug auf Regelmäßigkeit ein viel besseres Resultat liefert. Doch die Kohlen geben die größte Lichtintensität, es ist außerdem ein allgemeines Verlangen danach und folglich muß man sie haben.

Die Devil's konnte ich Dove nicht mitgeben da ich Ihren Brief zu spät erhielt, wollen Sie noch welche, so sende ich Ihnen durch Asher, vielleicht habe ich auch in 14 Tagen eine andre Gelegenheit. Schreiben Sie mir nur was Sie haben wollen. Die */*Devils*/* können aber nicht mit dem Wasser versendet werden. Am besten ist Flasche und Figur zu senden und Sie füllen sie dann dort.

Viele Grüße von meiner Frau */*&*/* von Anna die jetzt anfängt englisch fortzusetzen.

Von Herzen | Ihr | GM.

Berlin 12 Octob 1854.

Dear friend

You will have been surprised, or perhaps you will not have been surprised, that I have still not answered your kind letter of 11th Aug.[1] It arrived here after I had already left Berlin, and I only found it upon my return. If I am not mistaken, I told you already that I went to Carlsbad[2] because of my eyes, and I have to admit that, contrary to my expectation, using the water did me good. In the meantime, Dove has returned and has brought me your greetings, he is quite full of delight about his stay in England and about the reception that he found there. He has also told me about experiments that you have shown in Liv[er]pool;[3] however, our conversation up till now has only been very short, with the result that I could not understand him properly. Hopefully we shall get to read your work soon.

I am very pleased that M^r^. Gassiot is satisfied with the polytrope.[4] But Dove also told me in all haste that Foucault[5] has shown such nice lathes in Liverpool. Please tell me a little about it sometime or other.—The payment has arrived from Asher in very orderly fashion.

To my sincere regret, M^r^. Barlow came through here while I was still away. I am sorry that he did not call on Poggendorff or Riess, who were both in Berlin Give him my compliments and express my regret to him.

For the translation of my paper on black sulphur,[6] my most profound thanks.

You ask which individual items I have selected for the electric lamp, this information would not help you in any way because I have only selected what I had not already received earlier from Duboscq.[7] Incidentally, I am no friend of the electric lamp. It gives a very intense light but it is not regular enough; one would first have to train someone to deal with the lamp, for if one wants to regulate it, one cannot do anything else oneself at the same time, neither prepare other experiments nor talk much. You will remember that Dubosq had a young man who did nothing else but operate the apparatus. One will certainly get just as far with a very intense lamp, and I intend to construct myself a blower made of coal gas and atm. air that is supplied through a bellows, and to let this act on lime or magnesia. I am convinced that this will deliver a much better result, at least with regard to regularity. But the coals give the greatest intensity of light; there is in addition a general demand for them and consequently one has to have them.

I could not give Dove the devil's[8] to take with him as I received your letter too late. If you still want some, then I shall send them to you through Asher, possibly I shall have another opportunity in a fortnight. Just write to me what you want to have. The *[devils]* cannot be sent with the water, however. It is best to send the bottle and the figure, and you then fill it up there.

Many greetings from my wife *[&]* from Anna,[9] who is now starting to continue in English.

Sincerely | Your | GM.

RI MS JT/1/M/13

---

1. *your kind letter of 11th Aug*: letter missing. However, it was likely written in reply to letter 0956.
2. *Carlsbad*: Carlsbad, or Karlovy Vary, a spa town famous for its hot springs located in modern day Czech Republic.
3. *Liv[er]pool*: refers to the BAAS meeting at Liverpool (see letter 0950, n. 2). For Tyndall's lectures see letter 0967, n. 4. Tyndall described the meeting in letter 0975.
4. *polytrope*: see letter 0870, n. 7.

5. *Foucault*: see letter 0975, n. 17.
6. *my paper on black sulphur*: see letters see letters 0943, n. 2; 0948, n. 4; and 0961, n. 7.
7. *Duboscq*: see letter 0826, n. 10.
8. *devil's*: a Cartesian devil, or Cartesian diver, which is used to demonstrate buoyancy. A diver is 'a philosophical toy, consisting of a hollow figure, partly filled with water and partly filled with air, and made to float in a vessel nearly filled with water, having an airtight elastic covering. This covering being pressed down, the air inside the vessel is compressed, and more water forces through a small aperture into the figure, which consequently sinks, to rise again when the external pressure is removed' (*OED*).
9. *Anna*: likely Magnus's daughter, Anna Magnus.

From Thomas Henry Huxley 17 October 1854 0980

4 Rock House, Tenby[1] | Oct. 17th. 1854.

My dear Tyndall,

I suppose you and Francis have got back to work and that I may jam down the extinguisher over the small hopes I entertained of your paying me a visit here—where it is still jolly—much jollier than in Babylon[2]—though getting cold—as I found to day when out dredging with a vengeance.

However, my object in writing now is not to inform you of this interesting fact—but to ask if you will go shares with me in the 'Science' article in the 'Westminster.'[3] I had thought of giving it up altogether, principally on account of the pressure of other matters on my time—and because I feared there would be some interference in the management which I should not like. However the latter obstacle has come to nothing—but the former remains, unless the whole field can be divided. To give some account of the books in one's own department—is no particular trouble—and comes with me under the head of being paid for what I must, in any case, do—but I neither will nor can, go on writing about books in other departments, of which I am not competent to form a judgement even if I had the time. I wrote to Chapman, the publisher,[4] to tell him this and he urged me to suggest some way in which I could still assist. I told him that if Sciences were divided into Physico-Chemical and Natural History divisions I would still supply the latter half of the Science Article—and I recommended him before going anywhere (no more note paper, by Jove[5]) else, to apply to you to take the former. The total amount of stuff required is half a sheet;—the pay six guineas—(at least I used to get 12 for the whole sheet, but they may pay you extra as a Star—); the duty—to give an account critical, analytical, or anything else—you—please—ical, of the books sent you by Chapman. Vide[6] as models of style &c., the luminous articles on contemporary Science which have distinguished the last three months of Gt.[7] Westminster. Now he writes to ask me to apply to you—and so I have been and done it, by these presents. I need not say how much I should like to have you for a colleague in this or anything else—but don't hesitate to refuse as frankly as I have made the proposition—if you don't like it—Only, if you won't, keep the matter to yourself, for the sake of the next comer.

Remember me very kindly to Francis when you see him | and believe me | my dear Tyndall | ever yours faithfully | T.H. Huxley.

John Tyndall Esq.[8]

IC HP 8:16
RI MS JT/1/TYP/9/2866

1. *Tenby*: a walled seaside town in Pembrokeshire, Wales.
2. *Babylon*: the ancient city of Babylon, Mesopotamia. The term is often used to indicate a luxurious but corrupt city (*OED*).
3. *the 'Science' article in the 'Westminster*: in late 1853, Huxley took on the role of regularly writing the 'Science' column in the *Westminster Review* (P. White, *Thomas Huxley: Making the 'Man of Science'* [Cambridge: Cambridge University Press, 2003], p. 69).
4. *Chapman, the publisher*: refers to John Chapman (1821–94).
5. *by Jove*: a colloquial reference to the highest Roman diety, Jupiter; used to express surprise or emphasis (*OED*).
6. *Vide*: 'see, refer to, consult' (*OED*).
7. *Gt.*: LT notes: 'the?'. Likely an abbreviation of 'great'.
8. *Esq.*: a handwritten note appears on the bottom of the copy of the letter from Imperial College's Huxley Collection: 'J to H "I shd feel happy"'.

To Thomas Archer Hirst [18 October 1854][1] 0981

Wednesday night

My dear Tom

Here I am once more, my campaignings being ended for the present year. I will not boast of the amount of strength which I have collected as I know full well a day's imprudent work would send my brain all astray but I do hope to be able to husband the strength I possess so as to carry me victoriously over the duties of the coming year. I was not at all annoyed at the part Thomson acted in Liverpool.[2] I was surprised and nothing more. I believe he is very kindly disposed towards me, and you will see if you get the Athenaeum of Sept. 30 and of Oct. 7th.[3] from Fox[4] (to whom remember me kindly) that he did not hesitate to express a frank and favourable opinion regarding the experiments I brought forward. Your estimate of him is I think a correct one: One thing astonishes me, that he is able to do so much. He gave me two memoirs in Liverpool,[5] and I saw the proofs of a third large one in his hand on Thermo-electricity. This day I received a voluminous paper sent to me for my opinion[6] by the Council of the Royal Society[7] and containing an investigation by Thomson and Joule[8]—the most of it written by Thomson—I can only say that, if he thinks over his subjects as I am obliged to think over mine, that[9] nature has given him an immensely stronger brain than she has given to me. The two numbers of the Athenaeum to which I have referred will give you a far better account of the proceedings than I could possibly write. I did not see the papers myself until to day, otherwise I would have sent them to you.

After my trip in Wales had ended I went to Ireland and endured such a tossing in the passage that I almost swore never to cross the Channel again—on returning however it was so smooth that I quite forgot my former misery. I was not absolutely sick that is to say my sickness did not reach the <u>dynamic</u> stage it was sea sickness rendered <u>static</u> and the tension was far more disagreeable than if it had amounted to an open rupture.[10] On Friday night last while crossing the cabin of the steamer I thought I saw the countenance of a lady which was familiar to me—I turned my head aside to think who it could be and in doing so my eye met another face which assisted me—It was Dr Simpson and Mrs Simpson[11] who were going to Paris to meet John Martin[12] who has recently arrived there We had a good deal of pleasant chat together and I bade them good bye in Liverpool.

I do not think you can possibly estimate the value of having a garden to dig in,[13] for your strong brain has never been reduced to that state of utter prostration which would enable you to appreciate the inestimable value of an hour's digging. I hope you will continue to dig for depend upon it it is time saved eventually. For three hours imprudently occupied yesterday I have lost 6 to day. Had one of these three hours been devoted to digging I should not have lost a minute. I am learning more of my own constitution daily and I trust in the end to be able accurately to apportion the load which I am able to bear. I shall rejoice when I am permitted to seize a spade in your garden and to shew you how digging ought to be done. Little Tommy Grieve has written to me[14] and given me a cheery picture of Queenwood—I think arrangements ought to be made to have those little concerts pretty often—Any thing that diffuses a feeling of cheerfulness and happiness throughout such a number is of great value. Let play be liberal but let work be earnest. I am now engaged with the writing out of my memoir; it will be a long one and will take me a long time[15] as I wish to have it well done. Ruhmkorff is in London at present: he was with me to day. It would be a great blessing to me if I could withdraw myself from the thoughts of every body for a time as I find replying to letters and enquiries a dreadful draw upon my time, and what is worse, a dreadful distraction to my thoughts. I often long for a time similar to that which I enjoyed at Marburg when you were almost my only correspondent and when I felt a sweet refreshment in laying my work aside and writing to you. I have had a letter from Despretz[16] asking me to purchase some books from him. A letter from Magnus[17] who has been suffering from bad eyes but who is now better. A paper from the President of the Royal Irish Academy[18]—a letter from a Mr Sturm in Vienna[19] asking me to permit him to make use of me as a reference with regard to an invention of his, which I have declined doing—a book from Mrs Allen[20] containing an unusually good poem by a person named Bigg.[21]—Letters from sundry others too numerous to mention and all of which demand replies—Faraday returns to town tomorrow—He has been spending the last 9 weeks in the country. Barlow has been made Chaplain to the Queen, which is perhaps about the highest earthly reward that could be conferred upon him.

Good bye Tom. Remember me lovingly to Anna[22] | Yours as ever | John

RI MS JT/1/HTYP/366
LT Typescript Only

---

1. *[18 October 1854]*: Hirst wrote 'Oct. 20th' at the top of the letter, reflecting the date of receipt. It was presumably written on 18 October, which was a Wednesday.
2. *at Liverpool*: Tyndall had told Hirst that 'Thomson completely backed out of the position, which he had assumed in Belfast' (see letter 0975, n. 12).
3. *Athenaeum*: a British periodical on literature, science, and art established in 1828, edited by William Hepworth Dixon (1821–79) between 1853 and 1869. For the named issues see letter 0967, n. 4.
4. *Fox*: see letter 0800, n. 29.
5. *gave me . . . in Liverpool*: not identified, as Thomson wrote a number of papers in 1854 (see 'Thomson, (Sir) William', *RS Catalogue*, vol. 5, pp. 9779–81, on p. 979).
6. *for my opinion*: Tyndall wrote a referee report of Thomson and Joule's manuscript (see RS RR/2/239, 31 October 1854, letter 993).
7. *the Council of the Royal Society*: the Royal Society of London, which published the *Phil. Trans.*
8. *investigation by Thomson and Joule*: J. P. Joule and W. Thomson, 'On the Thermal Effects of Fluids in Motion.—Part II', *Phil. Trans.* 144 (1854), pp. 321–64. See letter 0994 for Tyndall's report.
9. *that*: LT appears to have penciled square brackets around this word, suggesting she may have added it to her typescript.

10. *Ireland . . . open rupture*: Tyndall noted this experience in his journal entry of 6 October, 'Heavy pitching through the night; lay awake in my birth with the wind playing on my face, not absolutely sick; but within a touch of it' (Journal, JT/2/13b/694).
11. *Dr Simpson and Mrs Simpson*: see letter 0816, n. 11.
12. *John Martin*: John Martin (1812–75), a Young Irelander, politician, and brother to Mary (née Martin) Simpson and Anna Martin. He had been indicted for treason in 1848 after publishing a paper advocating for an Irish Parliament. He moved to Paris after being pardoned in 1854 on the condition that he not return to England or Ireland; he was later unconditionally pardoned in 1856 (*ODNB*). See also Hirst, 'Journals', 15 October 1854.
13. *a garden to dig in*: Hirst had mentioned to Tyndall that his new home had a garden, which he was looking forward to spending time digging in (see letter 0977).
14. *written to me*: letter 0978.
15. *my memoir . . . long time*: while Tyndall did not specifically identify the paper, it must be J. Tyndall, 'The Bakerian Lecture.—On the nature of the Force by which Bodies are repelled from the Poles of a Magnet; to which is prefixed, an Account of some Experiments on Molecular Influences', *Phil. Trans.* 145 (1855), pp. 1–51. Although the *Phil. Trans.* stated that the memoir was received in October 1854 (see p. 1), Tyndall noted in his journal that he turned it in to the Royal Society on 2 November 1854 (Journal, JT/2/13b/697).
16. *Despretz*: see letter 0803, n. 10. Letter and books not identified.
17. *letter from Magnus*: letter 0979.
18. *paper . . . President of the Royal Irish Academy*: the Royal Irish Academy, founded in 1785, was established for the promotion and investigation of science, literature, and antiquities ('History of the Academy', *Royal Irish Academy*, 2016, https://www.ria.ie/about/history [accessed 22 December 2016]). The president of the Academy in 1854 was Thomas Romney Robinson (1792–82), an Irish astronomer and physicist known for inventing a device for measuring the speed of wind. The paper is missing.
19. *letter . . . Mr Sturm in Vienna*: may refer to Salomon Sturm (n.d.), who patented a lens grinding machine in 1854 (S. Sturm, 'Sturm's Machinery for the Manufacture of Optical Lenses', British Patent 2608, 10 May 1854, pp. 1–9); see also E. J. Atzema, 'Charles François Sturm's Writings on Optics', in J. C. Pont [ed.], *Collected Works of Charles François Sturm* [Basel: Birkhäuser Basel, 2009], pp. 49–65, on p. 49, n. 1). An article on the device and a discussion surrounding its novelty and provenance appeared in *Athenaeum*'s report of the BAAS meeting in Liverpool ('On a Method of Manufacturing Cylindrical Lenses', *The Athenaeum* 1406 [7 October 1854], pp. 1202–12, on p. 1204). Letter missing.
20. *book from Mrs Allen*: possibly S. Bigg, *Night and the Soul; A Dramatic Poem* (London: Groombridge and Sons, 1854). Not further identified.
21. *Bigg*: likely John Stanyan Bigg (1828–65), an English poet and journalist who spent time in Ireland.
22. *Anna*: Anna Martin.

To Thomas Henry Huxley 18 October 1854 0982

18th. Oct. 1854.

My dear Huxley,

I should feel happy and proud in being your colleague in any enterprise whatever; and were I not afraid of this poor miserable brain of mine which continually admonishes me that I am dilapidated used up mortal I would at once say 'yes' to your proposal regarding the Westminster.[1] The terms appear to me to be very fair. I will tell you how I am circumstanced at present: I have been working during the summer and have gathered materials for a heavy memoir[2] the writing out of which is only commenced. Some months ago I undertook, at his urgent request, to furnish Hughes of Greenwich[3] with 120 pages for his new schoolbooks[4] and there are but 30 pages as yet done. Before I undertake anything else this engagement must be fulfilled; then my lectures are looming upon me like a cloud. All these things combined render it impossible for me to undertake anything for the Westminster at present. But if you could drag on for another 6 months I should not mind trying what I could do. The engagement however looks rather formidable, to give 'an account critical, analytical or anything else of the books sent by Chapman.' Why he might send me more than I could possibly digest and I should not like to deal with matters I had not digested. Have you no voice in the choice of the books that you review? Why I thought that Chapman had been taken away by cholera[5] and here I find him, to my great satisfaction, turning up again. I was rather dismayed to read of what I conceived to be his death in the coffee room of a hotel in Llanrwst;[6] for, though I do not know the man, something of his that I have read has engendered within me a respect for him. I have read the articles to which you refer me[7] and have found great pleasure in them, I do not wonder at the reluctance of the publisher to lose the services of the writer of them. I shall see Francis tomorrow and convey your message to him—he is in capital condition—I envy his ruddy cheeks and copious digestion. You show your wisdom in keeping away from Babylon[8] as long as you can, and I shall certainly, follow your example next year if the gods spare me so long.

Believe me, dear Huxley | ever Yours most truly | John Tyndall.[9]

IC HP: 8.17
RI MS JT/1/TYP/9/2867

---

1. *your proposal regarding the Westminster*: Huxley had asked Tyndall to assist him in contributing to the 'Science' column of the *Westminster Review* (see letter 0980, n. 3).
2. *heavy memoir*: see letter 0981, n. 15.

3. *Hughes of Greenwich*: see letter 0900, n. 9.
4. *schoolbooks*: see letter 0900, n. 10.
5. *Chapman . . . taken away by cholera*: see letter 0985, n. 4.
6. *Llanrwst*: Llanrwst, Conwy County Borough, Wales.
7. *the articles . . . refer me*: not identified.
8. *Babylon*: see letter 0980, n. 2.
9. *Tyndall*: a handwritten note appears on the bottom of the copy of the letter from Imperial College's Huxley Collection: 'H to J "I am rejoiced"'.

From Thomas Archer Hirst 19 October 1854 0983

Vine Cottage | Broughton[1] | Stockbridge | Oct. 19th 1854

My dear John

I hear from Francis that you have returned[2] and are looking well and I write a few words of welcome to you before I go to bed. I wrote to you[3] at the post office Liverpool immediately after receiving your last letter.[4] Did you get it? I hear reports at Queenwood that Tommy Grieve has received a long letter[5] from you and that you make some enquiries about the 5th of November.[6] Will you come? And share my large house[7] with me, it will be a sweet harbour of refuge for you in times to come, it is so snug and quiet Come when you will its doors will open wide without creaking as if you were its lord and master. Everything about the house inside and out is in a transition state at present, from mere ruinous neglect to neatness and order. When will you come? At Queenwood there is not a very good spirit amongst the lads just now. Their dissatisfaction with McDougall and Blanche[8] went on increasing until it shewed itself in open defiance and ended in Blanche's dismissal. He was an empty-headed pedant and his dismissal was but a just concession to the lads. For McDougall there are hopes still but my faith in him rather decreases than otherwise Blanche had a bad influence on him perhaps he may now improve. For my part I never stood on a better footing with the lads than I do now & have done for the whole half year they never worked for me more willingly, yet it is not so with others Their dissatisfaction whose origin was with McDougall and Blanche vents itself at every weak point. Our Improvement Society[9] for instance is a capital thing when the lads are in a healthy condition and when every masters authority is based upon his own intrinsic merits and not propped up with mere appearances; but as late meetings have proved it is a perilous thing when discontent is abroad. When the relation between teacher and pupil is a healthy, firm one, it is refreshing to both to meet on common ground and let loose the reins of authority, feeling that you can gather them up again at will. But its constitution is too democratic to be safe when lads are seeking to set their masters at defiance. A Fortnight ago about a dozen members (among whom was our friend Alex,[10] I am surprised to say) sent in their resignation on the plea that the society no longer continued to interest them. The president could not do otherwise than receive it although he did so rather ungracefully. Next meeting another batch of 18 (mostly blockheads) present petition stating 'that they have resolved to leave the society seeing that members are punished for speaking' I happened to be present (I regret to say I was not there before owing to household matters) and I scarcely knew whether to be grave or to laugh. There was some justice in their plea, as I afterwards discovered Edmondson had in his short sighted way punished two lads for misbehaving (talking) at a meeting But what was significant the lads who were punished did not resign, shewing plainly that the 18 had made this a mere excuse for venting their displeasure and setting poor George[11] at defiance. Again he was hemmed in by the Societies[12] statutes and must accept their resignation though still more ungracefully than before. Poor George. This was not all the final stroke had yet to come which was to cut him most and leave him helpless and irresolute, not knowing on what ground to stand and confront them whether on his President-ship or his Principal-ship. Master Buchan[13] rises and with ridiculous pertness begs to move 'that members of this society may have free choice to attend its meetings or not as they may feel inclined.' The worthy president visibly trembling with rage turns helplessly to the Secretary to ask what the law is in this case to which the Secretary replies that the motion must be put on the books and decided next time by vote or ballot if necessary. All this took me by surprise for I had (as I said) been absent. Consequently I could but speak a word to them as a kind of friendly warning. I have since become possessed of all facts. I see clearly enough through the whole matter, and I think I dare predict that in a week or two it will be all straight again and on firmer footing than ever. The lectures & papers have been more than usually good lately so that I look upon all this nonsense as a mere symptom of a deeper malady in the school itself which latter of course must be at once attacked. You will hear. Let me know what you have been doing with yourself since you last wrote to me.[14] You will be glad to hear that John Martin[15] has arrived in Paris.[16] D^r^ & Mrs Simpson[17] have gone to see him If all be well D^r^ & Mrs Hirst[18] will go and see him shortly.

Write soon to | Yours affectionately | T. A. Hirst.

RI MS JT/1/H/206
RI MS JT/1/HTYP/367–368

---

1. *Broughton*: see letter 0872, n. 6.
2. *you have returned*: Tyndall returned to London on 16 October 1854 (Journal, JT/2/13b/696).

3. *wrote to you*: letter 0977.
4. *Your last letter*: letter 0975, as letter 0981 would not yet have been received by Hirst, as it was written on the evening of the 18th.
5. *a long letter*: letter missing. It was likely a reply to letter 0978.
6. *5th of November*: likely refers to a bonfire event (Guy Fawkes Night) at Queenwood that Tyndall mentioned in letter 0984.
7. *my large house*: Hirst had recently acquired a house in Broughton, where he would reside with Anna Martin after they married (see n. 18 and letter 0977).
8. *dissatisfaction . . . Blanche*: McDougall and Blanche were dismissed from their positions at Queenwood College in 1854. Blanche was dismissed prior to McDougall, as outlined in Tyndall and Hirst's discussion of the issue (see letters 0984 and 0987).
9. *Our Improvement Society*: the Queenwood Mutual Improvement Society (see letter 0816, n. 4).
10. *our friend Alex*: Alexander Cumming.
11. *poor George*: George Edmondson.
12. *the Societies*: the Queenwood Mutual Improvement Society.
13. *Master Buchan*: H. Buchan, a student at Queenwood College.
14. *last wrote to me*: see n. 4.
15. *John Martin*: see letter 0981, n. 12.
16. *arrived in Paris*: the Simpsons were travelling to Paris to meet Martin (see letter 0981, nn. 11–12).
17. *Dr & Mrs Simpson*: see letter 0816, n. 11.
18. *Dr & Mrs Hirst*: Hirst was referring to himself and Anna Martin, who he was due to wed in December 1854 (see letters 0974, n. 6 and 1018); see also Journal, 19 December 1854 (JT/2/13b/705).

To Thomas Archer Hirst 22[–23] October 1854[1] 0984

Sunday 22. Oct. 1854.

My dear Tom

I read your letter[2] with great concern last night: from Tommy Grieves epistle[3] I inferred that all was proceeding in a very happy and cheerful manner, but your note shews me that this is by no means the case. I am not in the least surprised, though I am sorry that things are so. The man who allies his fate with Queenwood and who wishes it to prosper must not expect a life of ease there, Mr E.[4] needs continual propping,[5] and if the problem had been laid before me a month ago as to the chances of the Mutual Improvement Society[6] under his sole superintendence I should have concluded that they were as 10 to 1 that it would get into confusion. During the last year of my sojourn at Queenwood I constantly attended its meetings and had a flood of questions to answer at each, and during this time I saw the Society not only expanding, but consolidating and becoming firmer and firmer each succeeding week. The elder boys need somebody there above them, to give them an interest in its proceedings—somebody from whose interest they will catch the same spirit: for it cannot be expected that they will take the same philosophic interest in the doings of the younger boys that you or I should if we were present. They tire, and seeing no life above them to which they can look and from which they can draw inspiration the affair soon becomes insipid to them. How is it that things were permitted to go so far with that 'pedant'[7] as to render his dismissal[8] a necessary concession to the boys. Such concessions ought never to be necessary. The case ought to have been foreseen; for it is a serious matter to let the boys feel that their clamour has been necessary to produce any proper result. It is in fact a premium to turbulence. I hope you will see that every just demand is satisfied—if you can feel this to be the case then it is quite an easy matter to quell any disturbance among boys, it is simply necessary to seek out those who are at the bottom of it and if they dont mend their manners send them off. You have not much time to reason with them where so many are requiring your attention. If I saw a boy who shewed any tendency to turbulence or disaffection I would give him a fair chance take all just cause of complaint away from him and if he proved still a disturber I would send him away on the ground that you could not agree together. Such boys are not always bad ones, and the best character possible ought to be sent with them to their parents. Edmondson would be the worst man possible to write such a character. For men of narrow minds are vengeful and look at the case only as it affects themselves. I can well imagine his rage as that foolish little Buchan[9] made his proposition.[10] I hope when you next write to me that you will be able to say that the storm is passed and you have sunshine once more.

How are you going to supply the place of Blanche?[11] There is a man in Manchester open I think to an engagement, and who would be of immense value to Queenwood. He was there before. Davidson[12] is his name. A first rate English teacher, and of sufficient force of character to keep things in order. He would prove I believe of inestimable value to Queenwood. I heard from Singleton[13] a few days ago that Davidson was at present free and he had some thoughts of engaging him himself. I should therefore say keep this matter to yourself until I hear from Singleton what he intends to do. I will write to him to day. If Edmonson heard of the thing he would consult his own interest first, and it would be unfair to cause what Singleton mentioned to me in conversation to redound to his own disadvantage.

On second thought I will defer writing to Singleton until I hear from you, for I do not know your present arrangements.

With regard to the 5th.[14] whether I can go or not will depend purely upon the amount of work before me. I shall probably defer a visit until Christmas. I simply asked Tommy the date of the bonfire and I suppose the little fellow has drawn the inference that I intend to be there.

My health is good, though at the present moment I am writing with a wet cotton sock tied against my forehead, a day's work after I came from the country set my brain in great disorder. I thought with regret upon the time when I could work for months from day dawn to night, for 16 hours a day. Well, I thought, I will use 30 per cent more water to my spongebath every morning. I will cut 30 per cent off my present amount of food and I will try a cold bandage—I did all, and lo! my ancient strength has in great measure returned to me and I have worked incessantly for the last three days and as yet feel no diminution of my vigour. The cold bandage I believe is the chief cause of my improvement for when I leave it off symptoms of headache soon shew themselves.

Goodbye Tom | Ever yours affectionately | John

Monday 12 o'clock I have worked hard for 5 hours to day without a bandage and feel very strong. I am writing my memoir[15]

RI MS JT/1/HTYP/369–370
LT Typescript Only

---

1. *22[–23] October 1854*: while this letter was dated 22 October, the postscript was clearly written on 23 October.
2. *your letter*: letter 0983.
3. *Tommy Grieves epistle*: letter 0978.
4. *Mr. E*: George Edmondson.
5. *needs continual propping*: Tyndall noted his frustration with Edmondson in his journal entry of 29 October, 'The same source of mismanagement and confusion is still there' (Journal, JT/2/13b/696).
6. *the Mutual Improvement Society*: see letter 0816, n. 4.
7. *that 'pedant'*: refers to Blanche, as Hirst had referred to him as 'an empty-headed pedant' in letter 0983.
8. *render his dismissal*: see letters 0983 and 0987.
9. *Buchan*: see letter 0983, n. 13.
10. *his proposition*: at the meeting of the Mutual Improvement Society, which Hirst and Tyndall were discussing, Buchan moved a motion 'that members of this society may have free choice to attend its meetings or not as they may feel inclined' (see letter 0983).
11. *Blanche?*: after Blanche was dismissed, there would have been a vacancy for a teaching position at Queenwood College (see n. 7).
12. *Davidson*: not identified.

13. *Singleton*: Josiah Singleton (see letter 0873, n. 19).
14. *the 5th*: Hirst had invited Tyndall to attend Queenwood College on 5 November for an event (see letter 0983, n. 6).
15. *my memoir*: see letter 0981, n. 15.

From Thomas Henry Huxley 22 October 1854 0985

Tenby, South Wales | Oct. 22nd. 1854.

My dear Tyndall,

I am rejoiced to find you entertaining my proposition[1] at all—No one believes how hard you work more than I, but I was not going to be such a bad diplomatist as to put that at the head of my letter—and if I had thought that what I want you to do involved any great accession thereto I think I could not have mustered up the face to ask you. But really and truly as long as it is confined to one's own department it is no great affair. You make me laugh at the long face you pull about the duties[2]—based on my phrase—The fact is you notice what you like, and what you do not, you leave undone, unless you get an editorial request to say something about a particular book. The whole affair is entirely in your own hands—at least it is in mine—as I went upon my principle of having a row at starting.

Ordinarily Chapman sends me such books as come to him; to these such as I ask for are added if he can get them; and then I select whatever I like for texts—omitting the rest or lumping them together.

I should have mentioned that all the books go back to him.

Now here is an equitable proposition—Look at my work: I have a couple of monographs; odds and ends of papers for journals; a manual and some three courses of Lectures to provide for this winter. 'My necessities are as great as thine' as Sir Philip Sidney[3] didn't say. So be a brick, and split the difference, and say you will be ready for the April number. I will then write and announce the fact to Chapman.

What idiots we all are to toil and slave at this rate! I almost repent me of tempting you, after all; so I promise to hold on—if you really think you will be overdoing it. With you I envy Francis his gastric energies. I feel I have done for myself in that line and am in for a life long dyspepsia. I have not now nervous energy enough for stomach and brain both; and if I work the latter, not even the breezes of this place will keep the former in order. That is a discovery I have made here and though highly instructive it is not so pleasant as some other physiological results that have turned up.

Chapman who died of cholera[4] was a distant relation of my man. The poor fellow vanished in the middle of an unfinished article which has appeared in the last Westminster,[5] as his forlorn Vale![6] to the world—After all that is the way to die—better a thousand times than driveling up into eternity betwixt awake and asleep in a fatuous old age.

Please to let me hear from you at once | and believe me | Ever yours faithfully | T.H. Huxley

IC HP 8.18
RI MS JT/1/TYP/9/2868

---

1. *my proposition*: Huxley had requested that Tyndall join him in writing the regular 'Science' column for the *Westminster Review* (see letter 0980, n. 3). Tyndall hesitated in accepting the offer in letter 0982.
2. *the long face you pull about the duties*: Tyndall was concerned that the column would require a great amount of effort (see letter 0982).
3. *Sir Philip Sidney*: Sir Philip Sidney (1554–86), an author and courtier, known as one of the most influential writers of the Elizabethan age. The quote by Huxley is a play on one attributed to a gravely wounded Sidney, who was believed to have offered another dying soldier his bottle with the words, '*Thy necessity* is yet *greater* than mine' (*ODNB*).
4. *Chapman who died of cholera*: John Chapman (1801–54), distant relative of the editor of the *Westminster Review*, was an English engineer and political writer. He had recently died of cholera on 11 September 1854 (*ODNB*).
5. *unfinished article . . . last Westminster*: the unfinished article was published as 'The Civil Service', *Westminster Review* 62 (July 1854), pp. 68–95.
6. *Vale!*: 'farewell; goodbye; adieu' (*OED*).

From Heinrich Daniel Ruhmkorff 22 October 1854 0986

Sontag den 22ten Octob. | 1854

Geehrtester Herr Tyndal

Zu dem Versuche welchen ich di[e] Ehre hatte Ihnen in Paris zu zeigen braucht mann zwei Condensatoren einen mit einer sehr groschen Fläche ungefähr 40 Centimeter viereck und einen sehr kleinen, man beruhrt mit einen Metalle den groβßen un[d] nach dem hebt man eine plat[t]e ab und bringt sie in Verbindung mit den kleinen, ich musß bis Mitwochen den 25ten noch in London bleiben den[n] die Administration will bis dahin mit den neuen Apparathe arbeiten und bin bis dahin ganz zu ihren di[e]nsten. mit meinen Fuss gehts es gut und ich konte wenn Si es wünschen zu ihnen kommen. I[c]h empfehle mich Ihnen ergebenst un[d] verbleibe mit aller Hochachtung

Ihr ergebenster Diener | Ruhmkorff.

Sunday | 22nd Oct. | 1854

Most honoured Herr Tyndal

For the experiment which I had the honour of showing you in Paris[1] one needs two condensers, one with a very large surface of approximately 40 square centimetres and a very small one, one touches the large one with a metal, and after this one picks up a plate and brings it into contact with the small one, I must still remain in London until Wednesday the 25th, for the administration[2] wants to work with the new apparatus[3] until then, and until then am entirely at your service. My foot is all right and I could, if you wish, come to you. I give you my most humble compliments and remain with all respect

Your very humble servant | Ruhmkorff.

RI MS JT/1/R/41

1. *in Paris*: Tyndall saw Ruhmkorff in Paris, noting the visit in his journal entry of 5 September 1853: 'Saw many experiments, those with the induction apparatus are particularly beautiful' (Journal, JT/2/13b/622).
2. *the administration*: not identified.
3. *the new apparatus*: likely Ruhmkorff's induction coil (see letter 0864, n. 1). See also letter 0884.

From Thomas Archer Hirst 25 October 1854 0987

Vine Cottage | Octr 25th 1854

My dear John

In answer to your question why was not Blanche dismissed sooner[1] I think if you had been here you would have agreed that there was not too much delay. After his manner with the boys had been seen and disapproved of he was told what it would lead to and recommended to change his tactics This he neglected to do and as soon as his neglect was apparent he was dismissed I do not think we had any solid reasons for dismissing him sooner than we did and it yet remains to be seen whether his dismissal will have the desired effect. When I used the term concession I did not mean that he was dismissed professedly to still their clamour, they do not even know the reason why he was sent away so that his dismissal cannot be considered by anybody as 'premium to turbulence'. The concession spoken of was silently made by the masters, here was a bad spirit abroad which must be quenched the question was whether to do so boys alone must be punished or whether in reality Blanche's mismanagement had not been the chief cause and therefore in justice whether he should not be first removed and all cause of complaint withdrawn. This was resolved upon and thus among ourselves a concession made to the boys. The method of treatment you advise[2] is just the one I have followed and am following throughout with the conviction that it will be successful. On Monday last came the crisis which you will see has been properly treated Between 5½ and 6½ P.M McDougall[3] was with the first class. Richards[4] was the only 1st class boy out at the time. Newton[5] and another in a suspicious way asked if the windows might be opened; McD.[6] suspecting something refused. Shortly after the door was gently opened and an arm threw in a cracker which exploding naturally caused the greatest confusion. Without delay Iliff went round to the several classes and found that Richards and Wesley[7] were the only two boys out at the time. Richards lives with Wright and went home directly afterwards so that without possibility of collusion Wesley was first examined. He denied having done it and offered to take an oath on it if required, so convincing was his denial to Wright and Haas who examined him felt sure Richards was guilty and Wright followed him home to ascertain. Richards denied it just as stoutly but on being told that there were only himself and Wesley out and that Wesley had denied it saying that he knew who had done it (which he did) Richards indignantly replied that he saw Wesley throw the cracker into the room! Next morning when Wright returned and the two boys confronted one another, Wesley confessed all, saying that he knew he should be expelled and therefore he did not care. We have before had trouble with him he was an unruly worthless fellow and had been previously threatened with expulsion, so that amongst ourselves there was no difference of opinion he must be sent away on the following day Wednesday. He was confined in Debus' old room all Tuesday but not sentenced. On Tuesday noon as all were at dinner he escaped (he ought to have been better imprisoned). All the masters except myself were engaged on Tuesday afternoon so that Wright came down to request me to go to Winchester and tell his father (the celebrated organist Dr Wesley[8]) I went by rail, at Romsey station I saw Wesley pass by the carriage window and he saw me I stepped out of my carriage immediately but he was no longer visible I knew he must have run round in front of the engine and looking between the carriages I saw him run towards the rear of the train. I ran also and just as he was about to step into the last carriage we met. 'Stop, my boy' said I 'we will return to Dunbridge if you please' He is a tall stout fellow nearly 17 years of age he looked at me for an instant irresolute & then turning on his heels ran off with all his speed along the platform towards Dunbridge. Of course I gave chase and my legs were longer than his; as he reached the end of the platform I was within arms reach of him and fearing he would spring down the embankment I sent him headlong on the floor with a sharp blow on the shoulder; before he could rise I had him by the collar fast and brought him thus ignominiously back towards the train which was waiting. The passengers and some Romsey people saw it all I am sorry to say, but although he protested and wriggled I held him firm by the collar told the guard he might leave me and I took him across into the waiting room where I kept him in charge until the next Dunbridge train came up. He was an awkward prisoner and with a good deal of determination he made several attempts to escape me. In a surly way he demanded his liberty saying he knew he should be expelled and I had no right to stop him for he was going home. I stood with my back to the door and quietly replied, that we had not done with him at Queenwood yet he had not received orders to go whereas I had received orders to take him back and take him back I would either quietly or by force whichever he thought proper. He began in an impudent way to reply but with a thundering 'silence' he sank frightened into a chair. When the train came up he still hesitated I walked up to him 'Now Sir shall I take you by the throat again or will you walk quietly by yourself' He was beginning to reply 'he did not see what right &c' but holding out my hand towards his collar I told him to answer my question and he promised he would follow me which he did. After a few similar attempts at escape all which were fruitless I brought him to Queenwood & gave him in charge to Edmondson with strict orders to see that the door and window of his room were safe. On arriving at Queenwood this morning I find that during the night he escaped. He was locked up in the bedroom next to Edmondson and how the fellow managed to get out of those awkward windows and jump out without Edmondson hearing him I cannot imagine. I am half inclined to think he did hear him but either did not wish or was afraid to stop him, for last night he told me he was sorry I had not taken him home instead of bringing him back for until he was gone he would cause nothing but trouble. I replied that in this case we ought not to take trouble into consideration. I wished to show him and all other boys that he could not walk off with impunity that even if he did anticipate his own expulsion he should be brought back to receive it and then <u>sent</u> away. I was angry that he had escaped at last though it was better than if he had not been brought back at all for last night whilst Wesley was still in prison Mr Edmondson told the lads what had happened since Monday and sentenced Wesley before them. Of course having expelled him we gave chase no more and Wright left early this morning to tell his father what had happened. It now remains to see whether a permanent check has not been applied. To day I spoke to Newton who was undoubtedly an accomplice and I have warned him. Newton is a mischiefous but still a high spirited lad and my talk may have some effect on him we shall see. For my part I feel sure enough that with care I can quell the whole disturbance—and I will do it. It is post time, no more at present from

Yours affectionately | T. A. Hirst

Edmondson has already written (before your letter arrived) to Davidson[9] with what result I do not know. Wright is against his coming for being a married man he cannot if his wife comes too take the Duty which is required.

RI MS JT/1/H/207
RI MS JT/1/HTYP/371–373

1. *why was not Blanche dismissed sooner*: see letters 0983 and 0984.
2. *the method of treatment you advice*: in letter 0984 Tyndall advised that 'it is simply necessary to seek out those who are at the bottom of it and if they dont mend their manners sent them off'.
3. *McDougall . . . first class*: McDougall was also dismissed from Queenwood College in 1854 (see 0983, n. 8).
4. *Richards*: W. Richards, a student at Queenwood College. Not further identified.
5. *Newton*: R. Newton, a student at Queenwood College. Not further identified.
6. *McD.*: refers to McDougall.
7. *Wesley*: Samuel Annesley Wesley (b. 1837), was the son of organist and composer Samuel Sebastian Wesley (see n. 8). He was expelled as a result of throwing a cracker (a type of fireworks) into the classroom.
8. *Dr. Wesley*: Samuel Sebastian Wesley (1810–76), a well-known English composer and organist, who had been appointed as the first Professor of Organ at the Royal Academy of Music in 1850 (*ODNB*).
9. *Davidson*: Tyndall had suggested him as a potential teacher to fill the vacancy at Queenwood College (see letter 0984, n. 12). He had initially offered to send a letter to Singleton regarding the matter, but changed his mind (see letter 0984).

To Thomas Archer Hirst 27 October 1854 0988

London 27th. Oct. 1854.

Dear Tom

Your long letter[1] has reached me and I now break in upon the writing of my long memoir[2] to reply to it. Your explanation of the term 'concession' deprives it I confess of the entire meaning with which my misinterpretation invested it and I have now no doubt that the matter was as wisely arranged as it could be under the circumstances. In fact I did not know that you had anything to do with the arrangement and I must confess that I doubted the decision of the others. You see Tom the matter is as plain as noon-day that the confusion arises through some incompetence, whether of talent or of character on the part of the teachers. Now we are all incompetent while we are learning, and if incompetence in the present instance be the result of inexperience it is just and right to place your extra force at the command of Mr. McD.[3] and thus virtually do his work for him. The great thing to bear in mind at Queenwood is I think to keep men there who shall be able to fulfil the tasks appointed to them. Imagine all of them like yourself, how smoothly would things go. This, I admit, is not easily attainable but an approximation to it ought to be aimed at. Davidson[4] I think is an unusually good man, and I don't think Wright's objection[5] can weigh for an instant against the advantage of having such a man: he is I believe a hard worker and as such a man is not easily found I hardly think it would be wise to let him escape if he can be secured. However you know your need of such a man better than I do and if you can get on prosperously with a single man it is perhaps all the better. You are learning more yourself at Queenwood than anybody else at present, and if an accession of strength, if power over your fellow men, if a knowledge of human character and the means of swaying it be at all desirable you are in a fair way of obtaining the desideratum. I have no doubt of the result: I have no doubt but you will set every thing right again. Endeavour to keep yourself from being exasperated with the boys—feel towards them in such a manner as to make <u>them</u> feel that you are their friend, and they cannot withstand you:—They may be impudent for a time, they may try to annoy you, but cultivate a little longsuffering and they will hug you in the end. I know that affair with that worthless liar[6] must have been unpleasant to you: but there are few boys at Queenwood who will be willing to risk what he encountered. Edmondson is a coward I know; but you must be courageous.

Ever affectionately yours | John

I almost forgot a question which I intended to ask you. Can you promise me a bed if I go down to Queenwood in the course of next week? If I go I will stop a week there. I shall have my memoir finished this week,[7] and afterwards I have to write an article for Hughes[8] which I could do pleasantly in the quiet of Broughton. Dont scruple to say you are not ready if it be the case. My health is bearing up wonderfully although I test it severely but it will be a grand thing when I am tired writing to be able to take a spade in my hand and dig in your garden.[9] That would be the very ideal of working. You will I trust let me hear from you by Monday.

J. T.

RI MS JT/1/HTYP/374
LT Typescript Only

1. *your long letter*: letter 0987.
2. *my memoir*: see letter 0981, n. 15.

3. *Mr. McD.*: McDougall.
4. *Davidson*: see letters 0984, n. 12 and 0987, n. 9.
5. *Wright's objection*: Wright objected to hiring Davidson on the grounds of him being married (see postscript to letter 0987).
6. *that worthless liar*: Samuel Annesley Wesley (see letter 0987, n. 7), a student who caused an incident at Queenwood then escaped punishment twice by running off; Hirst had been involved in apprehending him and returning him to Queenwood (see letter 0987).
7. *I shall . . . this week*: Tyndall noted in his journal that he gave the memoir to Faraday on 30 October and turned it in to the Royal Society on 2 November (Journal, JT/2/13b/696–97).
8. *write an article for Hughes*: see letters 0900, nn. 9–10 and 0982, nn. 3–4.
9. *dig in your garden*: see letters 0977 and 0981, n. 13.

To Thomas Henry Huxley 27 October 1854 0989

27th. Oct. 1854.

My dear Huxley,

In compliance with your desire I write "at once".[1] When you tell me the work you have before you I am ashamed to dwell upon mine—and it would be unkind in me to throw the weight of the April number[2] upon you. I am willing to do as you propose and try to make preparation for it.

But Mr Chapman must not forget that it is a mere matter of trial with me: my brain chances to be pretty tough at the present moment and this emboldens me to undertake what you propose—chiefly I must confess because it is you that propose it. I know not what time may bring forth—the brain may become tender, which the gods avert! and thus I might be forced to relinquish my task. Besides, as yet I do not know the nature of the said task, if it prove pleasant I may go on my way rejoicing but if otherwise the six guineas will form a very frail bond of connexion between me and it.

At all events if I do find myself unable to carry it on I shall endeavour to give Mr Chapman timely notice.

"What idiots we are all to toil and slave at this rate"[3] it all depends upon the spirit in which we face our task—I know numbers who do not slave or toil and who are miserable devils after all.

Goodbye dear Huxley | ever Yours sincerely | John Tyndall

Don't put "Upper" on your letters—They go astray.

IC HP: 8.19
RI MS JT/1/TYP/9/2869
Typescript Only

1. *In compliance . . . write "at once"*: this letter is a reply to letter 0985.
2. *the April number*: Tyndall was referring to the April issue of the *Westminster Review* (see letters 0980; 0982; and 0985).
3. *What idiots . . . at this rate*: Tyndall is quoting Huxley (see letter 0985).

From Tommy Grieve 28 October 1854 0990

Queenwood College | October 28th. 1854.

My dear Dr. Tyndall,

I hope you are quite well. We are going to have the bonfire on the 3[1] I think for one of the boys told me that some of the masters told him that we were going to have it on the third A great many of the boys asked me if you were coming down to see it when I had your last letter[2] and I said you did not say you were but I hope you are. On Wednesday I had a tooth taken out I had the tooth ache on Monday and Tuesday and my cheek sweled[3] very much after it was taken out and Power[4] had it worse than me and he had his tooth taken out at the same time and his cheek sweled as much as mine and we had to be in bed all Wednesday but I did not care I only missed a half-holiday and that is not much and it was very wet and Mrs Harrington[5] did not come but she came on Thursday, my bed was under a window that looked out into the play-ground and I could hear all the boys crying out and playing &c. my gums hurt a little now but it not much. I sleep in the same bed as I used too in number five so if you come you will know where I sleep. They play sometimes with nurs[6] now but I dont play when they play with nurs because I dont want to run the risk to have some of my teeth knocked down my throat like Henry Aster[7] had. James Coats[8] had one of his teeth knocked out and his lip very much hurt. The fourth and fifth classes changed class rooms a good while ago, Cozens[9] and W. Fox[10] send their love to you

I remain | Your affectionate Friend | Tommy Grieve.

RI MS JT/1/TYP/1/465
LT Typescript Only

1. *bonfire on the 3*: earlier it was discussed that the bonfire would take place on 5 November (see letter 0983, n. 6).
2. *your last letter*: letter missing.
3. *sweled*: misspelling of 'swelled'.
4. *Power*: A. B. Power, a student at Queenwood. Not further identified.
5. *Mrs Harrington*: not identified.
6. *nurs*: LT Notes: '?'. Not further identified.
7. *Henry Aster*: a student at Queenwood. Not further identified.

8. *James Coats*: James Coats (n.d.), a student at Queenwood. He was likely one of five brothers to Sir Thomas Glen-Coats (1846–1910), who was also educated at Queenwood ('Coats, Sir Thomas Glen', in *The County Families of the United Kingdom*, vol. 59 [London: Spottiswoode, Ballantyne, & Co. Ltd., 1919], p. 230). They were the sons of Thomas Coats (1809–83), a prominent Scottish thread manufacturer (*ODNB*).
9. *Cozens*: J. Cozens is identified as the Curator of Botany for Queenwood College's Mutual Improvement Society, listed in the *Queenwood Observer* 2:5 (31 August 1854), p. 35.
10. *W. Fox*: one of six sons to a local doctor (see letter 0800, n. 29) and a student at Queenwood. He was listed as the 'Curator of Entomology' for the Mutual Improvement Society in 1858 ('Mutual Improvement Society', *Queenwood Observer* 6:1 [27 February 1854], pp. 1–8, on p. 2).

From Thomas Henry Huxley 29 October [1854][1] 0991

4 Rock Houses, Tenby. | Oct. 29th.

My dear Tyndall,

I rejoice in having got you to put your head under my yoke[2]—and feel ready to break into a hand gallop[3] on the strength of it.

I have written to Chapman to tell him you only make an experiment on your cerebral substance—whose continuance depends on tenacity thereof.

I didn't suspect you of being seduced by the magnitude of the emolument[4]—you Cincinnatus of the laboratory[5]—I only suggested that as pay sweetens labour à—fortiori[6] it will sweeten what will be to you no labour.

I'm not a miserable mortal now[7]—quite the contrary—I never am when I have too much to do—and my sage reflexion was not provoked by envy of the more idle. Only I do sometimes wish I could ascertain the exact juste milieu[8] of work which will suit—not my head or will—them can't have too much; but, my absurd stomach.

Towards that organ my feeling is the costermonger's[9] towards his donkey—

'I gives him hay, I gives him corn,
As much as he can chump off,
And then—if he don't cut along,
Vy! I cuts his precious rump off.'[10]

Now I give him hay and corn of the quality he likes—but he won't always cut along in proportion—and you know I can't 'cut his rump off' that would be suicidal.

Excuse this abominable nonsense—the offspring of an extra cup of tea.

and believe me | Yours very faithfully | T. H. Huxley

I shall be home on the 1st—and right glad to see you.

IC HP: 8.20
RI MS JT/1/TYP/9/2870

1. *[1854]*: while the year is not included in the date of this letter, LT pencilled '[1854]' in the top of the typescript.
2. *I rejoice . . . my yoke*: Tyndall had agreed to start working on the 'Science' column for the *Westminster Review* in preparation for the April 1855 issue (see letter 0989, n. 2).
3. *hand gallop*: a controlled gallop (*OED*).
4. *emolument*: 'dues; reward, remuneration, salary' (*OED*).
5. *Cincinnatus of the laboratory*: refers to Lucius Quinctius Cincinnatus (*c.* 519–430 BCE), a Roman statesman and military leader known for embodying Roman virtues of manliness, civic virtue, and modesty.
6. *à—fortiori*: a Latin phrase 'used to express a conclusion for which there is stronger evidence than for one previously accepted' (*OED*).
7. *miserable mortal now*: Tyndall suggested he knew people with less work to do who are 'miserable devils' (see letter 0989).
8. *juste milieu*: 'a happy medium; judicious moderation' (*OED*).
9. *costermonger's*: 'a person who sells fruit, vegetables, fish, etc. In the street from a barrow or handcart'; 'also used as a term of contempt or abuse' (*OED*).
10. *I gives him . . . rump off*: this is reworked verse from the nineteenth-century song 'Whoppin' Ned' (*Library of Congress*, https://www.loc.gov/resource/amss.as114950/ [accessed 22 December 2016]).

To Thomas Archer Hirst 30 October 1854 0992

Royal Institution 30th. Oct 1854

My dear Tom

If Wright can afford me writing space I shall be very *[glad]* to avail myself of his kindness. What I shall want is a table, a chair, pen and ink. No fire, I don't use one in my own quarters. And then I can spend the night with you.[1] Just say whenever this can be managed without inconvenience to any body.

Your affectionate | John.

RI MS JT/1/HTYP/375

1. *And then . . . with you*: Hirst had recently acquired a house in Broughton, but was still working on preparing it (see letter 0977). Tyndall and Hirst had been discussing whether or not Tyndall might stay there should he come for a visit to Queenwood (see letters 0983, n. 6; 0984, n. 14; and 0988). It appears he was organizing for Tyndall to have a place to write, as he planned to work on his article for Hughes (Journal, JT/2/13b/698).

To James Coxe [31 October 1854][1] 0993

My dear Dr. Coxe

I am not to write to Mrs. Coxe—She has given me strict orders not to do so and it is my duty to obey—Therefore I submit though with the consciousness that her last note to me[2] would deserve a better reply than it is in my poor power to write. The warm kindheartedness breathing through every line of that note was sufficient to render it precious to me, but the fact of its having been written by her on her bed of suffering converts it into a kind of sainted relic in my estimation, and I shall preserve it as such. I am a bad hand at praying in the accredited fashion—if I did pray it would be in a peculiar dialect of my own which would doubtless sound heathenish to the orthodox—but if a strong silent desire partake at all of the nature of prayer, then I may be said to pray for the speedy recovery of Mrs. Coxe. I can truly say that it has added something to the pleasure of my life to have made her acquaintance, and therefore on the most selfish grounds I am impelled to wish for the long and happy continuance of hers—I trust sincerely that each succeeding day will add to her strength and that she will soon be restored—to both of us—in the full enjoyment of her vigour.

There is a little matter which has often crossed my mind of late and concerning which I intended to write to you in due time. A note I received from Alick[3] a few days ago has precipitated my intention. Alick is now far advanced in many respects as compared with his schoolfellows, new boys are continually streaming into a place like Queenwood and hence every class possesses its beginners which demand a large portion of the teachers attention. When therefore a boy has passed a certain limit, on account of the smallness of the number which stand on the same level he cannot have a great share of the teacher's attention. A higher school is needed for such a boy and I believe it will soon be needed for Alick,—a school in which he would find boys above him and teachers who can devote all their attention to such boys.

Has the thought of sending Alick for a year or two to Germany[4] ever crossed your mind? I believe what is now paid for him at Queenwood would amply cover his expenses in Germany. If the thought should ever occur to you do not forget that I have friends in Marburg, Berlin and elsewhere—men of intellectual eminence and moral worth, and that it will always give me the greatest pleasure to have their advice and invoke their assistance in behalf of Alick. I know that many of them would go some way to oblige me and I shall never ask them to do so with more earnestness than on Alick's account. I do not presume to give advice, but should you be uncertain what to do perhaps this may enter into your calculations,—and now good bye—present my warmest remembrances to Mrs. Coxe, and I trust that her own hand will soon be able to announce the glad tidings of her perfect recovery.

Believe me | Most sincerely yours | John Tyndall.

RI MS JT/1/TYP/1/1/261
LT Typescript Only

---

1. *[31 October 1854]*: LT suggested an approximate date of 31 October 1854, and this seems correct. In the letter, Tyndall refers to receiving a letter from Alick 'a few days ago', a letter he noted receiving 'yesterday' in his journal entry for 29 October (JT/2/13b/696).
2. *her last note to me*: letter missing. For the most recent reference to a letter from Mrs. Coxe see letter 0961, nn. 1 and 2.
3. *Note . . . Alick*: letter missing (see n. 1). Alexander Cumming, the Coxes's nephew (see letter 0781).
4. *sending Alick . . . to Germany*: Tyndall and Hirst had been contemplating suggesting to Cumming's family that he continue his education in Germany (see letter 0945, n. 14).

To The Secretary of the Royal Society[1] 31 October 1854 0994

The experiments recorded in the accompanying paper[2] were instituted for the purpose of ascertaining the thermal effect produced by the passage of air from a high to a low pressure across a porous plug. It is shewn that a cooling effect takes place, and that the amount of this effect is sensibly proportional to the difference of pressure on both sides of the plug. With carbonic acid the cooling effect is found to be greater than with air, whereas in the case of hydrogen the cooling effect, inferred from some preliminary experiments, appears to be only one thirteenth that of atmospheric air. A remarkable result has been obtained from the interdiffusion of two different gases—the thermal effect accompanying the interdiffusion of carbonic acid and air was found to be a diminution of temperature amounting to 2° Cent.[3] It is to be hoped that the authors[4] will fulfil their promise of making more exact experiments on this important point. The above experiments and the theoretic deductions drawn from them appears to me calculated to increase our knowledge regarding the nature of Heat & would therefore recommend their publication in the Philosophical Transactions.[5]

John Tyndall
31 Oct. 1854[6]

RS RR/2/239

1. *The Secretary of the Royal Society*: this letter was addressed to 'The Secratary of the Royal Society' (Tyndall misspelled 'Secretary'). At the time, the Secretary of the RS was George Gabriel Stokes.
2. *The experiments . . . accompanying paper*: see letter 0981, n. 8.
3. *Cent.*: centigrade.
4. *the authors*: Joule and Thomson.
5. *recommend . . . Philosophical Transactions*: it is unclear if Joule and Thomson undertook the further experimentation recommended by Tyndall. Their paper was eventually published in the *Phil. Trans.* (see 0981, n. 8).
6. *31 Oct. 1854*: the following was written in another hand that appears in the top right corner of the reverse side of the letter: '31. Oct. 1854 | Chpt. By D^{r}. Tindal | on Messrs Thomson & | Joule's paper.'

From Michael Faraday[1] 31 October 1854 0995

Royal Institution, | Oct. 31, 1854.

My dear Tyndall,

I send the enclosed letter[2] from MM. Van Breda[3] and Logeman[4] to you as an Editor of the Philosophical Magazine. If you should judge it proper for insertion in that Journal, I shall be very happy to see it there, but will beg you to accompany it on my part with the observation that it is not so conclusive in proving the negative (a thing very difficult to do) as to move me at present from the reserved condition of mind which I have recently expressed in respect of this matter.

Ever yours truly, | M. Faraday

MM. Van Breda and Logeman, 'On the Conductibility of Liquids for Electricity', *Phil. Trans.* 8:54 (1854), p. 465.
Faraday Correspondence, 4:2915

1. *From Michael Faraday*: this letter was originally published in the *Phil. Mag.*
2. *the enclosed letter*: this is a letter that was published in MM. Van Breda and Logeman, 'On the Conductibility of Liquids for Electricity', *Phil. Trans.* 8:54 (1854), pp. 465–69. It was also published in *Faraday Correspondence,* 4:2893.
3. *Van Breda*: Jacob Gijsbertus Samuël van Breda (1788–1867), a Dutch biologist and geologist who worked on magnetism and electricity in his role as director of the Physics Cabinet of the Teylers Museum in Harrlem, Netherlands.
4. *Logeman*: Wilhelm Martin Logeman (1821–94), a Dutch scientific instrument maker in Haarlem, Netherlands (see *Faraday Correspondence*, 4, p. lviii).

To [Peter Riess][1] 4 November 1854 0996

London 4th November 1854

My dear Sir.

My next course of lectures at the Royal Institution is to be on Electricity,[2] and I should like very much to be able to shew, if it were possible, the phenomena of the Nebenstrom,[3] and the dependance of the discharge upon the shape of the circuit. Will you kindly let me know whether there is any way of shewing these effects to an audience of one or two hundred. If you know any likely means I should feel under a great obligation to you if you would order the necessary apparatus and have it sent to me from Berlin. I should like to be able to shew the effect of the Nebenstrom in as many different ways as possible: by its heating power, its magnetising power, and its power of imparting a shock to the human body. If any other experiments which you deem likely to interest an audience such as ours—sehr gebildet aber nicht sehr wissenschaftlich[4]—should occur to you within the next two months, I should deem it a great kindness if you would communicate them to me.

I find that Matteucci has just sent his 'Cours Special Sur L'Induction, &c.'[5] to the Philosophical Magazine for review: now this brings to my mind that you also wrote a work on Electricity[6] which however has never, so far as I am aware of, been reviewed in this country. Few therefore know any thing about it—Now if you wish to have the work known in England there is no other way of doing this than by sending it to the principal reviews. If you like I will write a notice of it for the next number of the Philosophical Magazine,[7] and there is another periodical in which I think I could have it reviewed also.[8] Having gone to the trouble of writing a first rate work it is a pity not to take the proper means of making it known throughout the world.

With kindest remembrance to your wife and family, and many greetings to my friends in Berlin

Believe me | most sincerely yours | John Tyndall

Staatsbibliothek zu Berlin (Stabi) Slg Darmstaedter 1855 Tyndall Be. 1–58, 44–45

---

1. *[Peter Riess]*: while the letter does not explicitly state that it is from Riess, letter 1000 appears to have been written in reply.
2. *Lectures . . . on Electricity*: before Easter 1855, Tyndall delivered a series of eleven lectures on 'Magnetism and Frictional Electricity' (see 'General Monthly Meeting', *Roy. Inst. Proc.*, vol. 2, p. 5); after Easter he delivered a series of eight lectures on 'Voltaic Electricity' (see 'General Monthly Meeting', ibid., p. 56).

3. *Nebenstrom*: German for 'secondary current' (see letter 1000).
4. *sehr gebildet . . . wissenschaftlich*: German for 'well educated but not very scientific'.
5. *Cours Special Sur L'Induction*: see C. Matteuccci, *Cours Spécial sur L'Induction, Le Magnétisme de Rotation, Le Diamagnétisme, et sur les Relations Entre la Force Magnétique et les Actions Moleculaires* (Paris: Mallet-Bachelier, Gendre et Successeur de Bachelier, 1854).
6. *a work on Electricity*: see letter 0727, n. 14.
7. *I will write . . . Philosophical Magazine*: a notice of Riess's book appeared as 'Notices Respecting New Books:—M.P.T. Riess's Treatise on Frictional Electricity', *Phil. Mag.* 62 (February 1855), pp. 150–51.
8. *another periodical . . . reviewed also*: presumably the *Westminster Review*. Tyndall had recently agreed to co-author the regular 'Science' column with Huxley (see letter 0989). Riess's book does not appear to have been noticed there.

From Edward Sabine [6 November 1854][1] 0997

My dear Sir—

I send you Secchi's[2] original paper,[3] but will beg you to have the kindness to return it as soon as you have satisfied yourself that the diagram in the Phil Mag. is the same as in the Corr. Scien.,[4] and the text the same.

The diagram[5] does not appear to be well suited t[o] assist the conception; but of this I am & can be but an indifferent judge, because I am so familiar with the facts themselves—I take Secchi's illustration to be this—

At Toronto (T) the end of the needle acted upon is (according to his view) the upper end or S. Pole—The sun in his progress from E. to W. (East to West) causes the S. pole of the needle to be at 20" to the West of its mean position; at 2" to the East; at 9½" to the West; and at 15½" to the East of its mean position.

At Hobarton[6] (H) it is the N Pole (there the upper pole) which is acted upon, and the sun in his progress from E to W causes there the N. Pole to be to the East of its mean position at 20"; to the west at 2"; to the East at 9½, and to the west at 15½.

I repeat the illustration does not help me; but it possibly may others whose minds are in the state of Secchi's—About the facts however there is no difficulty (at least I hope not because I am there the responsible party)—He gives my statement in my own words,[7] and you or any one finds any difficulty in understanding these, pray tell me so.

You say that 'the arrow head is what we call the North Pole of the needle'. I say nothing against that in a general way, but certainly Secchi does not *[1 word illeg]* his arrows to a needle at all. An index hand projecting from the vertical SW, to the right at Toronto and to the left at Hobarton at 20", &c, would not have misled, as the arrow appears to have done.

If the arrows are to be regarded as needles, then the point of the arrow indicates at each station the pole which he conceives to act upon—i.e. the upper pole; i.e. the S. at Toronto, and the N. at Hobarton.

It is all clear in Secchi's mind because his mind is imbued with his hypothesis.

Sincerely yours, | Edward Sabine.

RI MS JT/1/S/14
RI MS JT/1/TYP/4/1304

---

1. *[6 November 1854]*: LT's typescript includes the date of 6 November 1854, though the original letter was not dated.
2. *Secchi's*: Pietro Angelo Secchi (1818–78), an Italian astronomer and Jesuit priest, who served as Director of Observatory of the Collegio Romano from 1850 until his death. He was a pioneer of spectroscopy who focused much of his work on the sun.
3. *original paper*: see 'Sulla maniera di determinare la flesione dei tubi nei grandi strumenti astronomici', *Corrispondenza Scientifica in Roma* 3 (1855), pp. 209–21; 261–67. It had appeared in the July issue of the *Corrispondenza Scientifica in Roma* and was translated for the *Phil. Mag.* as 'On the Periodical Variations of Terrestrial Magnetism', *Phil. Mag.* 8:53 (November 1854), pp. 385–99.
4. *Corr. Scien.*: *Corrispondenza Scientifica in Roma* was an Italian scientific journal.
5. *The diagram*: the diagram Sabine spoke of appeared on p. 394 of Secchi's article (see n. 3).
6. *Hobarton*: Hobarton (now Hobart), Tasmania, Australia.
7. *gives my statement... words*: pp. 394–95 of Secchi's article (see n. 3) quotes from E. Sabine, *Observations Made at the Magnetical and Meteorological Observatory at Toronto in Canada*, 2 vols (London: Longman, Brown, Green, and Longmans, 1845), vol. 1, p. 14.

From Thomas Charles Higginson 10 November [1854][1] 0998

Ipswich | Nov. 10th

My dear Tyndall,

I was very sorry at not having the pleasure of seeing you when in London on the 25th ult.,[2] but I fear our not having met was caused by my stupidity in putting a wrong date to my letter[3] I called twice at the Institution[4] on the 25th and left my card which the stupid porter seems not to have delivered. Non importa,[5] the pleasure is only deferred, if I had a bed to offer you I would ask you to come and spend a few days with me, but my sister[6] who keeps house (or rather lodgings) for me, occupies the only spare room in the small house I am in, in every other respect you should have the run of the house and an Irish welcome. I am now very busy providing food for powder[7] and at the same time lining my pockets, when we do meet what a chat we shall have recounting to each other all the events that have occurred since our boyish days

Believe me | Your sincere friend | T. C. Higginson.

RI MS JT/1/TYP/2/598
LT Typescript Only

1. *10 November [1854]*: in the typescript LT noted 'probly 1854'.
2. *ult.*: abbreviation for the Latin phrase, 'ultimo mense' or 'last month'.
3. *my letter*: letter missing.
4. *the Institution*: the RI. Tyndall noted in his journal entry of 29 October 1854, 'Higginson was to have come one day—I was not at the Institution and I know not whether he did come' (Journal, JT/2/13b/696).
5. *Non importa*: Italian for 'it's not important'.
6. *my sister*: possibly Mina Higginson (see letter 0963, n. 2).
7. *food for powder*: 'soldiers, esp. low-ranking recruits, collectively regarded as expendible in war' (*OED*). This is a direct reference to Shakespeare's *Henry IV, Part 1*, 'Tut, tut, good enough to toss; food for powder, food for powder. They'll fill a pit as well as better. Tush, man, mortal men, mortal men' (IV.ii.66–68).

From Michael Faraday 11 November 1854 0999

R. Institution, | 11 Nov. 1854.

Many thanks, my dear Tyndall, for your kind letter[1] which I have just received. I was anxious about you, thinking you might be confined at home by a little indisposition (as you would call it) and writing, and should probably have called today, in the evening. Now I shall rest, knowing how it is, and I hope you will enjoy the weather and the quietness and the time of work and the time of play, finding them all ministrants to your health and contented happiness.[2] Here we jog on, and I have just undertaken the Juvenile Lectures at Christmas,[3] thinking them the easiest thing for me to do. Reading Matteucci[4] carefully, and also an abstracted translation of Van Rees' paper,[5] is my weighty work; and because of the call it makes on memory, I have now and then to lay them down and cease to the morrow. I think they encourage me to write another paper on lines of force, polarity &c,[6] for I was hardly prepared to find such strong support in the papers of Van Rees and Thomson[7] for the lines as correct representants of the power and its direction, and many old arguments are renewed in my mind by these papers. But we shall see how the maggot bites presently, and as I fancy I have gained so much by waiting, I may perhaps wait a little longer.

Ever, my dear Tyndall, | Yours truly, | M. Faraday

RI MS JT/TYP/12/4137
Faraday Correspondence, 4:2921

1. *your kind letter*: letter missing.
2. *I was anxious . . . contented happiness*: Tyndall had been visiting Hirst at Queenwood for ten days (see letter 0992, n. 1).
3. *Juvenile Lectures at Christmas*: Faraday gave the RI's Christmas Lectures of 1854 on the topic of 'The Chemistry of Combustion'. See also letter 0747, n. 4.
4. *Matteucci*: possibly M. Matteucci, 'On the Electricity of Flame, with Observations', *Phil. Mag.* 8:53 (November 1854), pp. 399–403.
5. *Van Rees' paper*: Richard van Rees (1797–1875), a Dutch mathematician, physicist, and professor at the University of Utrecht. Van Rees's memoir *Over de Theorie der Magnetische Krachtlijnen van Faraday* (Amsterdam: C. G. van der Post, 1853), had appeared in German as 'Ueber die Faraday'sche Theorie der magnetischen Kraftinien,' *Poggend. Annalen.* 90:3 (November 1853), pp. 415–36. Faraday cited van Rees in M. Faraday, 'On some points of magnetic philosophy', *Phil. Mag.* 9:57 (February 1855), pp. 81–113.
6. *another paper . . . polarity &c.*: see Faraday, 'Experimental Researches in Electricity,' *Phil. Mag.* 142 (1852), pp. 25–56.
7. *Thomson*: see W. Thomson, 'On the Uniform Motion of Heat in Homogenous Solid Bodies, and its connexion with the Mathematical Theory of Electricity', *Phil. Mag.* 7:48 (Supplement 1854), pp. 502–15. Here Thomson directly engaged with Faraday's work on lines of force.

From Peter Riess 11 November 1854 1000

Geehrtester Herr

Es freut mich sehr, daß Sie in Ihren Vorlesungen über Elektricität Etwas von den neueren Forschungen beibringen wollen, die in England wol kaum bekannt sind, und es würde mir ein großes Vergnügen sein, Ihnen dabei irgendwie behülflich zu sein.—Alle Wirkungen des Nebenstromes laßen sich in beliebiger Stärke durch den einfachsten Apparat hervorbringen. Es genügt zwei Kupferdräthe, jeder etwa 50 Fuß lang 5/8 Linie dick, um einen breiten Holz- oder Glascylinder neben einander aufzuwickeln, so dasß zwischen den Dräthen überall eine starke Linie frei bleibt, und den einen Drath als Hauptdrath, den andern als Nebendrath zu benutzen. Doch ziehe ich es zur Demonstration vor, jeden der beiden Dräthe einzeln auf einer Holz- oder Guttapercha Tafel zu einer ebenen Spirale von 11 Zoll Breite zu winden, die beiden Spiralen etwa 3 Linien von einander entfernt, einander gegenüberzustellen und zwischen ihnen eine Glasplatte anzubringen, damit der Zuhörer nicht an ein Überspringen der El. von einer Spirale zur andern denken kann. Mit solchen Spiralen, die sich der Physiker recht gut selbst anfertigen kann, läßt sich durch den Nebenstrom Platindrath verbrennen, eine Stahlnadel ins maximo magnetisieren, ein unerträglicher Schlag ertheilen. Zu feineren Versuchen gehören freilich ein Thermometer, ein Funkenmikrometer etc, die von einem guten Mechaniker angefertigt sein müßen. Solche Instrumente, wenn Sie mir sie speciell bezeichnen wollen, will ich gern für Sie bestellen bemerke jedoch, daß nach der leidigen Sitte unserer Mechaniker, erst nach Verlauf von mehreren Monaten auf deren Ablieferung zu rechnen ist.—Die Abhängigkeit der Stärke des Nebenstromes von der Form der Nebenschließung läßt sich auch einem größern Publikum leicht aufzeigen. Man nimmt zwei ganz gleiche Dräthe, aber von nicht geringerer Länge, als der Drath besitzt in welchem der Nebenstrom erregt wird (also in dem obigen Beispiele nicht kürzer als 50 Fuß, und 5/8 Linie dick). Der eine Drath wird auf einem Brette zu einer ebenen Spirale von 11 Zoll größter Breite aufgewunden; der andere Drath wird auf einem Brette 50mal hin und her gewunden und die Hälfte der so gebildeten U aufwärts gebogen. Ich habe diese Vorrichtung in meinem Buche § 884 unter dem Namen Utafel beschrieben. Man bringt in der Nebenschließung einen dünnen 1 Zoll langen Platindrath und die Utafel an und sucht in vorläufigen Versuchen die Stärke der Batterieladung, bei welcher der Platindrath durch den Nebenstrom hell weißglüht. Bei meiner Batterie, wenn sie aus 5 Flaschen, jede von 2,6 quadrat Fuß Belegung, besteht, und den Apparaten von den angegebenen Dimensionen geschieht dieß, wenn bei ihrer Ladung 46 Funken von ½ Linie Länge in der Maaßflasche übergegangen sind, was durch etwa 24 Umdrehungen meiner Scheibe bewirkt wird.—Um nun die Abhängigkeit des Nebenstromes von der Form der Nebenschließung zu zeigen, bringe ich in dieser Schließung zuerst die ebene Spirale an und zeige, daß bei einer Ladung der Batterie mit 46, 48, 50, 52 Funken der Maaßflasche, der Platindrath durch die Entladung des Nebenstromes keine sichtbare Änderung erfährt (er bleibt, selbst im dunkeln Zimmer, vollkommen dunkel). Dann ersetze ich die Spirale durch die Utafel, wodurch ja die Dimensionen der Nebenschließung nicht geändert werden; jetzt wird der Platindrath bei einer Ladung der Batterie mit 46 Funken hell leuchtend und glühend, bei einer Ladung mit 48 Funken zerstiebt er in glühenden Kugeln.

Für Ihren Rath in Betreff meiner Elektricitätslehre danke ich Ihnen; ich weiß, daß der Rath gut gemeint, praktisch ist und von Vielen befolgt wird, aber ich kann keinen Gebrauch davon machen. Von jeher habe ich meine Abhandlungen und jetzt mein Buch (mit alleiniger Ausnahme der Sendung an Faraday und Humboldt) nur als Gegengeschenk versendet, oder wo mir ein Wunsch danach geäußert wurde. Ich denke so: von zwei Fällen einer;

entweder—mein Buch ist gut und erfüllt seinen Zweck. Dann ist es Pflicht des Redactors einer kritischen Zeitschrift, den Leser aufmerksam zu machen auf ein Compendium, das den jetzigen Stand der Elektricitätslehre darlegt, und ich habe nicht den Beruf, Jemand an seine Pflicht zu erinnern. Oder mein Buch ist schlecht—dann streitet es gegen mein Gewißen, den Redactor durch meine Höflichkeit zu verleiten, ein schlechtes Buch zu empfehlen. Was ich bei solcher Denkungsart an scheinbarer Anerkennung verliere, gewinne ich durch die verdoppelte Freude bei einer Anerkennung ohne mein Zuthun und wirklich haben mir bis jetzt zwei deutsche und eine americanische Zeitschrift diese Freude gebracht. Weit auffallender, als das Schweigen über mein Buch, deßen Beurtheilung ein längeres nur Wenigen gegöntes Studium voraussetzt, ist mir die Vernachläßigung, welche meine einzelnen Abhandlungen in England erfahren. Es ist, so viel ich weiß, seit langer Zeit weder eine der größeren, und schwerer verständlichen, noch eine der kleineren leicht übersetzbaren Abhandlungen die ich in Poggendorffs Annalen veröffentlicht habe, in England übersetzt worden. Ich würde den relativ geringen Werth dieser Arbeiten als Grund jener Vernachläßigung annehmen, wenn nicht Vieles, das in englischen Journalen übersetzt erscheint, diese Annahme zurückwiese, und ich darf hoffen daß, was ein Zufall verschuldet hat, ein Zufall wieder gut machen wird.

Mit dem Wunsche daß Sie Ihre schöne aber schwere Aufgabe in diesem Winter zu Ihrer Zufriedenheit lösen mögen, empfiehlt sich Ihnen

Ihr | ergebenster | P. Riess | Berlin 11 November 1854.

Most honoured Sir

I am very pleased that you want to teach something from the more recent research, which is probably hardly known in England, in your lectures on electricity,[1] and it would be a great pleasure to me to be of assistance to you somehow or other in the process.—All the effects of the secondary current can be produced in any strength using the simplest apparatus. It is sufficient to wind two copper wires, each about 50 feet long and 5/8 of a line thick, next to one another around a broad wooden or glass cylinder, so that a strong line remains free everywhere between the wires, and to use one wire as the main wire and the other as the secondary wire. However, I prefer, for a demonstration, to wind each of the two wires individually on a wooden or gutta-percha[2] tablet into a planar coil of 11 inches width, to place the two coils opposite one another about 3 lines apart, and to place a glass plate between them, so that the listener cannot think of the el. jumping from one coil to the other. With such coils, which a physicist can make quite well himself, platinum wire can be completely burned by the secondary current makes, a steel needle magnetized to the maximum degree, and an unbearable shock delivered. Of course, more sensitive experiments call for a thermometer, a spark micrometric etc, which must be made by a good instrument maker. I shall gladly order such instruments for you, if you want to name them specifically for me, however note that, in keeping with the tiresome custom of our instrument makers, their delivery can be expected only after the passage of several months.— The fact that the strength of the secondary current depends on the shape of the secondary circuit can also be demonstrated easily to a larger audience. One takes two completely identical wires, but of no lesser length than the wire has in which the secondary current is produced (that is, in the above example, not shorter than 50 feet and 5/8 of a line thick). One wire is wound on a board into a planar coil of 11 inches maximum width; the other wire is wound back and forth on a board 50 times, and half of the U formed his way is bent upwards. I have described this device in my book in § 884 under the name of the U-tablet.[3] One inserts a thin 1-inch-long platinum wire and the U-tablet in the secondary circuit and in tentative attempts looks for the strength of the battery charge, by which the platinum wire will glow brightly incandescent through the secondary current. With my battery, when it consists of 5 jars, each occupying 2.6 square feet, and the apparatus of the specified dimensions, this happens when 46 sparks of ½ line length have jumped in the unit jar during charging, which is achieved by about 24 turnings of my disc.—Now in order to show that the secondary current is dependent upon the shape of the secondary circuit, I first insert the planar coil in this circuit and show that when charging the battery with 46, 48, 50, 52 sparks in the unit jar, the platinum wire experiences no visible change through the discharge of the secondary current (it remains, even in a dark room, completely dark). Then I replace the coil with the U-tablet, by which, of course, the dimensions of the secondary circuit do not get changed; now, when charging the battery with 46 sparks, the platinum wire becomes luminous and glowing, when charged with 48 sparks it disintegrates in glowing balls.

I thank you for your advice with respect to my theory of electricity;[4] I know that the advice is well-meant, practical and is followed by many, but I cannot make any use of it. I have always sent my papers and now my book (with the sole exception of the copies mailed to Faraday and Humboldt[5]) only as a gift sent in return, or when a wish for them was expressed to me. I think this way: of two cases, one is; either—my book is good and fulfills its purpose. Then it is the duty of the editor of a critical journal to draw the reader's attention to a compendium that presents the current state of the theory of electricity, and it is not my job to remind someone of his duty. Or my book is bad—then it goes against my conscience to mislead the editor through my politeness into recommending a bad book. What I lose in apparent recognition by such a way of thinking, I gain through the redoubled pleasure of recognition without having a hand in the matter and indeed to date two German[6] and one American journal[7] have given me this pleasure. To me, what is far more striking than the silence surrounding my book, the appraisal of which presupposes lengthy university studies granted only a few few, is the neglect which my individual papers experience in England. As far as I know, for a long time now, neither one of the longer, and more difficult to understand, nor one of the shorter and easily translatable papers that I have published in Poggendorffs Annalen,[8] has been translated in England. I would assume that the relatively limited value of these works is the reason for that neglect, if much that appears in translation in English journals did not repudiate this assumption, and I may hope that whatever a coincidence has caused, a coincidence will put right.

With the wish that you may resolve your nice but difficult task to your satisfaction this winter, I commend myself to you as

Your | very humble | P. Riess | Berlin, 11 November 1854.

RI MS JT/1/R/28

---

1. *your lectures on electricity*: Tyndall had written to Riess asking for suggestions for experiments, specifically related to auxiliary currents, that could be conducted in front of a large audience (see letter 0996).
2. *gutta-percha*: a rigid, natural latex produced from the sap of the *Palaquium gutta* tree.
3. *the U-tablet*: Riess described the device, the 'Utafel' or 'U-tablet', in *Die Lehre von der Reibungselektricität*, 2 vols (Berlin: August Hirschwald, 1853), vol. 2, pp. 334–35.
4. *I thank you . . . theory of electricity*: Tyndall had suggested that Riess's book be submitted for review (see letter 0996).
5. *Humboldt*: see letter 0789, n. 4.
6. *two German*: see 'Riess, Pet. Theophil., Dr. phil., Prof. zu Berlin etc., die Lehre von der Riebungselekticität', *Literarisches Zentralblatt für Deutschland* 21 (21 May 1853), pp. 349–51. The other review has not been identified.
7. *one American journal*: see 'Die Lehre von der Riebungselekticität von P. T. Riess', *American Journal of Science and Arts* 17:50 (May 1854), pp. 273–74.
8. *published in Poggendorffs Annalen*: Riess published extensively in *Poggend. Annalen.* from 1832 to 1863 (see 'Riess, Peter', *RS Catalogue*, vol. 5, pp. 201–4).

From Heinrich Gustav Magnus 13 November 1854 1001

Berlin 13 Novemb 54.

Theurer Freund.

Sie sind, wie ich hoffe, wenn Sie diesen Brief erhalten, schon im Besitz der Teufelchen. Zwei meiner Neffen (Söhne meines ältesten Bruders) gehn nach Liverpool, und haben das Schächtelchen für Sie mitgenommen. Sie haben nun noch die Mühe die kleinen Figuren richtig mit Wasser zu füllen und in die Flaschen einzuschließen, aber ich hätte nicht gewußt wie ich sie Ihnen hätte bequemer zufertigen können. Zu entrichten ist nichts dafür, sie kosten alle zusammen eine Kleinigkeit.

Die Resultate Ihrer Arbeit erwarte ich mit großem Verlangen, was Sie über Feilitzsch als Experimentator sagen, ist auch meine Ansicht.

Regnault's neue Untersuchungen über die Dämpfe sind wieder sehr interessant, aber seine Sprache ist doch gar zu anmassend. Er sagt „Les physiciens n'ont pas cru necessaire de verifier la loi de Dalton par des éxperiences directes". Aber schon vor 18 Jahren habe ich das gethan, und die Gesetze, die er in dem 4^{tn} Abschnitt seiner Abhandl. Forces elastiques des vapeurs qui sont émises par les liquides volatils mélangés par dissolution réciproque, ou superposés erhalten hat, habe ich sämmtlich schon damals in einer Abhandl. „Über das Sieden von Gemischen zweier Flüssigkeiten und über das Stoßen solcher Gemische" in Poggend. XXXVIII. 481. bekannt gemacht.—Ich habe deßhalb an Regnault geschrieben, und ihn gebeten meinen Brief der Pariser Academie mitzutheilen und in den Comptes rendus abdrucken zu lassen. Ich zweifle auch nicht daß er dies thun wird, aber ich habe auch gleichzeitig an Poggend. eine Notiz gegeben; zumal eine der interessanteren Erscheinungen beim Kochen solcher Mischungen die getrennte Schichten bilden, von ihm gar nicht erwähnt ist. Daß nämlich die Temperatur des Dampfes um mehrere Grade niedriger als die der kochenden Flüssigkeit ist. Sobald die Notiz gedruckt sein wird, schicke ich sie Ihnen unter Kreuzband.

Mir ist die Polemik verhaßt und doch wird man dazu gedrängt. Nicht nur daß ich nun wieder mit Regnault zusammenkomme, sondern in Bezug auf meine Abhandl. über die Abweichung der Geschosse sind einige Bekanntmachungen erschienen. So wenig haltbar sie auch seien mögen so kann ich sie doch nicht mit Stillschweigen übergehn. Sie haben vielleicht einmal die französische Ubersetzung, welche von meiner Abhandl. erschienen ist, gesehn. Der Uebersetzer M^{r}. le Colonel Delobel à Liège, spendet mir in der Vorrede das größte Lob das man jemand spenden kann; und jetzt veröffentlicht er: „Memoires et notices relatives à l'influence de la rotation des projectiles sur leur trajectoire; und druckt darin die Uebersetzung nochmals ab, zugleich aber mit Gegenbemerkungen des Colonel Bormann, die zwar ganz schmeichelhaft für mich sind, in denen dieser aber sagt er habe die Sache früher gemacht. Er hat aber gesagt durch die Reibung einer Kugel gegen die Luft entstehe ein Druck gegen ihren Mittelpunkt der um so größer sei je größer die Reibung ist. Aber durch Reibung kann bei einem rotirenden Körper nur eine Verzögerung der Winkelgeschwindigkeit entstehen, sonst nichts.

Ob und wie ich hierauf antworten werde weiß ich noch nicht, aber unbeantwortet werde ich es nicht laßen dürfen.

Der einzige Trost der einem bei solchen Zänkereien bleibt sind dann die eigenen Arbeiten, und so hoffe ich denn daß meine Wasser Versuche bald fertig sein sollen. Grüßen Sie alle Freunde, von Allen D^r. Faraday, und M^ss F. so wie D^r. Bence Jones

Von Herzen Ihr | GMagnus.

Berlin 13 Novemb 54.

Dear friend.

You are, I hope, when you receive this letter, already in possession of the little devils.[1] Two of my nephews (sons of my eldest brother)[2] are going to Liverpool, and have taken the little with them for you. You will still have the bother now of filling the little figures correctly with water and sealing them into the bottles, but I would not have known how I could have forwarded them to you in a more convenient manner. You do not need to pay anything for them, all together they cost a trifle.

I await the results of your work with great desire; what you say about Feilitzsch[3] as an experimenter is also my opinion.

Regnault's new investigations on vapours[4] are very interesting once again, but his language is just quite too presumptuous. He says "Les physiciens n'ont pas cru necessaire de verifier la loi de Dalton par des éxperiences directes".[5] But I already did that 8 years ago,[6] and I published all the laws that he obtained in the 4th section of his disquisition[7] (Forces élastiques des vapeurs qui sont émises par les liquides volatils mélangés par dissolution réciproque, ou superposés[8]) even back then in a paper "Über das Sieden von Gemischen zweier Flüssigkeiten und über das Stoßen solcher Gemische" in Poggend. XXXVIII. 481.[9] I have therefore written to Regnault, and asked him to communicate my letter to the Paris Academy[10] and to have it printed in the Comptes rendus.[11] I do not doubt that he will do this, but at the same time I have also given a note to Poggend,;[12] especially as one of the more interesting phenomena with the boiling if such mixtures that form separate layers is not mentioned by him at all. Namely, that the temperature of steam is lower by several degrees than that of the boiling liquid. As soon as the note is going to be printed,[13] I will send it to you in a postal wrapper.

I hate polemics and yet but one is forced into them. Not only am I now rubbing shoulders with Regnault again, but some publications have appeared in relation to my paper on the deviation of projectiles.[14] As little tenable as they may be, I just cannot pass them over in silence. You have perhaps seen the French translation which has appeared of my paper. The translator, Mr le Colonel Delobel à Liège,[15] gives me the greatest praise in the preface that one can give someone; and now he publishes: "Memoires et notices relatives à l'influence de la rotation des projectiles sur leur trajectoire;[16] and reprints the translation[17] in it once again, but at the same time with rejoinders from Colonel Bormann[18] that are admittedly quite flattering for me, but in which the latter says that he has done the thing earlier. He did say, however, that through the friction of a bullet against the air pressure is produced against its centre that is all the greater the greater the friction is. But through friction only a declaration of angular velocity can be produced in a rotating body, nothing else.

If and how I shall respond to this, I still do not know, but I shall not be allowed to leave it unanswered.

The only consolation that remains for one with such squabbling then are one's own works, and so I hope then that my water experiments should soon be finished . Give my regards to all my friends, especially D^r^. Faraday and M^ss^ F.,[19] as well as D^r^. Bence Jones

Sincerely Your | GMagnus

RI MS JT/1/M/14

---

1. *little devils*: see letter 0979, n. 8.
2. *nephews (sons of my eldest brother)*: Magnus had several brothers, of whom the eldest surviving in 1854 was either Friederich Martin von Magnus (1796–1869) or Ludwig Immanuel Magnus (1790–1861). Ludwig Immanuel is sometime referred to as a cousin, making Friederich the likely brother Magnus mentioned. Nephews not identified.
3. *Feilitzsch*: see letter 0970, n. 6.
4. *Regnault*: see M. V. Regnault, 'Sur les forces élastiques des vapeurs dans le vide et dans le gaz', *Comptes Rendus Hebdomadaires des Séances de l'Académie des Sciences* 39 (14 August 1854), pp. 301–13; (21 August 1854), pp. 345–57; and (28 August 1854), pp. 397–409. The articles also appeared in English translation as M. V. Regnault, 'On the Elastic Forces of Vapours in vacuo and in Gases, at different Temperatures; and on the Tensions of the Vapours furnished by mixed or superposed Liquids', *Phil. Mag.* 8:52 (October 1854), pp. 269–80; and 9:56 (January 1855), pp. 4–25.
5. *Les physiciens . . . directes*: French for 'Physicists have never thought it necessary to verify Dalton's law by direct experiments'.
6. *I already did that 18 years ago*: see n. 9.
7. *4th section of his paper*: see n. 4, 28 August 1854 issue of *Comptes Rendus*, p. 397.
8. *Forces élastiques ... ou superposés*: French for 'On the elastic force of the vapours emitted by volatile liquids mixed by mutual solution, or superposed'.
9. *Über das . . . XXXVIII. 481*: G. Magnus, 'Ueber das Sieden von Gemengen zweier Flüssigkeiten und über da Stoſsen solcher Gemenge', *Poggend. Annalen.* 38:7 (1836), pp. 481–92.
10. *Paris Academy*: see letter 0737, n. 11.
11. *Comptes rendus*: see 0737, n. 12.
12. *Poggend.*: *Poggend. Annalen.*
13. *note . . . is printed*: the 'note' was published as G. Magnus, 'Ueber die Spannkraft der Dämpfe von Mischingen zweier Flüssigkeiten', *Poggend. Annalen.* 93:4 (November 1854), pp. 579–82.

14. *my paper . . . projectiles*: see G. Magnus, 'Ueber die Abweichung der Geschosse, und: Ueber eine auffallende Erscheinung bei rotirenden Körpern', *Poggend. Annalen.* 88:1 (January 1853), pp. 1–29. This paper also appeared in *Scientif. Mem.*, vol. 1, pp. 210–31.
15. *Mr le Colonel Delobel à Liège*: refers to Colonel L. Delobel (n.d.), the Colonel d'Artillerie and Director of the l'École de Pyrotechnie in Belgium.
16. *now he publishes . . . trajectoire*: 'Memoirs and Notes relating to the Influence of the Rotation of Projectiles on their Trajectory': see L. Delobel, *Revue de Technologie Militaire, ou Recueil de Mémoires, Expériences, observations et Procédés relatifs à cette science, choisis dans les meilleurs écrits Périodiques et non Périodiques, qui se publient en langues étrangères, ou empruntés à des documents officiels de dates régentes et provenant de tout autre pays que la France et la Belique; traduits, analysés et annotés*, 8 vols (Brussels and Leipzig: E. Noblet, 1854), vol. 1.
17. *the translation*: see L. Delobel, 'De la Déviation due a Mouvement de Rotation des Projectiles Sphériques et Oblongs; suivi de deux Appendices sur de remarquables phénomènes de physique générale qui se manifestent dans le mouvement gyratoire des corps par rapport aux alternatives de sabilité et d'instabilitié de leur axe de rotation', cited n. 16., pp. 371–413).
18. *with comments from Colonel Bormann*: likely Charles G. Bormann (1796–1873), a Belgian artillery captain and inventor of the Bormann fuse, used in the American Civil War (J. Kinard, *Artillery: An Illustrated History of its Impact* [Santa Barbara: ABC–CLIO, 2007], p. 186). See Borman, 'Notice Historique sur la partie de la Balistique qui se rapporte aux phénomènes de la rotation', cited n. 16. pp. 419–57.
19. *Mss. F.*: Sarah Barnard, Faraday's wife.

To May Anne Coxe 16 November [1854][1] 1002

London 16th Nov | before day in the morning.

My dear Mrs Coxe,

I returned last night from the country,[2] weary, draggled and grim with my speculations on the state of matters in the Crimea.[3] The plaid was here before me[4] and your note apprizing me of it had reached me in the morn[in]g. I commenced a note to you but my thoughts were too shifty or too dull to let me finish it. Well I have now commenced another, and what am I to say? I scarcely know. Your kindness is precious to me believe me but these expensive proofs of it are not needed. I want no other proof of it than your letters bring me. The first letter[5] you ever wrote to me is fastened in my Journal and the last[6]—the pencil one, is beside me. My constitution leads me to prefer these two simple documents to any number of plaids that you could possibly send me—Let me do justice to the plaid however—it is a magnificent one, but such things cannot add to that pure mental affinity which first prompted you to correspond with me. And then these postage stamps—

I shall forward them without delay to my friend Charlie[7] so that he may be able to write to me occasionally without invading his pocket money. I shall wear the plaid although I must invoke the aid of a northern friend to teach me how to put it on; but I trust from henceforth & forever you will permit the kindness of that large heart of yours to flow out through the nib of your pen alone.

Many thanks for Chambers's Journal.[8] To Mr Trevellyan is due the chief merit of that enquiry. He discovered it independently and sifted the matter to the bottom. This mode of experiment is hardly to be improved upon, and I have adopted it in lecturing upon the subject. all I have done is to rectify a few mistakes and to extend the matter a little beyond its former limits.[9]

I owe D^r^. Coxe many thanks for his letter regarding Alick[10]—no matter what may be his position Alick I doubt not will prove himself a superior man. I saw him several times last week: an accident placed me among a number of the teachers when his name was mentioned, and the character they one and all gave of him would have delighted you to hear. There can be no doubt that Alick's intellectual powers are of a very high order and I firmly believe the moral foundations of his character are just those upon which such intellectual qualities ought to rest. Take him all in all body, soul, and spirit I have rarely met a boy that has pleased me more.

And now I must end by thanking you for all your goodness to me | & believe me dear M^rs^. Coxe | ever most sincerely yours | John Tyndall

RI MS JT/1/T/252
RI MS JT/1/TYP/1/262

---

1. *16 November [1854]*: Coxe did not include a year in the date for the letter but LT's typescript includes the correct year of 1854.
2. *I returned . . . country*: Tyndall had been visiting Hirst at Queenwood for ten days (see letter 0992, n. 1).
3. *matters in the Crimea*: Tyndall was referring to the Crimean War (see letter 0873, n. 30). Tyndall noted a 'hand to hand conflict with the Russians before Sebastapol' in his journal entry, referring specifically to a battle that took place during the Siege of Sevastopol (Journal, 19 November 1854, JT/2/13b/698).
4. *The plaid . . . before me*: Tyndall noted its receipt in his journal entry of 19 November 1854 (Journal, JT/2/13b/698).
5. *The first letter . . . Journal*: possibly letter 0777.
6. *the last*: letter missing.
7. *my friend Charlie*: likely refers to Charles Cumming, brother to Alexander (or Alix/Alick) (see letter 0794, n. 7).
8. *Chambers's Journal*: *Chambers's Journal of Popular Literature, Science, and Art*, a low-priced, weekly journal founded by William Chambers (1800–83) and his brother Robert

Chambers (1802–71) in 1832 to provide 'educational and entertaining reading material for the "humbler orders"' (*ODNB*).

9. *To Mr. Trevellyan . . . former limits*: the article 'Music in Metal', *Chambers's Journal of Popular Literature, Science, and Art* 44 (4 November 1854), pp. 289–304, on pp. 303–4, in which the respective work of Trevellyan, Faraday, and Tyndall on the tones produced by heated rockers is outlined.
10. *his letter regarding Alick*: letter missing. It was likely a reply to letter 0993, as Tyndall had sent a letter to James Coxe regarding Alexander Cumming's academic progress.

From Edward Frankland 16 November 1854 1003

Richmond Grove | Longsight Nov. 16/54

My dear Tyndall

Dr. Gerland[1] has finished the analysis of that piece of shale.[2] It yielded the following results

SiO_3	41.29
Al_2O_3	22.00
FeO	13.46
CaO	.52
MgO	.42
KO_2NaO	3.01
Organic Matter & Loss	19.30
	100.00

I am glad to hear that you had such pleasant company to London,[3] it did not seem very promising on your entrance into the carriage.

I have just finished my first article for Hughes' 2nd Book on 'Chemical Nomenclature'[4] but have still two others to write one on 'Chemical Equivalents' & the other on 'Fuel' I expect the latter will cost me a good deal of labour to bring it down to the comprehension of boys. I am also making satisfactory progress with my research on Organo metallic bodies[5] & hope soon to have a memoir for the Royal ready.[6]

The chemical milk still answers admirably the little skeleton of a fellow[7] is getting quite fat & has never had a day's bad digestion since he began to use his lactochemic food.[8]

I understand that the Empress of Russia[9] exists exclusively upon 6 Wet nurses who accompany her wherever she goes & I have therefore some intention of making overtures to Old Nick[10] to the effect that if he will knuckle under & give up the war[11] I will relieve his wife of this intolerable bore!

Maggie[12] still talks about you & wishes now to repay your scotch kiss.[13]

With kindest regards from Mrs Frankland | Believe me very sincerely yours | E. Frankland

RI MS JT/3/45, interleaved in Volume

1. *Gerland*: Frankland's assistant. In his journal entry of 15 October 1854, Tyndall noted 'he is an intelligent young fellow but if he limited his speech to half its present amount it would be a great recommendation—he talks too much' (Journal, JT/2/13b/695).
2. *shale*: 'an argillaceous fissile rock, the lamnae of which are usually fragile and uneven, and mostly parallel to the bedding; often overlaying a coal formation' (*OED*). Tyndall had sent a piece of 'shell', possibly the very 'shale' Frankland spoke of here, to Debus for chemical analysis (see letters 0897, n. 2 and 0973, n. 1). The results here are similar to those in letter 0973.
3. *such pleasant company to London*: Tyndall had stopped to visit Frankland on his way back from Liverpool and Wales in October, as he stayed with Frankland on 15–16 October 1854. He must have written to Frankland about the young boy and his father seated opposite him in the carriage, and to whom he conversed during the journey (Journal, JT/2/13b/695–96).
4. *my first article . . . 'Chemical Nomenclature'*: see E. Frankland, 'Chemistry', in E. Hughes (ed.), *Hughes Reading Lessons. Second Book* (London: Longman, Brown, Green, & Longmans, 1855), pp. 204–16.
5. *Organo metallic bodies*: 'chemical compounds in which a metal atom is bonded directly to the carbon atom of an organic group' (*OED*).
6. *hope soon . . . the Royal ready*: Frankland published a series of memoirs on organometallic bodies. See E. Frankland, 'On a New Series of Organic Bodies Containing Metals', *Phil. Trans.* 142 (1852), pp. 417–44; 'Researches on Organo-Metallic Bodies.—Second Memoir: Zincethyl', *Phil. Trans.* 145 (1855), pp. 259–76; 'Researches on Organo-Metallic Bodies. Third Memoir: On a New Series of Organic Acids Containing Nitrogen. [1856.]', *Phil. Trans* 147 (1857), pp. 59–78; and 'The Bakerian Lecture: Researches on Organo-Metallic Bodies. Fourth Memoir', *Phil. Trans.* 159 (1859), pp. 401–15.
7. *little skeleton of a fellow*: Frankland's son, Frederick William Frankland (1854–1916), who had contracted diarrhea and was quite sick (D. Valenze, *Milk: A Local and Global History* [New Haven: Yale University Press, 2011], p. 172).
8. *lactochemic food*: in 1854, Frankland developed a formula milk for infants after his wife Sophie was unable to breastfeed and his son had fallen ill (ibid., p. 172, n. 36).
9. *Empress of Russia*: Alexandra Feodorovna (1798–1860), born Princess Charlotte of Prussia, was Empress consort of Russia and the wife of Tsar Nicholas I (see letter 0873, n. 32). She had one son, Alexander II Nikolaevich (1818–81), who later became Tsar (1855–81).
10. *Old Nick*: Tsar Nicholas I (see letter 0873, n. 32).
11. *Give up the war*: the Crimean War (see letter 0873, n. 30).
12. *Maggie*: Margaret Frankland (see letter 0885, n. 11).
13. *scotch kiss*: possibly sweets.

From Lyon Playfair 17 November 1854 1004

Marlboro House[1] | 17th Nov/54

My dear Sir

The Revd Mr Thomson,[2] a Scotch clergyman, will deliver to you this note.

He is about to take out a patent for certain appliances of a law depending upon the elasticity of fluids in pipes, and he is desirous to ascertain from you as a matter of professional advice whether his views are scientifically correct or whether there lurks some fallacy in them which would vitiate his applications as well the presumed Scientific result which he hopes to have attained.

Not possessing that knowledge of physics which would entitle me to express an opinion on his views, I have recommended him to endeavour to obtain your consideration of the subject, although I am not aware whether you do give professional opinions.

Mr Thompson is a most worthy member of the free Church of Scotland & a man of Scientific acquirements, and as his presumed discovery if Established will lead to very important practical results, it would be a great kindness to him & a favour to me, if you were to give this subject a professional consideration, if your time at all Enables you to do so.

Yours truly | Lyon Playfair

Dr Tyndall FRS. | Royal Institution

RI MS JT/1/P/114
RI MS JT/1/TYP3/979

1. *Marlboro House*: Marlborough House (see letter 0784, n. 12).
2. *Revd Mr Thomson*: not identified. Tyndall noted meeting the clergyman after being introduced by Playfair in his journal entry of 23 November 1854; he noted that they discussed Thomson's ideas on perpetual motion, but that after meeting with him for one hour and a half, 'He gave up perpetual motion but holds to another consequence equally absurd' (Journal, JT/2/13b/699).

From James and May Anne Coxe 20 November 1854[1] 1005

Edinburgh | Nov. 20th 1854

My dear Dr Tyndall

The death of poor Edward Forbes, which took place on Saturday from disease of the kidneys after a few days illness,[2] has suggested to me the possibility of your some day or other emigrating northwards. I suppose you know that James Forbes[3] has been for some time in very delicate health, and it seems to me very probable that before long he will feel himself obliged to resign his Chair. Should this happen, what would you think of being a Candidate?[4] The advantages of the Professorship are a certain status as professor, an income I should guess at about £800 or £1000, occupation that you delight in, and 6 months of leisure to follow out any investigation you were interested in. The drawbacks are giving up your old ties, and coming to town which is becoming every day more and more provincial, and where men's minds and actions are cramped, cabined and confined.

However pray consider the matter and let me know if I might speak of you as one that would occupy the chair if it were vacant. It is in the gift of the Town-council, and my conviction is that your chance of success would be so good as to make your election next to a certainty

always very truly Yours | James Coxe

Do Dear Dr Tyndall think of this—Edinburgh is stupid enough I admit, (but you would be entirely free for half the year) and you would be the one that would redeem Sodom![5]

Ask Dr Lyon Playfair—he knows this place well, and will I am sure tell you that there are some with whom you could 'fraternize'. At all events think of it.

Yours ever | M.A.C.

Dr Tyndall.

RI MS JT/1/TYP/1/263a–263b.
LT Typescript Only

---

1. *20 November 1854*: LT noted in the top left corner of the letter: 'Marked: "Private | J.T."'.
2. *The death of poor . . . few days illness*: Edward Forbes died suddenly on 18 November 1854 (*ODNB*). Upon learning of his death, Tyndall recorded in his journal that 'a genial manly soul is thus taken away from us' (Journal, 20 November 1854, JT/2/13b/699). See also letter 1006.
3. *James Forbes*: James Forbes had been Professor of Natural History at the University of Edinburgh since 1833. He was plagued by illness from 1850 onwards, often having his lecturing duties carried out by others. Forbes did not resign, however, as Coxe suggested he might.
4. *what would you think of being a Candidate?*: while Tyndall replied with letter 1006, he also drafted another letter that was copied in his journal but was never sent (Journal, 27 November 1854, JT/2/13b/701).
5. *you would . . . redeem Sodom!*: Sodom refers to 'an extremely wicked or corrupt place' (*OED*), as described in Genesis 13–14. In order to 'redeem' Sodom, Abraham and the Lord negotiated and determined that 10 righteous men, or even just one—Abraham's nephew Lot, would be needed (Genesis 18–19). The suggestion was that Tyndall may be able to redeem Edinburgh.

To James Coxe 23 November 1854 1006

My dear Dr Coxe

On Monday morning as I entered Faraday's room I noticed a shade upon his countenance—'Have you heard the sad news Tyndall.' was his first question 'No' I responded 'Forbes is dead' said he 'Edward Forbes—taken off last Saturday.'[1] I had heard of his illness on Thursday—heard of it at a little society[2] of which Forbes was the soul when he resided in London: but none of us deemed his illness serious. His health was proposed by Samuel Lover[3] who occupied the chair which he was accustomed to occupy—little did we think as we responded to it that at our next meeting we should have an opportunity of pledging the <u>memory</u> of our friend. Two short months ago I saw him at Liverpool,[4] apparently full of life and vigour, and heard his manly voice, which fell upon me like a refreshing dew, amid the soft sawder of the President's dinner. Exactly two months from the day in which he was finally smitten down he presided at the 'Red Lion' dinner,[5] and through his very jollity you could discern the substratum of a man.—a broad, genial, manly soul who could do honour to all the phases of human nature. We shall long remember him—for us who knew and loved him he is woven into the eternal texture of things and though dead yet speaketh.

Surely if there be one who more than any other has reason to bless the gods for raising up friends around him, I am he. By a kind of moral gravitation they move towards me and I towards them. I never made an arrangement to <u>catch</u> a friend in my life, but they have come without my contrivance, guided by laws which are apparently beyond my powers of investigation. In your note[6] which lies before me I have the most recent proof of this. I <u>will</u> do as Mrs Coxe desires me and 'think' of the matter[7]—meanwhile I may tell you my present thoughts concerning it. Faraday and I are very good friends and I have got a notion that I ought not to do anything that would harm him. This very notion prevented me from accepting a post in the Government School of Science[8] which was offered to me more than once and concerning which poor Edward Forbes wrote to me[9] expressing the wish of himself and his colleagues to have me among them. Other offers have been made to me but my feeling towards Faraday has induced me to decline them all. Now high as the position is to which you refer—higher indeed than I can hope to attain to in London—still if my accepting it would grieve Faraday I hardly think that I could accept it. But Faraday is a noble soul and will not ask me to do any thing contrary to my true interests—I will therefore talk the matter over privately with him and let you know the result.

I confess also that I should like to be a little <u>riper</u> than I am before undertaking duties of so responsible a character, but I suppose each succeeding year would diminish this drawback and render me less unworthy of being the successor of such men as Leslie[10] and Forbes. With regard to the position itself, it is I should say all that a man of science could wish for. It would set his mind at rest as regards the miserable 'thatch and fodder' of life and leave him free to work for his class and for science. After dealing with the known for half a year in his vocation as teacher he could refresh himself by a contact with the unknown during the next half year, and perhaps win something from it worthy of being incorporated with the scientific stock of the world.

With regard to the 'fraternizing' of which Mrs Coxe speaks[11] I should be perfectly content with the privilege of a cup of tea with herself at intervals—whether the thing is to come to pass is of course very uncertain, but whatever the result may be it will not I imagine modify to any great extent the feelings which I now entertain towards both her & you.

Believe me ever | most faithfully yours | John Tyndall.

23 November 1854.

I hope there are no religious tests in the University—though born in the Church of England I should pause before swearing to the 39 Articles.[12]

RI MS JT/1/T/243
RI MS JT/1/TYP/1/264

---

1. *On Monday . . . taken off last Saturday'*: Tyndall recounted this conversation in his journal entry of 20 November 1854 (Journal, JT/2/13b/699).
2. *A little society*: the Red Lions Club, of which Forbes was the founder (see letter 0900, n. 2).
3. *Samuel Lover*: Samuel Lover (1797–1868), an Irish painter and author who frequented many societies in London (*ODNB*), including the Red Lions Club (A. J. Symington, *Samuel Lover: A Biographical Sketch* [London, Blackie & Son, 1880]).
4. *I saw him at Liverpool*: see letter 0950, n. 2.
5. *'Red Lion' dinner*: see n. 2. Tyndall recorded in his journal dining with the Red Lions Club on 25 September (Journal, JT/2/13b/692).
6. *In your note*: letter 1005.
7. *'think' of the matter*: see letter 1005, n. 4.
8. *prevented me from . . . Science*: Tyndall had been offered, and declined, a position at the Government School of Mines. See letter 0863.
9. *Forbes wrote to me*: letter 0861.
10. *Leslie*: Sir John Leslie (1766–1832), a Scottish mathematician and natural philosopher known for his researches on heat. He held the Chair of Natural Philosophy at the University of Edinburgh from 1819 until 1832 (*ODNB*).

11. *the 'fraternizing' of which M*[rs] *Coxe speaks*: letter 1005.
12. *the 39 Articles*: refers to the Thirty-nine Articles of Religion, the defining statements of doctrine for the Church of England drawn up during the Reformation in the sixteenth century (*OED*). Adherence to them was required in order to hold civil office in England until 1824, and admission to and graduation from Oxford and Cambridge was restricted by this requirement until the early 1850s.

From Cornelius Hessel[1] 23 November [1854][2] 1007

Liverpool 23ten Nov. 1845 | Dane Street 153

Geehrtester Herr Professor!

Als ich vor einiger Zeit Deutschland verlies um mich mündlich wegen eines sehr vortheilhaften Engagements mit einem hiesigen Hüttenbesitzer zu benehmen, erhielt ich von unseren ehemaligen Lehrern Herrn Prof. Stegmann und Gerling, sowie von meinem Vater den Auftrag Sie in London zu besuchen und Ihnen die Grüße Ihrer Freunde zu überbringen. Zugleich übergab mir mein Vater einige Exemplare seiner jüngsten Abhandlungen um Ihnen dieselben zu überhändigen, da jedoch mein Aufenthalt in London zu kurz war, so konnte ich mich meiner Aufträge nicht persönlich erledigen, ich erlaube mir daher dies nunmehr schriftlich zu thun und bemerke zugleich, daß ich wahrscheinlich nächstens Gelegenheit finde Ihnen auch die Bücher bald zu übersenden, wenn ich nicht selbst nach London kommen sollte.—Da nemlich mein vorerwähntes Engagement sich fürs erste zerschlagen hat, indem die beabsichtigte Erweiterung der Hütte (durch Einführung eines neuen Entsilberungsprocesses auf nassem Wege) sich durch die gegenwärtigen ungünstigen Verhältnisse, verbunden mit einigen Unglücksfällen nicht ausführen läßt, so bin ich entschlossen mir ein Unterkommen als Lehrer zu verschaffen. Ich habe mich zu dem *[Zweck]* bereits umgesehen und werde mich auch ferner nach meinen Kräften zur Ausführung dieser Idee bemühen, da ich jedoch fest überzeugt bin, daß mir die Erlangung einer solchen Stelle nicht gelingen würde, wenn ich nicht der freundlichen Unterstützung einiger, angesehener Männer sicher bin, so erlaube ich mir, Sie sowohl um Ihre geneigte Beihülfe und Anwendung Ihres gewichtigen Einflusses zu meinen Gunsten zu ersuchen, als auch Sie um die Erlaubniß zu bitten mich bei etwa verlangten Empfehlungen auf Sie beziehen zu dürfen.

Ich habe in Marburg seit Frühjahr 47 studirt und, wie Sie wissen, mich mit Naturwissenschaften beschäftigt, habe alsdann in Mathematik, Physik und Chemie meinen Doctorexamen gemacht und im Frühjahr 1852 promovirt.—Alsdann habe ich eine Zeitlang (während der Abwesenheit des Assistenten) in Darmstadt die Assistentenstelle im chemischen Laboratorium versehen und ging von da auf Anrathen des preusischen Gesandten in die Bergwerksdistricte Preußens, Sachsens und Hanovers um durch practisches Arbeiten den dortigen Betrieb kennen zu lernen. Ich that dies um so lieber, als der preusische Gesandte für Centralamerika Geh. Rath Hesse mir versprach ein Engagement daselbst auszumachen, in mehreren Briefen, welche er mir hierüber schrieb, vertröstet er mich bis zur Beendigung der Revolution in Neu-Granada zu warten, da gegenwärtig aller Betrieb darniederliegt.—In Folge dieser Mittheilung knüpfte ich mit einem englischen Hüttenbesitzer Unterhandlungen an und befinde mit [sic] nun, da dieselben, wie bereits oben erwähnt wurde, zu Nichte geworden sind, in England.—Ich sehe mich daher genöthigt mir irgend ein passendes, anderes Unterkommen zu verschaffen und glaube dies am leichtesten zu erreichen, indem ich mich auf die Lehrer carriere werfe. Ich würde am passendsten wohl Unterricht in der deutschen Sprache, Mathematik und Zeichnen (Freihand- und Perspectivisch-Zeichnen) ertheilen können. Ich könnte ferner, wenn es gewünscht würde die Leitung der gymnastischen Uebungen übernehmen, wozu mich meine Bekanntschaft mit dem deutschen Turnen befähigte.

In der Hoffnung, daß Sie mich bei der Ausführung dieses Planes mit Ihrem freundlichen Rath und Einfluß unterstützen werden, habe ich die Ehre mich zu unterzeichnen, als

Ihr | ergebenster | Dr Heßel.

Liverpool 23rd Nov. 1845 | 153 Dane Street

Dearest honoured Herr Professor!

When I left Germany some time ago in order to discuss in person a very advantageous engagement with the owner of an ironworks here, I received from our former teachers Herr Prof. Stegmann and Herr Prof. Gerling, as well as from my father,[3] the task of visiting you in London and bringing you the greetings of your friends. At the same time my father gave me some copies of his latest disquisitions[4] to give these to you, but as my stay in London was too short, I could not discharge my tasks personally, and I therefore allow myself to do this now in writing, and note at the same time that I shall probably find the opportunity shortly to send the books soon too, if I should not come to London myself.—As, namely, my above-mentioned engagement has fallen through for the time being, in that the planned expansion of the ironworks (through the introduction of a new, wet process of desilverization[5]) cannot be carried out due to the present unfavorable circumstances, combined with some accidents, I am therefore determined to obtain employment as a teacher. I have already looked around to that *[end]* and shall further endeavour to carry out this idea to the best of my ability; however, as I am firmly convinced that I would not succeed in attaining such a position if I am not certain of the kind support of some respected men, I therefore allow myself both to ask you for your kind assistance and the bringing to bear of your substantial influence on my behalf, as well as to ask you for permission to be allowed to refer to you in the case that, for instance, I am requested for testimonials.

I studied in Marburg[6] from spring 47 and, as you know, devoted myself to natural sciences, then sat my doctoral exam in mathematics, physics and chemistry, and had my doctoral degree conferred in the spring of 1852.[7]—I then occupied the assistant's position in the chemical laboratory in Darmstadt[8] for some time (during the absence of the assistant) and on the advice of the Prussian envoy went from there to the mining districts of Prussia, Saxony[9] and Hanover[10] to become acquainted with the operations there through practical work. I did this all the more willingly when the Prussian envoy for Central America, Priv. Couns. Hesse[11] promised me that he would find an engagement there; in several letters which he wrote to me about this, he is feeding me hopes that he is waiting until the end of the revolution in New Granada,[12] as presently all operations are languishing.—As a result of this information I entered into negotiations with an English ironworks owner and now find [myself], as these—as already mentioned above—have come to nothing, in England.—I therefore see myself forced to obtain some suitable different employment or other, and believe this to be most easily achieved by throwing myself into a career as a teacher. I would probably be most suitably able to give lessons in the German language, mathematics and drawing (freehand and perspective drawing). I could further, if it were to be desired, take over the direction of gymnastic exercises, for which my acquaintance with German gymnastics would qualify me.

In the hope that you will support me in carrying out this plan with your kind advice and influence, I have the honour to sign myself, as

Your | very humble | Dr Hessel.

RI MS JT/1/H/121

---

1. *Cornelius Hessel*: Cornelius Hessel (1827–95), son of Johann Friedrich Christian Hessel (1796–1872), a German physician and Professor of Mineralogy at the University of Marburg where Tyndall had studied. Tyndall discussed the younger Hessel in letter 1008. Not further identified.
2. *23 November [1854]*: the original letter was mistakenly dated '1845', but it was written in 1854 as Tyndall noted receipt of 'a letter from the son of Prof. Hessel who is now in Liverpool' in letter 1008.
3. *my father*: see n. 1.
4. *his latest disquisitions*: see J. F. C. Hessel, *Die Anzahl ser Parallelstellungen und jene der Coincidenzstellungen eines jeden denkbaren Raulmdinges mit seinem Ebenbilde und mit seinem Gegenbilde, der Regelmässigkeitsgrad des Schwerpunctes und andere bei Raumdingen in Betracht kommende Zahlen, als Merkmale für den Begriff Familie von Raumdingen, nachgewiesen* (Cassel: Theodor Fischer, 1853).

5. *desilverization*: the process of extracting silver from lead (*OED*).
6. *Marburg*: the University of Marburg.
7. *doctoral degree . . . Spring of 1852*: see C. Hessel, *Beiträge zur Lehre über die von Foucault entdecke Drehung der Pendelebene. Inaugural–Dissertation, welce mit Genemigung der philosophischen Facultät zu Marburg zur Erlangung der Doctorwürde öffentlich vertheidigt Cornelius Hessel aus Marburg* (Marburg: Bayrhofferschen Universitäts–Buchdruckerei, 1852).
8. *the chemical laboratory in Darmstadt*: likely the Höhere Gewerbeschule in Darmstadt, which was known for having a good chemical laboratory (E. Homburg, 'The Teaching of Chemistry at the German Polytechnic Schools, 1803–1860', *Mitteilungen der Gesellschaft Deutscher Chemiker, Fachgruppe Geschichite der Chemie* 13 [1997], pp. 75–93, on p. 86).
9. *Saxony*: the Kingdom of Saxony, a state in the German Confederation.
10. *Hanover*: the Kingdom of Hanover, an independent territory until it was conquered by Prussia in 1866.
11. *the Prussian enovy . . . Priv. Couns. Hesse*: Franz Hugo Hesse (1804–61), a Prussian politician and diplomat.
12. *the revolution in New Granada*: the Republic of New Granada was established in 1831 and dissolved in 1856, in what is modern day Panama and Colombia.

To Thomas Archer Hirst [23–24 November 1854][1] 1008

Thursday

My dear Tom

I hope your lecture[2] will pass off to your satisfaction. Today a young German[3] called upon me with a note of introduction from Dove. He wishes to be a teacher in England. He speaks English pretty well. Chiefly as mathematician but has read physics and astronomy. He can teach arithmetic and all such matters. Has had several years practice as a teacher in the Gymnasium in Berlin,[4] was accustomed to take charge of the boys at the <u>Turnplatz</u>[5] and appears to be well acquainted with jumping, springing,[6] climbing and all those things that boys delight in on the playground. He is a decent steady-looking fellow and might be useful to you. I give you therefore notice of his existence. He says he knows what discipline is, knows its difficulties and is able to meet them; but here the proof is the only thing that would satisfy him.

This morning's post brought me a letter from the son of Prof. Hessel[7] who is now in Liverpool and also in search of a situation. So here are two, in case you do not know where to turn. Hessel I should like to serve on his father's account. He also is a first rate <u>Turner</u>[8] and a nice looking fellow, but I fear he lacks the practice necessary for Queenwood. I don't urge you to have either, but if you want a man you know where to look.

I had a note from Dr. Coxe[9] which I would enclose to you were it not just now in Faraday's hands. James Forbes of Edinburgh has been long ailing and he will probably soon feel himself obliged to resign his chair. Dr. Coxe thinks if I became a candidate that my election to it would be next to a certainty. The income is about £800 or £1000. If you were near me I would ask your opinion as to how I ought to act. 6 months of the year would be free. (Private)

I think either of those[10] that I have mentioned would accept a small salary, or go on trial for 3 months.

Affectionately yours | John

Many thanks for the magnet problem.[11] I have not yet read

RI MS JT/1/HTYP/375
LT Typescript Only

1. *[23–24 November 1854]*: Hirst dated this letter 'Nov. 24th' likely reflecting the date he received it. Tyndall likely started writing the letter on 'Thursday', 23 November, but finished it on 24 November, given the reference to 'This morning's post' in the second paragraph.
2. *your lecture*: Hirst gave a lecture at Basingstoke, Hampshire, England (see letter 1013).
3. *a young German*: not identified.
4. *the Gymnasium in Berlin*: not identified.
5. *Turnplatz*: a German outdoor exercise area.
6. *springing*: this was handwritten by LT as 'springing?'.
7. *Prof. Hessel*: see letter 1007, n. 1.
8. *Turner*: likely a reference to J. M. W. Turner (1775–1851), a well-known English landscape painter. Many of his works were studies of perspective. Hessel had noted in his letter to Tyndall that he was suited for giving lessons in 'drawing (freehand and perspective drawing)' (see letter 1007).
9. *a note from Dr. Coxe*: letter 1005.
10. *either of those*: Tyndall noted in his journal entry of 3 December that he had written to Queenwood about both of the potential teachers, but that there was no vacancy at Queenwood (Journal, JT/2/13b/702).
11. *the magnet problem*: not identified. The last time there is record of a problem sent to Tyndall by Hirst was in letter 0965.

From Heinrich Debus 26 November 1854 1009

Queenwood 26 Nov. 1854.

Lieber John!

Ein kalter Nordwind streicht durch das Land, die lebendige Welt zieht sich zusammen und fühlt den Herrscherstab der unorganischen Natur. Die kleinen Geschöpfe bergen sich in ihren Ritzen, die grossen in ihren Höhlen und der Mensch kriecht in Schaafwolle; die Pflänzchen hängen ihr Haupt als wollten sie sterben, ja sogar das flüssige Wasser wird fest und Alles scheint zu erstarren. Nur ich bleibe lebendig. Meine Haut zieht sich zusammen und treibt mein Blut nach meinem Herzen, mein Herz wird voll, mein Blut fliesst schneller und ich muss reden. Eben verschwinden die Wolken, die Sonnenstrahlen fallen auf die Erde und wie mit Zauberschlag liegt eine neue Welt vor mir. An jedem Blättchen, an jedem starren Punkte den ich um mich sehe hängt ein Diamant-Tröpfchen gleich Millionen von Sonnen. Ich finde es schön, werde stiller und ernst. Alles scheint lebend sogar der Stein auf dem ich stehe. Der Mensch liebt das Licht nicht um des Lichtes willen, sondern weil es ihm die Welt öffnet soweit sein Horizont reicht. Durch das Auge wird man allgegenwärtig, man wird Gott mehr ähnlich. Dort liegt Broughton in einem dünnen Nebelschleier mit seinem silbernen Bache.

Es ist Sonntag.

Die Menschen, nachdem sie 6 Tage gesündigt haben, wollen wenigstens einen Tag gut sein. Wohl, was man davon denken mag, es ist immer eine heilige Sache. Wenn der Mensch den Teufel zum Teufel jagt, seine irdischen und schlechten Qualitäten ablegt und seinem guten Genius einmal, wenn auch nur für einen Tag den Herrscherstab giebt, so achte ich ihn höher. Es ist die Grösse der Sache, es ist [dass] die ganze menschliche Körperwelt einen Feiertag hat, um der Seele ihren Lauf zu lassen, [was] mir erhaben erscheint. Der Teufel sagt, es ist Schein doch wie weit ist es vom Schein zur Wahrheit?

So ohngefähr dachte ich heute morgen, wenn es mir einfiel, dass ich Dir noch nicht geschrieben hatte um Dir für das schöne Zeugniss meinen verbindlichsten Dank zu sagen. Auch Hoffmann war so freundlich mir ein sehr gutes Zeugniss zu geben. Es ist jedoch ganz unwahrscheinlich dass ich einen Erfolg haben werde.

Mit herzlichen Grüssen | Dein | H. Debus.

Queenwood 26 Nov. 1854.

Dear John!

A cold north wind sweeps through the land, the animate world draws in on itself and feels the scepter of inorganic Nature. The little creatures hide themselves in their crannies, the big ones in their caves and Man cringes in sheep's wool; the little plants hang their heads as if they want to die, indeed, even liquid water becomes solid and everything seems to freeze. Only I remain alive. My skin draws in on itself and drives my blood to my heart, my heart becomes full, my blood flows more quickly and I must speak. The clouds are just disappearing, rays of sunlight are falling on the Earth and as if by a stroke of magic a new world lies before me. On every little leaf, on every frozen point that I see around me hangs a diamond droplet, like millions of suns. I find it beautiful, become quieter and serious. Everything seems alive, even the rock on which I am standing. Man loves the light not for the light's sake, but because it opens the world to him as far as his horizon reaches. Through the eye one becomes omnipresent, one becomes more like God. Over there lies Broughton in a thin veil of mist with its silver stream.

It is Sunday.

After sinning for 6 days, people want to be good for at least one day. Yes, whatever one may think of it, it is always a sacred thing. When Man chases the devil to hell, casts off his earthly and wicked qualities and gives his good guardian spirit the scepter once, even if for just one day, I regard him more highly. It is the greatness of the thing, it is [the fact that] the entire human corporeal world has a holiday in order to let the soul take its course, [which] seems sublime to me. The devil says it is an illusion, but how far is it from illusion to the truth?

This is more or less what I was thinking this morning whenever it occurred to me that I had not yet written to you to give you my most profound thanks for the nice testimonial.[1] Hoffman[2] too was so kind as to give me a very good testimonial. It is, however, quite unlikely that I shall be successful.[3]

With warm regards | Your | H. Debus.

RI MS JT/1/TYP/7/2411
LT Typescript Only

---

1. *the nice testimonial*: not identified.
2. *Hoffman*: misspelling of August Wilhelm Hofmann.
3. *It is nevertheless . . . be successful*: not identified.

From Walter Flight[1] 26 November 1854 1010

Queenwood College, | November 26th. 1854.

My Dear Dr. Tyndall,

Since you left us I have heard of a very extraordinary case of Electro-Biology[2] practised by a relation of one of my school fellows.[3] It was at a party near Fareham[4] and about 12 o'clock at night when the visitors were departing that a lady, one of the company, (who had never been in his house which was some 8 or 9 miles distant) asked him to mesmerize[5] her; he did so, and whilst she was sitting in a chair he did the following experiments. He placed a watch on her head and asked her what time it was; she said she could not tell him because there was only one hand to the watch (it happened to be exactly 12 o'clock so that the two hands coincided) He asked her again afterwards and she said 'Oh! I see; it is a little past twelve.'

He then asked her to go through his house in her imagination (his house was some 8 or 9 miles off in which she had never been) he told her to go up the stairs, she went up a little way and then gave a sudden start saying that she saw a man at the top of the stairs; he told her to look again and see what it was and she recognised it to be a picture; she then went through all the upper rooms, stating what they each contained, and every thing she said was quite true. He then told her to go down into the kitchen and tell him what there was there; she said that there was a cook, a house-maid and a man with a striped waistcoat who looked like a groom;[6] he asked her what they were doing, she said 'drinking brandy and water.'

The gentleman went home that night and calling the servants up the next morning, he asked them what they meant by staying up so late drinking brandy and water with Charles the groom;[7] they were greatly frightened and declared with tears in their eyes that they would leave the house, directly, if they had everything they did, pryed into in that manner.

At another time this gentleman mesmerized a man, who cried out with pain if anyone touched this gentleman.

Please to accept this letter from | Your old pupil | Walter Flight.

RI MS JT/5/15

1. *Walter Flight*: see letter 0818, n. 6.
2. *extraordinary case of Electro-Biology*: for definition of electrobiology see letter 0898, n. 6. Not otherwise identified.
3. *one of my school fellows*: not identified, but it would refer to another student at Queenwood where Flight also attended (see n. 1).
4. *Fareham*: Fareham, Hampshire, England, a town in Portsmouth Harbour.
5. *mesmerize*: 'to hypnotize' (*OED*).
6. *Groom*: 'a man of inferior position; a man-servant' (*OED*).
7. *Charles the groom*: not identified (see n. 6).

From Tommy Grieve 26 November 1854 1011

Queenwood College | November 26th. 1854.

My dear Dr. Tyndall,

There are 41 boys have got colds and I am one of them and I make 42. Mr Iliff gave us a very interesting lecture[1] on Christmas last night, we have the society[2] in the dinning room now for the lecture-room is too cold now. We are going to spend our Christmas holidays at Torquay in Devonshire for Scotland is too cold for mama but I wish I was going home for it is so cold and there is more skating there than at Torquay.[3] Papa tought me to skate last Christmas holidays the ice was about two feet thick. There is going to be a very grand concert at Broughton in the City Hall and we have to pay 6d to get in but the singers wont pay nowthing[4] I am not one of the singers and I will have to pay 6d but I am glad I am not, because I would not like to sing before such a lot of people. Mr Edmondson told me that there was a new master coming on Monday.[5] There are only three weeks and three days next Wednesday till the holidays. the time has passed very quickly this half has passed away the quickest of all the halfs I have been here. I think the time passes away quicker and quicker every half. some of the boys are learning oil painting they are Cunliff,[6] Dickenson[7] and Thomas Curtis[8] all three paint very nicely. There are a hundred and fifteen boys now. Last Tuesday all the school went out a walk between nine and twelve and each class had a separate master. some of the classes played at fox and hounds[9] but I was in bed and there was another boy that slept in the same room as I did he was in bed for a cold we had both colds one of the boys came up and told us that they were going out a walk. I lay in bed every morning now because of my cold for it would make it worse if I got up in the cold. Tompson Stephens[10] has made a moddle of the Laboratory and I think he is making one of the lodge but he has made the moddle of the Laboratory very nicely. There is a boy named Miles[11] he is a very clever boy he reads a paper nearly every Saturday night, he gave a very nice one about the four great Philosophers of England[12] and sometimes he gives resetations.[13] Frank Field[14] gave a resetation last night about the magpie calling his feathered friends together. Dr. Hirst comes into number nine every Saturday night after the society[15] and plays on the peoano but he did not last Saturday I think I must conclude now for it is nearly time that the boys came home now. With kind love to you

I remain | Your affectionate Friend | Tommy Grieve Esqr.

RI MS JT/1/TYP/1/466
LT Typescript Only

---

1. *interesting lecture*: not identified.
2. *the society*: the Queenwood Mutual Improvement Society (see letter 0816, n. 4)
3. *Torquay*: Torquay, Devon, a seaside town in southwest England.
4. *nowthing*: LT corrected this as 'nothing' in her typescript annotating it 'nowthing sic'.
5. *new master coming on Monday*: not identified.
6. *Cunliff*: J. Cunliff, a student at Queenwood. Not further identified.
7. *Dickenson*: A. Dickinson, a student a Queenwood. Not further identified.
8. *Thomas Curtis*: a student at Queenwood. Not further identified.
9. *fox and hounds*: 'a boys' game, in which the "hounds" chase the "fox"' (*OED*).
10. *Tompson Stephens*: likely a student at Queenwood. Not further identified.

11. *Miles*: not identified.
12. *four great Philosophers of England*: not identified.
13. *resetations*: misspelling of recitations.
14. *Frank Field*: a student at Queenwood.
15. *the society*: see n. 2.

To James Coxe 27 November 1854 1012

London 27th Nov. 1854.

My dear Friend[1]

I have thought over your note[2] and have conversed with Faraday upon the subject[3] of it. His feelings are just those which I expected from his high soul and kind heart. Here is my strongest bond of union with London—here the tie which it would grieve me most to sever:—my connexion with Faraday.

In the event of Prof. Forbes's resignation[4] many, I doubt not, would consider the Chair of Natural Philosophy in Edinburgh to be a prize worth fighting for. In a struggle of this kind I should like to bear no part. —I should not like to wade even to success through the gall and bitterness of an election contest.

But if the infirmities of its present distinguished occupant should induce him to resign his Chair—an event which I should be sorry to hasten—and if the Council should think proper to offer me the vacant post, I do not at present see any sufficient cause to prevent me from accepting it.

What I have done in science is, I suppose, sufficiently known: what I am to do is a matter of conjecture and of hope. Between the past and the future thus referred to stands my present self. People think that I am a worker,—at all events I feel the wish to do my duty, however far short of it performance may fall,—and the thought which rises predominant in my mind in connexion with the Chair at Edinburgh is, not the comforts which it will bring, but the sphere of action which it will open—the opportunity which it offers to the teacher and the investigator of making himself useful in his generation and leaving behind him an honourable name.

Believe me ever | Your sincere friend | John Tyndall.

Dr. Coxe.

RI MS JT/1/TYP/1/1/265
LT Typescript Only

---

1. *My dear Friend*: Tyndall had also drafted another version of this letter, dated '24th Nov. 1854', that was not sent. The version that was not sent was roughly the same with the significant exception that in the unsent draft Tyndall did not speculate about the great

opportunity that the position would offer to him as a teacher and investigator. He instead sought to further explain his seeming ambiguity about the position: "My independent way of writing when such a high distinction is at stake may perhaps cause you to smile. I write however exactly as I feel, and I do not think the state of mind which prompts me to write is in any way calculated to render me less vigorous in the discharge of my duties, or in the promotion of the cause of science by original research should Edinburgh ever become the sphere of my operations" (Journal, 23 November 1854, JT/2/13b/700). See also letter 1008.

2. *your note*: letter 1005.
3. *conversed with Faraday upon the subject*: as stated in his journal entry of 27 November, Tyndall had also consulted with Bence Jones on the matter (Journal, JT/2/13b/701). He had also written to Hirst about it (see letter 1008).
4. *Prof. Forbes's resignation*: see letters 1005, nn. 3–4 and 1008.

From Thomas Archer Hirst 28 November 1854 1013

Vine Cottage | Nov[r]. 28th. 1854

My dear John.

From your letter to Dr Coxe[1] I should say that you have taken the proper view of the Edinburgh professorship. I think you are quite right in not contesting the election and also in consenting to accept the professorship if offered to you On the one hand if you entered the lists as candidate it would appear as if you wished to get away from the Royal Institution which I am sure you do not and on the other hand I think if the place were offered to you that its superior advantages in a scientific point of view are sufficient to justify you in giving up the close communication with such a man as Faraday even. As I told you when you were here the situation at the Royal Institution is not exactly the one you could most usefully fill and in this respect Edinburgh would be far preferable. From D[r] Coxes letter[2] it would appear that the proposal is entirely his own. I do not know what influence he may have with the Town Council, but judging from the best kind of elections I scarcely think that the matter will be so easy as his letter would lead one to suppose. This does not affect you, however, your reply leaves you perfectly free and unaffected whatever may be the result I think I should omit the sentence about there being many who would consider the prize worth fighting for. It would seem to imply that you do not so consider it whereas on the very best grounds e.g. on those you mention in the conclusion of your letter it is worth fighting for. Again I think I would omit the sentence beginning thus 'People think me a worker &c.' and conclude the letter with the sentence beginning with 'The thought which rises predominant &c.' which conclusion is in my mind as manly and high minded as could be wished for. I appreciate as fully as most could do your present terms of intimacy with Faraday but upon mature consideration I think you ought to forego even such advantages were this situation offered you. Consequently I hope with all sincerity that the event may come to pass and should it do so or not I believe you will never regret the step you have taken I will bear in mind the two German applicants for situations[3] at present I do not see an opportunity to employ either at Queenwood. We require at present a young Frenchman and by Edmondson's order Haas is already on the look out. In a quarter of a year or so Taylor[4] will perhaps leave us (you recollect Wright and ourselves came to that conclusion when you were here) and then it is just probable that one of these Germans if at liberty would suit us. At all events I will bear them in mind when we arrange matters with Edmondson which we must do soon. My Basingstoke lecture came off satisfactorily the experiments were all good and successful. Our concert at Broughton comes off tomorrow our Queenwood one on Monday fortnight when we shall expect you.[5]

Yours affectionately | T.A. Hirst

RI MS JT/1/H/208
RI MS JT/1/HTYP/376

1. *your letter to Dr Coxe*: Tyndall must have shared his reply (see letter 1012) to letter 1005 with Hirst before sending it. It is possible that the version of the letter that was sent to Hirst is the one copied in Tyndall's journal (see Journal, 27 November 1854, JT/2/13b/701).
2. *Dr Coxes letter*: letter 1005.
3. *the two German applicants for situations*: Tyndall had recommended two German acquaintances for positions at Queenwood (see letter 1008).
4. *Taylor*: see letter 0927, n. 1.
5. *we shall expect you*: Tyndall did go to Queenwood and attend the concert (see letter 1020, n. 7).

From James Coxe 29 November 1854 1014

Edinburgh Nov. 29. 54

My dear Dr. Tyndall,

Your answer[1] quite embodies my own views and it is only in the event of the chair[2] being offered to you that I hope or wish ever to see you amongst us. What may be the course of events is hidden from our view, but I know how to act under certain contingencies, and this is all that is wanted in the meantime.

Whether you will be sorry or glad to hear that Alick[3] has got a Cadetcy[4] and goes to Addiscombe[5] after Christmas? I agree with you[6] that we shall yet hear of him in the world.

Always very truly yours | James Coxe.

RI MS JT/2/13b/702

1. *Your answer*: see letter 1012.
2. *the chair*: the Chair of Natural Philosophy at the University of Edinburgh.
3. *Alick*: Alick, or Alexander Cumming.
4. *Cadetcy*: refers to the role undertaken when entering the service of the East India Company, 'a cadetcy, which is in truth the first step in the military service of the [East India] company' ('Reg. v. Charretie and Others. Sale of cadetcy in the service of the East-India Company', *The Law Times and Journal of Property* 13 [14 April 1849], pp. 41–60, on p. 48).
5. *Addiscombe*: the East India Company Military Seminary at Addiscombe, Surrey, England, where young officers were trained for service in India.
6. *I agree with you*: Tyndall had discussed Alexander Cumming's education with the Coxes (see letters 0993 and 1002).

To Peter Riess 3 December 1854 1015

London 3rd Dec.1854

My dear Sir

Accept my best thanks for your very kind and useful letter:[1] I purposed when I wrote to you[2] to introduce the subject of induced currents in my regular course of lectures upon electricity, but the matter is so interesting and so new in England that I shall reserve it for one of our Friday evening meetings,[3] when I hope to have an audience fully capable of appreciating your researches. Faraday agrees with me that it is best to bring the subject forward in a Friday evening discourse.

Believe me I should consider the fact of my remarks regarding your book[4] being 'practical' a poor recommendation if they did not at the same time recommend a kind of conduct which was right and honourable. I did not mean to advise you to ask any editor to write a notice of your book—this I should never think of doing—But if editors are not informed of the fact they remain in ignorance of the existence of the book, and the usual way of informing them is to send them a copy. The matter quite escaped my recollection or I should have written a notice of the work long ago for the Philosophical Magazine.[5] I am not acquainted with any work on Frictional electricity which in point of usefulness is to be compared with yours, and therefore I think the English public ought to know that the work exists—acting on this conviction

I shall endeavour to prepare a notice for the January number of the Philosophical Magazine.[6]

Thanking you once more for the valuable information contained in your letter

Believe me | Most sincerely yours | John Tyndall

Herr Prof. Riess | *[Spandauer Straße]* | Berlin[7]

Staatsbibliothek zu Berlin (Stabi) Slg Darmstaedter 1855 Tyndall Be. 1–58, 46–47

---

1. *your very kind and useful letter*: letter 1000.
2. *when I wrote to you*: see letter 0996.
3. *I shall . . . Friday evening meetings*: Tyndall discussed Riess and frictional electricity in his Friday evening lecture of 1 June 1855 (see J. Tyndall, 'On the Currents of the Leyden Battery', *Roy. Inst. Proc.*, vol. 2, pp. 132–35).
4. *regarding your book*: see letter 0996, n. 6.
5. *I should have . . . Philosophical Magazine*: Tyndall included a notice of the book in February 1855 (see letter 0996, n. 7).
6. *I shall endeavor . . . Philosophical Magazine*: see n. 5.
7. *Berlin*: Riess's address was written on the reverse of the first page of the letter appears.

To James Coxe 4 December 1854 1016

4th Dec. 1854

My dear Dr. Coxe

The Scotsman[1] shall be posted to Hector[2] tomorrow and I know he will be gratified by your remembrance of him. Patterson[3] if I mistake not was a personal friend of his. I had been thinking of writing to Hector and the impulse imparted by your note will convert the thought into an act. Just as the tinkle of a bell will sometimes bring down an avalanche which is already on the point of falling.

With regard to Alick[4] I scarcely know whether to be sorry or glad.[5] I only hope that his comely character and clear head will find fitting exercise in his future vocation. Certain it is that the East India Company has rarely such a present made to it.

Ever yours most truly | John Tyndall.[6]

In a letter received sometime ago from a Professor of Mathematics in a German University[7] he mentions, if I mistake not, one of the Professors[8] at Addiscombe[9] as being an old friend of his. I set about rummaging for this letter to night but have not found it. During my search I stumbled on the enclosed documents[10] which perhaps will amuse you and Mrs. Coxe for five minutes. Many of my friends were angry with me for acting as I did,[11] but I have never been angry with myself. Perhaps you would kindly enclose the papers sometime when you are writing to me.

RI MS JT/1/TYP/1/267
LT Typescript Only

---

1. *The Scotsman*: a weekly liberal newspaper in Edinburgh, founded in 1817 and edited by William Ritchie (1781–1831) and Charles Maclaren (1782–1866) (*ODNB*).
2. *Hector*: Hector Tyndale.
3. *Patterson*: Robert Maskell Patterson (1787–1854), Professor of Natural Philosophy and Mathematics at the University of Pennsylvania, died on 5 September 1854.
4. *Alick*: Alick, or Alexander Cumming.
5. *I scarcely know . . . sorry or glad*: Coxe wrote to let Tyndall know about Cumming's new position as a cadet with the East India Company (see letter 1014).
6. *Tyndall*: LT notes below the salutation: '?', perhaps indicating that she was not certain if the following paragraph was part of this letter. The content, however, relates to the discussion of Cumming.
7. *Professor of Mathematics in a German University*: not identified.
8. *one of the Professors*: not identified.
9. *Addiscombe*: see letter 1014, n. 5.
10. *enclosed documents*: not identified.
11. *acting as I did*: incident not identified.

To Thomas Archer Hirst — 8 December 1854 — 1017

Royal Institution Friday | Dec. 8th. 1854.

My dear Tom

At the time you last wrote to me[1] I was still in doubt as to whether I ought to go to Queenwood or not, my convictions lay being rather on the negative side of the question. The fact is that I find I must draw in my expenses a little. My income next year will not be equal to that of last. Hence as I spent all my income last year, and more too, in order to make a less one answer I must be more economical than I was. I should like to go to Queenwood, but still as with many another liking I must put it under foot this time. But I anticipate matters. I said I was still in doubt when your last note came. Last night I was at the meeting of the Royal Society.[2] The Secretary[3] drew me aside and said that though the matter was not yet formally mentioned he knew that the paper I had sent in would be chosen for the Bakerian Lecture of the Society this year.[4] The lecture will probably come off on Thursday week next.[5] And up to that time I shall be busy with both hands and brain preparing experiments, and arranging my matter. This lecture, as its name imports, is due to a bequest left by one Baker,[6] which goes to pay some trifle for a lecture annually, but it is more a matter of honour than of profit—as I believe none but the best papers are chosen. You see then, whatever irresolution might have attached itself to my proceedings before last night, my fate is now fixed and going to Queenwood is out of the question. Your concert is on the 18th. but my lecture will be on the 21st.[7] The claims of both cannot be satisfied so I must rest contented with satisfying the claims of one.

Affectionately yours | John Tyndall

RI MS JT/1/HTYP/377
LT Typescript Only

---

1. *time you last wrote me*: letter 1013.
2. *Last night . . . Royal Society*: Thursday, 7 December 1854 ('December 7, 1854.', *Roy. Soc. Proc.*, vol. 7, pp, 175–82).
3. *The Secretary*: William Sharpey (1802–80), an English physiologist who served as a secretary of the Royal Society 1854–72 (*ODNB*). Tyndall noted being taken aside by Sharpey at the meeting in his journal entry of 10 December 1854 (Journal, JT/2/13b/703).
4. *the paper . . . this year*: see letter 0981, n. 15.
5. *Thursday week next*: Tyndall was told by Sharpey that his Bakerian Lecture would likely be read on Thursday, 21 December (Journal, 10 December 1854, JT/2/13b/703). The lecture would actually take place on 25 January 1855 ('January 25, 1855.', *Roy. Soc. Proc.*, vol. 7, pp. 214–19).
6. *This lecture . . . by one Baker*: upon his death, the English naturalist and popularizer of science Henry Baker (1698–1774) bequeathed the Royal Society £100 to establish a yearly lecture in his name (*ODNB*).
7. *21st*: see n. 5. As Tyndall's lecture was eventually pushed back, he was free to attend the Queenwood concern on 18 December, which he did (Journal, 18 December 1854, JT/2/13b/705).

From Thomas Archer Hirst 10 December 1854 1018

Broughton | Dec 10th 1854

My dear John,

My last week's work as a bachelor[1] is close at hand. I am sorry you cannot come on Monday week but glad at the cause.[2] On Tuesday week I start for Ireland on the 27th I shall be married. Now my boy decide to come after me.[3] After the 21st it will be well to give yourself a holiday. I shall not pass through London on Tuesday week write before I go and tell me your decision. Anna[4] asks me in every letter. When I get to Ireland I will write to tell you where to come to. If you are there on the 25th or the 26th it will suffice but sooner will suit us all better. Anna has been ill all the week with a violent attack of toothache which I am glad to say is now relieved.

Yours affectionately, Tom

(Turn over)

John![5] | Send the enclosed passport[6] to the French consul, 36 King William Street, City to be visèd.[7] Pay for it and send it me immediately telling me how much it has cost. | Tom.

RI MS JT/1/H/209
RI MS JT/1/T/HTYP/378

1. *My last week's work as a bachelor*: Hirst was to marry Anna Martin in December 1854 (see letter 0975, n. 25). See also letters 0983, n. 18 and 1021.
2. *cannot come . . . glad at the cause*: see letter 1017, nn. 5 and 7.
3. *decide to come after me*: Tyndall would in fact not attend the wedding of his closest friend (see letter 1020).
4. *Anna*: Anna Martin.
5. *John!*: this postscript appears on the back of the letter.
6. *enclosed passport*: Hirst had posted his passport to Tyndall and requested that he go to the French consul to have it viséd (see letter 1020).
7. *visèd*: misspelling of viséd, the 'term for an entry or note on a passport' (*OED*).

From Arthur Trevelyan 10 December 1854 1019

Tyneholm Pencaitland N. B.[1] | 10 Dec. 54

Dear Sir

Having been informed you wished my address—intending sending me a copy of your interesting paper, on the Vibrations of Metallic and other substances when in contact at different temperatures[2]—which I assure you will be accepted with pleasure[3]—I now write—and in the mean time beg your acceptance of the enclosed.[4]

Yours faithfully | Arthur Trevelyan

Dr. Tyndall | Royal Institution, London[5]

RI MS JT/1/TYP/5/1774
LT Typescript Only

1. *Tyneholm Pencaitland N. B.*: Pencaitland, East Lothian, Scotland, a village 12 miles south-east of Edinburgh.
2. *your interesting paper . . . at different temperatures*: see letter 0876, n. 14.

3. *which I assure you will be accepted with pleasure*: Tyndall must have sent the paper, as Trevelyan commented on it in a later letter to Tyndall (see letter 1023).
4. *of the enclosed*: not identified.
5. *London*: this is probably from the envelop as LT wrote 'To' above it.

To Thomas Archer Hirst 12 December 1854 1020

Dec. 12th. 1854.

My dear Tom

I went to day to King William Street and had your passport viséd.[1] You ask me to decide to come after you, and suggest that it would be a holiday for me[2]—it would I fear be a holiday which would render the doctor's aid necessary afterwards. My health is far from being strong at present. This day I was forced to lie in bed till 12 o'clock and the probability is that a journey to Ireland at this season of the year would completely upset me. But I would not say one word about this did I not see obstacles of a far graver character in my way. Faraday commences to lecture on the 28th.[3] and you are not married until the 29th.[4] Now though I am not bound by the Institution still I consider it my duty to be here during Faraday's lectures. I do not know what may happen to render my presence necessary. Faraday himself was poorly yesterday. The idea has glimmered upon me for some time that if the 26th were your day I might make a struggle to go—though it would be at the imminent peril of my health. But the 29th takes all choice out of my hands and fixes me inevitably here. I think if you weigh matters you will agree with me that I ought to remain. But in truth I do not care so much about satisfying you on this head as Anna.[5] I will write to her and state my case. My lectures commence on the 16th of January.[6] I suppose you will be in London by that time, and I shall do my best to render my little nest comfortable for your reception. If I am not very strong on Saturday I will run down to Queenwood[7] but if I do not need renovation I will remain here.

Affectionately yours | John

RI MS JT/1/HTYP/379
LT Typescript Only

1. *I went to day to King William Street . . . passport viséd*: Hirst had posted his passport to Tyndall and requested that he go to the French consul to have it viséd (see letter 1018, n. 7).
2. *You ask me . . . holiday for me*: Hirst had asked Tyndall to follow him to Ireland to attend his wedding to Anna Martin (see letter 1018).
3. *Faraday commences to lecture on the 28th*: Faraday delivered a series of six lectures 'On the

Chemistry of Combustion' that were aimed at a 'juvenile audience' and held during the Christmas holidays, one of which took place on December 28 ('General Monthly Meeting', *Roy. Inst. Proc.*, vol. 2, p. 5). See also letter 0747, n. 4.

4. *you are not married until the 29th*: Tyndall was mistaken on the date, as Hirst planned to marry Anna Martin on 27 December (see letter 1021, n. 3).
5. *Anna*: Anna Martin.
6. *My lectures . . . 16th of January*: Tyndall's series of eleven lectures was entitled 'Magnetism and Frictional Electricity' (see letter 0996, n. 2).
7. *on Saturday . . . Queenwood*: Tyndall travelled to Queenwood on Saturday, 16 December to visit Hirst and attend the concert on 18 December. He returned to London on 20 December 1854 (Journal, JT/2/13b/704–5).

From Thomas Archer Hirst 14 December 1854 1021

Broughton, | Dec. 14th 1854

My dear John.

I am very sorry you are yet so weak.[1] It is useless saying how sorry I am you are not going to Ireland for in fact I should not like you to go unless you could do so without all hesitation and scruple, or fear of consequences[2]—On Tuesday morning I start and as soon after I am married as I can I will write. Yours affectionately | T.A. Hirst.

Either you or I made a mistake in the date—I shall be married on Wednesday the 27th[3]

Anna's[4] address now (and mine too after Tuesday) is Robert Ross Todd Esq.[5] Solicitor Newry Ireland.[6] The lads are making a capital examination, our friend Alex[7] is doing his work magnificently, there is little doubt he will be 1st in the school and none whatever that he will carry away my half-yearly prize. He has got his appointment to the military school of the East India Company.[8]

RI MS JT/1/H/210
RI MS JT/1/HTYP/380

---

1. *you are yet so weak*: Tyndall mentioned that he was not feeling well (see letter 1020).
2. *not going to Ireland . . . fear of consequences*: Tyndall wrote that he would not be able to attend Hirst's wedding due to his commitments in London and his health (see letter 1020).
3. *Either you or I . . . Wednesday the 27th*: Hirst originally wrote to Tyndall that the date was 27 December (see letter 1018), but Tyndall remembered the date as the 29th (see letter 1020, n. 4).

4. *Anna's*: Anna Martin.
5. *Robert Ross Todd Esq.*: Robert Ross Todd (n.d.), an Irish solicitor who worked on the expansion of the railway between Newry and Armagh, Ireland ('Dundalk to Newry', in The Railway Companies [eds.], *The Tourist's Illustrated Hand-book for Ireland with Maps and Illustrations* [London: John Cassell, 1852], pp. 229–32, on p. 230).
6. *Newry Ireland*: a city in Northern Ireland, 34 miles from Belfast.
7. *our friend Alex*: Alick, or Alexander Cumming.
8. *appointment to . . . the East India Company*: see letters 1014 and 1016.

To Thomas Archer Hirst [15 December 1854][1] 1022

My dear Tom

I am anxious to look at you once more *[see]* you pass away into other hands. The Bakerian Lecture[2] is fixed for the 3rd Thursday in January.[3] This arrangement is made for the purpose of giving the Fellows[4] sufficient notice of the thing. Lord Wrottesley[5] announced it last night[6] at the Royal Society. This takes a little work off my hands at present. I will go down sometime tomorrow[7]—write & I will meet you at supper.

Tell Tommy Grieve that I will buy him the *[1 word illeg]*. He wrote to me[8] very dolefully when he heard that I was not to come. I am delighted with your assessment of Alick.[9]

ever yours | John

Elton Hall FAM 124

---

1. *[15 December 1854]*: while the letter is undated another hand has written the correct date of '15th Dec. 1854' at the top of the letter.
2. *The Bakerian Lecture*: see letters 1017, n. 4 and 0981, n. 15.
3. *3rd Thursday in January*: Tyndall initially told Hirst that the lecture would take place on 21 December, but it was not read until 25 January 1855 (see letter 1017, n. 5).
4. *the Fellows*: the Fellows of the Royal Society of London.
5. *Lord Wrottesley*: John Wrottesley (1798–1867), 2nd Baron Wrottesley and English astronomer, was president of the Royal Society from 1854 to 1857 (*ODNB*).
6. *Last night*: see 'December 14, 1854', *Roy. Soc. Proc.*, vol. 7, pp. 183–89.
7. *I will go down sometime tomorrow*: Tyndall left for Queenwood on 16 December (see letter 1020, n. 7).
8. *He wrote to me*: letter missing.
9. *I am delighted . . . assessment of Alick*: Hirst wrote that Alick (Alexander Cumming) was 'doing his work magnificently, there is little doubt he will be 1st in the school' (see letter 1021).

From Arthur Trevelyan 15 December 1854 1023

Tyneholm Pencaitland.[1] | 15 Dec. 54.

Dear Sir

Pray accept my thanks for the copy of your memoir; on the Vibration of substances at different temperatures,[2] when in contact—which I have perused with much gratification—and I may mention, that during my experiments I obtained vibrations in several instances, when assisting as you appear to have done, by 'shaking the rocker' but <u>not</u> considering such as fair experiments, I only note those as successful in which the vibration took place spontaneously—after the heated bar had been placed in the <u>steadiest manner possible</u> on its seat—or otherwise to insure spontaneity the vibrating temperature was raised by placing a lighted spirit lamp—beneath the end of the Bar over-hanging its seat, on which it had been previously placed when cold, and not afterwards touched during the exp^{t}.

The Balancing-bar I contrived for the purpose for enabling the vibration without the tone—to be seen by parties readily in a lecture-room at too great a distance*, to observe the movements in the vibrating-bar itself—and the longitudinal groove in the bar was only made during some of my earliest expts. as I soon found out its inutility.

Also thank you for the notice of mi theories.

Yours faithfully | Arthur Trevelyan

Dr. John Tyndall | Royal Institution | London

* from the Expts.

RI MS JT/1/TYP/5/1775
LT Typescript Only

---

1. *Pencaitland*: see letter 1019, n. 1.
2. *your memoir; on the Vibration of substances at different temperatures*: see letter 0876, n. 14. Trevelyan asked Tyndall to send him the memoir in letter 1019.

From Heinrich Gustav Magnus 26 December 1854 1024

Berlin 26 Decemb 54.

Verehrter Freund,

Entschuldigen Sie daß ich Ihren werthen Brief vom 3^{tn}. d. M. erst heut beantworte. Ich war bis zum Schluß meiner Vorlesung sehr beschäftigt, besonders weil ich auch meine Abhdl. über die Flüßigkeiten zu vollenden wünschte. Den von mir angegebenen kleinen Apparat, den Sie aus Riess Electricitäts Lehre ausgewählt, habe ich für Sie bestellt. Barreaud hat mir versprochen ihn in den ersten Tagen des Jahres zu liefern und werde ich ihn Ihnen dann sogleich durch Asher zusenden.

Poggendorffs Vorschlag die Magnetnadel zu beobachten ist in so fern verschieden von Gauß & Weber's Methode, als er den Spiegel parallel mit der Axe der Nadel anbrachte, während diese ihn bekanntlich senkrecht gegen die Axe befestigen. Dieser Vorschlag von P. ist früher als der von G & W. Poggendorff's findet sich im 7. Bd seiner Annalen *[pr.]* 1826. Gauß Methode ist im 28 Bd. p. 247. Anmerk. mitgetheilt und ist aus dem Jahre 1833.

Wenn Sie den Peltierschen Versuch mit dem Luftthermometer machen wollen, dürfen Sie nur einen ganz schwachen Strom anwenden. Ich pflege ein Grove'sches Element nur so weit einzutauchen, daß nur 2–3 Quadratzoll wirksam sind.

Die Vorstellung welche Sie mir mitgetheilt haben über die seitliche Bewegung einer Masse in Folge der ungleichen Reibung auf beiden Seiten, scheint mir nicht ganz begründet. Wenn der Punkt d festgehalten würde, etwa durch einen Faden, dann würde der Erfolg ganz so statt finden wie Sie meinen, aber so wirkt die Reibung nicht! Die Reibung ist nur eine tangentiale Kraft. Ich will versuchen die Schwierigkeiten die Sie veranlaßt haben eine Bewegung des Schwerpunkts anzunehmen, so weit es in meinen Kräften ist, fortzuräumen, aber ich weiß nicht ob ich werde hinreichend klar mich ausdrücken können.

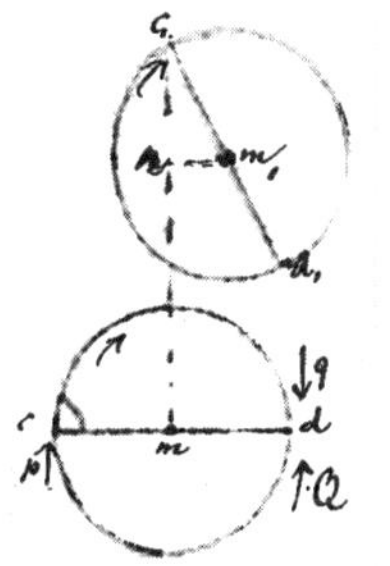

Denkt man sich die Scheibe cd zuerst in Ruhe, und es wirkt bei m, d.i auf den Schwerpunkt, eine Kraft ein, welche diesen nach n bewegt, so werden gleichzeitig auch alle andern Punkte der Scheibe sich parallel mit mn bewegen. Wenn ausserdem noch eine Kraft p bei c in Richtung der Tangente wirkt, so wird c mehr voraneilen als alle übrigen Punkte, die Scheibe wird rotiren, aber dabei wird doch die ganze Scheibe so fortrücken daß ihr Schwerpunkt sich in der Richtung mn bewegt. Ist bei d noch eine Kraft q=p vorhanden, welche die Rotation im gleichen Sinne hervorbringt, so wird die Scheibe mit doppelter Geschwindigkeit rotiren. Schafft man q fort, so wird sich nur die

Geschwindigkeit der Rotation wieder ändern, eine seitliche Bewegung des Schwerpunkts kann durch solche (parallel mit mn wirkende Kräfte nicht erfolgen. Die Schwierigkeit die Sie haben rührt, wie ich vermuthe, aus folgender Vorstellung. Sie sagen wenn bei d eine Reibung vorhanden ist, wenn z.B. eine Kraft Q dort wirkt die q gerade entgegen und genau so groß wie dieses ist; so geht der Punkt d in Ruhe über! Dies ist richtig aber nur in Betreff der Rotation um m. An der fortschreitenden Bewegung nimmt d noch immer unverändert Theil. Würde d durch die Einwirkung von Q fest im Raume, so daß es gleichsam als Hypomochlion für den Hebel cd dienen könnte, dann würde m durch die fernere Rotation von c nach m_1 gedrückt. Aber d kann niemals als /Stützpunkt/ dienen, da er selbst frei im Raume ist. Es wird fortfahren zu rotiren, auch nach Aufhebung der Wirkung von q durch Q, gerade so wie es rotirte, bevor q hinzutrat, als nur p die Rotation hervorbrachte. Durch die Reibung kann niemals eine Veränderung in der Richtung der Bewegung des Schwerpunkts herbeigeführt werden. Auch wenn die Reibung senkrecht gegen die Richtung der Fortschreitung verschieden ist. Gesetzt die Scheibe rotirte wie es die folgende Figur andeutet, und bei a [und] b brächten die gleichen Kräfte p [und] q die Rotation hervor; so wird wenn P=p, die Rotation bei a gehemmt wird, der Schwerpunkt m doch nach n fortschreiten. Die Scheibe wird nur langsamer rotiren da sie jetzt nur durch die Kraft q rotirt.

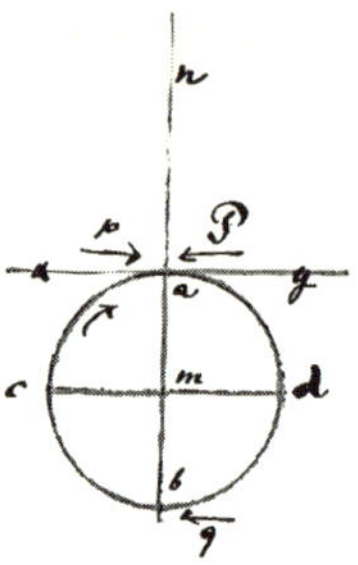

Die Reibung könnte eine seitlich Bewegung hervorbringen wenn sie gegen eine feste Fläche statt fände. Wenn xy eine feste Ebene wäre gegen welche die Scheibe drückte, während sie rotirt, dann könnte eine seitliche Bewegung eintreten. Vorausgesetzt daß der Druck groß genug ist. Aber solcher Druck ist gegen die bewegliche Luft nicht denkbar!

Viel mehr wüßte ich über diese Reibung nicht zu sagen. Übrigens hat Poisson dieselbe auch in seiner Abhandlung berücksichtigt, aber er wie alle andern Mathematiker, kommt zu dem Resultat daß die Reibung keine seitliche Abweichung hervorbringen kann.

Sie würden mir einen großen Gefallen thun wenn Sie mir einige Auskunft über die Lancaster Canonen geben könnten, vielleicht theilt Ihnen Col. Yorke etwas darüber mit. Alles was unsere Zeitungen darüber sagen ist mir unverständlich.

Grüßen Sie D[r]. Faraday recht herzlich von mir. Ebenso Mrs F. wenn dieselbe sich meiner noch erinnern sollte.

M[ss] Francis ist hier wohl und munter, ich habe sie vorgestern gesehn. Freitag ist Poggendorffs Geburtstag, da hoffe ich sie wieder zu sehn.

Gott schenke Ihnen ein glückliches Jahr! Ihr pupil Anna [und] meine Frau grüßen bestens.

Von Herzen | Ihr | GMagnus.

Berlin 26 Decemb 54.

Honoured friend,

Excuse me for the fact that I am answering to your kind letter of the 3rd of this month[1] only today. I was very busy up to the end of my lecture, particularly because I was also wishing to complete my paper on liquids.[2] I have ordered for you the small apparatus[3] indicated by me that you selected from Riess's study on electricity.[4] Barreaud has promised me that he will deliver it in the first few days of the year and I shall then forward it to you at once through Asher.

Poggendorff's suggestion to observe the magnetic needle differs from Gauss[5] & Weber's method, insofar as he inserted the mirror parallel to the axis of the needle, while the latter two are known to have fastened it perpendicular to the axis. The suggestion of P. is earlier than that of G & W. Poggendorff's can be found in the 7th vol. of his Annals *[for]* 1826.[6] Gauss's method is communicated in the 28th vol., p. 247,[7] footnote, and is from 1833.

If you want to do Peltier's experiment[8] with the air thermometer, you may use only a quite weak current. I am in the habit of inserting a Grove cell[9] only as far only 2–3 square inches are active.

The idea which you have communicated to me[10] about the lateral motion of a mass as a consequence of unequal friction on both sides does not seem entirely substantiated to me. If point d were to be fixed in place, for example by a thread, then the outcome would occur entirely as you think, but friction does not work like this! Friction is only a tangential force. I shall try, as far as it is in my power to do so, to clear away the difficulties that have caused you to assume motion of the centre of gravity, but I do not know whether I shall be able to express myself sufficiently clearly.

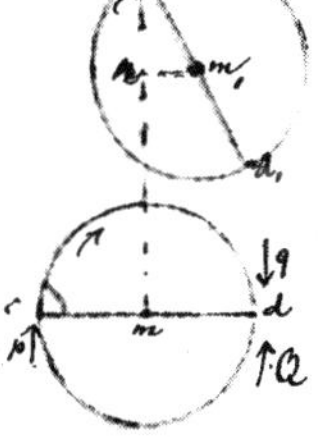

If one imagines the disc cd at rest to begin with, and a force acts at m, i.e. on the centre of gravity, which moves the latter towards n, then all other points on the disc will move simultaneously, parallel to mn. If, in addition, another force p acts at c in the direction of the tangent, then c will rush forward more than all the other points, the disc will rotate, but in the process the whole disc will move away in such a way that its centre of gravity moves in the direction mn. If, at d, another force q = p is present which produces rotation in the same sense, then the disc will rotate with double the speed. If one removes q, then only the speed of rotation will change again, a lateral motion of the centre of gravity cannot occur through such forces (acting in parallel to mn). The difficulty that you have stems, as I suspect, from the following idea. You say that if friction is present at d, when e.g. a force Q acts there that is directly contrary to q and precisely as large as this; then the point d will come to rest! This is correct, but only with respect to rotation around m. d continues to take part in the ongoing motion without any changes. Were d to become fixed in space through the action of Q, so that it could serve, as it were, as a hypomochlion[11] for the lever cd, then m would be pushed by the more distant rotation of c towards m_1. But d can never serve as a */fulcrum/*, as it itself is free in space. It will continue to rotate, even after the neutralisation of the effect of q by Q, exactly as it was rotating before q came along, when only p produced the rotation. A change in the direction of the motion of the centre of gravity can never be brought about by friction. Even if the friction perpendicular to the direction of the progression is different. Assuming that the disc was rotating as the following figure suggests, and the identical forces p [and] q were to produce rotation at a [and] b; then if rotation is inhibited at a when P = p, the centre of gravity m will still progress towards n. The disc will only rotate more slowly, as it is rotating now only due to the force q.

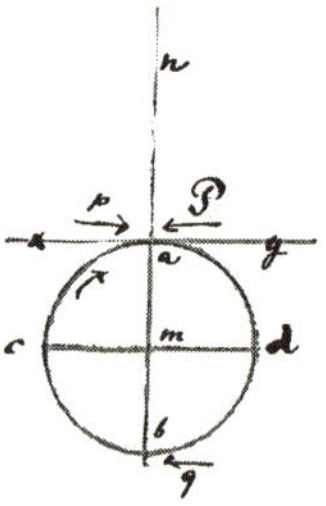

Friction could produce a lateral motion if it were to occur against a solid surface. If xy were a solid plane against which the disc pressed while it rotates, then a lateral motion could occur. Provided that the pressure is large enough. But such pressure is not conceivable against movable air!

There is not much more that I could say about this friction. Incidentally, Poisson[12] has considered it also in his paper,[13] but he like all the other mathematicians comes to the result that friction cannot produce a lateral deviation.

You would do me a great favour if you were able to inform me about the Lancaster cannons;[14] perhaps Col. Yorke will tell you something about them. Everything that our newspapers say about them is incomprehensible to me.

Give Dr. Faraday my very warm regards, and also to Mrs F,[15] if she should still remember me.

Miss Francis is well and cheerful here, I saw her the day before yesterday. Friday is Poggendorff's birthday and I hope to see her there again.

May God give you a happy year! Your pupil Anna[16] [and] my wife[17] send you their best regards.

Sincerely | Your | GMagnus.

RI MS JT/1/M/15

---

1. *your kind letter . . . this month*: letter missing.
2. *my article on liquids*: not identified.
3. *small apparatus*: Tyndall mentioned an apparatus used in Riess's experiments on auxiliary currents that he wanted Riess to order for him (see letter 0996).
4. *Riess's study on electricity*: see letter 0996, n. 6.
5. *Gauss*: Johann Carl Friedrich Gauss (1777–1855), a German mathematician who together with Weber built one of the first electromagnetic telegraphs.
6. *Poggendorff's . . . 1826*: see J. C. Poggendorff, 'Ein Vorschlag zum Messen der magnetischen Abwichung', *Poggend. Annalen.* 7:1 (1826), pp. 121–30.
7. *Gauss's method . . . vol. 28, p. 247*: see C. F. Gauss, 'Die Intensität der erdmagnetischen Kraft, zurückgesührt auf absolutes Maass', *Poggend. Annalen.* 28:6 (1833), pp. 241–72.
8. *Peltier's experiment*: Jean Charles Athanase Peltier (1875–45), a French physicist who researched thermoelectricity and discovered the so-called Peltier effect—'an effect whereby heat is given out or absorbed when an electric current passes across a junction between two materials' (*OED*).
9. *Grove cell*: an early fuel cell, invented by William Robert Grove.
10. *The idea which you communicated to me*: not identified.
11. *hypomochlion*: fulcrum, or 'the point against which a lever is placed to get a purchase or on which it turns or is supported' (*OED*).
12. *Poisson*: Siméon Denis Poisson (1781–1840), a French mathematician and physicist known for the eponymous Poisson's ratio, equation, bracket, and distribution (*OED*).
13. *in his paper*: not identified. For papers by Poisson see 'Poisson, Simon Denis', *RS Catalogue*, vol. 4, pp. 964–69.
14. *Lancaster cannons*: the Lancastor cannon, or 68-Pounder Lancaster gun, a rifled, muzzle-loaded cannon with a slightly oval bore invented by Charles Lancaster (1820–78), which was used in the Crimean War (see letter 0873, n. 30) (*ODNB*).
15. *Mrs. F.*: Sarah Barnard, Faraday's wife.

16. *Your pupil Anna*: Anna Magnus, daughter to Heinrich Gustav and Bertha Magnus (see letter 0852, n. 17).
17. *my wife*: see n. 16.

From Hermine Ballin [December 1854][1] 1025

Stamford Hill[2] | Saturday night.

Dear Sir,

Let me thank you for your kind letter[3] and for your friendly wishes to the first Christmas I spend in England. Though I cannot help feeling deeply the separation from my dear parents and sister[4] on this first Christmas of my life I pass without them, still I feel that I have so much reason for being grateful and happy that I do not allow my sad thoughts to dwell in my mind. Dear Miss Hornblower[5] is more than kind, she is quite m<u>other</u>ly to me and this makes me doubly happy as I know what a source of joy it is for my dear parents. The fact that the year I fixed for my stay at Miss Hornblower's house is now soon elapsed causes me many a sad hour; I feel that it will be now doubly difficult for me to accustom myself to the treatment governesses generally meet with in England. When I came first here I was prepared for the worst and I think I should have had strength of mind enough to bear it; now I fear all the love and kindness with which I am treated here has spoiled me and I shall feel every change the more. However I will not lose courage I know that Miss H.[6] will do every thing to promote my interest in another way, if she cannot manage to keep me in her family, and my first step into the wide world had been such a favoured one that it would be wrong to lose confidence now.

I was very sorry not to have seen you when you kindly came to Stamford Hill to see me;[7] I should have told you so and thanked you for your kind visit long before this, had I not been afraid of troubling you with a letter now I know how little time you have to spare. It is very fortunate for me that I did not know this in Germany or I should never have troubled you with my business and then of course never have known my beloved Miss Hornblower. I am sorry to learn from your letter that you do not feel quite well, indeed you ought not to sacrifice your health to your science though I can understand how difficult it must be for a true lover of science to yield to other claims. I am glad that you have not yet quite forgotten our cheerful trips to the woods of dear Marburg; they are fresh in my recollection, though my thoughts are now taken up too much with different pursuits and duties to dwell often upon those times long gone by. Since my two friends Tette and Sophy[8] are married and have left Marburg I never hear anything from the old town, which though I know now nobody there in whom I feel an interest will always keep a favourite place in my recollection. My friends are both very happy; the letters from Mrs. Bromeis[9] give me true pleasure because every word in them is expressive of her exceeding happiness.

But I am afraid I am trying your patience too hard and you will think this letter endless. Take my kind wishes for your health and a happy new year

Very truly yours | Hermine Ballin.

RI MS JT/1/B/12
RI MS JT/1/TYP/1/141

---

1. *December 1854*: as this letter refers to Ballin's first Christmas in England, and mention of her arrival at London in late February 1854 first appears in letter 0885, it must necessarily be from December 1854. Tyndall likely replied to this letter on 31 December, as he noted in his journal, 'I thought of writing to Miss Ballin and the duty was so evident that I did not dare to shirk it' (Journal, JT/2/13b/708–9).
2. *Stamford Hill*: the school for girls at which Ballin was a governess was located in Stamford Hill, an area in north London, England (see letter 0880, n. 6).
3. *your kind letter*: letter missing.
4. *my dear parents and sister*: not identified.
5. *Miss Hornblower*: Governess Jemima Hornblower, who governed a school for girls in Stamford Hill (see letter 0880, n. 6).
6. *Miss H.*: Jemima Hornblower (see n. 5).
7. *When you kindly came . . . see me*: not identified.
8. *my two friends Tette and Sophy*: not identified.
9. *Mrs. Bromeis*: wife of Johann Konrad Bromeis (see letter 0880, n. 1). Not further identified.

BIOGRAPHICAL REGISTER

This register contains the names and biographical details of people who are mentioned three or more times in the letters included in this volume. Biographical information about people mentioned once or twice is contained in the notes appended to the appropriate letter(s).

Conventions

Tyndall and some of his correspondents were often unaware of the correct spelling of the names of people whom they met; in particular, it is clear that they spelled many German names phonetically—and incorrectly. In transcribing the letters the editors have generally retained the spellings of family names given in the letters or LT's transcripts. In this register persons for whom we have a preferred spelling—for example, as used in biographical dictionaries—are entered under that spelling and any alternative usages within the letters are placed in "( )" parentheses: thus, "Anderson (Anderton), Charles." Where we do not know the preferred spelling, the alternatives are given in alphabetical order, separated by a "/": thus, "Dickenson/Dickinson, Carl." Nicknames and pseudonyms were widely used by Tyndall and his friends; these nicknames have been entered, with cross-references, in the register below. Where such nicknames occur in the letters, the person they refer to is often given in an accompanying note. These nicknames are given in this register and are placed in "( )" parentheses: thus, "Phillips (Philips), George Searle ("January Searle")." Where known, the dates of birth and death are given in "( )" parentheses to the right of the part of the name for which the nickname is given; that is, "Plücker, Julius ("Plucker") (1801–68)."

Along with material from the Tyndall correspondence and from the journals of Tyndall and T. A. Hirst, the sources for the biographical entries are standard biographical dictionaries, such as the *Dictionary of Scientific Biography*, the *Oxford Dictionary of National Biography*, Poggendorff's *Biographisch-literarisches Handwörterbuch der exacten Wissenschaften*, and the *Dictionary of Nineteenth-Century British Scientists*. Where additional sources add important material, they are given in "( )" at the end of the entry, often in abbreviated form.

Adie, Richard (1810–81), maker of scientific instruments. He was the third son of instrument maker Alexander Adie (1775–1858), who set up a shop in Edinburgh, which Adie took over after the death of his eldest brother. He also published in scientific journals, including the *Phil. Mag.* and the *Edinburgh New Philosophical Journal.* He and Tyndall had a running argument in the *Phil. Mag.* in 1852–3 concerning the temperature produced at the joint between antimony and bismuth.

Allen, Charles (Charley) Edward (d.1853), son of Margaret and Robert Copeland Allen. Charles died from consumption on 11 January 1853 at about three years old. He was buried in Trinity Church in the town of Preston next to his sister Emily Ellen Allen who had passed away at the age of one in 1852.

Allen, Margaret (née Smith) (*c.* 1828–?), wife of Robert Allen, daughter of a merchant named Samuel Smith. Three of her children (Charley, Sophy, and another daughter) died young of consumption. She and Tyndall exchanged several letters in 1853–54. After her husband, Robert, moved to America, Margaret was entirely dependent on his parents for support. Tyndall occasionally gave her money and helped her obtain a position in October 1854.

Allen, Robert (Bob) Copeland (*c.* 1823–?), civil assistant on the Ordnance Survey of Ireland, who Tyndall met in Preston. He was the son of Robert Allen, an excise officer. He married Margaret Smith on 6 March 1845. Robert, who had a drinking problem, later moved on his own to America in order to reform himself.

Anderson (Anderton), Charles (1790–1866), lecturer and laboratory assistant at the RI.

Asher, Adolphus (1800–53), a Berlin bookseller and author. Asher spent five years in England in 1820–5, before he moved to St Petersburg where he worked in the diamond trade before he became interested in bookselling. Due to the religious persecution of the Jewish community in Russia, Asher moved to Berlin around 1830. His shop in Linden Walk attracted many international visitors, and he also worked with the British Museum's library. Asher was connected to Tyndall through Heinrich Gustav Magnus.

Atkinson, Edmund (1831–1900), chemist, teacher, author, and translator of many German and French scientific works. Atkinson received a PhD in Chemistry at the University of Heidelberg in Germany in 1854. After that he taught at Queenwood College before becoming assistant to Benjamin Brodie. He then became lecturer on chemistry and physics at Cheltenham College. After spending some time at the Royal Military College, Sandhurst, he later became professor of experimental science at the Staff College in Surrey.

Barlow, John (1798–1869), Anglican clergyman and the secretary of the RI in 1843–60. Together with Michael Faraday, he undertook the day-to-day running of the RI. When Faraday's health declined in the 1840s, Barlow took over many of Faraday's duties. In 1854, Barlow was made Chaplain-in-Ordinary at Kensington Palace.

Barreaud, a scientific instrument maker who was associated with Heinrich Magnus. Magnus commissioned him to produce a polytrope for Gassiot. While it appears to have reached Gassiot eventually, Magnus was unhappy with how long it took.

Best, Samuel (1803–73), rector of Abbots Ann near Andover. He was a mutual friend of Hirst and Tyndall.

Bevington, James Buckingham (1804–92), a wealthy and benevolent manufacturer of leather goods, later described as a leather merchant, of Neckinger-Mills in Bermondsey, south London. He enjoyed wide-ranging religious discussion with Tyndall. They were good friends in the 1850s and remained in contact throughout their lives. ("Wills and Bequests," *Standard*, 13 October 1892.) Bevington was a member of the Entomological Club and invited Tyndall and Hirst to their annual social dinner in July 1854, which they were unable to attend.

Blanche, a teacher at Queenwood College who was dismissed in 1854. This matter, together with the dismissal of McDougall, was the subject of several letters between Tyndall and Hirst.

Booth, Francis (d.1853), a mutual friend of Tyndall and Hirst. He was employed in an important position in a wool mill near Halifax. Tyndall had hoped Booth might be able to advance himself through school mastering. In 1851, he became ill with tuberculosis and his health deteriorated quickly. Both Tyndall and Hirst helped him monetarily during his illness, as did Bevington.

Buchan, H., a student at Queenwood who was involved in an internal row over Queenwood's Improvement Society.

Bunsen, Robert Wilhelm (1811–99), German chemist. He received his PhD from the University of Göttingen in 1830. He taught chemistry first at Göttingen in 1833, then in the Polytechnic School in Kassel in 1836, before accepting a professorship at the University of Marburg (extraordinary in 1838 and ordinary in 1842). In 1846, he also traveled to Iceland as part of a scientific expedition sponsored by the Danish government. At Marburg, he was an extremely popular teacher, mentoring both Tyndall and Frankland, and a renowned laboratory researcher. Tyndall's initial experiments in diamagnetism relied heavily on crystals that he secured from Bunsen's laboratory. Tyndall and Bunsen had a mutual interest in steam boilers, and discussed this in letters in 1854.

Carlyle, Thomas (1795–1881), Scottish essayist and historian. Tyndall and Hirst discussed and made references to Carlyle's work in many of their letters.

Carter, Richard (1818–95), a surveyor in Halifax and former employer of Tyndall (1844–47) and Hirst (1845–50). Later in life, Carter purchased significant property in the South Yorkshire town of Barnsley, where he was eventually elected mayor. In the 1880s he retired to Harrogate. (J. Forrest, ed., *Minutes of the Proceedings of the Institution of Civil Engineers*, vol. 123 [London: Institution of Civil Engineers, 1896], pp. 445–48.) Tyndall and Hirst discussed Carter's views on the relationship between science and religion in April 1853. Carter stated that religion and science were equally noble endeavors, while Hirst assessed, and Tyndall agreed, that Carter "bows politely to Revealed Religion /but that is all./"

Chapman, John (1821–94), a British publisher and physician. In 1851 he became proprietor and editor of the *Westminster Review*, which was a quarterly liberal publication initially founded by Jeremy Bentham. In 1854, Thomas Henry Huxley invited Tyndall to coauthor his regular "Science" column for the *Westminster Review*.

Christie, Samuel Hunter (1784–1865), an experimental philosopher and mathematical master at the Royal Military Academy, Woolwich. He served as a secretary of the RS until 1854. While Christie taught mathematics, his later research was related to magnetism and he was closely linked to the establishment of colonial magnetic observatories during the so-called magnetic crusade. However, from the 1840s onward his duties with the RS, at Woolwich, and with the Admiralty prevented him from carrying out his own research.

Clausius, Rudolf (1822–88), renowned German scientist and one of the founders of modern thermodynamics. His famous paper on the theory of heat "Ueber die bewegende Kraft der Warme," was published in the *Annalen der Physik* in 1850. He taught physics at the Royal Artillery and Engineering School in Berlin in 1850 before moving to a professorship in mathematical physics at the Polytechnicum in Zurich in 1855. Tyndall translated many of his major works into English for publication in the *Phil. Mag.* and the *Scientif. Mem.* in the early 1850s.

Coxe, James (1811–78), a Scottish physician, married to May Anne Coxe. He was elected a fellow of the Royal Society of Edinburgh in 1854. Coxe was keen for Tyndall to find employment at the University of Edinburgh and both he and his wife sought education and employment advice from Tyndall about their nephew, Alexander (Alick) Cumming.

Coxe, May Anne (née Cumming) (1806–75), married to James Coxe. She was the daughter of Robert Cumming and Lesley (Leslie) Baille who was famously known as Robert Burns's "Bonnie Lesley." She sought advice from Tyndall about the future educational and employment opportunities for her nephew Alexander (Alick) Cumming.

Craven, James (Jemmy/Jimmy) (1829–86), a young man who had trained as a surveyor in Halifax, very likely under Richard Carter, and, like Hirst, a young protégé of Tyndall. Craven and Hirst were good friends and supported each other's educational pursuits. Under Tyndall's guidance, Craven studied philosophy, geology, and chemistry; however, Tyndall did not feel that Craven was as dedicated to the work as Hirst. Craven set up independently as an "agent and surveyor" in Halifax late in 1850. He had a brother, possibly named John.

Cumming, Alexander (Alick/Alix), nephew of May Anne and James Coxe. Alick was one of Tyndall's students at Queenwood College, who later trained at the East India Company Military Seminary in Addiscombe, Surrey. Tyndall took a great interest in Alick's education and was keen for him to continue his studies in Germany.

Cumming, George S., a student at Queenwood College and brother of Alexander (Alick/Alix) Cumming. He began writing to Tyndall in 1853.

Daubeny, Charles Giles Bridle (1795–1867), a chemist, botanist, and geologist who spent time in the United States studying its thermal springs and geology (*ODNB*).

Davy, Humphry (1778–1829), chemist and inventor. Davy was given a post at the RI soon after its foundation and was appointed to full lecturer at the RI in 1802. Davy was the first to isolate a series of substances, including potassium and sodium. Davy's employment conditions became a useful point of comparison when Tyndall was offered a professorship at the RI.

De la Beche, Henry (1796–1855), geologist and paleontologist. He was the first director of the Geological Survey of Great Britain. In 1853 De la Beche was keen for Tyndall to take up a position at the Government School of Mines.

De la Rive, Auguste Arthur (1801–73), Swiss physicist who experimented with magnetism and electricity. His book *Traité de l'électricité théorique et appliquée* was translated into English as *A Treatise on Electricity, in Theory and Practice* in 1853.

Debus, Heinrich (1824–1915), chemical assistant in Bunsen's laboratory in Marburg until early 1851 and a friend to Tyndall, Frankland, and Hirst at Marburg. Debus took up a position teaching chemistry at Queenwood College, where he succeeded Frankland. He was an excellent teacher. After leaving Queenwood in 1867, he spent three years at Clifton College, Bristol, then returned to London as lecturer in chemistry at Guy's Hospital Medical School (1870–88), adding to this the position of professor of chemistry at the Royal Naval College, Greenwich (1873–88), under Hirst, who was founding director of studies at the college.

Dickenson/Dickinson, Carl (b.1833/4), an English student who arrived in Marburg in mid-1851. He had probably been a student of Friedrich Schnackenberg, who accompanied him on his journey to Marburg, and was therefore probably from Lancashire (possibly Preston or Blackburn). Hirst mentored him, and at the beginning of the 1852–53 academic year, he and Hirst moved to Berlin where they lodged together. He had a younger brother at Queenwood.

Dirichlet (Derichlet), Peter Gustav Lejeune (1805–59), a German mathematician known in particular for his contributions to number theory. Dirichlet lectured at the University of Berlin until he was offered a position at the University of Göttingen in 1855. Tyndall spent time in Berlin in the summer of 1853, where he saw Dirichlet.

Dove, Heinrich Wilhelm (1803–79), Prussian meteorologist and experimental physicist. In 1829, Dove moved to Berlin where he became an ordinary professor in 1844 and director of the Prussian Institute of Meteorology beginning in 1849. He also worked to expand the coordination between meteorologists internationally. In 1853, Dove received the RS's Copley medal. Tyndall saw Dove while visiting Paris in September 1853, and again in September/October 1854 when Dove visited London where they, among other things, dined at the Red Lions during the annual BAAS meeting in Liverpool.

du Bois-Reymond (du Bois/Du Bois/Dubois), Emil (1818–96), German physiologist. He began his studies at the University of Berlin, where he attained his degree in 1843 and began lecturing in 1846. He is known for his pioneering work in electrophysiology, particularly his ability to measure the electrical currents produced by nerves. Du Bois-Reymond had a conflict with the Italian Carlo Matteucci over their views on animal electricity. Tyndall doubted the validity of either position and got involved in the dispute when he wrote an article that referred to du Bois-Reymond's work on animal electricity that was published as the "Rise, Progress, and Present Condition of Animal Electricity," *British and Foreign Medico-chirurgical Review* 25 (1854), pp. 126–42, which soured their relationship.

Edmondson, Anne (née Singleton) (d.1863), wife of George Edmondson. She was active in the management of Queenwood College and, like her husband, was a committed Quaker. Tyndall and Hirst regarded her as the stronger character of the two.

Edmondson, George (1798–1863), Quaker, educationist, and headmaster of Queenwood College from 1847 until his death. His Quaker beliefs contributed to his innovative educational practices, for example the inclusion of manual pursuits in the curriculum. (For his earlier life see volume 2.) In 1853–54, Tyndall and Hirst had a drawn-out dispute with Edmonson concerning Queenwood teacher John Mummery's purchase of a spurious German doctorate.

Edmondson, George William, son of George and Anne Edmondson. Edmondson studied at Queenwood.

Emerson, Ralph Waldo (1803–82), an American poet and essayist and a leader of the Transcendentalist movement. Tyndall greatly admired Emerson's writings and urged his protégés Thomas Hirst and James Craven to read Emerson's work.

Faraday, Michael (1791–1867), natural philosopher and Sandemanian. Faraday married Sarah Barnard in 1821, and the same year he was appointed acting superintendent of the RI. In 1833, Faraday became the first holder of the RI's Fullerian professorship of chemistry. Faraday's research focused particularly on electromagnetism, and he discovered electromagnetic induction. Faraday was an important influence on Tyndall and assisted him in navigating the scientific scene of London. He was also a key actor in getting Tyndall his position at the RI in 1853.

Faraday, Sarah (née Barnard) (1800–79), wife of Michael Faraday. Sarah's parents, Edward Barnard and Mary Boosey, were Sandemanian, and she met Michael through her elder brother Edward Barnard, who knew him from a self-improvement essay writing group. After Michael's death, Sarah donated many of his manuscripts to his early biographers, including Tyndall.

Forbes, Edward (1815–54), naturalist. Forbes undertook extensive geological survey fieldwork. He started the Red Lions Club in 1839. He was appointed lecturer in the newly expanded Government School of Mines in 1851, where he taught on subjects such as zoology and paleontology. Throughout 1853, Tyndall corresponded with and about Forbes regarding the possibility of him joining the School of Mines. In 1854, Forbes was appointed to replace Robert Jameson as the Regius Chair of Natural History in Edinburgh. He died later that same year.

Forbes, James David (1809–68), Scottish physicist and geologist. As the Edinburgh professor of natural philosophy, Forbes undertook research on a range of topics such as radiant heat and light, sound, as well as glacial motion about which he later had a conflict with Tyndall. In 1853, Tyndall challenged Forbes's views on the vibration of heated metals. Forbes suffered from recurring illness, and in 1854 there was speculation that he would have to vacate his position at Edinburgh, potentially freeing it up for Tyndall. Forbes, however, did not resign.

Francis, William (1817–1904), chemist, publisher, and author. Along with his father, Richard Taylor, Francis established the Taylor & Francis publishing house in 1852. It was a prolific venture and published many scientific journals, including the *Phil. Mag.* Tyndall and Francis corresponded extensively (see volume 3, in particular), and worked together on multiple publishing ventures, including the *Scientif. Mem.*, and the *Phil. Mag.*, of which Tyndall became a conductor in 1854.

Frankland, Edward (1825–99), chemist and teacher at Queenwood College. In 1848, he accompanied Tyndall to the University of Marburg in Germany, where they both studied for doctoral degrees. Frankland married Sophie Fick in 1851 and had several children, including Margaret (b. 1853), Frederick (b. 1854), Sophie (b. 1855), and Percy Faraday (b. 1858).

Galvani, Luigi (1737–98), Italian physician and physicist. Galvani investigated the effects of electricity on animal tissue, and discovered what he termed "animal electricity." Animal electricity was, Galvani believed, an innate force of animal tissue. Galvani had a longstanding disagreement over his findings with Alessandro Volta.

Gassiot, John Peter (1797–1877), electrician, wealthy wine merchant, and a founding member of the London Electrical Society (f.1837) and of the Chemical Society (f.1845). Gassiot was moreover a manager of the London Institution, and throughout 1853 sought to secure Tyndall's employment as a lecturer there. He was also a member of the Council of the RS in 1853 and acted as an intermediary between Tyndall and the RS in regard to Tyndall's declining of the Royal Medal.

Gerling, Christian Ludwig (1788–1864), German mathematician. He became professor of mathematics at Marburg in 1817. He was noted for work in geodetics, for his correspondence with Gauss, and for being doctoral adviser of Julius Plücker. Letters between Gerling and Tyndall mention a wife, Christiane Wilhelmine (née Suabedissen), and possibly daughters as well.

Ginty, William Gilbert (*c.* 1820–66), Irish civil engineer, was a close friend of Tyndall from his Irish Ordnance Survey days and an ally in the campaign for better treatment of the civil assistants employed in the English Ordnance Survey (see volumes 1 and 2). He married Margaret Roberts, and the couple relocated to England, first in or near Wakefield then to Manchester in 1846. During his trip to England in the summer of 1850, Tyndall often visited the Gintys. Ginty became a prominent Manchester businessman and was involved in various civil engineering projects, including the construction of the Liverpool-Bolton railway. He served briefly as honorary secretary of the Manchester Athenaeum (elected late 1853) before moving to Rio de Janiero in April 1854 to manage the city gas works. He was awarded a Brazilian knighthood—Knight of the Imperial Order of the Rose—in 1857 for his contributions to major construction projects. He died suddenly from unspecified causes in mid-1866.

Graham, Mrs. Her son, Newton, attended Queenwood, and she wrote to Tyndall about possibly sending another of her sons to the school.

Grieve, Tommy W., student at Queenwood, who kept a regular correspondence with Tyndall after Tyndall left his position as a teacher there in 1853. His father may have been H. M. Grieve.

Grove, William Robert (1811–96), natural philosopher. During the 1850s, Grove researched on electrical conduction through a vacuum and gave a Friday evening lecture at the RI "On the Transmission of Electricity by Flame and Gases" in 1854. From 1852 until 1854, Grove was one of the managers of the RI.

Haas, John, Swiss educator and author. Haas taught modern languages, in particular French, at Queenwood from 1848 onward before he became headmaster of the College of Morat in Switzerland. Haas had several publications, including a French language textbook.

Hansen, son of the Danish-German astronomer Peter Andreas Hansen. Heinrich Magnus wrote the young Hansen a letter of introduction for when he visited London in 1854.

Helmholtz, Hermann Ludwig Ferdinand von (1821–94), German physicist and physician. In 1847 Helmholtz published his paper "Über die Erhaltung der Kraft," which discussed the mathematical principles of the conservation of energy. When Helmholtz visited England in 1853, he discussed this research with Tyndall.

Herschel, John Frederick William (1792–1871), first baronet, mathematician, and astronomer. He was the son of the famous astronomer Frederick William Herschel and himself became one of the most eminent scientists in Britain in the early Victorian era. His *Preliminary Discourse on the Study of Natural Philosophy* (1830) was widely read and greatly admired. He became Sir John in 1838 when, on his return from mapping the southern skies from Capetown, he was created a baronet. In 1849, he published his *Outlines of Astronomy*, a popular presentation of astronomy that went through eleven editions by 1871. He also edited the *Admiralty Manual* in 1849, a field guide in science for those traveling throughout the British empire. Tyndall was introduced to Herschel by Faraday in 1854.

Higginson, Thomas Charles (*c.* 1821–98), Major in the Royal Artillery. Higginson knew Tyndall from their time with the Ordnance Survey. In 1853–54, Higginson was a recruiting officer with the British Army and was based both in Ipswich in England and County Kildare in Ireland.

Hill, Henry. Hill had financial difficulties and corresponded with Hirst about them. This eventually led to his gainful employment with Bevington at Neckinger Mills.

Hirst, Anna (née Martin) (1831–57), wife of Thomas Hirst, sister of Mary Simpson (née Martin). Anna met Hirst in Marburg and they were married in 1854. Anna became ill with symptoms of tuberculosis soon after their wedding.

Hirst, Thomas Archer ("Tom") (1830–92), a mathematician and one of Tyndall's closest friends. Hirst was the son of a successful Yorkshire businessman, also named Thomas Hirst, and Hannah Oates Hirst. He was twelve when his father died and seems to have been deprived of most of his inheritance under questionable circumstances. Tyndall and Hirst met in 1845 when Hirst moved to Halifax to become an apprentice to Richard Carter on the West Yorkshire Railway. Tyndall took an interest in Hirst's development and served as a mentor to him, even contacting Thomas Carlyle on his behalf. Hirst later studied mathematics and chemistry, first at the Halifax Mechanics Institute, and later at Marburg, Germany, where he followed in Tyndall's footsteps. Hirst then replaced Tyndall at Queenwood College in 1853 when the latter took up his position at the RI. Hirst left his position at Queenwood in 1856, when his wife Anna became very ill with tuberculosis.

Hofmann (Hoffmann), August Wilhelm (1818–92), German chemist. Hofmann's research focused on organic chemistry, and he made several discoveries including formaldehyde and developed the "Hofmann reaction." From 1845 onward Hofmann served as the director of the Royal College of Chemistry until he moved to Bonn in 1864. Throughout 1853 and 1854, Hofmann delivered "A Course of Lectures on Organic Chemistry" at the RI.

Hopkins, William (1793–1866), mathematician and geologist. In 1853–54, Hopkins undertook research on the effects of pressure on the melting points and conductivities of solids, with colleagues such as James Prescott Joule and William Fairbairn. As a member of the Council of the RS, Hopkins appears to have opposed the decision to award Tyndall the Royal Medal behind the scenes. Tyndall later sent Hopkins his published papers.

Hughes, Edward, author and educator. Hughes was the headmaster of the Royal Naval Lower School, Greenwich. He edited the series "Reading Lessons," which brought together introductory chapters on a wide range of topics for general readers. Tyndall wrote a series of articles on "Natural Philosophy" that appeared in successive volumes of *Hughes Reading Lessons* from 1855 to 1858.

Huxley, Thomas Henry (1825–95), biologist. Between 1846 and 1850, Huxley acted as scientific assistant surgeon on the *Rattlesnake* expedition. While in Sydney, Australia, Huxley met Henrietta (Nettie) Anne Hearthorn and began an eight-year long engagement while he attempted to establish himself on the scientific scene in London. Huxley received the Royal Medal in 1852 for his *Rattlesnake* papers and became lecturer in natural history and paleontology at the Government School of Mines in 1854. Huxley and Tyndall met in 1851 and quickly became friends.

Iliff, teacher at Queenwood College.

Jones, Henry Bence (1813–73), chemist, wealthy physician, and a manager (later secretary) of the RI. Bence Jones organized Tyndall's first lecture at the RI and was central in helping to secure Tyndall a position there in 1853.

Joule, James Prescott (1818–89), physicist. Joule examined the relation between heat and mechanical power, which laid the foundation for the law of conservation of energy promoted by figures such as William Thomson, James Clerk Maxwell, and Peter G. Tait. In 1854 Joule and Thomson submitted a paper to the *Phil. Trans.*, which Tyndall reviewed and recommended substantial changes.

Knoblauch, Karl Hermann (1820–95), German physicist. In 1853, Knoblauch left the University of Marburg for a professorship at the University of Halle. Knoblauch studied under Heinrich Magnus and researched radiant heat. Tyndall coauthored two papers with Knoblauch on diamagnetism, which appeared in both German (in Poggendorff's *Annalen*) and English (in the *Phil. Mag.*) in 1850.

Kohlrausch, Rudolph Hermann Arndt (1809–58), German physicist. In 1853 Kohlrausch became Extraordinary Professor of Physics and Mathematics at the University of Marburg. He is best known for his work in electricity and electromagnetism, including work with Weber showing the ratio of the electrostatic unit of charge over the electromagnetic unit of charge.

Magnus, Heinrich Gustav (1802–70), German chemist and physicist. Magnus was professor and later dean at the University of Berlin where he maintained an expansive scientific laboratory. Magnus undertook research in a wide range of areas, such as on thermoelectricity and the behavior and nature of gases. Tyndall and Magnus had an extensive surviving correspondence, which included discussions on scientific subjects as well as on personal and professional matters.

Matteucci, Carlo (1811–68), Italian physicist. Matteucci was professor of physics at the University of Pisa, where he experimented on the nature of animal electricity and nerves. In the early 1850s, Matteucci and du Bois-Reymond engaged in a heated public conflict over their views on animal electricity that also involved Tyndall.

McDougall (McD/McD), teacher at Queenwood. McDougall had conflicts with the students at Queenwood and was dismissed from his short-lived position in 1854 soon after Blanche was also dismissed.

Moigno, François-Napoléon-Marie (1804–84), French Catholic priest, physicist, and author. Moigno acted as scientific editor of several journals and founded the scientific journal *Cosmos* in 1852.

Mummery, John S. (Mr. M), teacher of classics and history at Queenwood. In 1853, it was revealed that Mummery had purchased a German doctorate in London. Tyndall and Hirst were particularly displeased with Mummery's purchase, as they were concerned it would negatively influence the reputation of Queenwood as well as other German doctorates. Mummery gave his notice of resignation from Queenwood soon after.

Newman, John (*c.* 1783–60), instrument maker. Between 1827 and 1860, Newman ran a shop on 122 Regent Street, London. Newman supplied scientific instruments to the RI, where he also acted as laboratory assistant.

Newton, Isaac (1642–1727), natural philosopher and mathematician. Newton made important contributions to a range of scientific subjects, including the development of the law of universal gravitation.

Parsons, William (Lord Rosse), the third Earl of Rosse (1800–67), astronomer. Parsons served as president of the RS between 1848 and 1854.

Perkinton, Thomas ("Tom"), land surveyor. Friend of Tyndall and Hirst dating back to their surveying days in Halifax. The 1850 *Directory of Halifax, Huddersfield, Holmfirth, and Surrounding Villages* lists Perkinton as a surveyor and land agent. He also went into business with George Searle Phillips, though the nature of the business is not known.

Phillips (Philips), George Searle ("January," "January Searle") (1816–89), a journalist and writer who befriended Thomas Hirst in 1849. Phillips worked as editor of the *Leeds Times* beginning in 1844. He also edited the Transcendentalist publication the *Truth Seeker* along with F. R. Lees, and was a lecturer at the People's College in Huddersfield. Hirst and Tyndall often called him "January" or "January Searle."

Playfair, Lyon (1818–98), politician and chemist. In 1851, Playfair was appointed professor of chemistry at the Government School of Mines, and in 1853 was made secretary for science in the Department of Science and Art.

Plücker, Julius ("Plucker") (1801–68), German mathematician and physicist. Plücker worked on diamagnetic and magnetic properties of crystals and gases at the University of Bonn from 1847 until he turned his attention to mathematics in 1865.

Poggendorff, Johann Christian (1796–1877), German chemist, physicist, and editor. He was professor of chemistry at the University of Berlin, and edited the *Annalen der Physik und Chemie* between 1824 and 1876.

Powell, Baden (1796–1860), the Savilian Chair of Geometry at the University of Oxford. He was both a physicist and theologian and worked in particular on radiant heat and the wave theory of light. He was the vice president of the RS in 1853.

Price, Bartholomew (1818–98), mathematician. In 1853, he became Sedleian Professor of Natural Philosophy at the University of Oxford.

Regnault, Henri Victor (1810–78), French chemist and physicist. Regnault was professor of chemistry at the Collège de France, where he researched the properties of gases.

Riess, Peter Theophil (1804–83), German physicist. Riess was the first Jewish member of the Berlin Academy. Riess is best known for his work in electricity and developed the spark micrometer, or Riess micrometer, and the Riess spiral coils. Tyndall contacted Riess in regard to experiments on auxiliary currents that could be conducted in front of a large audience during a Friday evening lecture at the RI. Tyndall also assisted Riess with publishing a translated letter "On the Generation of Heat by Electricity" in the *Phil. Mag.* 7:46 (May 1854), p. 348.

Ruhmkorff, Heinrich Daniel (1803–77), German instrument maker and engineer. Rumkorff is known for his induction coil, or Ruhmkorff coil, which he first patented in 1851. Ruhmkorff was based in Paris, and Tyndall saw him during a visit there in September 1853.

Sabine, Edward (1788–1833), army officer and physicist. Sabine participated in several exploratory missions, including to the Arctic, during which he undertook scientific research on magnetism. Sabine was an active member of the BAAS, where he acted as general secretary for twenty years and president in 1852. Sabine was also a valuable patron to Tyndall. As president-elect of the BAAS in 1852, Sabine offered Tyndall one of the secretaryships at the Belfast meeting in charge of several of the sessions and the order of papers. He also proposed Tyndall for election to the RS, using his influence to help garner support and gather signatures. And he was a member of the Council of the RS that awarded Tyndall's Royal Medal, which Tyndall subsequently declined.

Sauerwald (Sauerbach/Sauerback), F. Possibly a Berlin instrument maker who produced instruments for Tyndall.

Schneider, a Berlin-based shipping agent whom Hirst often used to send Tyndall scientific instruments.

Schwerd, Friedrich Magnus (1792–1871), German astronomer and physicist who is known for developing a prism-photometer. He published a treatise on diffraction and wave theory in 1835, which Tyndall read in 1853.

Simpson, Mary (neé Martin) (d.1900), wife of Maxwell Simpson. Together they had six children. Mary and Anna Hirst (née Martin) were sisters, and Thomas Hirst spent time with the Simpsons as he was getting to know Anna during 1853 leading up to their wedding in December 1854.

Simpson, Maxwell (1815–1902), organic chemist. Simpson taught chemistry in the Peter Street school of Medicine in Dublin. He spent three years in Germany with his family between 1851 and 1854, where he met Hirst, who would become his brother-in-law.

Stegmann, Friedrich Ludwig (1813–91), German mathematician. Stegmann was professor of mathematics in Marburg and was adviser to both Tyndall and Hirst. Tyndall wrote his dissertation at Marburg under the supervision of Stegmann.

Steiner, Jakob (1796–1863), Swiss mathematician and professor of mathematics at the University of Berlin.

Steuart/Stewart, Elizabeth Dawson (née Duckett) (1802–93), born into a very wealthy family with an estate of 12,000 acres and a gothic mansion (Duckett's Grove) situated several miles north-east of Carlow. She was married to William Richard Steuart (1792–1852), with whom she lived at Steuart's Lodge, Leighlinbridge. She took a keen interest in Tyndall's career and was probably his most consistent and long-lived patron.

Stokes, George Gabriel (1819–1903), physicist and mathematician. Stokes became Lucasian professor at Cambridge in 1849, a position he held the rest of his life. He also took up a lectureship at the Government School of Mines in 1854. Stokes was the secretary of the RS from 1854 to 1885, where he took an active role in the publication of the *Phil. Trans.*

Sylvester, James Joseph (1814–97), mathematician. Sylvester was Jewish and as such was barred from completing a degree at the University of Cambridge, where he studied mathematics. He unsuccessfully applied for the professorship of mathematics at the Royal Military Academy in Woolwich in 1854. He was, however, successful when he applied again in 1855. Sylvester had also been proposed as a recipient of the Royal Medal the same year as Tyndall.

Taylor, Richard (1781–1858), prominent English publisher, especially of scientific journals, including *Annals of Science*, *Annals of Natural History*, and the *Phil. Mag.*, where Tyndall submitted his first scientific papers for publication. He also founded and edited seven volumes of *Scientific Memoirs* between 1837 and 1852, a collection of translations of major foreign language scientific works. Taylor married Hannah Corke in 1807, with whom he had one daughter, Sarah. After Hannah was struck down by severe psychiatric illness, he had a long-term relationship with Frances Francis, his daughter's governess, with whom he had two children, including his only son, William Francis. Taylor suffered a nervous breakdown in 1852 and was sent to an asylum at Sydenham, where Tyndall often visited him. He went into partnership with his son William at that time, founding the Taylor & Francis publishing firm, an imprint that first appeared on the cover of the *Phil. Mag.* in June 1852. Francis and Sarah moved to Richmond the following year to look after their father, and Tyndall moved into Francis's former lodging in London. (Brock and Meadows, *Lamp of Learning.*)

Tennyson, Alfred Lord (1809–92), first Baron Tennyson. Poet laureate of Great Britain and Ireland. Tyndall purchased one of Tennyson's collections of poems in 1853, which he described as "wine to the intellect."

Thomson (Thompson), William (1824–1907), brilliant young mathematician and physicist. He was appointed professorship of natural philosophy at Glasgow University in 1846, a post he held for fifty-three years. He published significant papers on the dynamic theory of heat in the late 1840s and early 1850s. For much of his career Tyndall regarded Thomson as a scientific rival.

Trevelyan, Arthur (1802–78), a fellow of the Edinburgh Geological Society. He discovered a phenomenon whereby sounds are produced by heated metals during the cooling process. He presented his experiments on the phenomenon to the Royal Society of Edinburgh on 17 January and 21 March 1831 (A. Trevelyan, "Notice regarding some Experiments on the Vibration of Heated Metals," *Trans. Roy. Soc. Edin.* 12:1 [1834], pp. 137–46). Tyndall conducted similar experiments for a Friday evening lecture on 27 January 1854.

Tyndale, George Hector (1821–89), American military officer and member of a family business based in Philadelphia that imported pottery and porcelain. Hector was a cousin of Tyndall, who Hector described as his "nearest Atlantic relative." The two had a mutual interest in getting to know each other, and Tyndall requested Hector to have his portrait taken for him to keep. Hector visited Tyndall when he was in Europe for business in 1854.

Verdet, Marcel Émile (1824–66), French physicist. Verdet was professor of physics at l'École Normale, where his research focused on optics and magnetism. The Verdet constant is named after him. Tyndall met Verdet during his visit to Paris in 1853.

Weber, Wilhelm Eduard (1804–91), German physicist and a professor of physics at Göttingen from 1831, and then Leipzig from 1843, and then Göttingen again from 1849. Along with his colleague Gauss, he established the Göttingen Magnetische Verein to coordinate magnetic observatories, first in the Germanic states and then across Europe. He collaborated with Kohlrausch, establishing the ratio between the electrodynamic unit of charge and the electrostatic unit of charge. He also became involved in the experimental and theoretical development in diamagnetism.

Wertheim, Guillaume (1815–61), German-born French physicist who worked on the polarization of light.

Wheatstone, Charles (1802–75), an English experimental physicist and prolific inventor. In 1834 he accepted a professorship of experimental physics at King's College, London. In the same year he used a revolving mirror to estimate the speed of electricity moving through a wire, a similar experimental design that he later used to estimate the speed of light. Tyndall was especially interested in Wheatstone's invention of electro-magnetic clocks, which could be used to measure small intervals of time.

Whewell, William (1794–1866), English tidologist and historian and philosopher of science. He held professorships first in mineralogy (1828), then in moral philosophy (1838), and ultimately he accepted the Mastership of Trinity College, Cambridge (1841). Whewell is best known for his multivolume *History of the Inductive Sciences* (1837) and *Philosophy of the Inductive Sciences* (1840). His own research in physical astronomy focused on what he termed "tidology." He wrote fourteen major papers on the study of the tides, the last appearing in 1850.

Wiedemann, Gustav Heinrich (1826–99), German physicist and editor. Wiedemann was married to Clara Laura (née Mitscherlich) (1827–1914), the daughter of the chemist Eilhard Mitscherlich (1794–1863). When Poggendorff died in 1877, Wiedemann became the editor of *Annalen der Physik und Chemie*, and Clara assisted by translating the scientific papers. Wiedemann wrote his dissertation on organic chemistry under Magnus at the University of Berlin. In 1854, Wiedemann became professor of physics at the University of Basel. Together with Rudolph Franz, Wiedemann developed the so-called Wiedemann-Franz law in 1853, a fundamental law of physics which holds that the "ratio of the thermal conductivity of any pure metal to its electrical conductivity is approximately constant at a given temperature" (*Oxford Dictionary of Physics*). However, in his letters in this volume Tyndall primarily discussed the translation of one of Wiedemann's papers, presumed to be "On the Motion of Liquids in a closed Galvanic Circuit," in Tyndall and Francis (eds.), *Scientif. Mem.*, pp. 232–56.

Wright, Richard Sears (1829–92), English mathematician and teacher who taught engineering and surveying at Queenwood College from 1851. He would later write *The Elements of Plane Geometry for the Use of Schools and Colleges* (1868), which would include a preface by Thomas Hirst.

Wrightson, Francis, son of a Birmingham manufacturer. He studied chemistry in Marburg. While he was under the supervision of Hermann Kolbe, and obtained his PhD in 1853, he was also tutored and mentored by Hirst, who wrote about him occasionally in letters to Tyndall.

Wynne, George (1804–90), a former boss, friend, and patron of Tyndall. Wynne was from a military family, although his father was a clergyman in Killucan. From 1825 he was a Lieutenant in the Royal Engineers and from 1835 until 1840 employed in the Irish Ordnance Survey. When he and Tyndall met, Wynne was in charge of the division in which Tyndall worked (see volumes 1–2). He was a military officer until 1877, when he retired with the rank of general. Over the period of this volume he had the rank of captain, and the position of inspector of railways (August 1847 to April 1858) under the Board of Trade.

Yorke, (Col.) Philip James (1799–1874), a British army officer, chemist, and FRS. He was president of the Chemical Society from 1853 until 1855 and he also served as a manager of the RI.

INDEX

The index refers to the letters using their serial numbers. The letter 'n' denotes a note reference and appears if an item is not otherwise mentioned in the letter to which the note pertains. A number in **bold** signifies a letter to or from a correspondent.